THE BIG BLACK BOOK OF ELECTRONIC SURVEILLANCE

THE BIG BLACK BOOK OF ELECTRONIC SURVEILLANCE

EDWARD TEACH

Copyright © 2022 by Edward Teach

All rights reserved. No part of this book may be used or reproduced or transmitted in any form or by any means electronic or mechanical, including photocopy, recording, or any information storage and retrieval system now known or to be invented, without permission in writing from both the publisher and author, except by a reviewer who wishes to quote brief passages in connection with a review written for inclusion in a blog, website, magazine, newspaper, broadcast, or digital media outlet; and except as provided by United States of America copyright law.

The information presented in this book represents the views of the author at the date of publication. This book is presented for informational purposes. The author reserves the right to alter or update his opinions based on new information obtained after the publication of this book. While every effort has been made to verify the information in this book, neither the author or publisher or any distributor assume responsibility for any errors, inaccuracies, or omissions.

*I dedicate this book to the love of my life –
my gypsy woman, Veronica.*

Table of Contents

Introduction 1

Chapter 1: Public Policy 5
Fundamentals:
    Lawful Intercept the American Way 5
    Bulk Metadata Collection in the USA 8
    The UK's Investigatory Powers Act 18
    France's 2015 Intelligence Act 21
    Russia's SORM: Lawful Intercept KGB Style 26
    China's Counter-Terrorism Act vs. Western Laws 28
    U.S. National Cybersecurity Strategy 35
    The Wassenaar Arrangement 40

Chapter 2: Packet Monitoring 43
Fundamentals:
    Overview 43
    DPI "Versus" IP Flow Monitoring 50
    Field-Programmable Gate Arrays 53
    Network Functions Virtualization 57
Surveillance Players and Solutions:
    ALBEDO Wire-Speed Packet Capture (Spain) 64
    Amdocs VoLTE Interception (Israel) 69
    Cisco and the Great Firewall of China (USA) 73
    Decision Group DPI (Taiwan) 77
    DESOMA Packet Extraction (Germany) 85
    Flowmon: Flow Monitoring (Czech Republic) 88
    Glimmerglass Photonic Surveillance (USA) 91
    Incognito Broadband Command Center (Canada) 94

Keysight Network Monitoring (USA)	97
Netcope: Scouting 100GB Ethernet (USA)	101
Netronome: Driving SS8 and Blue Coat (USA)	105
NETSCOUT: Silent Spyware Vendor (USA)	110
Packet Forensics: DPI for the NSA (USA)	112
Protei Deep Content Inspection (Russia)	117
Radisys Wire-Speed Packet Capture (USA)	121
Riverbed and Wireshark Packet Capture (USA)	125
Rohde & Schwarz's ipoque DPI (Germany).	128
Savvius Network Forensics (USA)	131
Silicom Denmark: VoLTE Interception	138
Sovereign Intelligence and Sixgill (USA/Israel)	142
VSS Monitoring: Network Packet Brokers (USA)	146

Chapter 3: Analytics　　151
Fundamentals:

Overview	151
Big Data Analytics Top to Bottom	157
Artificial Intelligence in ISS	160

Surveillance Players and Solutions:

Axilent ACE Profiler (UK)	168
ComWorth SwiftWing SIRIUS (Japan)	172
Fifth Dimension Crime Prevention (Israel)	175
IBM i2 Safer Planet (USA)	178
Neo4j and Linkurious Visualization (USA, France)	182
MemSQL: Escape from Hadoop Batch Mode (USA)	185
NORSI-TRANS Vitok 3-X, Vitok-CLUSTER (Russia)	188
Ockham Solutions: Mobile Forensics (France)	193
Sqrll: Cyber Analytics with Machine Learning (USA)	197
Verint: OMNIX Intelligence Fusion Center (Israel, USA)	200

Chapter 4: Offensive Cyber　　203
Fundamentals:

Overview	203
Cracking Cryptocurrency	206
DNS Hijacking: The Art and Science	210
Man-in-the-Middle Attacks	214
Mobile Hacking via SS7 and Diameter Protocols	217
NetFlow Versus Tor: Conquering Anonymity	220
Tor OONIprobe: Outing Surveillance	225

Surveillance Players and Solutions:
   China's Move on Quantum Cryptology (PRC)   228
   Dark Mail vs. Offensive Cyber (USA)   232
   Darktrace: AI Versus Offensive Cyber (UK)   236
   DARPA Memex Dark Web Intelligence (USA)   239
   Endgame Zero Day Exploits (USA)   245
   FBI Offensive Cyber Powers (USA)   248
   FBI Network Investigative Techniques (USA)   251
   FinFisher FinSpy Rootkit Infection (Germany)   254
   Gamma Group: FinFisher's Parent (UK)   259
   The Hacking Team (Italy, Switzerland)   266
   NSA Equation Group and TAO (USA)   269
   NSA/GCHQ Hack of Gemalto SIM Card Keys (USA, UK)   273
   NSO Group: Offensive Cyber King (Israel)   277
   Ouroboros: Assault on Critical Infrastructure (Russia)   282
   Wintego Cracks WhatsApp Encryption (Israel)   286
   ZERODIUM: Zero Days for Cybersecurity (USA)   289

Chapter 5: Mobile Location and Monitoring   293
Fundamentals:
   Overview   293
   LTE Protocol and Chipset Vulnerability   298
   MegaMIMO 2.0 Wi-Fi Interception   305
   Stumping IMSI Catchers via Multi-IMSI Phones   308
Surveillance Players and Solutions:
   Boeing DRT Box (USA)   313
   cellXion LTE IMSI Catchers (UK)   316
   Cisco Hyperlocation: Tracking Wi-Fi (USA)   320
   CyberSeal IMSI Catchers and Detectors (Israel)   323
   Harris StingRay IMSI Catchers (USA)   326
   Innova GPS Tracking Device (Italy)   328
   Micro Systemation Mobile Forensics (Sweden)   331
   Polaris Wireless: Vertical Location (USA)   334
   Pro-Solve IMSI Catchers (UK)   338
   Rafael PowerSpy: Amping-Up Location (Israel)   341
   Rayzone Piranha LTE IMSI Catcher (Israel)   347
   Rayzone InterApp for Wi-Fi Interception (Israel)   350
   Saab Medav: 5G Blind Mobile Location (Germany)   354

Septier Mobile Location (Israel)	358
Wintego WINT for Wi-Fi Interception (Israel)	360

Chapter 6: Open Source Intelligence 363
Fundamentals:
Overview	363
Social Media Monitoring and GEOINT	365

Surveillance Players and Solutions
AGNITIO: Pioneer of Voice Biometrics (Spain)	372
BrightPlanet Tor Cracking (USA)	374
Expert System: Deep Semantic Analysis (Italy)	377
Group 2000 Facial Recognition (The Netherlands)	380
Knowlesys KIS: Chinese OSINT (PRC)	384
Kofax Dark Web Monitoring (USA)	387
NEC Neoface Facial Recognition (Japan)	391
Phonexia Voice Biometrics (Czech Republic)	395
Recorded Future: Real-Time GEOINT (USA)	397
SYSTRAN: OSINT Machine Translation (S. Korea)	400
SciEngines Custom Hardware Attacks (Germany)	406
S2T: Finding Foes via Cyber-HUMINT (Singapore)	412

Chapter 7: Lawful Intercept Multi-Play Vendors 417
Fundamentals:
Overview	417

Surveillance Players and Solutions
Aqsacom: Lawful Intercept Used by Verizon (France)	425
ATIS UHER: Visualizing Lawful Intercept (Germany)	429
BAE Systems Applied Intelligence (UK)	433
Cisco Routers' Wiretap Role (USA)	438
ClearTrail: Lawful Intercept and Offensive Cyber (India)	441
Elbit Systems CYBERBIT Multi-Play (Israel)	444
iPS: From Lawful Intercept to Dark Web (Italy)	448
Iskratel: SORM & ETSI Hardware (Slovenia)	453
NETI: Hacking and Mobile Location (Hungary)	455
NetQuest: Policing 100GB DWDM Fiber (USA)	457
NICE: All-in-One ISS (Israel)	461
Paladion: From Malware to Cybersecurity (India)	465
Pen-Link: Moving Beyond Pen Registers (USA)	468
RCS Lab MITO^3: Unified LI Management (Italy)	471
Roke Manor Research: LI with a Military Twist (UK)	473

Russia's SwitchRay Masquerades as U.S. Company	476
Sinovatio: China's Global Surveillance Player (PRC)	480
SS8: Evolution from LI to Commercial Cyber (USA)	485
TelcoBridges Lawful Intercept for VoIP (Canada)	489
Telesoft Technologies: Probes for Voice and Data (UK)	493
TraceSpan: Where Fiber and LTE Monitoring Meet (Israel)	496
Trovicor: Spyware's Black Panther (Channel Islands, UK)	499
Utimaco: Flexible, Top Quality Interception (Germany)	504
Vehere Takes on The West (India)	507
Verint's Big Iron STAR-GATE (Israel)	512

Chapter 8: Military Intelligence — 517

Fundamentals:
Overview	517
Airborne ISR: SAR, Electro-Optical/Infrared Fusion	519
Signal Jammers and Anti-Jamming Systems	524
Military Alternatives for GPS Location	527

Surveillance Players and Solutions:
Airbus Zephyr: Satellite Surveillance Supplement (UK)	531
Elbit Systems Airborne and Ground Surveillance (Israel)	533
ELTA Systems Defeats Drone Bombs (Israel)	536
ELTA Systems LTE and SATCOM Interception (Israel)	539
Keysight Signal Analyzers and RF Sensors (USA)	541
Lumacron: Lighting Up Enemy Forces (Scotland, UK)	544
Parsons OGSystems: Disruptive GEOINT (USA)	548
Palantir: Visualizing the Future of Warfighting (USA)	551
PLATH Group: Military and Malware (Germany)	554
Providence Group: SOCOM Tracking (UK)	558
Raytheon 3G Forward Looking Infrared (USA)	561
Raytheon Visual Analytics Fiasco: DCGS (USA)	565
Sepura COVERT SRC3300 Terminal (UK)	570
Thuraya: Terrorists' Favorite Satellite Network (UAE)	574
Ultra Electronics C5ISR (UK)	578

Glossary — 583

Introduction

As *The Big Black Book of Electronic Surveillance* goes to press, nearly a decade has passed since Edward Snowden leaked secret documents disclosing the extraordinary surveillance powers of the United States and aligned nations. While Snowden has largely faded from view since then, the surveillance capabilities he exposed have grown far beyond the scope and sophistication of that era.

Its purveyors are technology companies and government service contractors that serve the interests of national and homeland security. In the U.S., Snowden worked for one, Booz Allen Hamilton. In addition to well-known U.S. government contractors, some of the largest players are global brands best-known for commercial tech products and services: IT giants including Cisco, IBM and Microsoft. Worldwide, hundreds of other companies drive advancements in surveillance technology. Even in authoritarian and communist nations, surveillance is a business; for example, surveillance companies abound in Russia and the People's Republic of China.

In the United States, some 80 percent of the people deployed at Intelligence Community (IC) secured facilities are employees of government contractors: Booz Allen, BAE Systems, CACI, General Dynamics, Leidos, Northrop Grumman, Raytheon and other multi-billion dollar behemoths whose offices ring the nation's capital. This army of some one million highly-skilled private sector personnel serves as a "shadow government" responsible for deploying and managing surveillance solutions used by U.S. Intelligence agencies and the intelligence branches of the Department of Defense.

Contractors also play an important role in training government personnel in the art and science of spyware. Budding hackers at the National Security Agency's Maryland facilities learn offensive cyber techniques not from fellow employees but from contractors. These public/private sector relationships are close in the U.S., where government surveillance is very much like a family business.

A great many contractor employees have IC or military intelligence backgrounds. Managers and technical experts who leave government employers such as the Department of Homeland Security (DHS), National Security Agency (NSA), National Reconnaissance Office (NRO), National Geospatial Intelligence Agency (NGA) and Central Intelligence Agency (CIA) commonly move to jobs with contractors, for whom they serve their former government agency employers.

Similar symbiotic relationships between government/military intelligence and industry counterparts thrive in other nations. Intelligence support systems (ISS) vendors such as Tel Aviv-based NSO Group have deep roots in the Israeli Army's 8200 unit. In Great Britain, the GCHQ (the United Kingdom's version of the NSA) is just down the road from Cheltenham, where numerous tech experts – many of them former GCHQ employees – work for private companies in roles that support national security in the UK.

Curiously, in the United States, government contractors perform very little original spyware research and development of their own. For the most part, these corporate giants act as systems integrators of other companies' work. When a U.S. contractor "introduces" a new solution it is more than likely a re-branded version of another innovator's work.

Given concerns and restrictions about using technology solutions made in foreign nations, spyware sold by contractors to U.S. government agencies is nearly 100 percent made in America. Even more remarkably, discussions with U.S. contractors' leadership and technology experts often reveals that they are oblivious to the work of surveillance innovators based off-shore.

The Big Black Book opens the door to this world of intelligence arms merchants whose work shapes modern surveillance. These companies hail not just from the U.S., but also from Austria, Australia, Canada, China, the Czech Republic, Denmark, Dubai, France, Germany, India, Israel, Italy, Japan, Korea, The Netherlands, Singapore, Slovenia, South Africa, Spain, Sweden, Switzerland, Taiwan, Ukraine, the United Arab Emirates, the United Kingdom, and many other nations.

This volume presents the market leaders and the surveillance solutions they sell to governments: packet monitoring, analytics, offensive cyber, mobile location and forensics, lawful intercept, social media intelligence

(SOCMINT), facial recognition, voice biometrics and other forms of open source intelligence (OSINT) – as well as relevant forms of artificial intelligence that automate their performance. Also covered: military-focused technologies that deliver or intercept intelligence at the tactical edge: forward-looking infrared (FLIR), RF monitoring, Electro-Optical/Infrared, eLoran, and systems with the power to take control of space assets vital to national security and critical infrastructure.

The Big Black Book is at once: a textbook; a manual for law enforcement, IC and military agencies committed to safeguarding national and homeland security; an encyclopedia for human rights advocates committed to protecting privacy; and a resource for journalists who seek the truth. This book does not take sides or indulge in political debate or favoritism. It simply provides the facts. Among the conclusions that readers may draw from the intelligence contained herein:

- Virtually every advance in electronic communications is accompanied by a way to monitor or attack it, as often as not, produced by the same enterprise. Some of the most vocal critics of government surveillance are companies that quietly profit from it.
- Electronic surveillance is ubiquitous. Laws that govern its deployment, both in-country and against other nations, are quite similar from one country to the next. Democratic nations such as the USA, UK, France, Germany, Italy and The Netherlands are no more constrained in deploying surveillance solutions than are Russia, China, Iran, North Korea and other authoritarian regimes.
- The power that nations exercise via current modes of electronic surveillance will be dwarfed by what follows: advances in artificial intelligence and quantum computing, cryptology and cryptography that take surveillance to the next level.
- The classified electronic surveillance technologies that Edward Snowden revealed were not the product of the NSA. These and similar government agencies worldwide are merely end-users of solutions developed and deployed by Intelligence Systems Support (ISS) vendors.

What lies ahead? In the wake of the Coronavirus pandemic and its impact on the availability of safe human talent, governments will inevitably come to rely on automated, outcome-based solutions to preserve national security, law and order. This transformation raises possibilities once confined to the realm of science fiction – a world where technology makes the rules.

Finally, although respected pundits now deem hostile nation states the greatest threat to national security in free nations, it is short-sighted to overlook, as so many do, the ever-present danger of terrorism. It is precisely this failure of focus that opens opportunities for political extremists of the Far Left, Far Right and fanatics of all stripes.

Any disregard for this threat mirrors the foolish oblivion of a quiet, peaceful "day before" an event that will forever live in infamy. September 11, 2001.

– Edward Teach

CHAPTER 1

Public Policy

Lawful Intercept the American Way

Lawful intercept (LI) is the process of obtaining signaling data, call records or full content of communications via court order or subpoena to support a criminal, terrorist or hostile nation state investigation. With the rapid growth of technology, lawful intercept has leapt from the simple voice "wiretap" of yesteryear to include the capture of data, signaling and content from mobile networks, as well as from IP, broadband and cloud services and IT infrastructure.

Laws governing electronic surveillance are equally complex. In many nations there is no single law with oversight. Rather, like the technologies involved in U.S. lawful intercept, laws have evolved to have integrated functionality. United States surveillance law is a case in point. In the United States, law enforcement agencies (LEAs) conduct lawful intercept via an interconnected framework of laws and policies:
- Title III of the Omnibus Crime Control and Safe Streets Act of 1968 ("The Wiretap Act");
- The Foreign Intelligence Services Act (FISA) of 1978;
- The Electronic Communications Privacy Act (ECPA) of 1986;
- The Stored Communications Act (SCA), enacted as Title II of ECPA in 1986;

- The Communications Assistance for Law Enforcement Act (CALEA) of 1994;
- The Patriot Act of 2001, which amended FISA and ECPA;
- The 2005 Federal Communications Commission (FCC) Order extending CALEA to apply to facilities-based providers of Internet, broadband and Voice over IP (VoIP) services that connect to the public switched telephone network (PSTN) – but exempting "over the top" services, i.e., those not connected to the PSTN;
- The FISA Amendments Act (FAA) of 2008, principally Section 702 on requirements for NSA surveillance of non-US targets within the U.S. under PRISM (downstream collection from ISP servers of Google, Apple, Yahoo, Facebook, Skype, etc.) and upstream (directly from cables and other network infrastructure).
- The USA Freedom Act of 2015, reauthorizing most aspects of the Patriot Act but eliminating Section 215, and ostensibly ending bulk metadata collection by the NSA.

In the United States, law enforcement agencies use these laws to conduct lawful intercept over networks operated by communications service providers (CSPs), which in turn are required to deploy and maintain technology solutions and expert staff to support court orders for lawful intercept. Court interpretation and administration of surveillance laws may vary state-to-state. For example, the use of mobile location technologies varies widely depending on local jurisdiction and applicable case law.

Failure to comply with surveillance laws carries stiff penalties – in the case of CALEA, fines of up to US $10,000 per day. As yet, however, there is no known instance of a CSP being fined for non-compliance. With the emergence of broadband and over-the-top services, a growing number of service providers believe they are exempt from the law. Quite often they are wrong. The rule of thumb: If a carrier's service interconnects at any point with the public switched telephone network (PSTN), it is subject to the rules of CALEA and must have a technology solution in place to facilitate lawful intercept upon receipt of a valid court order. CALEA applies not only to wireline and wireless carriers, but to broadband and VoIP providers whose networks touch the PSTN.

CSPs may either purchase, deploy and manage the technology solutions required for compliance themselves – as the largest tend to do – or hire a Trusted Third Party (TTP) to help. In either event, when a CSP has in place a solution that meets the technical standards of CALEA, it is considered to be in "safe harbor," i.e., in compliance with law.

Putting and keeping CSPs in "safe harbor" with CALEA is the principal mission of TTP lawful intercept vendors. Some TTPs operate as service bureaus that provide end-to-end CALEA compliance systems that: (1) deploy, test and maintain the technology solution; (2) provide in-house legal counsel to confirm the accuracy of court orders received by a CSP; (3) implement the intercept as detailed in the court order; (4) ensure that the intercept follows the strict privacy protections outlined in CALEA; (5) employ former law enforcement officers to liaise with the LEA in charge of the investigation; and (6) shut down the intercept when the court order's time stamp expires.

New regulations add to the complexity and cost of lawful intercept. For example, in the U.S., federal authorities require vendors and CSPs to partition LTE technology solutions so that they will not intercept Voice over LTE (VoLTE) and VoIP unless specifically targeted by a court order. The upshot: Vendors must reconfigure legacy LTE lawful intercept technology solutions to follow the rules.

CALEA is generally perceived as the principal law governing U.S. lawful intercept. That said, the bulk of activities undertaken by TTPs often pertain to FISA and Wiretap Act court orders requested by the U.S. Federal Bureau of Investigation.

Bulk Metadata Collection in the USA

When President Barack Obama signed the USA Freedom Act into law in June 2015, his action was widely promoted as the end of one of the most controversial aspects of government-run "bulk metadata collection" under Section 215 of the Patriot Act. The USA Freedom Act was hailed for curtailing mass interception of U.S. citizens' call data records, including all data records for voice and data traffic. But as close analysis proves, that assumption is far from true.

Bigger and more productive forms of bulk metadata interception most certainly survived, albeit in ways that by and large have escaped public attention. To understand how and why this happened, it is essential to begin at the beginning – way back beyond the USA Patriot Act that supposedly saw the first authorization of bulk metadata collection – and examine the full history of this issue. Many unanswered questions remain about the origins, evolution and current status of this ISS capability:

- How and where did bulk metadata collection begin?
- What exact powers did Patriot Act Section 215 endorse?
- How were these powers interpreted and implemented by government agencies?
- What communications services were included in U.S. bulk metadata collection?
- When and why was the program extended to voice call data records (CDRs)?
- What communications companies were involved in cooperation with CDR orders?
- What about Internet services – were they under a separate program or programs?
- How have IP intercept programs evolved? How broadly were these IP intercept programs deployed?
- What key mass surveillance programs, if any, survived the USA Freedom Act?

One can find some of the answers if willing to dig deep and far enough among a wide variety of sources. But sometimes these sources conflict. Here we will make sense of the pieces and put them together in the context of their implications for the USA Freedom Act.

Did Bulk Metadata Collection in the U.S. Pre-Date 9/11?

Few can forget the image of a shaken President George W. Bush sitting in a classroom of second graders in Sarasota, Florida as he received word of the events of September 11, 2001. The lingering picture is of a President and nation totally unprepared for the devastating attacks in New York, Washington, D.C., and over the skies of Pennsylvania by a cadre of Al-Qaeda operatives – ironically, primarily made up of natives from the U.S.'s strongest ally in the Arab world, Saudi Arabia.

For the most part, this derogatory view is a purpose-driven fabrication, but let's see where it leads.

As the days passed, so the story goes, there was much behind the scenes finger-pointing at and between members of the U.S. Intelligence Community. Typical of the rumored dialogues: "If any member agency had advance intel on the attacks, why wasn't it channeled upstream immediately for action?" and, "If no member of the IC had the intelligence beforehand, then why not?"

President Bush, aided by a uniformly supportive Congress, supposedly went into high gear to fix an alleged intelligence gap by creating new laws that granted unprecedented powers to U.S. national security and law enforcement agencies. In truth, this story wasn't even right on the "unprecedented powers" part.

Nonetheless, as documented in the public record, members of the IC, the U.S. Department of Defense and its principal foreign intelligence arm, the National Security Agency, ostensibly jumped into action to create new technology solutions that would preempt future sneak attacks.

If so, they had good reason to be optimistic from the outset. The technology already existed. For that matter, its immediate precursor was a proven solution.

DEA's Precursor – USTO

In 1989, then-President George H. W. Bush expressed his interest in a program that would exploit "advanced technology" to pursue the War on Drugs against criminals in the U.S. and abroad. Three years later in 1992, that wish came true for the Clinton Administration: a sophisticated system developed by the Drug Enforcement Agency (DEA) to track and apprehend drug lords based on intelligence revealed in their call detail records. They called it USTO, short for tracking metadata on calls from the "U.S. to" foreign nations.

USTO was a first of its kind program. It required all telecom and wireless carriers to hand over call detail records of individuals located in the U.S. and placing calls to offshore parties, a huge undertaking and completely novel for that era, for unlike other forms of lawful and intelligence intercept of the time, USTO was warrantless. Another twist: The DEA could and did execute the requests without a subpoena.

While some carriers voiced momentary qualms, none raised objections – at least not for years. AT&T, Verizon, Sprint and many others provided full cooperation. The few that later disagreed with DEA on USTO received firm letters from the U.S. Department of Justice instructing them to comply.

Initially confined to drug investigations in a few countries, USTO was ultimately expanded to pursue major drug offenders in 100+ nations, often without notifying those countries' law enforcement officials. In certain instances, USTO was also applied to terrorism investigations by agencies including the FBI and NSA. Among the extracurricular success stories linked to USTO: its use in investigating the 1995 Oklahoma City bombing to quickly ascertain that this event was in no way linked to foreign terrorists.

USTO remained in operation until the fall of 2013, when U.S. Attorney General Eric Holder shut it down following a leak on the DEA program by Edward Snowden. But well before then, USTO had established what some consider a "blueprint" for programs by other members of the Intelligence Community, notably the NSA.

The one drawback of USTO: It relied on call data records that were often several days to a week old. That wasn't good enough for the NSA, the DoD – or the Office of the Director of National Intelligence reporting directly to the President. "Bush II" era intelligence strategists wanted a system that worked in real-time, or very close to it. Such help was readily available in Silicon Valley long before the first airliner crashed into the World Trade Center.

Role of Narus and AT&T

With its founding in 1997, Sunnyvale, California-based Narus Corporation had begun work to advance the field of packet filtering, specifically the development of Semantic Traffic Analysis (STA) technology. By late 1999 Narus had a product, STA 6400, a platform of "standalone traffic analyzers" connected to an Internet Service Provider (ISP) network to collect customer usage information. Today we call this capability deep packet inspection (DPI), the capture, inspection and analysis of the full packet payload of single or unlimited numbers of packet streams, both for metadata and content. Among the first clients was AT&T, which deployed the STA 6400 at its network operation center (NOC) in Bridgeton, Missouri. Bridgeton was no backwater, but rather the key NOC supporting AT&T WorldNet, then the largest ISP and operated by the nation's still-dominant long haul telecom company. In tandem with Bridgeton, another site opened in San Francisco. From there the work of the STA 6400 and similar solutions began to spread in a way peculiar to the Internet: virally.

AT&T, as a common business practice, had peering arrangements with all other ISPs – from Abovenet up through the alphabet list of Internet providers to XO Communications – as well as to Mae West, one of the two key Internet internodal centers in the U.S. From secret NOCs outfitted with more Narus STA 6400s, AT&T quickly gained the ability to track not only all the traffic on WorldNet, but all traffic crossing the Internet in the U.S.

The Narus STA 6400 was just one of several similar DPI solutions used by AT&T for filtering IP traffic on the Internet backbone. And AT&T wasn't working in a corporate vacuum. The company had partners in high places. AT&T hosted "secret rooms" in the service of the DoD and NSA.

When the Intelligence Community Wrote the Law

In mid-2002, Admiral John Poindexter, then Director of the DARPA Information Awareness Office (IAO), had begun to "out" the government's involvement in DPI in public speeches. Remember: This was a time well before Edward Snowden when any and all means of preventing terrorism were viewed favorably by the populace.

IAO itself had an interesting story. It sprang from the Admiral's own vision. Long an expert on national security matters, Poindexter suggested the creation of such a group right after 9/11. President Bush bought in. IAO's mission: to develop information systems to detect and prevent terrorist threats under the "Total Information Awareness" (TIA) program, later renamed "Terrorist Information Awareness." IAO was quickly formed and Poindexter was appointed as its head in early 2002.

By summer he was on the record describing a government system that would "provide intelligence analysts and law enforcement officials with instant access to information from Internet mail and calling records to credit card and banking transactions and travel documents, without a search warrant." It is no coincidence that these capabilities map to those of the AT&T/Narus system.

Poindexter's list also lines up with features enumerated in another important document: the statutory list of document and record types opened to mass surveillance under Section 215 of the Patriot Act.

Congress was hardly pulling the active mechanism of Section 215 out of thin air. They charted the law to surveillance capabilities already recognized, available and dictated to them verbatim by the Intelligence Community. Call it a point-by-point case of law written to support a set of technology specs.

Consider the language of Section 215. The statute reads that the Director of the FBI or a designee of the Director "may make an application for an order requiring the production of any tangible things (including books, records, papers, documents, and other items). . ." It goes on to detail examples of the types of records eligible for surveillance: "library circulation records, library patron lists, book sales records, book customer lists, firearms sales records, tax return records, educational records, or medical records containing information that would identify a person. . ."

One restriction: The statute specified that the grounds of an investigation using Section 215 would be "to obtain foreign intelligence information not concerning a United States person or to protect against international terrorism or clandestine intelligence activities, provided that such investigation of a United States person is not conducted solely upon the basis of activities protected by the first amendment to the Constitution."

Of note, Section 215 did not specify conducting the investigation using evidence intercepted on the Internet. But it didn't rule out the Internet or public switched telephone network, either.

In short, Section 215 provided broad latitude. The list of "tangible things" left ample room for interpretation, and the circumstances for pursuing suspects wide open. Equally remarkable for a law providing almost carte blanche powers: an incredibly short "window" between conception and execution.

Typically, when a law is passed, months of wrangling follow over the regulations governing every detail of its implementation. Not so with Section 215. Reason: the surveillance program with the ability to conduct bulk metadata intercepts preceded the law. The system was in place on October 4, 2001 – three weeks ahead of the Patriot Act becoming law on October 26, 2001. In effect, Section 215 of the Patriot Act codified a vast surveillance system that was already in place.

When Admiral Poindexter subsequently referenced an intelligence gathering system that could provide "instant access" to any type information, he wasn't projecting into the future. He was revealing the present.

The system's name: STELLAR WIND, quite appropriate given the original astronomic definition. A stellar wind is the flow of gas particles surrounding the stars in the universe. For each star the stellar wind is different, as is the data emanating from every human "particle" on earth.

STELLAR WIND: The NSA Takes Over

Unique in the history of U.S. intelligence capabilities, STELLAR WIND was the code name of a suite of secret intelligence activities that comprised the "President's Surveillance Program" (PSP) introduced by former President George W. Bush following 9/11. Under STELLAR WIND, this suite was run as a "Sensitive Compartmented Information" or SCI operation reporting to the Director of National Intelligence, under final authority of the President and administered by Executive Order.

What exactly was STELLAR WIND, and what did it do?

STELLAR WIND was the first iteration of government bulk metadata collection in every sense of the term. True, the NSA, CIA and FBI had managed large databases of suspect information before, but nothing on the scale or with the sophistication of STELLAR WIND.

Depending on the source, STELLAR WIND has been variously described as a "data mining" or "email interception" tool, capabilities that in fact barely scratch the surface of what the system could do. STELLAR WIND was AT&T/Narus on steroids – what the NSA describes as a series of "Upstream" programs, capable of intercepting all traffic from backbone networks, whether Internet or telephony.

As importantly, even as new and more sophisticated solutions such as big data analytics and machine intelligence emerged, STELLAR WIND remained the model for succeeding generations of bulk metadata and content collection, both in the U.S. and with its "Five Eyes" partners – Canada, the United Kingdom, Australia and New Zealand – which quickly came to share network connections and intelligence. The model was classic DPI: packet capture, mirroring, scoring and analytics, but on an ever-enlarging scale, all in next-to-real-time. With STELLAR WIND, this power was focused on the U.S.

Initially, STELLAR WIND comprised DPI solutions backed by analytics and the near-limitless processing power of the NSA's Fort Meade, Maryland facilities for tapping U.S. Internet backbones, thus able to track, mirror and analyze every packet that transited a network.

The list of "tangible things" eligible for surveillance under Section 215 of the Patriot Act was nearly limitless, like the universe itself: email, SMS, files, downloads, uploads, websites visited, marriage and divorce records, pornography preferences, passport records, arrest records, travel plans, hotel stays, restaurant reservations, credit card and debit transactions, cash transactions, real estate records, automobiles purchased or rented, medical records, education records, gun ownership, professional licenses, club memberships, political donations, political affiliations – and the same information and more on spouses, children, family and friends.

The immense power granted intelligence agents via STELLAR WIND didn't come without controversy. Highly-placed law enforcement officials within the Bush Administration, including then Attorney General John Ashcroft and FBI Deputy Attorney General James Comey, questioned the legality of mass Internet surveillance. As later revealed in a report on an investigation conducted by the Inspector General's office, Ashcroft in March 2004 refused to sign an order extending authorization of STELLAR WIND. Flummoxed by the AG's adamance, President Bush was forced to shut down STELLAR WIND for a period of three months while finding a way around internal opposition within his own Administration. The solution came on July 14 of that year: a secret signed order by Colleen Kollar-Kotelly, Chief Judge of the Foreign Intelligence Surveillance Court (FISC). After a brief hiatus, bulk metadata collection was back in business.

In the aftermath of this intra-Administration flap, management of STELLAR WIND underwent organizational changes. In 2006 the "powers that be" folded the program under a new initiative called Special Source Operations (SSO) at the NSA. SSO had many functions including the collection of data and intelligence from fiber optic networks – international, transoceanic and domestic – on its own and in partnership with major CSPs and ISPs. It was a good home for STELLAR WIND, and for what came next.

Taking a Page from DEA's USTO

Controversy surrounds the question of whether bulk metadata collection of telephony – wireline and wireless – call data records from communications service providers (CSPs) began with STELLAR WIND in 2001 or came later. Some sources insist that STELLAR WIND programs as applied to voice CDRs began in the 2001-2002 time frame, while others vow that mass CDR interception waited another five years.

The baseline: CDR metadata collection was inspired by DEA's USTO program: warrantless bulk metadata collection that could easily have included CDRs. Even before 9/11, the technology existed to capture all CDRs, IPDRs and the content of both, regardless of type.

Whichever origin date one agrees with – pre- or post-9/11 – it is clear that STELLAR WIND operated strictly under Presidential authority for years. But a big change came shortly after mid-decade in the form of official court authorization.

In May 2006, the FISA Court ordered CSPs to maintain detailed phone call detail records. Really, the order outlined little more than carriers already did for billing and other purposes. But now CSPs had to maintain the CDRs in full format for a set period of time, and provide the metadata in a continuous stream to government agencies and law enforcement by court order from FISC. Instead of specifying a FISC order with an individual suspect's name in order to obtain legal authorization, LEAs could simply cite the name of the carrier and proceed to access all CDRs in bulk metadata form. "Warrantless" went to "warranted" on a grand scale.

Under DEA's USTO, telecom and wireless operators were required to comply – and without a warrant – with requests for CDRs on calls from the U.S. to foreign destinations. All major carriers are known to have cooperated with USTO. But the FISC's interpretation of Section 215 of the Patriot Act took bulk metadata collection to a new level. Instead of helping law enforcement agents track just suspected drug trafficking and money-laundering activities revealed in CDRs on calls from the U.S. to offshore, FISC now applied bulk metadata collection authorization to the entire panorama of CDRs generated in the U.S.

The change didn't sit well with communications service providers, who typically like to jettison call data records after they've served their purpose, and when they do store them tend to do so in truncated format that is less costly – and less useful to members of the IC. To carriers, data retention is a cost center. As a result, CSPs were among the slowest commercial enterprises to embrace big data analytics. Some failed to even retain sufficient data required to use the tool.

The same principle applied to Internet-focused companies. Always looking for cheaper alternatives, many CSPs abandoned the once firm principle of keeping customer data close at hand, and instead farmed it out to lower cost offshore data centers. Not that CSP lethargy mattered to the NSA in 2006. FISC's order meant that CSPs were required to store all CDRs and IPDRs and provide access to the files to the NSA, on demand. NSA's data access powers were further strengthened two years later by amendments to the Foreign Intelligence Service Act (FISA) of 1978. Under Section 702 of the 2008 FISA Amendments Act (FAA), the NSA could filter and copy telecom and phone traffic from networks, and store even incidental traffic of U.S. parties for analysis and investigation.

In addition, the NSA found that it could use Section 702 to advance surveillance measures in the U.S. under PRISM, developed for intercepts on fiber backbones outside America's borders. As long as a FISA Court order was directed at a carrier with traffic originating 51 percent from foreign nationals, Section 702 allowed the NSA to move forward, compelling a CSP or ISP to hand over call data. Remember FAA Section 702 as we proceed.

STELLAR WIND In Its Sails

In the succeeding years, bulk metadata collection capabilities grew in sophistication owing to more advanced packet capture and DPI capabilities, improvements in storage and search functions, and improved analytics. Foremost among these advancements was the NSA's SHELLTRUMPET, which from 2008 – 2012 combined classic collection of "Upstream" (network backbone) traffic with real-time analytics. Used by multiple agencies of the U.S. Intelligence Community, SHELLTRUMPET processed more than one trillion (1,000,000,000,000) documents through 2012, more than half of them in the last year alone.

The NSA in 2011 launched RAMPART-A, a program that opened foreign cables to American surveillance via the hosting of U.S. equipment in-country, not just amongst the "Five Eyes" nations but many other countries, with the full cooperation of those nations' intelligence agencies. On the list of cooperative parties: some 33 foreign countries. In keeping with NSA nomenclature practice, each location had a separate "all caps" name such as MOONLIGHTPATH, SPINNERET and AZUREPHOENIX. These three sites, the largest, by 2013 provided full data access across more than 70 cables. In all, RAMPART-A provided a data feed of more than 3 terabits per second.

Speed continued to be a major driver. As SHELLTRUMPET wound down, along came EVILOLIVE in late 2012, using a sophisticated new packet filter that doubled SHELLTRUMPET's performance.

Crunching EVILOLIVE's input: sophisticated search and analytics capabilities provided by X-KEYSCORE, used with all Upstream collection programs.

Together, the Intelligence Community's commitment to bulk metadata collection in the U.S. and abroad constituted a sizable investment of manpower and resources.

IP Rules

One contradictory element that cut against the grain of U.S. mass data collection efforts through at least 2012: The Obama Administration reportedly shut down bulk Internet metadata interception in the U.S. in 2011. Which leads to a new set of questions:

- If mass IP intercepts were forbidden by President Obama in 2011, then how did NSA EVILOLIVE get off the ground the very next year?
- Why did Congress go to the trouble of including already banned mass IP intercept in the list of the USA Freedom Act's taboos?
- Because NSA CDR requests focused only on the top three carriers, never comprised more than 30 percent of their network traffic, at most, and classic wireline telephony was even then a declining business anyway, why would the NSA fret over loss of the ability to access this set of CDRs?
- And what about Section 702 of the FISA Amendments Act – didn't that give the IC a way around USA Freedom Act restrictions?

Short answers:
- IP traffic is all that matters now and in the long run. For that reason alone, bulk metadata collection of Internet traffic never went away, and never will.
- The USA Freedom Act does not proscribe the types of traffic forbidden to bulk metadata interception. It simply states that those agencies applying for FISC authorization may not do so by "term" – a reference to the type of company whose traffic is to be intercepted in bulk. That is what the legal profession might call flawed language wide open to a loophole.
- The NSA doesn't care so much about the loss of access to voice network CDRs. Circuit-switched telephony is in its death throes. "Call" metadata has long been deemed as not particularly useful to intelligence.
- Section 702 of the 2008 FISA Amendments Act "giveth" what the USA Freedom Act "taketh away" – an automatic green light for capturing the IP and call traffic of greatest concern.

With the USA Freedom Act, Congress succeeded in banning a program that delivered limited intelligence from a network resource that average Americans, let alone suspected terrorists or their affiliates, are less and less apt to use.

The real gems of U.S. intelligence gathering –- programs that collect massive amounts of IP data and from global resources that overlap into the U.S. –- were untouched by the new law, and remain so.

How did U.S. legislators commit such oversights, and conclude by characterizing the USA Freedom Act as a "win" for privacy? One might fall back on the old "policy always lags technology" argument, except that the technologies used to collect IP data in bulk have been around for years.

The UK's Investigatory Powers Act

With passage of the Investigatory Powers Act (IPA) by the British Parliament in 2016, the UK implemented one of the most aggressive surveillance laws on earth. The IPA allows for "equipment interference" – a legal euphemism for the right to hack any network or device, at multiple locations. In addition, all communications service providers including ISPs are required to retain complete customer Internet data records for 12 months and allow law enforcement and government agencies to access the metadata without a warrant.

Specific codicils spell out the near limitless power of the government to monitor communications.

Section 136, Part 5 authorizes government agencies to "interfere" with any equipment "producing electromagnetic, acoustic or other emissions "or any device capable of being used in connection with such equipment. " In other words, any device on either end of a link, or any hardware in between, is fair game for government monitoring.

Section 264 sets forth the parameters:

In this Act, "systems data" means any data that enables or facilitates, or identifies or describes anything connected with or facilitating, the function of any of the following:

- *A postal service;*
- *A telecommunications system (including any apparatus forming part of the system);*
- *Any telecommunications service provided by means of a telecommunications system;*
- *A relevant system (including any apparatus forming part of the system);*
- *Any service provided by means of a relevant system.*

Another aspect of the IPA: the ability to "interfere" with equipment whether network-based or the end user's, *outside* the UK, provided the case involves British parties or interests.

The lever that determines whether a targeted warrant may be issued is the Section 108 requirement that "there is a British Islands connection." Here, too, the application is broad. The UK connection may be established if:

- *Any of the conduct authorized by the warrant would take place in the British Islands (regardless of the location of the equipment that would, or may be interfered with);*
- *Any of the equipment which would, or may be interfered with would, or may, be in the British Isles at some time while the interference is taking place.*

The phrase, "regardless of the location" means that the IPA can be used to justify a warrant to hack into equipment of systems data located anywhere – including outside the nation – if the intended crime or act of terrorism pertains to or might occur in the British Isles.

The IPA is but the latest chapter in a long series of UK surveillance laws including the country's Telecommunications Act of 1984, the Intelligence Services Act of 1994, the Terrorism Act of 2000, the Regulatory and Investigatory Powers Act (RIPA) of 2000, the Anti-terrorism, Crime and Security Act of 2001, and the Data Retention and Investigatory Powers Act (DRIPA) of 2014.

Up until enactment of the IPA, the principal laws governing lawful intercept in the UK were RIPA and DRIPA.

In general terms, RIPA read much like the U.S.'s CALEA. RIPA began by defining unlawful interception (done without knowledge of the target), lawful interception without a warrant (allowed only when the target has agreed), and lawful intercept, which could only be conducted under authority of a warrant issued by the Secretary of State. To obtain a warrant, the relevant LEA or agency was required to show that surveillance was essential to protecting national security, preventing serious crimes, safeguarding the country's economic well being, or in the interest of providing mutually beneficial international assistance.

Typically a RIPA warrant named the target, identified his or her location, and the types of communication to be intercepted within a set time frame – just like a CALEA court order. One significant extension: Under RIPA the Secretary of State had the latitude to overrule including such items in the warrant.

RIPA also applied to any telecommunications provider, meaning all communications services operators, not just those designated as being connected to the PSTN. Just as in the U.S. under CALEA, service providers who received a court order under RIPA were reimbursed for the cost of each instance of lawful intercept.

RIPA warrants were primarily issued to collect "telecommunications data," i.e., call metadata identifying the target and those who communicated with him, plus the date and time of their communications, and location. However, content access was not expressly ruled out, nor was the use of technologies to decipher encrypted data. Lawful intercepts were conducted without informing the target, and notification was required only after the intelligence gathered had been formulated into evidence for a court case. The British government's collection of cypher keys to obtain evidence was protected and non-disclosable unless the warrant expressly ordered disclosure. In a word, RIPA was very much like CALEA.

Passage of the UK's DRIPA in 2014 added a twist. Under DRIPA, UK telecommunications firms were required to retain all metadata "or any description of data" for a period of up to 12 months. Included were not just telecom, Web and social media data, but financial information, medical records, legal documentation and the "confidential" communications of Members of Parliament – potentially every bit of data generated in the UK.

RIPA and DRIPA seemed simple and straightforward enough. However, the picture changed dramatically when another UK law, the Intelligence Services Act, came into play. The opening line of the law under "Warrants: general" stated:

"No entry on or interference with property or with wireless telegraphy shall be unlawful if it is authorized by a warrant issued by the Secretary of State under this section."

And:

"The Secretary of State may, on an application made by the Security Service, the Intelligence Service or GCHQ, issue a warrant under this section..."

Under the Intelligence Services Act, as long as the Secretary of State gave formal approval, anything was allowed from the standpoint of surveillance capabilities. The basic rule of thumb in the UK as of 2015: If GCHQ or other intelligence services wanted to use whatever technology was required to pursue a suspect, they could do so.

Why did Parliament return to the subject in 2016 and pass the Investigatory Powers Act? Arguably, to thumb their noses at the EU. The EU Court had challenged RIPA in 2015 in a case that remained unresolved the following year when the British Parliament approved BREXIT. Passage of the IPA combined all powers of prior surveillance laws into a single Act, rendering the EU court's move on RIPA irrelevant.

France's 2015 Intelligence Act

France's national surveillance law came into being as a direct result of the January 7, 2015 attacks on the editorial offices of *Charlie Hebdo*, where staff of the publication were targeted by Islamic fundamentalist gunmen for printing humorist depictions of the Prophet Mohammed. Twelve journalists were killed by automatic gunfire and another 11 seriously injured. The terrorists responsible for the murders: a pair of brothers affiliated with Ansar-al-Sharia, a branch of Al-Qaeda in the Arabian peninsula. The incident brought to public attention the full force of radical Islamists' strategy to export terrorism beyond the Middle East to where European opponents lived. The same strategy would soon be adopted by ISIS, to offset rising losses on the battlefield. But France did not wait for ISIS. The *Charlie Hebdo* attack gave proof to law enforcement's conviction that French surveillance law was out of date and needed significant revision to keep pace with new communications technologies used by Islamic fundamentalists to facilitate terrorism.

On March 9, 2015, the French Prime Minister introduced the Intelligence Act, a law that significantly expanded the powers of law enforcement to monitor communications networks, and the requirements of all industry players – wireline and wireless communications service providers, ISPs, VoIP companies, social media services and web hosting companies – to aid in assisting investigations. The law led the way to staggering change and a new world order for French surveillance.

Service providers were, for the first time, required to install "black boxes" enabling deep packet inspection (DPI) and analytics capabilities in their networks, with the ability to collect customer content and metadata in real time.

In direct contravention of European Union edicts banning data retention in member states, the Act extended the time period for data retention in France. Service providers would be required to store content and metadata – both voice and IP – for up to six years.

To ensure secrecy, the law introduced new restrictions. All service providers came under strict non-disclosure rules forbidding any public mention or discussion of surveillance actions undertaken on their networks or facilities.

The penalty for non-compliance would be severe: up to €75,000 and imprisonment of up to 1 year for violation of the non-disclosure rule; and up to €750,000 and an imprisonment term of up to 1 year for failure to cooperate with law enforcement network monitoring, or failure to produce requested data.

After debates, the vote in favor was overwhelming. France's National Assembly voted 438 in favor versus 86 against and 42 abstentions. The Senate voted 252 for the Act, 67 against, with 26 abstentions. The Intelligence Act became law of the land on July 24, 2015.

Sweeping Change by a Law Long Overdue

The Hebdo incident provided the spark to update France's antiquated surveillance statutes. At the time of the attack, the country was still operating under the 1991 Wiretapping Act, a law implemented well before the advent of the World Wide Web or broad acceptance of mobility as the dominant form of portable voice/data communications or of VoIP, SMS and social media. The 1991 law was designed purely to support conventional telephone wiretapping. While French law enforcement had modernized surveillance methodologies to the extent possible with new mobile location and monitoring capabilities, they had done so without clear legal authority.

The Intelligence Act brought French law enforcement up to speed with criminal and terrorist elements who had been taking advantage of advanced communications services for years. Now law enforcement agencies had the law behind them.

Prior to the Hebdo attack, public opinion in France ran strongly negative on the issue of issue of surveillance. During the Arab Spring of 2010-2011, public outcry against the involvement of French ISS companies in the Middle East led to the dismantling of Amesys, a DPI specialist exposed for doing business with autocratic Middle Eastern regimes.

With the fate of Amesys fresh in mind, other French companies specializing in DPI and lawful intercept – notably Qosmos and Aqsacom – deliberately limited their business to export sales. Qosmos was so concerned about public backlash that when questioned about Russia's use of the Qosmos ixEngine DPI solution for surveillance purposes, the company denied it – though the facts to the contrary were plain.

The bloodbath of Charlie Hebdo turned public opinion around. The Intelligence Act broke new ground in France. Among the many major changes positively impacting national security via judicious use of surveillance:

- New Management. The Act created a new body, the Commission nationale de contrôle des techniques de renseignement (CNCTR), to provide direct oversight of surveillance. The CNCTR consists of nine individuals: four members of Parliament assigned by both houses of the legislature; two judicial judges chosen by the Court of Cassation, France's highest judicial court; two administrative judges picked by the Council of State; and one technical expert designated by the French National Regulatory Authority for Telecommunications.

The CNCTR reviews and opines on surveillance orders issued by the Prime Minister, except in exigent circumstances. The CNCTR also holds responsibility for all intelligence gathered during surveillance activities.

- Geotagging. In addition to conventional wiretaps, the Act authorized LEAs to "geotag" suspects via mobile location devices and techniques such as beepers and geofencing.
- Hacking. Law enforcement agencies may use intrusive techniques including Zero Day exploits, malware and Trojans to infect target computers and access, intercept and retain real-time communications, stored data, and keystroke logging generated by a computer and received by other equipment including "audio visual devices," a term which may be broadly interpreted to mean a feature or smartphone. Hacking is permitted by permission of the CNCTR in circumstances where evidence cannot be obtained by other means, and is limited to a 30-day period.
- International Surveillance. When targets roam across borders, French law enforcement is authorized to extend surveillance outside France including "communications emitted from or received abroad." The primary agency in such cases is the Directorate General for External Security (DGSE), France's foreign intelligence service organization. Under the Act, any foreign surveillance may be conducted with full impunity for the agents involved.

The Act also established the ability of French citizens to voice opposition to specific instances or capabilities of surveillance. Redress must first take place before the CNCTR, and if unsuccessful may proceed to the Council of State, France's equivalent to the U.S. Supreme Court, which also serves as principal advisor to the Prime Minister on legal matters of national import.

Legal Challenges to the Intelligence Act

Almost immediately following passage of the Act, "les "exégètes amateurs" [amateur exigists or interpreters], a group including French Data Network, La Quadruture du Net and other pro-privacy leaning entities challenged the government's right under the new Intelligence Act to monitor radio frequencies.

The pro-privacy group took their case directly to the French Constitutional Council, the country's highest judiciary group for judging constitutional issues. In so doing, the amateurs followed a comparatively new legal precedent. From its founding in 1798 by Napoleon through 2009, the Council had reviewed the constitutionality of proposed laws prior to their enactment by French legislative bodies. The thinking behind this approach: Permitting anyone to challenge the constitutionality of a law after its enactment would infringe on the authority of the legislature.

However, beginning in 2009, opponents of extant French law began to present their challenges to the Constitutional Council after the fact and challenge existing laws. One of the key drivers behind this approach: belief that European Union law, which tends to be politically liberal and privacy-focused, should hold precedence over nation state law.

In the case of "the amateurs" versus The Intelligence Act, the Constitutional Court on October 21, 2016 decided in favor of the litigants, ruling that the object of their litigation – radio surveillance as permitted under the Intelligence Act – was unconstitutional. Since that date, monitoring of mobile networks in France has been off-limits.

But French intelligence and law enforcement agencies arguably had one far greater concern at the time: a contractor that turned the nation's plan for an efficient, reliable country-wide surveillance system into a total fiasco.

How Thales' Bungled Handling of PNij Left France Vulnerable

Few nations have suffered more horrific terrorist attacks than France. Yet the two worst incidents, in Paris Nov. 13, 2015 and Nice July 14, 2016, actually followed France's deployment of a system designed to prevent such attacks – the PNij (Plateforme Nationale des Interceptions Judiciaires). According to French law enforcement officers, the PNij is still so bug-ridden that law enforcement officers forced to use it are "exasperated."

Launched by military contractor Thales on October 12, 2015, the PNij was intended to create a single unified source of intelligence gathering and sharing that would accelerate the process of identifying targets, put local and federal law enforcement on the same page, and help preempt both terrorist and criminal acts. PNij was intended to replace a quilt of multiple vendors serving various ISS needs, yield vast improvements in performance and results, and cut the state's national security budget to a fraction of its former sum. All communications service providers – wireless, wireline and Internet – would be connected, making it simple to access communications metadata and content. Altogether a grand idea.

Initially, users noticed significant improvements in several areas. Call metadata for a lawful intercept could be retrieved in seconds – versus the weeks or months delays experienced with prior vendors. Agents were generally satisfied with PNij's ability to quickly identify target phone numbers and IP addresses or determine mobile location. But user satisfaction was premised on the availability of these and other features on-demand and with total reliability, areas where PNij earned low marks.

Routine system failures required regular intervention by Thales. Among the innumerable problems cited: faulty connections, disconnects, features that didn't work or were missing. Just as bad, when investigators conducting 24X7 surveillance encountered a glitch, they discovered to their dismay that Thales tech support for the PNij only worked on weekdays, never on weekends.

French police didn't take to the system well. In November 2015, the Union of Officers sent a list of complaints about PNij to the Director General of National Police, stating, "This software seems unworkable and even could compromise investigations." Chiming in, an official with SCSI, France's Homeland Security Agency, warned of "catastrophic losses for investigations" if law officers were required to use PNij. Thales offered no immediate response. Then came the great PNij crash.

At 9am on Monday Feb 29, 2016, systems operators began receiving panicked calls from police who could not access the system to pursue lawful intercepts and investigations. Within hours, other branches of government were impacted: France's Anti-Terrorist Sub-Division (Sdat) and the Internal Security Directorate (ISDB). Was the problem simply a systems bug or a case of network penetration? Neither Thales nor the Chancellery in charge would comment. The PNij shut down for a week while Thales worked overtime to make it operable. But the damage was done. Some 1,800 wiretaps were lost. Asked to account for the debacle, Thales promised to "increase the application performance" of the PNij.

Some expressed doubt whether any further improvements by Thales would be worth paying for. In April 2016 a Court of Auditors pointed out that when Thales won the bid, PNij's budget was set at 17 million Euros and would thus drive a significant savings when compared to the estimated 50 million Euros paid annually to various ISS vendors. The auditors then asked why PNij's price tag had soared to nearly 90 million Euros. Thales' answer: lengthy implementation delays owing to a variety of factors – lawsuits by ousted ISS vendors, disagreements within the Ministry of Justice over management control of PNij, and an endless string of technical problems.

Little has improved since the 2016 audit. In late September 2019, the Ministry of Justice threatened to move PNij's operations under direct government management. Meantime, in the Paris prefecture, where the bulk of lawful intercept activities are centered, agents rarely use the PNij and instead rely on solutions by vendors such as Elektron, Foretec, Amecs and Azur. In other words, France is paying twice – once for the Thales PNij and again for smaller ISS vendors that police already know and trust.

Russia's SORM: National Surveillance KGB Style

SORM (System for Operational Investigative Activities) is the system under which Russia's Federal Security Service (FSB) conducts in-country surveillance.

To understand SORM, the state's system of in-country monitoring and surveillance, it is best to begin with an anecdote.

At a 2016 meeting of the International Telecommunications Union (ITU) held in Dubai, members heard a proposal from Russia's delegation: to switch management of Internet domain names and IP addresses from U.S.-based ICANN to an international body such as the ITU. While the measure sounded harmless enough, it concealed a very different motive: Russia's goal of Balkanizing the Web by transferring management from global companies to local or national control – in Russia's case a vast in-country intranet under management by the government. The goal was to move international Internet and social media companies that operate in Russia under strict Russian jurisdiction. SORM is the principle vehicle.

SORM, which began in the early 1990s and evolved over time, is not one, but actually three sets of rules:

- SORM-1: for surveillance of wireline and mobile communications.
- SORM-2: for monitoring Internet communications.
- SORM-3: for collecting all communications from all media and storing it for up to three years.

Effective August 2014, SORM-3 applies to all video, social media, Wi-Fi networks and websites.

How SORM Works

In principle, "lawful intercept" in Russia begins much as it does in the United States and European Union nations: The relevant law enforcement or intelligence agency, usually Russia's state police force – the FSB – is first required to obtain a court order. Other government agencies with the authority to use SORM include the Ministry of the Interior, and the Federal Prisons Service and Anti-Drug Agency. However, it is the FSB that defines interception procedures.

The communications service provider or ISP is required to pay for and deploy a technical solution that will accommodate the court order. Failure to comply can cost a provider its license.

SORM and Western laws diverge in one critical way. Under SORM, the FSB is not required to present court orders to the communications company involved or notify them that a surveillance investigation is underway. The carrier has no control over the surveillance device or knowledge of the suspect targeted or information intercepted. Furthermore, unlike in the U.S. and many other Western nations, under SORM there is no time limit on intercepts.

As in other countries, the mediation device, probe or server handling an intercept is linked by secure line to a monitoring center operated by the law enforcement agency for analysis, forensics and storage. Unlike in Western nations, all communications companies, ISPs and media companies are hard-wired by cable to a local FSB monitoring center. At any given time, an FSB agent can enter his region's monitoring center and check the status of a current intercept, or initiate a new one.

They keep busy. According to Russia's Supreme Court, the number of phone conversations and emails intercepted doubled from 265,937 in 2007 to 539,864 in 2012. Much of the increase came after 2011, following mass protests over Vladimir Putin's campaign to return to the Presidency. Russia's Supreme Court has not issued a report on wiretaps since 2012. One glib analyst commented that with SORM, Putin has created a surveillance state that would be the envy of his former employer, the KGB.

Ostensibly, SORM applies solely within country where it is leveraged by the FSB, which is responsible for counter-intelligence, counter terrorism, border and internal security, and criminal investigations. However, SORM's influence extends farther.

Former Soviet bloc countries that now belong to the Commonwealth of Independent States (CIS), notably Belarus, Uzbekistan, Kyrgyzstan, Kazakhstan and even Ukraine, which remains at war with Russia, often follow the same surveillance practices as Russia, under laws that mimic SORM. Kyrgyzstan draft legislation updating its surveillance laws copies SORM virtually word for word. CIS countries often favor Russian surveillance equipment, as well.

Customers interested in buying Russian equipment have a wide variety of products to choose from: active mediation and passive probes; mobile location; deep packet inspection; social media monitoring; voice biometrics; malware and IP video equipment and networks. Many of these surveillance solutions come with high recommendations by the FSB.

China's Counter-Terrorism Act vs. Western Laws

Of all the nations that have committed atrocities against mankind – including Nazi Germany, Stalinist Russia and the Empire of Japan before defeated in World War II – the Peoples Republic of China owns the title of "Most Murderous Regime in Human History." As disclosed in research published by the *Washington Post* on January 1, 2020, of the 100 million people killed by autocratic regimes in the 20th century, 65 million died at the hands of the PRC.

Bear that in mind as you read this comparative analysis of 21st century Chinese and Western surveillance law, and for that matter, whenever you purchase products made in the PRC, or stop and consider the vast number of Western companies that disregard all moral principle by doing business in or with a nation governed by slave masters.

Several years have passed since the 2016 implementation of China's Counter-Terrorism Act, a law that drew fierce criticism from privacy organizations, tech companies that do business in China, and former U.S. President Obama. Heading the list of complaints were objections to Chinese authorities' plans to require "back doors" into network equipment and end user devices, as well as access to all encryption keys by Chinese law enforcement and government intelligence agents.

The final law was toned down by Chinese authorities to assuage these concerns. In the interim, the world has not come to an end due to China's new law, as many fearmongers predicted would happen.

And for good reason. China's Counter-Terrorism Law would, on close inspection, appear to be little different from similar laws long in place in the United States – and in the United Kingdom and most countries in Continental Europe, where surveillance laws are typically far more aggressive than in the U.S.

For that matter, the Great Firewall of China so often denigrated by the Electronic Freedom Foundation, ACLU and other U.S.-based advocacy groups, continues to rely heavily on U.S. technology. Remove a brick from the Great Firewall and one invariably finds hardware by Blue Coat (now part of Symantec) and Cisco, which remain the bulwark of Internet surveillance and censorship in China, notwithstanding Cisco's public stance condemning surveillance by Western governments.

The facts: The Chinese draft law's original insistence on back doors and device encryption keys was not so far removed from common practices of the U.S. Federal Bureau of Investigation and Britain's Government Communications Headquarters (GCHQ), that nation's principal intelligence agency.

As illustrated in Chapter 4 on offensive cyber, the FBI has carte blanche powers to use any available technology in the quest for evidence, and wields massive parallel processing power sufficient to have hacked the San Bernardino terrorist's iPhone in 2015. Contrary to public speculation, Israel's Cellebrite had nothing to do with that successful hack. According to inside sources at the FBI, the agency handled mobile forensics on their own.

In the U.K, passage of the Data Retention and Investigatory Powers Act 2014 (DRIPA) gave the government power to acquire bulk metadata of every imaginable type and require ISPs to retain all consumer/business Web browsing histories indefinitely in case needed for future investigations. An earlier law, the Anti-terrorism, Crime and Security Act of 2001, might be best described as the Chinese Counter-Terrorism Act on steroids. Most recently, Britain's Investigatory Powers Act allows law enforcement to surveil citizens without limit, and commit privacy violations commonly seen in totalitarian states.

The seemingly limitless reach of Western law enforcement and government agencies on intelligence matters raises a legitimate question: Are the Chinese law and the actions taken under its authority so different from established law and common practice in other nations?

What China's Counter-Terrorism Law Requires of Carriers

The Counter-Terrorism Law of the People's Republic of China was enacted on December 27, 2015 and went into effect on January 1, 2016. The law consists of 97 provisions contained in 10 chapters. Compared to most U.S. legislation, the law is – for the most part – simple, straightforward and so "to the point" that a child could read it and understand every passage. Even translated from the original Han, the law is crystal clear.

U.S. lawmakers should take this irony to heart: policymakers in the world's most severely totalitarian regime compose policy more clearly and succinctly than policymakers in the "Land of the Free."

The authors of China's law state its intent right up top. The purpose is "to prevent and punish terrorist activities, strengthen counter-terrorism efforts and to safeguard the security of the state, the public, and the lives and properties of the people."

The Law is equally lucid about the compliance requirements placed on Chinese communications service providers.

Article 19: Telecommunications operators and internet service providers shall, according to provisions of law and administrative regulations, put into practice network security systems and information content monitoring systems, technical prevention and safety measures, to avoid the dissemination of information with terrorist or extremist content. Where information with terrorist or extremist content is discovered, its dissemination shall immediately be halted, relevant records shall be saved, and the relevant information deleted, and a report made to public security organs or to relevant departments.

Network communications, telecommunications, public security, state security and other such departments discovering information with terrorist or extremist content shall promptly order to the relevant units to stop their transmission and delete relevant information, or close relevant websites, and terminate relevant services. Relevant units shall immediately enforce [such orders] save relevant records, and assist in conducting investigations. Departments for network communications shall adopt technical measures to interrupt transmission of information with terrorist or extremist content that crosses borders online.

What China's Law Allows in Surveillance

If a surveillance technology has proven results, you can bet that China is using it. China is also in the forefront of tech advancement, among the first to explore and deploy quantum computing to recognize, identify, monitor and quickly capture terrorists.

Chinese law enforcement and government agency powers are essentially unlimited in this regard, although the law does stipulate the requirement of advance authorization:

- Article 45: *As needed for counter-terrorism intelligence information work, and on the basis of national provisions, public security organs, state security organs and military organs may employ technological investigative measures upon strict formalities for approval.*

Note that China has its own version of counter-intelligence organizations consisting, as in the U.S., of multiple agencies that span central intelligence, the military and law enforcement:

- Article 47: *The national counter-terrorism intelligence information center, local leading institutions for counter-terrorism efforts, and also public security organs and other relevant departments shall screen, scrutinize, verify and monitor relevant intelligence information, and where finding a threat of a terrorist incident occurring, and that it is necessary to employ corresponding safety precautions and response measures, shall promptly report this to the relevant departments and units, and may issue alerts according to the situation. Relevant departments and units shall complete safety precautions and response handling efforts in accordance with the report.*

China's Law vs. CALEA, ECPA and FISA

Article 19 is in at least one way reminiscent of the U.S. Communications Assistance for Law Enforcement Act (CALEA), which requires a telecom or wireless carrier and any broadband or VoIP service that interconnects with the public switched telephone network (PSTN) to be in technical compliance with the law, i.e., to have the appropriate technical solution in place to perform lawful intercept upon receipt of a court order.

The equipment must have the capability to intercept both call related data (metadata) and content. In the U.S., passive probes and active mediation systems perform both functions. Furthermore, probes, which detect targeted traffic via deep packet inspection could, if so configured, not only intercept but censor targeted content.

CALEA does not go that far, but many of the technology devices used to ensure an operators' reaching "safe harbor" CALEA compliance most certainly have the capability to do so. DPI is a rules-based engine, and the user can configure the rules however he or she sees fit, be it for more purely commercial purposes such as monitoring traffic flows in order to supply adequate bandwidth, or to intercept targeted communications, or to trap and censor specific types of traffic.

One key difference between China's law and CALEA is that the latter is not applicable to Internet, broadband or social media services that do not touch the PSTN. Also, as mentioned, U.S. surveillance laws are not in the business of censoring content.

CALEA lumbers on in general ignorance of social media, Internet and cloud communications that do not intersect the PSTN. Given the generally hostile attitude of U.S. lawmakers to surveillance, the odds of expanding CALEA to include these missing services still look fairly dim.

Neither CALEA, the Electronic Communications Privacy Act (ECPA) nor any other U.S. surveillance law provides the authority to order service providers to shut down terrorist or criminal content, as the Chinese law does. Thanks to the USA Freedom Act, carriers are not required to save the offending data, either.

But the U.S. Foreign Intelligence Service Act (FISA) does kick in on cross-border communications, just as the Counter-Terrorism Law of the PRC does. If a U.S. citizen sends or receives communications to or from an offshore entity, a U.S. intelligence agency may apply to the FISA court for an order to monitor and capture this traffic for investigatory purposes.

What the EU Says and Does

In 2016, the European Union joined forces with leading tech players including Microsoft, Facebook, Twitter and YouTube in executing a new "code of conduct" that will censor so-called "hate speech" on social media in Europe. Any message or language deemed objectionable now must be expunged from European comms within 24 hours.

The move is well-meaning and politically correct, but from the standpoint of intelligence gathering, sadly misguided. Hate speech is detestable, to be sure, but it can provide vital clues as to the identity and plans of dangerous elements including terrorists. The EU and leading tech companies are, in effect, wiping the slate clean and replacing offensive posts with "counter-narratives," whatever that means – they are not defined by the EU.

Otherwise, surveillance laws in European nations are in general significantly more comprehensive in scope than their U.S. counterparts, and on par with the 2016 Chinese law. Notwithstanding the new "hate speech" censorship initiative, laws in the EU nations provide the same powers now available in China – and have done so for a longer period of time.

Most EU member states empower law enforcement to perform full intercept of Internet comms, including Web surfing, email, chat, texting, VoIP, FoIP, ftp and telnet. Law enforcement agencies can demand decryption support and cypher keys of secured email, HTTPS web surfing, virtual private network (VPN) links, and encrypted telephony. Articles 20 and 21 of the Council of Europe's Convention on Crime permit real-time collection and storage of metadata and content.

Granted there are variations from one country to the next. France, for example, was comparatively lax on ISS until the January, 2015 Charlie Hebdo attacks inspired legislators to enact and enforce more aggressive surveillance laws. However, as discussed, the national surveillance network that was to have been established by Thales remains a technical train wreck. In lieu of such a consistent approach, a loose federation of smaller surveillance companies throughout France continue to operate with the same abandon and lack of oversight as a band of gypsies, a situation that became evident with the November 2015 ISIS attacks in Paris.

The Netherlands, in contrast, has always been in the forefront of European nations insisting on strong surveillance measures. Well before the events of 9/11 in the U.S., The Netherlands had a set of laws outlining service provider responsibilities in assisting law enforcement electronic surveillance and streamlining criminal and terrorist investigation procedures, both legally and technically. Moreover, the Dutch have operated a National Interception Office and a Central Bureau of Interception since the early 2000s.

The Netherlands Transport of Intercepted IP Traffic (TIIT) was for years put forth as a model by the European Telecommunications Standards Institute (ETSI) for state-operated surveillance activities. TIIT is a set of specifications for communications interception on IP networks. TIIT outlines the precise mechanism for handover interfaces at ports including HI1, HI2 and HI3, as well as access to crypto keys, email, login and logout and all content and metadata.

Initially TIIT relied on equipment supplied by Verint, SS8, Accuris, Pine and Aqsacom. However, use of Verint interception hardware may have "gone south" at least temporarily following revelations of the company's tight relationship with Israeli intelligence, alleged to have funded up to 50 percent of the company's R & D programs. This relationship raised the specter of an ISS vendor potentially leveraging "backdoors" in its own hardware to spy on friendly nations including The Netherlands and U.S. However, following the initial public embarrassment over this possibility and subsequent investigations, the spying charges were dropped for lack of evidence. TIIT went on and The Netherlands system of intelligence regained its luster as a standard setter, first as the National Signals Intelligence Organization (NSO) beginning in 2003, then centralized in 2014 under military authority as the Joint SIGINT Cyber Unit (JSCU).

Worth mentioning: With the increase in cyberattacks, JSCU is exercising its mandate to execute cyber offensive missions. The group's areas of authority include:

- Infiltration and monitoring of computers and networks to acquire data, to better understand the technology involved in offensive cyber work.
- Real-time predictive analytics to glean intelligence valuable for advance warning against attacks.
- Counter-intelligence on the cyber front.
- Close coordination of cyber intelligence with human and signals intelligence.

Counter-terrorism efforts in China are very close to the Dutch model: government-run but based on commercial products, with extensive integration amongst diverse intelligence sources. Many of the older ISS vendors present in the days of TIIT remain actively involved in The Netherlands. Some, like Verint, have grown and moved on to open global offices, with stations in Hong Kong and mainland China.

Global Perspective

In all, some 140 nations have enacted anti- or counter-terrorism laws since the events of 9/11 in the U.S. There are a few exceptions. Norway, for example, has resisted all such efforts at new legislation on the grounds that anti-terrorism laws threaten basic rights to privacy. Japan, which has experienced little terrorist activity, nonetheless assisted in the war on terror by providing troops to fight in Middle Eastern wars against Al Qaeda and ISIS, but has no counter-terrorism law per se.

With the death of two Japanese soldiers in combat action, the country began to take a more serious look at what it might need to do to prevent terrorist attacks in the home country. Japan's steps thus far: creation of a counter-terrorism task force to monitor any threats. Total headcount: 20 persons, which to some might appear a tepid response. Although Japan went on "high alert" following the November 2015 attacks in Paris, incidents of terrorism in Japan have been virtually nil since the sarin poison attack on a train in 1995.

The rest of the world has not been so fortunate. In retrospect, it would seem reasonable that any nation considered a target for terrorists must take whatever steps necessary to safeguard national security and public safety. Is Chinese law any more strict than similar laws of the U.S., UK or other Western nations? Not really. When critics condemn China's Counter-Terrorism Act, it's a classic case of "the pot calling the kettle black."

U.S. National Cybersecurity Strategy

On May 12, 2017, the Trump Administration issued its first Executive Order on Cybersecurity, outlining the President's strategy and mandate to improve security for federal agencies, critical infrastructure industries and other enterprises. Depending on one's interpretation, it was either perfect timing for President Donald Trump, cyber-wise – or Friday the 13th struck a day early.

At roughly the same moment, hackers unleashed the WannaCrypt (aka WannaCry or Wanna Decryptor) malware, attacking some 250,000 computers in 150 countries. We shall not digress into technical details on WannaCry here, as Wannacry is typical ransomware used to take over operating systems, and not particularly sophisticated. The important aspect is the impact on national policies and how the U.S. government has responded under the current Administration. The WannaCry episode proved to be Act I – Scene I showcasing Donald Trump's ineptitude in handling threats to national security such as the COVID-19 pandemic.

In this case, WannaCry demonstrated the extreme vulnerability of federal agencies and critical industry infrastructure to cyberattacks. To date, not much has improved. Despite the 2017 Executive Order, and the followup National Cybersecurity Strategy issued in September 2018 by the White House, federal networks by large remain as vulnerable as ever.

At the outset, the 2017 Executive Order had the earmarks of an aggressive plan of action, requiring all federal agencies to thoroughly review their networks, discover and disclose all vulnerabilities, and create concise plans of action to patch flaws and improve cybersecurity. Agencies were given 90 days to submit their analyses/plans to the Department of Homeland Security (DHS) and Office of Management and Budget (OMB).

Similarly, the DHS and other "relevant agencies" such as the Commerce Department and Federal Trade Commission were ordered to make their own evaluation of critical infrastructure sectors such as electric utilities, health care, communications, and public transportation, and report back to the President.

As a result, the Trump EO on Cybersecurity was viewed by most as a positive step forward. But well after the fact, there remain unresolved issues that the Administration either chooses to ignore, or is oblivious to.

Problems with the NIST Cybersecurity Framework

First, in two of the key sectors outlined by Trump's Executive Order – federal agencies and critical infrastructure – the yardstick for improvement was the government's "CyberSecurity Framework." The Framework was then, and still is a set of voluntary guidelines established by the National Institute of Standards and Technology (NIST) years before under President Barack Obama.

The NIST Framework contains the expected verbiage on the goal of the NIST program: to help agencies and companies identify risk within their infrastructure, develop and implement "appropriate safeguards," detect cyber threats, respond safely and institute "resilience programs" that expedite recovery from an attack. Nothing startling, it was typical high-minded government rhetoric.

The weak point of the Framework remains its set of four "Implementation Tiers" that describe "an increasing degree of rigor and sophistication" for managing cyber risks. The tiers range from "Partial" up to the most demanding level of readiness, "Adaptive." When issued in 2014, the Tiers allowed different bodies, be they federal or private, to choose the tier of cybersecurity deemed most appropriate based on their risk assessment, varying organizational needs, and regulatory requirements. Though updated in 2019, this phased approach to cybersecurity still rules the NIST Framework and is deeply flawed.

Just how weak is the Tiered concept? The Framework's description of the "Tier 1 – Partial" lays it out clearly:

"Organizational risk management practices are not formalized, and risk is managed on an *ad hoc* and sometimes reactive manner. . .There is limited awareness of cybersecurity risk at the organizational level". . .and "an organization may not have in place the procedures to participate in coordination or collaboration with other entities," i.e., experts more enlightened on safeguarding against cyberattacks.

While it is hard to conceive a federal agency having the audacity to report back to the DHS with a Tier 1 – Partial plan, particularly in the wake of WannaCrypt, that tepid level of cybersecurity is still in the NIST guidebook, and allowable.

Another key factor that has led to delay, curiously unmentioned by the Framework's authors, is obvious: cost. While not so much a concern for federal agencies funded by taxpayer dollars, cost may well prove a major obstacle erected by the private sector.

What Trump Deleted from the Executive Order

On October 3, 2016, a month before the Presidential election, Mr. Trump vowed to make cybersecurity a top priority and to issue an Executive Order "within 100 days of taking office." Reality set in quickly as the President's new team fathomed what a truly convoluted issue they were up against. A signing ceremony for a cybersecurity Order was scheduled for January 31, 2017, and cancelled the morning of the event. Then a first draft of the Order was leaked in early February, giving a good indication of what must have been the major challenge for the President: balancing the need to propose strong cybersecurity measures against fear of alienating the business community with new regulations – and the cost of implementing those rules.

To end-run business objections, the Administration's initial ploy was to command a federal report "on options to incentivize" critical infrastructure industry cooperation with "effective cybersecurity measures." The goal was to "encourage investment," but the draft Order did not specify how. Speculation ran rampant that the Trump team might seek tax incentives. Another rumored option: discounts on cybersecurity insurance with the added benefit that those companies so covered would not be penalized in the event they were taken down by a successful cyberattack.

But when the final Executive Order went public on Friday May 12, 2017, the incentives language was missing.

Also absent: language from an earlier draft provision encouraging cyber training for employees of critical infrastructure and commercial enterprises. The specific language in the draft called for "workforce development review . . . to understand the full scope of U.S. efforts to educate and train the workforce of the future."

Modernization of federal IT, long a key concern, was initially part of the plan. Karen Evans, then cybersecurity leader of the Trump transition team, was quoted saying, "We're only fooling ourselves if we believe that you're going to take $80 billion of legacy systems and lift, shift and re-engineer it into a 21st century solution." Modern networks were seen as key to effective cybersecurity.

The cost of maintaining government's antiquated IT systems has since soared to $90 billion per year. On a positive note, a new law – the IT Modernization Act – made its way into the 2017 Appropriations bill, providing fresh funding for advances such as migration from reliance on data centers to cloud infrastructure beginning in 2018. But two years later, there are still major gaps in the Administration's thinking. One such overlooked area: What is the U.S. doing on the critical infrastructure front? Answer: not much, largely through fear of drawing swords with big business.

Enemies in Strange Places

Overlooked by the Trump Administration is the fact that other key agencies have a role in monitoring cyber readiness. The Federal Communications

Commission (FCC), charged with regulating much of the U.S. telecom and broadband industries, is one such body.

Following public distribution of the Framework in February 2014, the FCC took review of telecommunications industry's cybersecurity preparedness under its wing. That same year, the Communications Security, Reliability and Interoperability Council (CSRIC), a joint study group including FCC officials and industry experts, undertook a study of telecom's state of cybersecurity readiness. Members then included senior executives from the giants of the network business such as Verizon, as well as a smattering of federal agency and public safety representatives. Their recommendation: Companies should follow the guidelines in the NIST Framework and adopt appropriate levels of security.

Again, the recommended strategy was all voluntary. Reason: Many of the same companies involved in the CSRIC study actively opposed new regulation and any form of mandatory investment in cybersecurity. Years later, telecom companies still may choose to follow the NIST guidelines – or not. Verizon's customers have since paid the price: a major hack discovered in July 2017 that exposed the records of 14 million customers.

Lack of mandatory cybersecurity requirements will continue to plague both federal agencies and the private sector in the U.S. Until such rules and penalties are in place, any so-called cyber strategy issued by the Administration will be policy without teeth.

2018 National Security Policy

In September 2018 the White House issued a new "National Cybersecurity Strategy" detailing the President's long term approach to safeguarding critical assets from cyberattacks. Responsibility is divided between the Department of Homeland Security (DHS) for federal civilian agencies, and the Department of Defense for the military branches – remember that members of the U.S. Intelligence Community such as the NSA report to the DoD, and thus are military agencies.

The DHS approach relies on a four-phase system known as Continuous Diagnostics and Mitigation (CDM). The phases: What is on the network?; Who is on the network?; What is happening on the network?; and How is data protected? The rationale behind CDM is sound, setting uniform standards for how scores of federal agencies implement cybersecurity. But there are problems.

First, agencies may choose the phase they need and graduate to higher phases at their own pace. Second, some agency cyber leaders dislike the idea of relinquishing authority to a central administrator and may slow-roll adoption. Finally, the term, "uniform standards," does not mean uniformity of hardware and software across all agencies – it is inevitable that overlooked vulnerabilities will arise in some brands, leaving the compromised agency open to attack.

As is the case with its polar opposite, ISS technology solutions, deployment of CDM is left to government contractors, but has not been nearly as effective. In many instances, federal civilian agencies are long wed to ineffective cyber defenses and reluctant to replace them. Office politics interferes, as well: high-ranking department heads are reluctant to accept help that might inadvertently expose the unacceptable vulnerability of cybersecurity systems they have lorded over for years. As a result, contractors that have won major contract awards from DHS to support multiple agencies often find themselves in the position of having to sell once again – to each of these end user communities.

The final step in this ballet of mis-steps is by DHS itself. The agency proudly acknowledges that it has tested and approved some 56,000 cybersecurity solutions for use by federal civilian agencies. Asked if this is a daunting number for agencies to consider, and if approval of so many doesn't render such rating utterly meaningless, DHS had no comment.

The Wassenaar Arrangement

The Wassenaar Arrangement is a "multilateral export control regime" (MECR) established to encourage transparent, responsible transfer of conventional weapons and "dual purpose" goods and prevent their being obtained by rogue states or any nation with questionable human rights practices. The Arrangement is named after a suburb of The Hague where it was first signed by 30 nations in 1996.

The three requirements for membership are: (1) manufacture of arms or "sensitive" industrial equipment; (2) adherence to a nuclear nonproliferation treaty; and (3) effective export controls. The current 41 nation states that follow the recommendations of the Wassenaar Arrangement meet twice annually to exchange information on the sale of conventional weapons to non-members. The recommendations within the Arrangement are often the model of or influential to members' export rules. In addition, while certain nations such as China and Israel are not members, they are reported as having modeled their export controls on the Wassenaar Arrangement.

Of note, the Arrangement is not a treaty and thus not legally binding on members. As set down, the Arrangement allows a degree of latitude:

"Export controls are implemented by each individual WA Participating State. Although the scope of export controls in Participating States is determined by WA lists, practical implementation varies from country to country in accordance with national procedures."

However, though not binding, the Wassenaar Arrangement remains influential as a joint multinational effort to set the rules for export of conventional weapons, weapons of mass destruction (WMDs) and "dual-purpose" goods and technologies.

"Dual-purpose" defines technologies that can be used not only for commercial purposes but also in ways that per the Arrangement's new conventions require strategic export controls. In 2015, the Wassenaar Arrangement expanded the definition of "dual purpose" to include surveillance technologies.

Including conventional weapons or WMDs on the list is a cut and dried issue. But in many instances, "dual purpose" technology is a gray area.

For example, SS7's primary use is to create a separate signaling path for communications transiting the network. Deep packet inspection (DPI) provides the means for Internet companies to monitor the bandwidth requirements of traffic by author and type and set usage-based market rates. The two technologies' respective deployments as location and content surveillance technologies are a secondary application. But to the technology industry as a whole, not just the electronic surveillance sector, the Arrangement's new conventions could undermine the sale of any device or solution capable of "dual purpose" use – whether or not its innate surveillance capabilities come into play.

The move to including surveillance tools in the WA's list of dual-purpose technologies was initiated and driven by the United Kingdom.

In the UK, responsibility for formulating export rules falls to the Department for Business Innovation and Skills (DBIS), in charge of the UK strategic export control lists. New export rules for intrusive surveillance solutions, classed under "arms and controlled goods," went into effect in late 2014. Thus 2015 was the first year for which full disclosure on the export of these technology solutions by the UK was made publicly available. Evidently, efforts to restrict sales of surveillance solutions to proscribed nations did not go as planned.

According to the records, in 2015, the UK government granted export licenses valued in the millions of British pounds for solutions that include IP monitoring systems, ethical malware and IMSI catchers. Buyers included Egypt, Saudi Arabia and the United Arab Emirates (UAE), i.e., some of the very authoritarian regimes considered off-limits by the Wassenaar Arrangement.

The act of continuing sales of these items would appear to fly in the face of the DBIS's 2014 export rules based on the Wassenaar Arrangement.

A possible reason for this discrepancy may lay with the UK Home Office. The HO oversees country-internal areas related to national security: law and order, immigration and security – the latter including MI5. MI5 is predominantly concerned with police, internal security and counter-intelligence, working alongside departments that handle foreign threats: MI6, Government Communications Headquarters (GCHQ) and Defense Intelligence. However, much like the FBI, MI5 does have authority to carry out its mission offshore.

The Home Office also crosses the line from government activities into the commercial arena of national security and law enforcement – as sponsor of the Center for Applied Science and Technology (CAST) Security and Policing event held every March in the UK. At CAST, surveillance solutions, including the type proscribed by the Wassenaar Arrangement without strict export governance, cover a sizable swath of products and services exhibited. Granted, FinFisher, NSO Group and Trovicor are not on the list of exhibitors. But a great many other major brands that offer "intrusive" surveillance systems are, such as Hewlett Packard, iPS, NeoSoft, Polaris Wireless, Rohde & Schwarz and Telesoft. These companies sell their spyware under the wing of the Home Office in a unspoken but nonetheless very tight partnering relationships.

In a word, while the DBIS goes about the business of regulating the export of intrusive surveillance products, the HO actively assists in their marketing. By authoring rules that tightly control the UK's export of intrusive software and surveillance products for IP networks under the Wassenaar Arrangement, the Department for Business Innovation and Skills is butting heads with the all-powerful Home Office, and specifically with MI5. Considering the evidence – the volume of export licenses granted for intrusive surveillance systems – DBIS would appear to have lost that battle.

The situation is similar in the United States and other nations that adhere to the Wassenaar Arrangement. The requirement for export licenses has not in any way deterred the sale of surveillance solutions to proscribed nations. However, the rules do provide insights on the companies that partake of this activity, the solutions they sell, and to whom. To the extent that the Wassenaar Arrangement promotes visibility for tracking such sales, this unusual international accord has proved a success.

CHAPTER 2

Packet Monitoring

Overview

Packet monitoring technologies are an outgrowth of network management and analysis, subsequently applied to intelligence support solutions (ISS). Current thinking revolves around three approaches: deep packet inspection (DPI), IP Flow Monitoring, and Network Packet Brokers. All are used for commercial applications as well as lawful intercept and intelligence collection activities.

In the ISS arena, packet monitoring technologies are commonly used in lawful intercept solutions for the purpose of monitoring IP networks. DPI, IP Flow Monitoring and Network Packet Brokers may also be used standalone for other surveillance needs.

Although we treat the three categories of packet monitoring as separate capabilities here for the sake of simplicity, in the ISS world they all serve the same purpose: collection of packets for the purpose of identifying, tracking and capturing targets. Many rely on hardware accelerators such as field programmable gate arrays (FPGAs) to ensure rapid screening and capture of packets. Finally, the three monitoring solutions may be combined in various ways. One often sees IP Flow Monitoring paired with DPI, the former to sample anomalous packets on high-speed networks, and the latter to perform full inspection of packet content.

Deep Packet Inspection

DPI is a multi-purpose tool, a primary means of intercepting and examining suspicious data communications, as well as for improving the efficiency of networks via prioritization of traffic. In both disciplines, DPI has the power to determine traffic and type of application, protocols, origin, content and destination of every packet that transits a network, or alternately, of specifically targeted data streams.

There is little agreement on when, where or how DPI began, or even if it qualifies as a technology. Like many developments in the networking field, DPI represents an evolution and agglomeration of capabilities over time. Many features common to DPI, such as packet sniffing and packet inspection as part of the routing process, existed many years before the term "DPI" emerged to define the combination of these and other monitoring functions.

DPI, like its predecessors, focuses on live data. But unlike classic packet inspection used in facilitating data transmission, DPI does not stop at un-packing the header to determine its destination. DPI opens the entire packet including the payload.

DPI also does far more than simply decide where to route packets. It analyzes bit streams to perform two core functions essential to surveillance: recognition and notification of patterns. A third DPI function, manipulation, is widely deployed on the commercial side to determine consumer interests and purchasing habits, as well as by nation states to censor or block access to outlawed content.

Exactly what a specific DPI application recognizes depends entirely on the patterns or trends it is programmed to identify. This function, controlled in the DPI engine, might direct DPI apps to search for and single out particular protocols, urls, content, applications, text strings, malware, or specifically formatted data such as a credit card number or bank account.

The rules that are programmed into the DPI engine have various names such as expressions, rules sets, signatures and fingerprints. Moreover, DPI apps will only act on the rules dictated by the engine, nothing else. Omit a rule that might track evidence critical to detecting a potential criminal, terrorist or hostile nation state act, and the operator is out of luck. DPI cannot find what it is not instructed to look for. Because the criteria of an investigation might change constantly, the DPI rules needed to support it may require almost continuous updating, or new rules. In short, DPI acts very much like a service.

In action, the recognition process moves from the simple to the complex quickly. Core rules might tell the DPI app the basics to look for in potential malware. A more sophisticated layer of rules will outline the patterns to watch for. A final set could be very specific, detailing precise identifiers to apply to a target's traffic at exact points in the network.

Doing it all in real-time is the trick. The use of DPI to scan and assess vast amounts of data "in the moment" can quickly lead to scalability issues, depending on the complexity and quantity of predefined patterns it is asked to search for. Scalability might not be a problem for NSA facilities at Fort Meade, Maryland, or Camp Williams, Utah, but it can be an issue when DPI is used in routine lawful intercept scenarios.

A central feature of all DPI solutions is the ability to send notifications or alarms, depending on the priority of findings set by the user's predefined rules. In the commercial world, the alert might pertain to a customer approaching the halfway mark on his or her monthly texting allotment. For government agencies, an alarm might be triggered when individuals with known or suspected criminal, terrorist or foreign intelligence affiliates set a meeting or give some indication of an imminent attack or other event. Otherwise, the DPI solution simply does its job monitoring assigned data communications traffic and producing reports.

As used in surveillance, DPI is passive. It performs recognition and notification, but not manipulation, which for obvious reasons would make targets aware that their traffic is under scrutiny. Depending on the user's available processing power, DPI may be used to capture traffic in-line in real-time – a huge scalability issue given today's data network volumes – or off-line, in which case copies of network traffic are captured and stored for subsequent analysis. The chief benefit of the in-line option is total invisibility to the target. The challenge: processing power behind the DPI must be on par with the speed of the network being monitored, either through massive parallel processing, FPGAs or both.

As a profiling tool, DPI is enormously useful to government agencies, the military and law enforcement. By ferreting out patterns associated with dark or malicious activities, or to spot the use of Tor and other encryption tools, DPI solutions work continuously to protect national and homeland security and public safety.

The components found in DPI hardware and software frequently cross nation state borders. For example, makers of lawful intercept device manufacturers such as Germany's Utimaco use DPI products made by Procera Networks of California, USA. France's Qosmos ixEngine, a DPI solution widely used by U.S. communications service providers, also powers Russia's Protei, an ISS solution used for SORM lawful intercept and government intelligence, not withstanding Qosmos' public stand that it forbids the use of ixEngine for surveillance purposes. This latter point illustrates another fact-of-life where DPI is concerned. Once it leaves the manufacturer's hands, DPI may be used for virtually any type of network monitoring that its owner wishes to pursue.

IP Flow Monitoring

The concept of IP Flow Monitoring surfaced in 1996 with Cisco's commercial launch of the NetFlow protocol, roughly concurrent to when DPI emerged. Like DPI, NetFlow was developed as a network management tool, in this case to help users of Cisco routers better understand traffic flows and support customer needs, or to signal the need for more scalable hardware or virtual (software) network facilities to serve rising bandwidth demands.

NetFlow quickly inspired competition as other makers of IP routers introduced their own versions of the protocol for their devices. At the same time, NetFlow began to see use as a surveillance capability. Because NetFlow is a generic term and not trademarked by Cisco, it has become interchangeable with "IP Flow Monitoring" in both the commercial and surveillance arenas.

At one level, IP Flow Monitoring appears to act exactly like DPI. It collects data at a network device interface and exports this data to an analysis engine. While the two methods of network monitoring thus seem quite similar, the fundamental operating principles of each are radically different from the other.

DPI can be configured to mirror, open and analyze selected packets or every packet that enters or exits a router interface.

In contrast, IP Flow Monitoring *samples* traffic and via a different methodology – IPFIX, an international standard for formatting selected traffic streams, and forwarding them to network management, lawful intercept mediation devices, or to sophisticated analytics systems used by government intelligence agencies. To underscore the importance of IPFIX, makers of IP Flow Monitoring solutions often couple the protocol's name with descriptions of product capabilities, e.g, "NetFlow IPFIX. As network speeds soar beyond 100GB, IP traffic sampling has proved an effective means to measure network activity at each node monitored. However, critics contend that IP Flow Monitoring falls short for government surveillance purposes because it inevitably misses packets. While dropped packets may be of little concern to network operators charged with overseeing general network conditions, to LEAs and government agents the same prospect raises concerns that essential evidence might go missing. Advances in a sister technology have all but eradicated such fears.

Hardware accelerators known as field-programmable gate arrays (FPGAs) today are often used with both DPI and IP Flow Monitoring systems. FPGAs facilitate massive parallel processing with zero packet loss even in the highest-speed networks. Because FPGAs are programmable, they can be configured differently to meet the objectives of changing network or surveillance scenarios, in contrast to ASICs-based hardware, which is "wired" at the factory to perform specific functions. With FPGA-enabled IP Flow Monitoring tools, the user can perform most of the same jobs typical of DPI: target and identify specific IP addresses, MAC addresses and other identifiers, as well as examine packet content.

Network Packet Brokers

Network Packet Brokers (NPBs) are packet filtering systems capable of monitoring IP traffic flows at line rate speeds, providing dynamic visibility into the network, both for commercial and surveillance purposes. NPBs have grown in popularity with the evolution of software defined networks, virtualized networks, the cloud, and the attendant rise in the volume, bandwidth, complexity and variety of network traffic.

Traditionally, NPBs have been hardware-based, using either ASICS or FPGAs. Increasingly, however, vendors are looking to "virtual" or software-centric NPBs to provide the same functionality at lower cost. As usual, a battle has ensued between hardware and software vendors.

Software proponents contend that NPB hardware is too complex and expensive, meddles with the network, and is slower and harder to reconfigure than software. NPB hardware proponents fire back that its competitors' virtual NPB products are less reliable. Over the horizon, some foresee network equipment that comes packaged with NPB capability. Meantime, the two primary forms of NPB, virtual and hardware-based, co-exist in the marketplace and find buyers, depending on applications and budgets.

Any discussion of how NPBs operates begins with "SPANs" and "TAPs." SPANs are Switched Port Analyzer ports, interfaces on network switching devices that copy and aggregate traffic, then send it on to monitoring devices or software for analysis. TAPs are more sophisticated mechanisms for monitoring two-way IP traffic on a pass-through basis.

SPAN ports, which emerged in the legacy network era, maintain a role in communications service provider networks today. They have good and "not so good" characteristics to consider. At the link level, SPAN ports may interface with copper, fiber or both types of line connectivity at various speeds. As a result, SPAN ports offer the advantage of low cost compared to more advanced alternatives. But they come with built-in issues.

One problem is that most network switching hardware is equipped with only one or two SPAN ports, meaning that multiple lines must compete for connectivity to access traffic for monitoring and analysis. Before long, port availability is overwhelmed and can lead to data loss during a network monitoring scenario. Traffic that consumes more bandwidth, such as social media and streaming video, may contribute to data loss.

Another issue: As network speeds advance, legacy monitoring tools designed for 1G networks must be replaced by tools for 10G, then for 40G and 100GB. With each installation of new monitoring tools, any savings generated by reliance on SPAN ports tends to diminish.

Finally, SPAN ports may encounter hurdles such as switches or routers that do not support port mirroring at all, or that slow traffic flows when port mirroring is introduced.

The next generation of data collection and aggregation came about with development of the TAP, a term sometimes deemed an acronym for "test access port," though there is no firm proof to support that claim. Originally designed as a portable test device, the TAP evolved into hardware for permanent deployment in the network, or in an enterprise database, to monitor traffic on a pass-through basis in "dual mode," i.e., both directions simultaneously.

At present there are three types of TAPs: (1) basic TAPs where the number of ports on the TAP hardware is equivalent to the number of network ports; (2) aggregation TAPS that use one port to monitor multiple network ports; and (3) regeneration TAPs wherein a single TAP is used to mirror traffic from one part of the network, then passes the data along to multiple monitoring and analysis systems.

A TAP may be either in-line (active) or off-line (passive). When a TAP is deployed in active mode, network traffic flows through the device and may be modified or blocked. Passive TAPs, in contrast, simply mirror the traffic and send alerts to users when the device observes targeted communications. Note that a network TAP may be configured in either active or passive mode for data collection, but monitoring is strictly passive or off-line. In other words, any data missed in the collection stage cannot, for obvious reasons, be subjected to deeper analysis.

The arrival of Network Packet Brokers or NPBs represented, to many, a breakthrough. An NPB literally sits between a network switching device's SPAN ports or routers to take charge of traffic flows. In addition to performing the essential monitoring function of filtering high-bandwidth traffic flows, an NPB can do much more: filter, de-duplicate and time-stamp packets, and be coupled with deep packet inspection (DPI) capabilities from Layers 2 – 7 of the OSI stack. In other words, an NPB can provide end-to-end inspection of targeted packets through the OSI Layer 7 application level – where much intelligence resides – together with the target's IP address, MAC address and content.

The NPB can scale to monitor entire networks while at the same time drilling down to provide visibility and analysis of the link layer of IP traffic. It can see and identify any packet on the network and provide full packet capture with accurate time and port stamping. With an assist from DPI, the NPB can also open and view full packet content.

Given its capabilities, speed, intelligence and redundancy, the NPB is perfect for network forensics when connected to probes, mediation devices and other intercept hardware. But again, NPBs don't necessarily operate in isolation. NPBs are often found in combination with DPI capabilities. The same goes for NPBs and NetFlow. The Cisco Series 9000 router uses both Cisco "NetFlow IOS" IP Flow Monitoring and the "Cisco Network Data Broker," an FPGA-based NPB for accelerating packet capture.

DPI "Versus" IP Flow Monitoring

Widely portrayed as a dystopian technology with the power to penetrate any/all IP communications and manipulate traffic, DPI is often cast as the "Big Brother" of modern society. There may be an element of truth to that assertion. DPI has a variety of key capabilities: the ability to target specific individuals or groups and identify IP address, MAC address, and location. But DPI's primary benefit is the capture and analysis of packet payloads at the individual packet level. DPI is first and foremost a means of accessing IP content.

As discussed, DPI is not infallible. A system is only as useful as the rules it is based on. In every instance where DPI is deployed, it must be programmed with predefined rules that determine the information elements (IEs) and trends the system should look for. Moreover, each DPI system must be continuously updated to stay current with evolving threat scenarios.

Variable parameters including operating systems and line speed scalability profoundly influence accuracy. In North America and Europe, for example, a DPI system's level of accuracy in determining applications at Layer 7 might reach as high as 95 percent, but elsewhere can drop as low as 60 percent.

In terms of OSI/TCP-IP stack penetration, DPI capability depends on what the user purchases. Some DPI solutions used for lawful intercept only penetrate as high as Layer 2, the Data Link, responsible for the transfer of data across networks. In many instances, Layer 2 Access identifies packets only by originating and terminating hardware addresses: who's sending and receiving data. Moving upstream to the packet payload held in Layers 4 – 7 involves greater sophistication and cost, particularly in a high-speed networking environment. A general rule of thumb with DPI is that the more bandwidth a network offers, the higher the cost of DPI system elements that perform capture, storage and analysis. As networks reach to 100G and beyond, some would argue that such costs have become prohibitive for all but the largest LEAs and government intelligence agencies. Indeed, the cost of massive DPI is one factor behind the rapid rise in popularity of flow monitoring, which is more scalable at lower cost. But, as always, the user gets what he or she pays for. Flow monitoring alone cannot match the depth or detail provided by DPI.

Flow Monitoring

Flow monitoring developed on a separate track from DPI as a method of sampling IP traffic flows with specific predesignated identifiers. Flow monitoring checks and analyzes the flows of traffic at the packet header level only, versus examining every packet. Given the scalability of flow monitoring, and the high speeds at which it processes data, Flow Monitoring is designed to operate in the big data realm of today's networks.

Commercial flow monitoring works in four stages. First, an Observation Point is established using a probe powered by an FPGA line card or other high performance network interface controller (NIC) such as Application Specific Integrated Circuits (ASICs) hardware. FPGAs accelerate hardware performance in the packet capture process, are accurate to within 100 nanoseconds, and less costly and more flexible than ASICs – thus are the preferred method. Alternately, the user may opt for a forwarding device that is connected to a probe. Whatever the type of flow monitoring vehicle chosen, the results are the same: packets are captured and pre-processed using key flow monitoring protocols that encapsulate flow records into messages. Cisco NetFlow v5 and v9, and IPFIX (Internet Protocol Flow Information Export) are the dominant protocols.

In the second stage, a metering and export process within the probe identifies flows by factors such as port numbers, IP addresses, packet and byte counters, and then time stamps the results.

The third stage is data collection, done by Flow Collectors that format and store data. Storage may be either "Volatile," meaning done in real-time by flash memory, or "Persistent" for long term storage, a process that is more time-consuming.

Finally, in the fourth stage, flow monitoring performs high-level traffic analysis, looking for primary targets that appear with greater frequency and might indicate any behavior considered abnormal, e.g., a brute force cyber attack or DDoS attack.

There are technical issues to consider before proceeding. Flow monitoring works almost flawlessly on wireline networks, as well as virtual networks, which act as wired LANS. However, wireless networks do present one challenge. While it is possible to use a NIC card interface for flow monitoring of a wireless network, such cards can only capture one frequency at a time. Therefore, it is advisable when monitoring a wireless network via flow monitoring to place the NIC card or FPGA at the wireless LAN controller.

As mentioned, DPI and Flow Monitoring initially were viewed as diametrically opposed approaches to network monitoring. DPI was considered all-encompassing for its ability to access and analyze the full packet payload at Layers 2 – 7 of the OSI stack, and thus reveal not only the application used but the exact content. Flow Monitoring was seen as less comprehensive for its packet header-only focus, but as a direct result of such refinement – more scalable, faster and less expensive.

However, the two technologies have always borne similarities. Flow monitoring provides critical data on targeted flows including originating and terminating IP addresses, port numbers and metadata. Flow monitoring systems may be deployed in in-line mode to intercept traffic between two hosts, or in mirror mode to make exact copies of packets transiting between ports. DPI captures similar types of data (and more), and may be deployed in the same ways. But again, the sticking point with flow monitoring is its focus on "packet sampling," capture of all packets or examination of packet payload as is the case with DPI. With flow monitoring, packet sampling is accomplished by means of filters that restrict capture criteria.

While filtering certainly reduces the amount of information gathered, flow monitoring has grown more robust over time thanks to a key addition in the form of deep packet inspection. For example, any flow monitoring system using the IPFIX protocol can work at Layer 7 of the OSI stack to gather data on applications used by a target. Such "application awareness" is made possible by integrating DPI with the flow monitoring platform.

Field-Programmable Gate Arrays

Field-programmable gate arrays (FPGAs) are circuit boards that can be programmed by the user for a variety of applications including lawful intercept and government surveillance. FPGAs are important to ISS because they accelerate the capture of "unstructured" data, i.e., any type that does not follow a predefined data model, from voice to SMS, video, HTML web pages or text.

As unstructured data continues to dominate traffic, and at ever higher velocities, new methods for capturing it become all the more critical. Advanced FPGAs have grown into this role. Without FPGAs, much data essential to an investigation might go missing simply because the massive amount of traffic involved overwhelms conventional collection systems.

When first invented in the 1980s, FPGAs were so large that it was difficult to fit more than a few onto a single chip, meaning the engineer designing an end user system had to create an interface between the processor and an external FPGA or an entire subset of multiple FPGAs. The processing unit itself was still based on ASICS (application-specific integrated circuit) built-in at the factory, and thus not programmable in the field.

Designing external FPGAs was time-intensive. The concept of high power FPGAs that act as "coprocessors" so small that multiple units could reside in a chip – and allow individual logic elements (LEs) within the FPGA to be changed without affecting other systems functions – was years away.

Today a variety of companies offer tiny, high-performance FPGAs that are or can be embedded in quantity in surveillance systems. The principal vendors behind FPGAs are Xilinx and Intel (through its 2014 acquisition of Altera), though other important brands are also actively involved: IBM via its SPSS and Netezza products, Fusion IO, GiDEL, and ALBEDO and EZChip (now a property of Israel's Mellanox), to name a few. Why is their work so important and where did it all begin?

At least partial credit goes to a team of researchers from MIT, the University of Glasgow and Hewlett Packard who joined forces in 2012 to prove the merit of FPGAs for unstructured data capture and analysis. Their seminal work helped fuel the popularity of products made by a number of vendors that continue to improve FPGA circuit boards today.

Initial Lab Testing

In a data-centric networking environment, the major considerations of data capture are high-speed performance, energy requirements and heat, all of which combine to fuel a fourth overriding issue: cost. Increase the amount of data flowing into a data center or through a traditional surveillance system and the classic solution was to add more processing sets, which in turn elevated energy demands, system heat and higher costs for cooling the system.

The problem increased geometrically as big data arrived on the scene. Much of the influx was attributable to unstructured data littered with ambiguities that were not friendly to typical machine processing, or partially unstructured in ways unique to each type of data. Classic meta-tagging and data mining techniques for structuring quickly proved to be out-classed and obsolete – unable to keep pace with big data in real time.

The MIT/Glasgow/HP team set out to test the feasibility of real-time processing unstructured data using first generation advanced FPGAs provided by Altera – the Altera Stratix IV – on a GiDEL PROCstar platform. The goal was to match massive amounts of data against specific pre-selected topic profiles, including terrorist activity.

Each GiDEL PROCstar platform was outfitted with five Altera Stratix IV FPGA chips. Each chip operated at 150 MHz, supporting three RAM memory blocks and two SDRAM (software defined) memory blocks. The system also included a Bloom filter to delete false negatives and provide a more reliable cache filter vs. that of traditional systems.

The team observed at the outset that the diverse functions performed in data capture range from collection and distribution to organization, maintenance and analysis. The academics focused solely on analysis – the success of queries on specific topics such as keywords or topic signatures pitted against unstructured or semi-structured big data flows. Any document or other type of formless data that scored above a predefined profile threshold algorithm would be deemed relevant to the search.

In operation, the platform processed unstructured data through the SDRAMs and FIFO (first in, first out) buffer. The Bloom Filter then discarded false negatives so that the system focused solely on measuring profile weights to positives.

Even with conservative design, the academic team was able to achieve a capture rate of 772 million terms per second at power usage of just six watts per FPGA chip. The FPGA platform and chip were so powerful that they ran at this rate using only four percent logic utilization and 22 percent memory utilization. For typical usage, FPGA outperformed baseline servers by a factor of 20 to 40 percent on both performance and energy usage. As importantly, FPGA was 10 times more cost efficient.

Conquering Unstructured Data

Today the use of FPGAs has essentially taken over the world of unstructured data capture and the acceleration of complex transactions in industry and government. Applications range from military avionics, missile and drone guidance to automotive networking and repair, medical ultrasound, digital displays in consumer electronics, routers and servers, industrial imaging, video image processing in security, wireline connectivity interfaces, mobile backhaul configuration, search engines, real-time financial transactions on Wall Street, and of course, rapid-fire SIGINT used in surveillance. Where fast computing processes are required you will almost invariably find FPGAs at work.

Altera (Intel) and Xilinx, respectively the first company to patent FPGA technology and the first company to market a commercial FPGA product, still retain control of the marketplace, with combined market share of 67 percent and Xilinx slightly out ahead in winning customers. But as with any hot market, a multitude of competitors have arisen to fight for what remains.

The key to FPGA's success is reliance on the use of multiple logic elements (LEs) that can be wired together like gates to perform an array of complex, combinational processing tasks as fast as ASICs, and with the added feature of being able to reprogram any single LE without disrupting the performance or tasks of another. In addition, FPGA chips and LEs can be programmed at far lower cost and in less time than attempting to reconstruct any chip built with ASICS. Production is fast, and energy requirements and heat generated significantly reduced. Modern FPGAs are even produced with embedded processors to provide a complete system that is 100 percent programmable on one chip. In some instances, FPGAs may also be constructed to self-tune and reprogram on their own, depending on the mission required at that moment.

With ASICS complexity and costs rising, FPGAs would appear to have a secure future – at least until something better emerges.

Toward Software Configurable Microprocessors

Just as communications service providers are looking to the future with plans for network functions virtualization (NFV), the concept of software-centrism can be expected to play an important role in high performance processing, as well. "Software configurables" (SCFs) by companies such as Stretch are already positioning as viable alternatives to FPGAs. The basic concept: Rather than place processors inside FPGAs, SCF companies are making the processor itself programmable. Instead of using "gates" as FPGAs do to link diverse logic blocks or elements, an SCF is based on an extensible architecture that uses embedded software that can be reprogrammed to accelerate hardware – without involvement by or need for an FPGA.

The SCF provides pre-defined and optimized acceleration functions for capturing, processing and analyzing unstructured data including audio, video and text. Granted, SCF has its detractors who question whether this new technology can do everything possible with FPGAs. Some raise doubts about ease of use for software engineers grappling with reconfiguration via SCF for the first time. In all fairness, however, FPGA had its critics in the beginning, as well. Whereas FPGA is by now a well-recognized and mature technology, SCF has its entire future ahead of it.

Network Functions Virtualization

Of the many disruptive technologies to emerge in recent years, few have aroused greater fanfare – or confusion and even anxiety in certain quarters – than Network Functions Virtualization (NFV), the proposed migration of network tasks from proprietary hardware to more economical "virtual" systems comprised of software and white boxes. Stripping away the tech jargon that invariably submerges the basics in any discussion of the topic, NFV in its simplest terms is a revolt by communications service providers caught in the middle between rising bandwidth demand and network hardware vendors who have for years held carriers by the throat. With NFV, CSPs hope to achieve their strategic ambitions in quadruplicate: lower costs, higher performance, simplicity and independence from equipment vendors plying high-end proprietary hardware as the end-all/be-all solution to managing today's bandwidth explosion.

Can't the decades-old hardware centric approach handle the explosion in data traffic led by video? Most certainly it can, but not without problems. One is the aforementioned high cost, which in turn helps drive Problem #2: long development times typically required for new applications. NFV is widely viewed as a tool that will sweep away both issues *and* the hardware *and*, ultimately, dependence on hardware vendors.

That vision has alarmed the world's leading network hardware brands: Cisco, Huawei, ZTE and others that by some estimates stand to see margins on proprietary boxes collapse as much as 50 percent when and if NFV takes hold. In desperation, the vendors have adopted an attitude of faux-cooperation, masking their true intent of tacking proprietary barnacles onto an NFV "ship of state" envisioned by all others as completely open source.

Either way, the prospects are grim for the hardware houses: Losing by half is not so far removed from losing it all, particularly when CSPs have their minds set on replacing the big brands with no-name routers made cheaply in third world countries, ironically, by the self-same sweatshops that make proprietary equipment. In the end, NFV exposes network hardware vendors for what they are – purely middlemen – and eliminates the price mark-up that always accompanies market participation by third parties.

Surprisingly, a critical area of NFV that has attracted little attention is the impact of the movement on the surveillance sector. A number of key ISS companies are mapping out an active role in NFV's development, for example, Mellanox EZChip and Xilinx in the area of data plane acceleration. Still others, notably those reliant today on hardware, might be negatively affected unless they adapt. Deep packet inspection is often cited among the victims of NFV – often by analysts who in the same breath forecast 22 percent CAGR for DPI through the 2020s. Whatever the final outcome, it stands to logic that any major trend in networking will involve or influence ISS. But long before that happens, the marketplace must first agree on a definition of NFV and how it differs, if it does, from parallel movements that appear to do much the same thing – in the process confounding even industry experts.

Confusion surrounding NFV stems from the tendency to use the term interchangeably with related development such as Software Defined Networks (SDNs) and OpenFlow. While related and in many instances complementary to one another – and backed by many of the same tech enterprises – NFV, SDNs and OpenFlow have separate origins and purposes.

OpenFlow

Work on OpenFlow began in 2001 at Stanford University and UC Berkeley as a way to introduce the concept of programmable networks in a campus environment. The six-year program was shepherded by Drs. Nick McKeown, Scott Shenker and other academics, culminated in 2007, and is described in detail in their paper published in April, 2008.

OpenFlow was seminal for two reasons: (1) The researchers set out to build "an open, programmable virtualized platform" to deploy new protocols; and (2) they never had any intention of involving "brand name" equipment vendors. As opposed to the old hardware-centric way, where packet forwarding and high level routing decisions are made in the switch, OpenFlow divided the work between two boxes: packet forwarding still done in a switch, but routing now handled in a separate "controller," i.e., a simple router.

The authors are clear on why the team pursued an open source versus proprietary approach:

The practice of commercial networking is that the standardized external interfaces are narrow (i.e., just packet forwarding), and all of the switch's internal flexibility is hidden. The internals differ from vendor to vendor, with no standard platform for researchers to experiment with new ideas. . .And, of course, open platforms lower the barrier-to-entry for new competitors.

At the time, the researchers' ambitions were modest: line-rate processing for a college wiring closet without the technical overkill, high cost and limitations of commercial hardware. Their solution emerged from a simple observation: While Ethernet switches and routers vary vendor-to-vendor, each with its distinctive "flow tables" that manage packet forwarding in the data plane, all share common sets of functions and run at line-rate. OpenFlow was designed to take advantage of these commonalities, providing an open protocol to manage flow tables in different devices and in essence, act as a uniform OpenFlow flow table.

OpenFlow proved itself immediately capable as a high performance, "extensible" (multi-function), flexible and low cost alternative to reliance on proprietary boxes. Instead of centralizing packet forwarding (the data plane) and high-level routing decisions (control plane) in the switch, OpenFlow separated the functions. This distinction was so innovative for the time that it bears repeating in italics:

The switch still handled data plane packet forwarding, but routing decisions were managed by a separate controller, i.e., a simple router.

With OpenFlow, researchers could now install any variety of innovative applications in the network quickly, easily and inexpensively. Savings resulted from the ability to power the campus system with a single switch – at the same performance level as the prior use of multiple switches, regardless of data traffic loads.

In short order, OpenFlow became a standard for achieving these goals by dedicating separate devices for the data plane and control plane for efficient management of Layer 2 (data link) and Layer 3 (network switching and routing) of the OSI model.

Did OpenFlow permit users to do anything they couldn't do before? No. But now they had a single programmatic interface to build and customize networks to specific requirements. In 2011, OpenFlow jumped from academia to the commercial arena with its own "Open Networking Foundation" (ONF) backed by major service providers including founding members Google, Facebook, Verizon, Deutsche Telekom, Microsoft and Yahoo.

Network hardware vendors quickly piled on, too: Broadcom, Cisco, Citrix, Dell, Ericsson, HP, Juniper Networks, NEC and NTT, to name a few that began producing their own OpenFlow switches to meet demand. In essence, market pressure forced hardware makers into catch-up mode. OpenFlow had broken the mold of classic networking and today is viewed as the antecedent and complement to yet another school of virtualization, the software-defined network (SDN).

Software-Defined Networks

The software-defined network is a network architecture defined by the advent of the softswitch, an abstraction of high-level network tasks performed by software instead of hardware. SDNs are often confused with OpenFlow because both work by separating management of the data plane from that of the control plane. But let us be clear on one point: OpenFlow and SDNs are related – they are not twins.

In an SDN the interworking of the two planes requires a specific method of communication. OpenFlow is one such, but not the only method or standard that may be applied to facilitate the SDN architecture. Alternatives to OpenFlow are now widely available from hardware vendors. Options such as the Cisco Open Networking Environment (Cisco ONE) add the benefit of extending network programmability through Layers 4 – 7 (network services and apps) of the OSI stack, moved from its former home on network devices onto commodity servers.

Cisco also provides a kit that lets users build an API that makes all Cisco hardware software programmable. The advantage for Cisco, of course: open source becomes "Cisco source." Other vendors have followed suit with their own clones of OpenFlow.

As a network architecture the SDN offers distinct advantages over reliance on multiple types of proprietary hardware including: the ability to configure multiple boxes at once versus doing so one-off; full programmatic management of the network; greater speed-to-market with new services; and, simplified integration of services following a merger.

In many ways, SDN represents the fruition of work by those researchers who brought OpenFlow to the networking community – separation of the forwarding plane (dumb switches and routers) and control plane (where the intelligence resides for routing decisions), centralized programming and control, and distinct interfaces to monitor and manage the network – as an embedded architecture that deploys OpenFlow or similar standards branded by equipment vendors. Underscoring that tight relationship, SDN is actively promoted by the ONF, the same group behind the OpenFlow standard.

Network Functions Virtualization

Network Functions Virtualization represents a step up from SDN: the use of virtual servers in virtual networks, *with the power to convert physical infrastructure into software code.* It is an open, competitive ecosystem where independent software vendors and virtual appliances "run the show."

NFV is unique not only for what it does but from whence it arises: the service provider community operating under the umbrella of the European Telecommunications Standards Institute (ETSI) Industry Specification Group (ISG) for NFV.

The development of NFV was first presented to the public at the Software-Defined Network and OpenFlow World Congress in October 2012.

Standards development was assigned to ETSI ISG NFV the following month. From an in initial roster of seven founding member companies, the group has grown to include nearly 40 network operators such as AT&T, Verizon, Deutsche Telekom, NTT, NEC and Telecom Italia, as well as another 270 individual companies. Listed supporters include entities well-recognized in the ISS community: Aqsacom, BAE Systems, Mellanox EZChip Technologies, IXEA, Qosmos, and Yaana Technologies.

Missing from the list: top network equipment vendors such as Cisco, Ericsson, Juniper, Huawei, Oracle and ZTE.

The operating premise of ETSI ISG NFV is that any network function can be abstracted and virtualized. That in mind, the goal of the group is to translate any/every appropriate network function to a common commodity platform for the express purpose of reducing network infrastructure costs. Virtual ports, servers and routers will replace their current physical counterparts. Functions previously managed by a hard box will be taken over by virtual appliances. Network services and applications – all virtualized.

Such is the vision. The reality is troubled by contention and the tendency of key participants to either hedge their bets by backing multiple initiatives, or to spin off from groups and go solo on NFV. Nearly eight years after being assigned the responsibility of setting standards for NFV, ETSI has yet to produce a single one.

As a result, new and more aggressive schools of thought have surfaced in competition with ETSI ISG NFV. Among the foremost is the Open Platform for NFV (OPNFV) consortium, formed in late 2014 due to members' impatience with what they deemed as slow progress by ETSI. It is a fair criticism. On January 13, 2015 ETSI published "agreed definitions of the key concepts of NFV" and ETSI-endorsed "management and orchestration" (MANO) stack for NFV. Then in July 2016, ETSI formed "a new working group, Solutions (SOL), charged with the task of delivering a consolidated set of protocols and data model specifications to support interoperability."

The membership of ETSI ISG NFV grew, ultimately winning over dissidents such as Cisco, Huawei and Juniper who initially gravitated to OPNFV. The ETSI group has moved on track to make NFV open source, in line with OPNFV. However, NFV standards remain a work in progress. The profusion of players going off in their own directions brings to mind the American saying about "too many chiefs," except in this instance there are too many Indians, as well. A movement that set out to free service providers from over-reliance on network hardware vendors would appear to have devolved into its opposite: a catch-all where CSPs are every bit as dependent on vendors as before.

Not content to abide by standards (but eager to give that impression they are), equipment vendors have set out with their own proprietary NFV standards, hoping a commercial version will become de facto.

Spyware's Move to Virtualization

In a cloud or virtual environment, lawful intercept and other forms of intelligence gathering can face significant new challenges that arise with the change from physical to abstracted network architecture. Lawful intercept, for example, has for many years relied on "tapping" services carried on physical network connections. As networks move to NFV, wiretapping encounters problems:

- Services operating on virtual ports, routers and machines are not linked to specific servers.
- As service demands change, so may the number of virtual hardware devices, which may be created or turned down in accord with demand. The number of virtual machines thus may be in a constant state of flux, versus in a fixed number as is the case in physical networks.
- Finally, physical connections between a target and other parties disappear because they are located on the identical server.

With NFV, classic wiretapping is now "going virtual." One approach is to create a virtual tap that duplicates network traffic at the Ethernet interface. The tap applies to all virtual machines in the network for monitoring purposes, and automatically replicates itself in each new virtual machine that is created. The virtual tap can intercept any traffic from any company operating on cloud networks and has sufficient intelligence to identify the virtual machine or machines from which the traffic has been tapped.

Another challenge in an NFV environment is data plane acceleration. Essentially there are two bottlenecks in the NFV data plane: (1) the virtual switch on the server platform; and (2) performance of virtual network functions (VNFs). Standard virtual switches of the type envisioned for use with NFV often do not meet expectations for delivering high-bandwidth traffic on a continuous basis, thus any degradation in speed negatively impacts performance. At the same time, if a VNF does match the performance of the same function on a physical network, the cost benefits of NFV quickly evaporate. Neither the switch nor the VNFs must be allowed to devolve into bottlenecks.

In the physical network arena, data plane acceleration is achieved through means that include field programmable gate arrays (FPGAs) provided by companies including Altera (now part of Intel), Albedo, Mellanox EZChip

Technologies and Xilinx. Altera, Mellanox EZChip and Xilinx are contributing members of the ETSI ISG NFV. In addition, all three are deeply involved in creating virtual solutions to NFV data plane roadblocks.

NFV data plane acceleration enables improved resource management and performance using FPGAs, either across integrated chips or in the NFV servers. VNFs can be outfitted with application program interfaces (APIs) that quickly determine which data plane accelerators to access in the respective hardware resource (chip or server). The accelerators ensure consistently high performance for VNFs and full interoperability with any underlying hardware regardless of make or model. Will NFV bring an end to deep packet inspection? Not necessarily. Market leaders such as Qosmos are far along in adapting DPI solutions for a virtual network environment. Qosmos' core product ixEngine is already used as a VNF component, as are many other DPI solutions. DPI will continue to be a winner during the transition from hardware to NFV, and when networks are purely virtual.

ALBEDO Wire-Speed Packet Capture

What if a lone police officer or agent had access to a simple, hand-held device every bit as sophisticated as the big boxes or their virtual alternatives, but designed strictly for tactical purposes? In that event, he or she could monitor a target's IP traffic at close range and short duration – only as long as was needed to make an informed decision to take action.

The solo investigator can do just that thanks to ALBEDO Telecom, a Spanish enterprise that has distinguished itself not once but many times with market "firsts" for making hand-held devices that perform targeted packet capture at wire speed.

Founded in Barcelona in 1983, ALBEDO Telecom over the last three-plus decades has branched out to establish a presence on five continents. ALBEDO Telecom's steady growth is a reflection of the broad importance of its work to the US $2 trillion telecom industry.

Let us make this clear from the start: ALBEDO Telecom is not confined to the lawful intercept/intelligence niche. The company designs and manufactures a wide variety of products deemed essential to all manner of communications service providers (CSPs):

- Equipment for field testing and maintenance of telecom circuits.
- Network monitoring solutions to back up a CSP's service level agreements (SLAs) to ensure contractual requirements for quality of service and end user experience.
- IP network emulation systems that verify the performance of new CSP products and applications.
- Systems for measuring network transport SLAs.
- Hardware for configuring best practice network configuration validation and auditing.
- Solutions that test emerging network technologies to address operational and functionality issues.
- And, finally, "Quality of Timing" devices to ensure network synchronization for LTE operators across multiple technologies at the transmission layer and timing plane. ALBEDO Telecom, by the way, is a thought leader in this mission-critical space, regularly demonstrating its solutions at commercial events including Mobile World Congress and more esoteric scientific forums such as the ITSF (International Telecom Sync Forum).

Not surprisingly, some tools developed for monitoring or improving performance of commercial networks are pertinent to lawful intercept and other areas of surveillance, where ALBEDO Telecom is a strong player.

A factor that sets ALBEDO Telecom apart from many competitors is its focus on compactness and solo user functionality. The company certainly makes its share of rack-space designed hardware playing an important role in intelligence gathering. But ALBEDO Telecom really stands out for its focus on user-friendly hand-held (literally, with handles set left & right) or even palm-sized devices designed primarily for field use. ALBEDO Telecom has in many ways pioneered the field of small, discrete monitoring devices. Here is a condensed view of the company's R & D output from early days portable devices progressing to today's tiny ones:

- 1989: First portable multiplexer instrument from 64 Kbits up to 140 Mbit/s.
- 1996: First hand-held device incorporating a touch-screen.
- 1999: World's first hand-held tester for SONET/SDH.
- 2009: World's first portable wire speed tap with active filters.
- 2012: World's first hand-held wire speed WAN emulator.
- 2014: World's smallest 10Gb/s Ethernet, PTP, SyncE tester.
- 2015: PTP (Precision Time Protocol) synchronization assurance.
- 2016: Hold-over with built-in atomic rubidium clocks optimizing 4G access.

To field agents and officers, words and phrases such as "hand-held" and "world's smallest" have an almost magical effect, indicating ease of concealment in tactical scenarios versus "portable," which all too often translates to "suitcase-sized" and easy to detect by those being monitored.

However, the real razzle dazzle comes from what lies inside an ALBEDO Telecom device: FPGAs that are ideal for wire speed data capture, and SSD (Solid State Drives) for storage and rapid processing.

Add it all up and you have: easy access to a designated traffic stream; connection to the target's network via "bridge mode" that transmits through the device completely unnoticeably; applicability to any network environment at whatever speed; and, lower cost.

ALBEDO Telecom offers two devices that define the world of compact packet capture for field operations: NetHunter and NetShark.

In the Hunt with NetHunter

NetHunter is ALBEDO Telecom's mainstream packet capture device. It comes in two formats: traditional rack mounted or battery-operated hand-held version. This is a powerful tool for pre-filtering, capturing, recording and analyzing a target's data streams in tactical field operations. It works with any/all types of formats including VoIP, data, email, SMS, TCP/IP, IPTV and others in either IPv4 or IPv6.

A common feature of all ALBEDO monitoring devices is the company's "Zero Delay" technology that captures every packet, live in full duplex (FDX) at wire speed without a single packet lost, even when there is temporary loss of power.

NetHunter is the first of the line to leverage Zero Delay. Working in either pass-through or mirror mode, the device provides interception that is completely undetectable to the target and his/her network service provider. No MAC or IP address will ever appear, ensuring silent, secretive interception. The same feature makes the device virtually impervious to hackers.

NetHunter immediately captures all packets that meet a specified condition or any of a wide range of filters. NetHunter's multilayer trigger may be set to diverse logical conditions, or simply set to "on" to capture all traffic. For a specific target, suspicious packets can be saved to a built-in 120 GB solid state drive, then for analysis may be forwarded to WireShark, a free, open source protocol analyzer for Unix and Windows, or routed to data retention for subsequent analysis.

Filtering is done by 16 simultaneous filters that capture Ethernet source and originating MAC addresses, IP source, destination, and IP address group.

The first step, pre-filtering, is important because it lets the user decide which types of traffic to capture and store, and which to eliminate from the investigation. This selective approach to packet capture refines the focus of the intercept, for example, to VoIP traffic or email, thus extending the duration of packet capture time. Pre-filtering also permits the marking of individual packets per the filter applied to each, to simplify subsequent analysis, providing the ability to peg packets to a specific IP address.

Similarly, other filters offer their own discrete advantages. Port filters key to specific words, phrases or sentences. Then come event filters that trigger the device to begin allowing all packet frames to pass through for capture. Specific to the LEA tech specialist, NetHunter's "Lawful filter" performs a 64-byte pattern match at any point in the packet payload.

Unlike large DPI systems, NetHunter is not intended to capture huge amounts of data, but rather, finite traffic relevant to known or suspected targets. A "rule" of such devices that really turns out to be an advantage: NetHunter must be configured on the device itself, thus is completely at the command of the user. ALBEDO Telecom provides ample training to ensure that the user is completely familiar with every aspect of the device before setting out to monitor and catch evildoers. And once in the field, the agent will find it simple to configure NetHunter to the task at hand via the device's extremely user-friendly GUI.

With NetHunter, the user can literally be within yards of the target, intercepting every VoIP call, message, website visited, file downloaded or uploaded, as well as the affiliate party or parties with whom the target is communicating – all without being observed. From the standpoint of executing on tactical needs from the field, what could be better? Possibly another ALBEDO Telecom product.

Fins to the Left, Fins to the Right

Welcome to NetShark, touted as the "world's first hand-held, battery powered and 100% autonomous TAP." Using FPGA cards, NetShark connects in pass-through mode to filter and capture packets. As with NetHunter, captured data can be relayed to Wireshark for subsequent protocol analysis.

One distinction to be clear on. Packet capture and packet protocol analysis, though related, are separate functions. Wire speed packet capture, by definition, must be real-time to ensure no packet loss. Packet protocol analysis is not real-time. Ergo the handoff of one function (capture) to the other (protocol analysis).

FPGA technology offers definite advantages, particularly to field operatives on tactical missions. Being programmable on the fly, NetShark with FPGA capability can be quickly adapted to track elusive targets switching communications devices, or to new targets entering the field of view. The programmable parts of FPGA vastly simplify the process of rapid-fire reprogramming. The parts are "logic blocks" that implement complex digital circuitry, and "logic gates" that handle complex computations at very high speed. With FPGA, the user can reprogram one or a set of logic blocks that handle specific functions, without impacting or causing any need to reprogram the other logic blocks.

Of equal importance, FPGA acts as a "hardware accelerator" that can meet the demands of 10 GB, 40 GB, 100 GB and higher network speeds – a key feature as networks switch to higher bandwidth facilities to meet rising broadband demands for LTE, VoLTE, streaming video and 5G. Finally, FPGA offers definite cost advantages over competing technologies.

Note that many of NetShark's features are quite similar to NetHunter's: capture of VoIP, data, email, SMS, TCP/IP or IPTV packets; works with both IPv4 or IPv6; offers 16 simultaneous filters; tracks IP source source-and-destination; and offers the "Lawful Filter." Where NetShark and NetHunter differ is in the interception mode. NetHunter operates either by port mirroring or pass-through. NetShark is strictly pass-through.

Each capture mode has its advantages. Port mirroring is typically used for low-speed applications where packets are forwarded to a third party such as Wireshark for analysis. The downside of port mirroring, given the slower speed, is that packets may be dropped. In contrast, pass-through mode captures all packets, and at wire speed.

Granted, although both NetHunter and NetShark are small, their look is distinctive and might arouse curiosity among sophisticated targets who have "studied up" on ALBEDO wares. If that is a concern, the company has yet another solution: its VNC (Virtual Network Computing) Remote Control unit, which lets agents use their own tablet, laptop or smartphone to tap NetHunter and NetShark from afar. Yet another smart idea from a company where the focus is on "small" and the results always impressive.

Amdocs Bulk Metadata and VoLTE Interception

When news of the NSA bulk metadata collection scandal broke in 2013, some liked to joke that carrier billing would be far more accurate and reliable if service providers turned the job over to the NSA. As it happens, billing for many of the same carriers was then and still remains in the hands of a vendor once alleged to have close ties to the Intelligence Community – Amdocs – though the alleged spies were Israeli, not American.

Why that is significant: Amdocs "customer experience solutions," marketing jargon for back office systems spanning business support (BSS) and operational support (OSS), are the basis of billing information used in call data records (CDRs) and Internet data records (IPDRs) for metadata collection whether of the bulk or targeted variety. As the dominant BSS/OSS platform on the market, Amdocs is used by more than 250 communications service providers (CSPs) worldwide.

Another noteworthy achievement of Amdocs: With LTE now the dominant standard in mobile communications worldwide, Amdocs is a leading force behind platforms that eliminate a major headache for consumers, law enforcement and government agencies alike: dropped calls on VoLTE networks. Per one source, VoLTE radio access networks (RAN) drop calls "four to five times more often" than legacy 2G and 3G networks.

The author of that quote is, by no coincidence, Amdocs, which highlighted the remark in research on the issue of VoLTE network dropped calls, a problem that can be resolved by using an Amdocs solution.

But for the momentary taint of scandal in 2002 when a small group of its employees were exposed as Israeli intelligence agents using the company's products to spy on the U.S., Amdocs has steered clear of any public involvement in intelligence support systems. The company was subsequently absolved by U.S. investigators of any direct participation in that episode, though suspicions lingered in some quarters.

Today, when warrants for call data records (CDRs) and Internet Protocol Data Records (IPDRs) are issued to service providers by a law enforcement agency, the odds are very high that such records are created and stored on systems made by Amdocs. Thus, Amdocs' involvement in surveillance is no different from that of network hardware and software vendors: They make systems that may be primarily used for commercial purposes, but can equally well serve the needs of government surveillance, even if the company's involvement is at arm's length.

It is a long arm. Amdocs CES solutions and their resident metadata collections capabilities have for all practical purposes reached ubiquity.

Clients include AT&T, Verizon Communications, Verizon Wireless, Sprint, T-Mobile in multiple countries, BT, Bell Canada, Comcast, Telefonica, Cox, Deutsche Telekom, Etisalat, Vodafone, Tata Communications, Yellow New Zealand, and many more Tier 1 operators.

Inside Amdocs

Amdocs did not start out as a telecom billing company, at least not immediately. Launched in Israel in1982, the company was originally part of an Israeli telephone directory service owned by Aurec Group. Fairly quickly the directory subsidiary developed billing software for telcos and was named Aurec Information & Directory Services, which was changed to Amdocs when Southwestern Bell purchased a 50 percent interest in 1985. Two years later Amdocs was spun off to independent investors for $US 1.0 billion, with a focus on billing services for wireline and wireless companies. Today Amdocs is a multinational company based in Chesterfield, Missouri, USA – but its leadership and roots remain strongly Israeli.

In the mid- to late 1990s, telecom operators took advantage of "convergent billing," integrating charges for diverse services on a single invoice. Companies such as the UK's Geneva Technology and Israel's Amdocs were in the forefront of this movement. Communications service providers were attracted by the ability to incent usage, cross-pollinate between services, increase user satisfaction and loyalty, and most importantly, monetize rising network investment essential to supporting the services.

Integrated convergent billing had the effect of elevating billing from a cost center to a profit center. As new capabilities such as inventory-based network element management were folded in, vendors such as Amdocs made it possible to speed order fulfillment time, better manage product lifecycles, pinpoint and repair network problems before they became issues that negatively impacted customer experience, and in general improve the customer experience. Soon "customer experience management" became the mantra of these solution providers. At Amdocs, multiple technology products are now offered under the heading of "Customer Experience Solutions."

From BSS/OSS to Streaming Customer Data

Even in the early days of convergent billing, Amdocs always cut a large swath in its sector. At trade shows, the company's booth was always among the largest, displaying life-size images of clients with glowing testimonials. The company advanced its technology leadership through internal innovation and strategic acquisitions: Clarify, Comverse and Actix on the BSS side; and in OSS, the UK's Cramer, which had carved a unique niche in the field of automated inventory management. Cramer took Amdocs to a new level, giving CSPs the ability to accurately identify, track and deploy network assets essential to service delivery, at a sizable savings over prior manual processes. With Cramer on board, Amdocs made OSS an integral, reliable profit generator in tandem with convergent billing.

Competitors such as Geneva Technology never grew beyond their roots. Geneva was purchased by Convergys Corporation with great fanfare in 2000 for $700 million in stock, immediately began to founder, and was subsequently sold off to Japan's NEC at a loss. Amdocs, in contrast, has grown into a company with $US 4.5 billion in revenue, 25,000 employees in 85 countries, and a global client base.

The lingua franca of Amdocs CES is data. Long before the word "smart" became attached to phones and other devices and data became big, Amdocs was empowering its clients with data to make more intelligent and lucrative decisions based on network assets, product innovation and an intensely detailed understanding of the needs and interests of customers. The same 0s and 1s generated by customers – detailing who they called, texted or emailed, when, where, for how long, using what services on which devices – are also valued by the law enforcement community for lawful intercept of content and metadata. Although Amdocs itself is simply a technology provider and not in any way involved in intercepts, it is, by extension via the capabilities it offers CSPs, law enforcement's best friend for intelligence gathering of target intelligence from the network.

Amdocs supports the management of this intelligence through solutions such as the company's TeraScale, "a carrier-grade, big data solution designed to cost-efficiently store, manage and extract real-time charging usage data." Based on Apache Software Foundation big data lake Hadoop, TeraScale can stream the massive amounts of data generated by Amdocs CES billing solutions, providing modeled, scored data in clusters for analytics and visualization tools. While this solution is designed to help the CSP improve the customer experience, there is also an inherent value to law enforcement and government intelligence. The object is the collection and analysis of metadata, after all. A system that stores, manages and analyzes "massive" amounts of data is dealing in bulk metadata.

Amdocs' Cure for Dropped VoLTE Calls

Companies never produce research in a vacuum or for its own sake. When Amdocs delved into the problem of dropped calls on VoLTE, they were motivated by purely business-driven reasons: They have a commercial solution. That's good news not only for customers but for LEAs that experience gaps in surveillance evidence due to the very same shortfalls still found in VoLTE call quality.

The Amdocs research was solid – based on a year-long study of 25 million voice and data connections hosted by 80 CSPs throughout the globe. Key findings:

- Hasty construction of radio networks often leads to a five-fold increase in dropped calls, particularly on newer networks. What they did not say but which is implied: As the quality of service declines, the quality of mobile surveillance follows. Operators need to be more "aggressive," says Amdocs, in tuning networks.
- Broad Wi-Fi offload by consumers, and hence targets, is a myth. While the volume of mobile data usage has grown 60 percent per annum, as little as 5 to 15 percent of traffic transiting Wi-Fi networks may be classified as offload from carrier networks. As a result, the broader measure of data traffic stays on the network, where the ratio of dropped LTE calls is highest.
- During periods of network congestion – and higher numbers of dropped calls – indoor users contend with a 25 percent higher incidence of problems than users who call out-of-doors. CSPS are doing little by way of network optimization to resolve the issue. The inference is that LTE mobile interception also suffers in these conditions.

Amdocs' answer to these problems is VoLTE Controller, a policy control and charging rules function (PCRF) that "de-risks" VoLTE networks by establishing rules that ensure VoLTE traffic flows, a quality experience for user, accurate billing and, not incidentally, accurate mobile interception results on LTE networks. Readily deployable and highly scalable, VoLTE Controller manages up to 12 times more VoLTE transactions than do networks without the solution. Users can roam with guaranteed quality of service, and LEAs can monitor their every byte and syllable. VoLTE Controller provides the same QoS in rare cases when LTE still defaults to circuit-switched fallback on 2G or 3G networks. As with CES and metadata collection, VoLTE Controller is mainly geared toward improving the customer experience. But for government intelligence agencies, the side benefit of an improved surveillance experience is a positive consequence.

Cisco and the Great Firewall of China

As the security and censorship platform for monitoring nearly 1.4 billion people, the Great Firewall of China is at once the most complex and little understood in-country electronic surveillance system on the planet. Often overlooked: Much of the Great Firewall's core technology is American made, and Chinese developers themselves often get an assist from U.S. tech leaders.

The principle application used to block access to sites and services deemed unfriendly to the Chinese government is deep packet inspection (DPI). Much has been written about China's use of DPI solutions made by the former Blue Coat in tandem with routers from Cisco. What remains surprising: company claims by Cisco and Blue Coat that neither was ever involved in providing technology to China for the Great Firewall. Both companies have indeed been accomplices in this activity for years.

Blue Coat has since ceased independent operations. In 2016, Symantec (now a property of Accenture) acquired the DPI specialist, folding it into its operations. Of note, Blue Coat CEO Greg Clark became CEO of Symantec, and Blue Coat COO Michael Fey was named President and COO of the integrated US $4.0 billion parent company.

Symantec has since gone through two ownership changes in four months. Broadcom acquired the company for $10.7 billion in September 2019, then spun it off to Accenture in January 2020 for an undisclosed amount. While it was still independent, Symantec operated in China for more than 20 years with offices in Beijing, Shenzhen and other major Chinese cities. Its DPI solutions retained the Blue Coat name, and were sold wherever the company conducted business, either directly or through third party vendors.

Key Targets of the Great Firewall

China is intensely sensitive to the political activities and unrest caused by the non-Han elements of society, principally of Tibetan and Uighur origin. The PRC also believes it essential to block major search engines, social media sites, as well as sites that provide information on ways around DPI, such as virtual private network (VPN) service to all citizens of the PRC.

High-ranking websites blocked in mainland China via DPI include Google, Facebook, Wikipedia, Youtube, Twitter, Instagram, Tumblr, Flickr, Bloomberg, Amnesty International, Reporters Without Borders, *The New York Times,* the *BBC* and many others. All VPN services are monitored and blocked whenever possible. The Chinese government also uses DPI to to surveille Taiwan and monitor communications of pro-democracy activists in and Hong Kong.

The sheer volume of communications monitoring required to root out a comparative handful of individuals of groups is staggering. To assume that the products of a single company such as Blue Coat handle the full payload is an exaggeration, to say the least, not giving "credit" where it is due to the multiple players involved. But U.S. technology companies definitely play a role in assisting repression in the most autocratic and murderous regime in history.

Cisco: China's Favorite Router Supplier?

In 2005, Cisco and Blue Coat were the subject of unfavorable publicity in the West, the former for selling 200 routers to China, and the latter for providing its ProxySG and PacketShaper DPI boxes. Subsequent investigations by privacy groups claimed to have located both types of Blue Coat hardware in the PRC. Cisco at the time denied any involvement in supporting the Great Firewall's DPI prowess.

Through 2015, Blue Coat DPI products were still actively sold in China via any number of third parties, with prices shown in Yuan currency. Online links to vendors selling the ProxySG vanished after Symantec's buy-out of Blue Coat, and other factors.

In the year prior to Edward Snowden's release of NSA records, tensions between the U.S. and Chinese governments ran nearly as high as they do today, resulting in cross-accusations about both countries' use of communications equipment to spy on the other. The Obama Administration subsequently raised the bar for any American company using Chinese equipment that wasn't first tested and approved by the FBI. Similarly, China's anger over the NSA scandal raised concerns about U.S. equipment used to the same ends, implying threats to Cisco's dominance in router sales to China and possible openings for Huawei and ZTE to gain the lead.

To date, Cisco's sales in China continue to rise, totaling in excess of $US 2 billion per year. China does not want to abandon the tech innovation of the world's most advanced routers, or be in the position of relying solely on Chinese-made products, nor does Cisco wish to forfeit a favorite cash cow. As a result, Cisco equipment also remains an important part of the Great Firewall.

Cisco's SCE8000

One well-known challenge to DPI hardware/software is that the larger the volume of traffic to be analyzed, the greater the processing power required. Mastering that challenge is the Cisco SCE8000, a high capacity DPI product.

The Cisco SCE8000 is programmable and extensible for commercial applications such as traffic management, as well as surveillance and censorship.

How a customer uses the device or modifies its use depends on the rules he or she decides to program in.

Using DPI to monitor and censor the communications of a population as large as China's is no small feat. Processing demands are significant. The risk of packet loss is high. But Cisco is up to the challenge.

A key feature of the SCE8000 is the principle of "stateful" DPI. Rather than processing packets one-by-one, the unit reconstructs individual traffic flows and the Layer 7 application state of each individual session.

Application-level classification of IP traffic helps ensure real-time analysis and control of content-based services for a given subscriber or group of subscribers. Real-time advanced control functions include granular bandwidth shaping, quota, and redirection using protocol-specific traffic flow analysis.

With the SCE8000, DPI can run 7X24X365. High availability is achieved by stacking devices in a cascading configuration that overlays dual 10-gigabit links. One unit processes the IP traffic while sharing stateful information with the secondary engine as a backup. In the event of any interruption in the operation the second device fills in. Cisco recognized early on that "the cloud" would be the future of DPI as well as many other network functions. Cisco helped pioneer this movement with its Meraki "cloud" solution.

Cisco Meraki "Cloud Networking" DPI

Cisco Meraki cloud networking products are at the core of over 20,000 customer networks – meaning the built-in DPI capabilities of Meraki are there, as well. The Meraki packet processing engine performs traffic analysis using Cisco's Layer 7 technology to identify, classify and report hundreds of applications. Like the SCE 8000, Meraki cloud systems go far beyond IP addresses, host names, and ports, with the ability to classify traffic based on application, including evasive, dynamic, and encapsulated apps.

With the Meraki cloud, users can quickly spot trends, track the target's app utilization on graphical display or drill for more data. The beauty of Meraki cloud: centralized management and high throughput, without the cost of stacking multiple high end devices.

Huawei SIG 9800 Series

Not to be outdone in its home country, Shenzhen-based Huawei offers its own "big iron" DPI suite – the SIG 9800 series, with SIG standing for "service inspection gateway."

While it is primarily marketed as a network management and efficiency enhancement system, the SIG 9800 performs equally well at surveillance.

The SIG 9800 suite supports advanced identification technologies to single out multiple categories of traffic including P2P, VoIP, IM, video, as well as some 850 protocols and 1,000 applications: Skype, Facebook, MSN and many others. Highly flexible, it can be customized to accommodate new protocols and to increase the knowledge base of what it needs to monitor in the network.

The SIG 9810 and 9820 come with a library of over 65 million urls in 40+ categories and in multiple languages including not just Chinese, but also English, Russian, Arabic, French and Spanish.

Huawei's system is also adept at quickly identifying suspicious targets whether the individual is using fixed, wireless or Wi-Fi networks. SIG 9800 devices quickly pin down all subscriber information: IP address, IMSI number, location and roaming status, in the event the target is on the move.

URL filtering lets either system eliminate access to undesirable websites including those that promote pornography, criminal activity, violence or any questionable content.

Semptian DPI Suite

Also of interest is a smaller company that is a rising star in the Chinese DPI market, even though it is best known for its lawful intercept products: Semptian, based in Shenzhen and with offices in Beijing.

Semptian is active in multiple solutions areas of lawful intercept and intelligence: mediation devices, probes, social media monitoring and ethical malware. The company also offers DPI and IP Flow Monitoring hardware assisted by field programmable gate arrays (FPGAs) to ensure wire speed packet capture.

Semptian DPI and IP Flow Monitoring products provide mass interception of IP traffic, capturing data at all seven layers in the OSI stack. Semptian decodes email content, IM and website visits in real-time.

One solution relevant to China's "Golden Shield" is SempScope National Firewall, which can be installed either in an uplink or international link to block domestic Internet customers from visiting selected offshore sites. SempScope examines every request for a DNS IP address and checks keywords to decide, based on how it is programmed, whether access is permitted. SempScope can also be programed to block any and all visits to out-of-country websites.

As a small and comparatively young company, Semptian relies on American technology to support its role in the Great Firewall of China: the OCTEON II processor produced by Cavium, a California-based maker of high-throughput semiconductors expressly designed for DPI solutions.

Decision Group DPI

Three AsiaPac DPI vendors dominate the market for surveillance-focused deep packet inspection in AsiaPac: Taiwan's Decision Group, Singapore's Expert Team and Korea's Hanvit. Although small compared to vendors of multipurpose standalone DPI solutions such as those made by Sandvine, Allot, Cisco, Procera, Qosmos and others, each has carved out a healthy niche in the DPI for lawful intercept and intelligence realms within region, and in at least one instance globally, as well. The greater issue is how well the AsiaPac threesome and their giant DPI colleagues will hold up against a mounting competitive threat from new technologies including Flow Monitoring and Network Functions Virtualization (NFV).

Forecasts are alternately promising or clouded. According to sources the global market for DPI will reach a value of US $12.5 billion by 2023. AsiaPac customers will represent nearly 40 percent of market demand. NFV, premised on decoupling network services from hardware, will soar to US $36.3 billion.

If analyst speculation on NFV is true, then forecasts on the inevitability of virtualization could foretell dark days ahead for classic DPI. The software component of traditional DPI solutions could be transferred to the cloud, where it would decrypt packets at Layers 4 – 7 of the DPI stack, while the hardware piece moves to small, low-cost devices handling Layers 2 – 3. Such reduction in reliance in special high-end hardware, the end game of NFV, is every network equipment maker's worst nightmare. Is that day just over the horizon? Perhaps, but in the meantime critics of the purported coming NFV boom are quick to point out that standards for the technology remain wanting. Nearly a decade after announcing its plans to issue uniform standards for NFV, the European Telecommunications Standards Institute (ETSI), is still at the drawing board on this topic. As ETSI itself somewhat embarrassingly admits on its website:

"This large community (300+ companies including 38 of the world's service providers) is still working intensely to develop the required standards for NFV as well as sharing their experiences of NFV implementation and testing." Assuming that NFV does arise as a standards-based capability at some point, it is reasonable to ask which entities are best positioned to retain control of the DPI sector in the AsiaPac market where demand is and will likely remain strongest. Also worth considering: which vendors have the flexibility to take the next step into new technologies when and if DPI theoretically becomes a gray-haired also-ran in packet forensics. A rough consensus:

- **Decision Group**, with more than 30 years of experience in the DPI for LI/intelligence arena and hundreds of law enforcement and government agency clients worldwide, would appear to hold a decisive edge in AsiaPac DPI, with toe-holds in global markets, as well.
- **Expert Team** has come on strong since its founding in 2012, offering its own line of products showcased regularly at events such as Interpol World, Milipol and ISS World, the latter including conferences in the company's native Singapore, as well as Dubai, Kuala Lumpur, South Africa, and Europe.
- **Hanvit** of Seoul, Korea prefers to partner with and resell the work of others such as: the Czech Republic's Flowmon for Flow Monitoring; the USA's Keysight for Ixea solutions that perform packet filtering on high density 100 GB mobile networks; and a suite of TAP, DPI and Dense Wave Division Multiplexing (DWDM) fiber optic hardware made by Israeli vendor Allot Communications through its acquisition of Optenet. As a result, Hanvit has a formidable presence not only in DPI but also in IP Flow Monitoring and fiber network monitoring and interception.

DPI's Evolution for Interception

Increasingly, DPI products expressly designed for lawful intercept and government intelligence collection come equipped with capabilities that go beyond commercial DPI used for network optimization. Foremost among these capabilities are:

- **HTTPS/SSL Network Forensics:** Provides real-time access to full metadata and content of IP communications secured by these protocols, and uses packet reconstruction to make content visible in the exact format viewed and used by the target. DPI products designed for surveillance get around HTTPS/SSL security either by conducting Man-in-the-Middle (MITM) attacks or by obtaining the target's private key and working offline to open encrypted data. Since private keys are not typically readily available, MITM attacks are the more common mode. The DPI solution poses as the target and obtains the requisite certificate from the SSL server, then generates its own "valid" certificate that is indistinguishable from the original for authentication purposes.
- **Tactical Wi-Fi and LAN Interception:** Lets the agent conduct close location/short term surveillance of targets in public areas by tapping into Wi-Fi and Wireless LAN networks, then decoding and reconstructing traffic.

Decision Group

Founded in 1986, Decision Group is arguably the best-established AsiaPac-based provider of DPI. Headquartered in Taiwan, the company draws 40 percent of revenue from the AsiaPac region, and also maintains a strong, lucrative business in Africa, the Americas and Asia. Decision Group sells through offices in Malaysia, Japan, Canada, Africa, Europe and Taiwan (but not the PRC), as well as through partners. Aggressive product development keeps sales humming.

Heavily committed to its own R & D, Decision Group maintains a large staff of engineers focused on software and hardware development. What they produce is impressive: separate DPI modules for wireline and mobile – the latter with full WEP and WPA (algorithms for secure IEEE 802.11 mobile communications) cracking capability – and master systems that do both; decoding and reconstruction for VoIP, HTTPS and SSL decryption showing content in the exact digital format seen and used by the target; digital forensics; a mediation device for wireline, and for wireless including LTE; a data retention and management system (DRMS); and a Centralized Management System (CMS) that can be integrated with third party systems and with other data sources such as OSINT.

While the products may be purchased one-off, Decision Group treats the individual modules in its suite as integral components of a grander vision where each solution makes a unique contribution to the big picture of surveillance in service to public safety and national security. That vision is end-to-end, beginning with a master "does it all" system and extending through modular DPI units designed for: tactical Wi-Fi,; defeat of secure (HTTPS/SSL) IP protocols: storage and analysis; centralized management; and finally, forensics. Whether the client chooses the master product or a module, access to data is shared securely among assigned agents, then cleared and continuously monitored for aberrant internal user behavior by the administrator.

Decision Group DPI products have mastery of Levels 1 through 7 of the OSI stack, and the ability to identify over 200 protocols. For the user, the end result is full insight into metadata and content, revealing not just every data bit transported across the network for suspects, but their identity, intent and planning, affiliations, sphere of operation and location, all in real-time. Complementary modules build on the intelligence the master Decision Group DPI product delivers.

A high-level overview of the Decision Group portfolio:
- **E-Detective:** This is Decision Group's core DPI product. Designed for LAN and WLAN monitoring, it sniffs, decodes and reconstructs

packets in their original viewed format from wireline, wireless and social media networks for analysis. E-Detective's baseline functionality may be augmented with modules such as: the HTTPS/SSL module, to apply the same intercept/decode/reconstruct capabilities on secure packet traffic (Gmail, Hotmail, SMS, social media, including the target's user name and password); and Decision Group's DRMS data retention system, which receives captured data from E-Detective in ftp or iso format for purposes of archiving and analysis. E-Detective may also be integrated with third party tools including OSINT collection systems. The system can be deployed as the strategic centerpiece of a lawful intercept or government intelligence solution, or used for tactical operations of short duration. E-Detective works off-the-shelf with minimal systems integration required, features a user-friendly GUI, and requires only minimal training, though Decision Group recommends and offers product training to ensure optimal performance, and provides 24X7X365 customer support to handle questions/issues that might arise. Sold widely throughout AsiaPac and globally, E-Detective is often provided as an integral component of other vendors' surveillance suites, as well.

- **Wireless-Detective:** a comprehensive mobile intercept product that intercepts, decodes then reconstructs wireless traffic in 802.11a/b/g/n networks and cracks both WEP and WPA target traffic in either the 2.4 GHz or 5.0 GHz frequency, within a 500-meter range of the target. Recognizes over 200 protocols. The solution is built on a commercial laptop so that the user can "blend in" during tactical surveillance, and includes USB ports for adding data from other sources, as well as network ports for interfacing with other Decision Group products such as E-Detective, HTTPS/SSL and DRMS.
- **Law Enforcement Management Solution Suite:** a passive LI solution for the LEA, or for the communications service provider (CSP) required to support lawful interface technical standards set forth in CALEA or ETSI. DPI for data & VoIP decodes packet data associated with protocols at all 7 OSI stack levels. The system identifies the target by specific device or account, exports metadata and content and reconstructs it in the exact format seen by targets. Includes the Decision Group "iMediator" mediation device, which configures the targeted intercept, connects to network hardware lawful intercept software modules, then receives and formats the data in designated LI protocols for routing by VPN to Decision Group's purpose-built "iMonitor" for archiving and analysis. Connecting to one or multiple

iMediators, iMonitor provides access to diverse security-cleared users and provides "deep content inspection" of intercepted traffic, with the option to present the data in visualized format.
- **DRMS:** data retention that inputs and lets users search, scope and view intercepted/reconstructed data collected by multiple front-end systems including E-Detective. With DRMS users can import large amounts of data and view multiple files simultaneously. Decision Group's DRMS data integration capability also integrates with 3rd party data sources or with other data retention systems such as Java-based HDFS (Hadoop Distributed File System developed for the Apache Hadoop "data lake").
- **Decision Group Centralized Management System (CMS):** The Decision Group CMS is useful when managing IP interception across one or multiple large networks. The CMS can be managed remotely. Another plus: CMS can be configured to permit access by diverse users in different locations through secure VPN connectivity.
- **E-Detective Offline Centre:** provides offline packet decoding and reconstruction and can be integrated with other intelligence systems to add narrative and detail to an investigation. The administrator has the flexibility to set up separate paths of investigation, or to allow designated users to pursue different cases, or even to assign designated aspects of the same case as a pre-step to digital forensics. E-Detective Offline Centre decodes any/all types of content: email, Webmail, SMS, file transfers, or HTTPS/SSL protocols applied to streaming video, VoIP and Webcam.
- **HTTPS and SSL:** sidesteps these commonly used Web security protocols to provide access to packet headers, footers and payloads. Operates at all 7 OSI stack levels. Modes of the Decision Group HTTPS/SSL module include transparency proxy for MITM attacks, forward proxy and passive capture. The module also provides the option of customized certificate replacement.
- **VoIP-Detective:** intercepts, captures (by mirror or tap), decodes, and exactly reconstructs the target's VoIP calls. Decision Group offers VoIP-Detective in a complete hardware/software unit, and also sells the software-only piece for use on a client server.
- **Network Investigation Toolkit:** For field agents, the Toolkit is an all-purpose wireline/mobile system for interception, capture and reconstruction of Ethernet LAN or WLAN traffic including HTTPS/SSL in real-time via MITM attack. Allows for offline decoding and reconstruction and provides initial forensics analysis.

- **Forensics Investigation Toolkit:** for "after the fact" analysis, a Windows-based system that can import IP or other network traffic for forensics analysis.

Which solution is best for the client, be it a network service provider, LEA, defense or intelligence agency? And once that choice is made, is the client sufficiently trained? Decision Group helps on both counts.

The company begins by providing a detailed consultation to assess needs, then recommends the most appropriate solution, how and where it is best deployed, and provides insight on legal issues and procedures relevant to the client's geopolitical sphere of operation.

Next up is comprehensive training on: product installation and optimal use; the fundamentals of conducting investigations with DPI; analysis of data; integration with third party systems including social media monitoring and OSINT; and, lawful intercept training for systems that adhere to CALEA and ETSI standards.

Decision Group's investment in all facets of DPI – robust product suite, close analysis of client needs and the full scope of training – underscores the company's view that deep packet inspection is "a" and perhaps "the" most important component of surveillance.

However, excellent though its products may be, Decision Group needs to look over its shoulder as new competitors emerge to offer very similar capabilities. Features touted as "unique" by Decision Group only a few years past – packet reconstruction in original format and HTTPS/SSL decoding – are now offered by a rising generation of DPI providers such as Expert Team of Singapore.

Expert Team

Expert Team's arrival in the DPI market in 2012 roughly coincided with the "great expectations" of the technology's evolution to embrace NFV functionality, for example as a component of network virtualization on a software-defined switch. Among the leaders in this movement is France's Qosmos, whose ixEngine is among the more popular and respected DPI products, albeit – or so it claims – strictly for enterprise communications service provider network optimization, bandwidth prioritization, customer experience management, network/service QoS, targeted marketing based on segment/individual usage scores, and cybersecurity.

Expert Team follows a similar mold, but with no qualms on DPI for law enforcement and government intelligence or cyber censorship included.

Close examination of Expert Team's product serves as evidence of the "DPI for LI" niche's growing commoditization. Indeed, Expert Team's DPI and supporting products for law enforcement would appear virtually identical to those of its Taiwan-based neighbor's portfolio. Expert Team's package spans products that individually perform: Layer 7 packet filtering; decoding and real-time packet reconstruction; data retention; centralized management; HTTPS/SSL interception, decoding and reconstruction; a tactical unit for both wired and WLAN Wi-Fi; and finally, a mediation device.

The key difference from Decision Group: Expert Team markets its products as piece parts rather than as modules in support of a core product. Otherwise, the portfolio is a carbon copy of DPI offered by Decision Group, as well as by many competitors located worldwide.

The emergence of vendors with essentially lookalike products and services is not a negative. On the contrary, it is extremely beneficial, introducing market opponents that should, in turn, stimulate innovation and price competition. However, owing to the fact that Expert Team aggressively promotes the contribution of extensive in-house R & D, it is reasonable to query what exactly these software engineers spend their time doing by way of creating new and different solutions.

The answer to that question initially seemed to be Expert Team's 3i-Tactical System, which provides features such as Deep Web OSINT, using a rules-based engine called IRGO (Intelligent Reconstruction Gear OS) developed for its earlier 3i-Web product.

3i-Tactical System is a small, lightweight laptop version of 3i-Web that may be used in field operations, completely unobserved. Interested buyers may choose from models ranging in input speed up to 160 GBs. However, what is new for Expert Team with its 3i-Tactical System has for some time been available in Decision Group's Wireless E-Detective, which is equally compact, provides integration with OSINT sources and is just as fast.

Like Decision Group, Expert Team is for the moment steering clear of NFV, an area where Korea's Hanvit may have the edge – simply by hedging its bets.

Hanvit's Connections

Seoul-based Hanvit appeared on the DPI scene with its founding in 2002. From the standpoint of R & D, the company is a non-starter: Hanvit is strictly a reseller of products made by its partners, which include Allot, Flowmon Networks, Keysight (Ixia), and in the not-too-distant past the network packet broker vendor, VSS Monitoring.

The characterization as reseller by no means impugns Hanvit. On the contrary, a strategy based on resale of others' best-in-class network optimization and surveillance systems may be the ultimate guide to market survivability. Hanvit is not wedded to any technology in which the company itself has invested. The Koreans thus can move from one-to-the-next category of technology with every shift in the marketplace. The same flexibility flows through to Hanvit's customers, as well.

As explained in the chapter on Flowmon Networks, this Czech vendor is among the leading forces behind Flow Monitoring, which is differentiated from DPI by its focus on distinct types of packet flows at the packet/header level versus DPI's concentration on packet payload, an important distinction as carrier networks evolve to 100 GB and above, and thus beyond the back office processing capabilities of most DPI systems. Flow monitoring can segment IP packets by specific identifiers including protocol, application, originating and terminating points and addresses, and thus learn the identity and location of the targeted parties. More granular detail may be obtained by routing the refined data to DPI systems.

Hanvit offers the full Flowmon Networks Flow Monitoring package comprising: FlowMon collectors for interception of IP traffic; FlowMon probes for export of designated IP flows; data retention and analytics; and plug-ins for additional behavior analysis of suspicious traffic.

The Hanvit MIDAS suite is a set of OEM'd versions of same-named products made by Optenet, which was acquired by Allot Communications of Israel in February, 2015. What is interesting here: the spread of Allot/Optenet products that allow Hanvit to go head-to-head with competitors in three markets – with companies such as Keysight (Ixea) and Gigamon for passive and active network TAPs, DPI by Decision Group and Expert Team, and Silicom Denmark (Fiberblaze) for fiber optic network TAPs.

In studying the various DPI solutions available, the client has multiple criteria to consider in matching products to needs – standalone or integrated option, system-wide or modular approach, proven performance in the field, evolutionary path (if any) to NFV, and availability of complementary solutions. The AsiaPac DPI scenario addressed in this analysis serves as a case study whose lessons may be applicable to the selection process for DPI products offered by vendors situated in other geographic regions, as well as in the broader commercial DPI market.

DESOMA Packet Extraction

DESOMA GmBH, a provider of advanced packet extraction capabilities, is almost as quiet as the field it plays in: mass IP Monitoring. DESOMA initially set out to provide its unique skills via products specifically designed for government intelligence and law enforcement, then realized that the same talents apply to protecting enterprise digital assets, as well. Today the company serves both markets globally through business alliances that range from its European home base to the Middle East and South Asia. But beyond a very occasional public appearance at public forums such as ISS World Europe, DESOMA does very little marketing. Instead, the company relies on word of mouth from partners like FinFisher, cybersecurity firms such as CSDINT (Switzerland), systems integrator COMINT (India) and others.

DESOMA's solutions would almost appear to sell themselves, a fact that portends a rare degree of uniqueness in a field that by now is nearly as common in ISS, IT and networking as is water to the earth's surface. Certainly, DESOMA believes it has something special. Company founders in 2013 applied for and in 2016 received a patent from the U.S. for new advances in data extraction technology.

DESOMA is a smallish company of some 30 – 35 employees. It was formed in 2010 in Rosenheim, Germany, where it remains to this day. What sets the company apart and contributes to business continuity? The answer may reside in understanding the limits and challenges of conventional Mass IP Monitoring and how DESOMA claims to have advanced the marketplace.

First Stake in the Ground with Gamma Group/FinFisher

Within a year of opening its doors for business, DESOMA had already won a key partnership with one of the most renowned vendors in the ISS space: offensive cyber leader Gamma Group/FinFisher. The winning product that catapulted DESOMA to such prominence so quickly was its "DAISY" solution, a packet extraction engine that connected to the network on either the CloudShield CS-2000 high-speed hardware or the IBM PN41 Deep Packet Inspection Blade.

Company-internal Gamma Group/FinFisher documents show that DESOMA's DAISY engine took on the leadership role in both the front end (administration, data capturing/handling, target identification) and back end (decoding/demodulation, storage/archiving, reconstruction) in Gamma's master design for capturing data from public switched telephone, UMTS and LTE networks. For IP networks, however, Gamma handed off the same functions to a different partner, DreamLab. That was 2011. DESOMA's claim of "We are IP" apparently didn't yet fully resonate with Gamma. But that would soon change as DESOMA set out on an ambitious plan to own the IP monitoring market. DESOMA saw and understood issues that DreamLab had overlooked.

Real-time Extraction on a Budget

For competitive reasons, ISS companies rarely if ever disclose details of their proprietary technologies in public documents. However, disclosure is inevitable to a degree when a company seeks to protect its intellectual property via patent. DESOMA did just that in its 2013 U.S. filing for a new packet extraction technology, a patent that was subsequently granted in April 2016.

DESOMA set out by pointing to issues with high processing systems that capture massive amounts of data then store it, as well as real-time systems that analyze data in the moment. Both common approaches have negative offsets. High-power processing systems are expensive. When data is stored for a long period of time, it loses its "real-time" value. Finally, reliance on storage of bulk data can lead to memory overflow. DESOMA's rating of such systems: not ideal for permanent IP monitoring.

DESOMA took issue with typical real-time analytics systems, too:

"When using these systems, it is usually not possible to reconstruct the entire content of a network session, because the data packets are often transmitted twice or not in chronological order via the network. This occurs particularly in load-balanced and redundant networks. These kinds of real-time capable analysis systems therefore cannot guarantee completely secure monitoring, because for example, key words whose common occurrence in a message or a network session is searched for, may be distributed among different data packets, so that these key words are then untraceable."

DESOMA's alternative: a method providing analysis of both metadata and "useful content" from packets, enabling a permanent architecture packet monitoring system that guarantees detection and capture of targeted content. DESOMA's system operates as follows:

1. Extract packets targeted for analysis.
2. Copy each packet.
3. Assign a unique hash code or signature to each packet so that duplicates have the same #code and are easy to spot. [The unique hash code or signature of a packet prevents repeat storage and saves on memory costs.]

4. Store coded packets in a buffer two ways: metadata in a first memory and useful content data in second memory, using pointers to show when they are related.
5. Conduct at least a partial reconstruction of the network session, sorting packets either chronologically by time of arrival or by sequence in the OSI stack, in a ring buffer, then conduct the first search for targeted fingerprints within the stored metadata.
6. When the search is successful it is "flagged" in the metadata. Because the metadata from Memory #1 produces a result, so will the related code-linked "useful content" stored in Memory #2.

As DESOMA explains, "Because the search term can also occur in another packet of the network session, it is ensured in this way that all data packets for the reconstruction of the network session are available, and that the network session can therefore be fully reconstructed."

For further analysis, DESOMA's engine moves to the next level with the ability to replicate exact copies of emails, documents, videos or other content, but doing so economically by starting with search terms that succeed in the metadata and from there proceeding to the content. After a network session reconstruction is completed, DESOMA "marks" it so that it can be readily identified as a completed decode and stored for further analysis or built upon with the arrival of new data stream analysis.

One advantage of DESOMA's packet extraction and analysis solution is that it frees up resources so that key results – not the data dump but information that matters – may be identified and seen in real-time. The process gains speed because the solution automatically discards irrelevant data from the specific search and analysis being conducted at that moment, though that "useless" data may still be stored for future analysis. The solution can also search for specific types of files, using a MIME (multi-purpose Internet email extension) search to extract specific data from a network session and include it in the analysis.

The key advantages of DESOMA: efficient, economical use of memory and storage; and, packet extraction and analysis in close to real-time. Once the first search is successful, the entire network session is available.

Flowmon Networks: DPI with Flow Monitoring

Until splitting into two companies in 2015 – Flowmon Networks and Netcope Technologies – the Czech Republic's INVEA-TECH was for nearly a decade recognized as a leader in packet flow monitoring and Field Programmable Gate Array (FPGA) solutions that helped law enforcement and intelligence agencies track targets on high-speed networks. With the rapid growth of each marketplace, whether for surveillance or commercial network performance requirements, the parent company felt it was time to dedicate separate corporate resources to each.

Here we will concentrate on the work of Flowmon Networks, delving into the union of technologies formerly considered technological adversaries, or at least incompatibles – flow monitoring and deep packet inspection. While packet flow monitoring and DPI were for years deemed polar opposites, the two have always borne a similarity. Now they are seen as complementary by companies like Flowmon Networks, which uses DPI to add greater depth to the data revealed by flow monitoring.

Flowmon Networks relies to a great extent on FPGAs, just like its sister company, Netcope Technologies. FPGAs play an important role in Flowman's suite as hardware accelerators that reduce CPU load during packet capture to help ensure comprehensive monitoring without packet loss on 100GB networks.

One such device is the Flowmon Probe originally developed by INVEA-TECH and now offered in updated versions by Flowmon Networks.

The Flowmon Networks probe is a passive device that connects to a network and monitors all packets. At the outset, Flowmon uses Cisco NetFlow to collect information by ingress interface, source and destination IP addresses, IP protocol, source and destination ports, and type of service.

The solution is based on a pair of FPGA cards called "COMBO" that can process near-limitless amounts of data at wire speed with no input sampling required and zero packet loss. Flowmon inspects all packets, and with DPI embedded captures the entire packet payload in Layers 4 – 7, creating flow records which are then ready for storage or analysis.

The probe is available in models that collect IP data at speeds beginning with 10 Mbps, and going up to 100 GB. The probe uses the NetFlow v 5/v 9 and IPFIX protocols and FPGA cards to generate a complete copy of packets traversing an IP network. Probes are deployed at locations in the network with the highest traffic, both ingress and egress, with connection made either by mirrored port or Ethernet splitter. Because the probe is a passive device, it operates invisibly and without detection.

Flowmon offers multiple versions of its probe, each tailored to meet the operational and budget requirements of the user. Pricing moves up a sliding scale in tandem with higher performance capabilities and the number of interface ports per unit. On the low end, the Flowmon 1000 Probe with a single 10/100/1000 Mb Ethernet interface can intercept 500,000 packets/second. At the top, the Flowmon Pro 100000 CFPR can capture 1.5 million packets/second via a single 100GB Ethernet interface.

The models in-between offer significant choice, with a few other differentiators. For example, the comparatively high-end Flowmon 80000 comes with a Quad Small Form-Factor Pluggable (QSFP) transceiver that permits "hot swapping" to intercept traffic on fiber optic cables without changing the underlying intercept system. The tradeoff: In QSFP mode the unit does not support DPI capability to capture key application data at Layer 7 such as VoIP and HTTP stats.

Complementing the probe is the Flowmon Collector, a single-purpose server that provides data retention. Each Collector takes in NetFlow and IPFIX-formatted data collected from probes connected to the network. The Collector comes with 500 Gbs of resident storage capability supplemented by virtual storage from RAID multiple disk drive components. Note that with the "smaller" Flowmon probes, which collect lower volumes of data, storage is not part of the package and is sold as an add-on.

In addition to hardware-based flow monitoring/DPI, Flowmon also offers a selection of six virtual Flowmon products that take packet capture "to the cloud." Flowmon virtual appliances are designed for a VMware environment. Performance is significantly lower than that of the hardware versions, beginning with the Flowmon VA 1000 which provides 0.3 Mps packet capture and capping out with the top-of-the line Flowmon VA 20000 with 0.7 Mps. Another caveat with these virtual appliances is that the Collector function is available only as an add-on.

The key takeaway with Flowmon is choice. In addition to hardware and virtual Flowmon products, buyers can opt to go the do-it-yourself route by purchasing and building on the COMBO-100G, an FPGA card that delivers full throughput of data transfers. With the Collector added, users also can leverage Flowmon analytics tools or those developed by third party analytics systems vendors such as InterSystems.

For law enforcement agents, Flowmon Networks provides a specific package that meets the full requirements of ETSI standards-based lawful intercept: the INVEA LI System. With three exceptions – absence of the standard Flowmon 40,000 Pro and the two virtual systems with the highest throughput in that class – the LI probes are identical to units offered for commercial clients. The key differentiator of the Lawful Intercept System is the addition of a mediation device. Flow data is routed from the probe to the mediation device for standard formatting prior to forwarding to a law enforcement agency (LEA) monitoring center.

In instances where the LEA or government agency already knows the IP address of the target, or up to 100 targets in a single location, Flowmon offers a tactical version of the Lawful Intercept System. The tactical system may be deployed for finite interception needs and quickly removed when the intercept is completed. While in operation the tactical unit provides both probe and mediation functionality, capturing call data and content, then converting the data to the correct LEA format.

For data retention, LEAs may use one or more Flowmon Collectors. As on the commercial side, Collectors store every byte of data collected, for real-time or subsequent analysis. Flowmon's data retention system can be designed to comply with the data collection rules of any country/client, and to provide stored data in the formats required by different LEAs and government agencies.

Speed-wise, INVEA LI System does not disappoint. Using FPGA-powered flow monitoring it can monitor 100G IP networks and deliver the right flows to Flowmon's IP engine in the probe for analysis, thence onward to mediation and a law enforcement monitoring center.

Glimmerglass Photonic Surveillance

Based in Hayward, CA, Glimmerglass Cyber Solutions is a pioneer in fiber optic interception and monitoring technology, with a robust product offering that today serves clients including the U.S. Intelligence Community, all branches of the U.S. military, the State Department and a Who's Who list of defense systems integrators – Harris Corporation, Leidos, Northrop Grumman, Raytheon and Lockheed Martin, to name a few – and an equally impressive roster of customers throughout Europe, Asia and the Middle East. Add major communications service providers, too.

That is quite a coup for a company that began operations at the least auspicious moment in telecommunications history: the great telecom bust of 2000, an event that crippled the industry for months and raised serious doubts about the heady days of widespread fiber optic deployment.

In that year, *Bloomberg Business Week* magazine ran a cover story on the debacle, observing that during the boom (or bubble) of the late 1990s, operators had sunk some $6 trillion worldwide into fiber optics. In the U.S., when urban markets became congested with competitors, eager players spread into the countryside, laying or stringing fiber in even the tiniest Tier III and Tier IV markets – and basing their valuations not on customers and revenue won, but on their cash outlay for network infrastructure.

The ensuing crash of this latter day glass menagerie wiped out more than $US 2.0 trillion in stock market value. Pessimists observed that all the fiber capacity created during the infamous telecom bubble would never be used.

At the time, of course, few could foresee the vast amounts of data that would be generated by the convergence of the Internet, mobility, smartphones, social media and video. By mid-2010, investment in fiber optics was roaring once again, and by 2012 the U.S. reached its pre-crash level of 19 million miles of fiber optic cable installed per year.

Among the insightful few early-on were the founders of Glimmerglass who knew, as all industry insiders do, that bean counters always bewail network investment as too high, while operations folk understand that capacity is priceless and tends to fill up months and years ahead of official projections.

Today, fiber optic cables dominate core networks in developed nations, are making significant inroads into developing nations, and by far dominate international communications, responsible for 99 percent of transoceanic traffic.

The challenge, for intelligence and law enforcement agencies, is sifting through all the data carried by fiber networks to find targets and intercept useable data. What they need: intelligent optical signal access and monitoring that can select defined signals on demand, at any data rate and protocol and then select, extract and monitor data for intercept purposes, all at the speed of light. It's a demanding field calling for special technologies and skills. Which is precisely why Glimmerglass technology has long held such a strong interest for the U.S. Intelligence Community and many others.

Glimmerglass Intelligent Optical System (IOS)

Glimmerglass arguably offers the leading solutions in transparent optical switching: Glimmerglass Intelligent Optical Systems (IOS). All technology is developed in-house. The company faces competition in commercial niches from market opponents such as Calient Networks, but in photonic switching for the intelligence and defense sectors, Glimmerglass is a standout player.

The Glimmerglass Intelligent Optical System (IOS) uses a combination of hardware and software to let the user quickly target and access optical signals and route them to signal processing and monitoring equipment, in real-time, and without disrupting the network.

IOS is offered in several models – The System 100, 500 and 600 – that move up the scale in port management capability from 16 X 16 fiber ports to 192 X 192 fiber ports. Each system supports optical data rates up to 100 GBs, and any protocol including SONET/SDH, Ethernet, C/DWDM, video, ESCON, FICON and many others – an important feature as new protocols are always being added. Standard 20 millisecond switching can create an optical path between any fiber input and output.

IOS uses Glimmerglass proprietary 3D microelectromechanical systems (3D MEMS) technology, coupled with micromirrors, to capture selected light signals from among the thousands transiting a fiber optic cable. Optical signals are monitored by photo detectors before entering a tiny switch, and from there are directed into a 3D MEMS array with scores of micro-mirrors.

The system uses this micromirror array as a lens so that each beam is directed to a specific mirror. Via remote software, it then sends individual light beams to selected processing points for analysis. The system can also split signals into multiple copies and route each one through a separate output, if the user needs to distribute the captured data to different locations.

The appeal of IOS with 3D MEMS is two-fold: automation and ubiquity. A user clicks once for an input and then again for output to analysts in multiple locations – no manual processing required. The reason: customers in the intelligence space typically need mission-critical data sent not to one or a handful locations, but to hundreds of sites simultaneously.

Glimmerglass CyberSweep Sapience

The company's core product for acquiring actionable intelligence from dynamic signal selection is dubbed Glimmerglass Cybersweep with "Sapience" threat insight for real-time analytics capabilities.

CyberSweep entails a 3-part interception process: Select, Extract and Monitor.

Out of the billions of conversations, data transmissions, video and financial transactions that cross fiber optic networks, where do you begin? This part is hands-on. The agent can, if he wants, grab data in bulk. More commonly, agents will single out communications from obvious trouble spots such as Yemen and Syria. Again, with Glimmerglass IOS at the core of CyberSweep, they can select any type or combination of communication signals to be extracted.

Extraction is the process of converting the selected optical signals via optical layer mux/demux and application protocol processing to the end form of communication, be it mobile, text, social media, etc.

Finally, monitoring provides files for analytics to create a target profile with ID and image, messages, frequency and distribution of communications, full content and the relationships that reveal networks of affiliate targets.

A Power Behind the NSA?

Where in the network is photonic monitoring deployed? There are three options, and Glimmerglass has them all covered: central office; international gateway; and submarine cable landing station, a favorite work site of the NSA. The NSA is expressly forbidden to do domestic surveillance. However, gateways and landing stations can be offshore where U.S. law does not apply.

In 2013, as the news broke of NSA's capability of sneaking into and monitoring transoceanic fiber optic cables, some observers drew up an elaborate conspiracy theory wherein credit went to NSA's crack hacking team, variously called Tailored Access Operations (TAO) or The Equation Group, together with the QUANTUM alert system and FOXACID servers and malware that could take over networks.

But was NSA using Glimmerglass? The intelligence and law enforcement agencies of leading nations in Europe, Asia and the MiddleEast do. Given that the NSA employs numerous government contractors to work in secured facilities and even to train its own people, it is highly unlikely that the agency goes to the trouble and expense of doing photonic surveillance entirely on its own.

Incognito Broadband Command Center

Canada's Incognito Software Systems is well regarded as a provider of operational and billing systems support (OSS/BSS) for the communications service provider (CSP) industry. But Incognito touts other skills, as well: the ability to collect customer IP data using a method the company claims to be more effective than deep packet inspection.

The vehicle is a product called Broadband Command Center (BCC). BCC is marketed as a multi-network/environment provisioning device that can speed multi-play service rollouts, cut service downtimes and fulfill every carrier's dream: the ability to deliver a superior experience that builds loyalty while simultaneously cutting costs and boosting profits.

Not incidentally, BCC also helps cops catch bad guys. As such, Incognito markets BCC as a dual-purpose tool: fully functional on the service provisioning/back office side of the business, while also capable of helping CSPs meet their legal requirements for compliance with lawful intercept in multiple nations.

The Download on Incognito

Founded in 1992, Incognito operates from its Vancouver headquarters and – by a different name, "Interactive Systems" – from offices in Dublin, Ireland. The company's primary customers are cable MSOs, VoIP and video providers.

Since its early days, Incognito has built a sizable global clientele numbering some 250 companies including "five of the 10 largest cable companies in North America." Named clients include Bright House Networks, BT Group, Mediacom and Suddenlink, as well as cable companies in Brazil, Chile and Hong Kong. One notices right away that the identified cable companies skew toward the smaller half of the top 10. The big players – Comcast, DirecTV, Dish, Time Warner and AT&T U-Verse – aren't on the list. However, this may be an intentional omission. The Top 5 are notoriously loathe to acknowledge the need for help in the areas Incognito specializes in. Probable reason: reluctance to admit the outrageous cost, passed along to customers for decades, of maintaining error-ridden manual network inventory management and service delivery processes in the decades preceding automation.

For the first 22 years of its existence, Incognito boot-strapped the business. Then in September 2014 the company made the judicious decision to be acquired by Volaris Group. Volaris is a Toronto holding company that invests in software firms and is itself one of four operating groups of

Constellation Software, Inc. The acquisition was a good deal for both sides. Incognito's management stayed in place managing operations more or less independently, with backing from a deep-pocketed owner. Volaris, noted for picking winners, has oversight but does not interfere in the day-to-day running of Incognito.

Broadband Command Center

Broadband Command Center, Incognito's anchor product, is Dynamic Host Configuration Protocol (DHCP) software that can be used to capture, record and transfer customer data. DHCP automatically distributes configuration parameters including lease records and IP addresses during an Internet session. In a commercial scenario, BCC connects to DHCP servers to collect IP data on a client's customer base and determine bandwidth use by specific customers.

BCC is also the core of Incognito's "Lawful Interception Reporting Solution." In a lawful intercept, the BCC follows the same process to collect similar data of targets specified in a court order.

BCC comes equipped with a RADIUS (Remote Authentication Dial In User Service) feature which also shows metadata records including start time, duration and stop time of an IP connection. RADIUS updates may then be pushed to a DPI solution to provide refined intelligence. In the event DPI fails, BCC automatically buffers the collected data, which can be forwarded after the DPI solution resumes service. Essentially, the Command Center is designed to take DPI out of the loop on IP data collection.

Incognito cites the Command Center's RADIUS function as a competitive advantage over data collection by DPI alone, noting that high levels of broadband traffic at times overwhelm DPI, leading to lost packets. The principal benefits touted for BCC, then, are more accurate validation of target info and back-up.

With BCC onboard, Incognito says clients can comply with lawful intercept requirements by:
- Generating subscriber IP communication reports on any device or IP address;
- Retrieving subscriber records; and
- Storing information.

BCC in its Proper Place

In general, Incognito overstates its case for Broadband Command Center as a DPI replacement on the data collection side.

It is true that traffic volumes can impact DPI and increase the risk of packet loss for functions such as measuring broadband volumes. However, passive probes configured for lawful intercept use DPI to capture the data of designated targets, not to consume all network traffic. Most probes also provide buffering to prevent any loss of packets. Where DPI is used in broader surveillance capacities, users know to back it up with sufficient processing power.

RADIUS is hardly unique to Incognito's product. Many DPI products also integrate with both RADIUS and DHCP engines to collect IP data, which is then mirrored and analyzed at Layers 2 – 7 of the OSI Stack to decrypt and view the full packet payload including content.

It is in this latter realm – content intelligence – that the Incognito BCC comes up short. For metadata collection, the BCC is a fine performer. But for content analysis, it is a non-starter. BCC simply hands off raw data to a DPI engine, which performs the all-important job of deciphering content. In its primary role as a provisioning and customer experience management system, the Incognito Broadband Command Center is a fine tool for broadband companies. As a lawful intercept solution, it is just a supplement.

Keysight Network Monitoring

When Keysight Technologies purchased Ixea in 2017 for US$1.6 billion, the deal captured attention as a folding of two impressive network test and measurement competitors into one giant with the power to dominate competitors in the T & M marketplace. Under the radar: The pairing created a separate type of behemoth combining Keysight's formidable RF Monitoring for military clients (see "Keysight Signal Analyzers" in Chapter 8) with Ixea's equally impressive line of network packet brokers (NPBs) and DPI engines used by government intelligence agencies to capture and decode data from high-speed networks.

Often forgotten: Ixea's strength in the NPB arena gained momentum from M&A activity of its own – the US $190 million acquisition in 2013 of Net Optics, whose NPB and deep packet inspection prowess added weight to Ixea's product line, both T & M and ISS-focused, from that point on.

At the time, Net Optics was itself a newcomer to the network monitoring business, and solely by acquisition of a pair of private jointly-owned Australian vendors, TripleLayer and nMetrics in January 2012.

Keysight's acquisition of Ixea was actually its second stab at the network monitoring business. Before spinning off its "measurement" business as Keysight in January 2014, former parent company Agilent sold the company's entire network monitoring business to JDSU in 2010 for US$160 million.

Keysight's purchase of Ixea might be cast as the return to a business its former management abandoned and the current team evidently missed. The irony is that Keysight's payout for Ixea in 2017 was nearly 10 times more than its former parent Agilent received for *exiting* the business seven years earlier in 2010. Keysight could afford it, and judging from its soaring stock price in the years since purchasing Ixea, shareholders heartily approved.

The value add for government agencies is what this new & improved Keysight brings to the table in network monitoring capabilities. Packet interception tools are one part. But there is more in play than just conventional network monitoring hardware and software. Ixea also brought mobile monitoring tools that deliver fresh power to Keysight's core RF sensor capabilities via "design and test systems" for 5G mobile networks. How the mobile T & M side of Ixea's business impacts Keysight's ISS niche remains to be seen.

Ixea Network Visibility Platforms

Make no mistake, Ixea solutions are primarily dedicated to network test and measurement purposes for network analysis and optimization, to ensure peak performance for network operator customers. That said, several key products are definitely dual-purpose and overlap into the ISS area for government intelligence work. Among these, a "network visibility platform" called the Ixea Net Tool Optimizer (NTO) or Vision 5200 Series can focus more than the usual set of technical eyeballs on the network.

The Vision 5288 is designed for both network management and monitoring, providing a simplified user interface that renders both tasks comparatively simple. From the ISS standpoint, think of the Vision 5288 as a control point that intersects with key tools that perform network monitoring – including network packet brokers, DPI solutions – and then on to analytics. With the Vision 5288, the same level of visibility available to network managers within a communications service provider (CSP) or enterprise is available to an investigative agent, and may be tailored to capture highly granular detail.

High-speed is the rule. With network packet brokers in the mix, Vision 5288 devices can readily manage vast amounts of data whether the source network is 10GB, 40GB 100GB or beyond. The system provides a comprehensive view of all network activity in real-time, and can drill down to specific targeted packets for investigation, leveraging filters that eliminate extraneous packets not relevant to the case. Refined data is then handed off to analytics tools for Layer 2 – 7 OSI stack examination including not just metadata but apps and content.

Remember, though, that the Vision 5288 is a management system, not a collection system. Other modules do this latter job. As Ixia describes it, the Vision 5288 "sits between" devices such as SPANs, TAPs, network packet brokers (NPBs) and DPI products that do the real work of data collection. Ixea does a knock-out job on both NPBs and DPI.

Appliance vs. Virtual Network Packet Brokers

Recall that Network Packet Brokers (NPBs) are packet filtering systems capable of monitoring IP traffic flows at line rate speeds, providing dynamic visibility into the network. NPBs have grown in popularity with the evolution of software defined networks, virtualized networks, the cloud, and the attendant rise in the volume, bandwidth, complexity and variety of network traffic.

An NPB sits between a network switching device's SPAN ports or routers to take charge of filtering high-bandwidth traffic flows. In addition, an NPB can do much more: de-duplicate and time-stamp packets and be coupled with DPI capabilities from Layers 2 – 7 of the OSI stack. In sum, an NPB with

DPI in tow can provide end-to-end inspection of targeted packets through the OSI Layer 7 application level – where much intelligence resides – together with the target's IP address, MAC address and content.

The star of Ixea's data collection family is the Vision 7300, described as a "net tool optimizer," marketing lingo for a network packet broker. The Vision 7300 comes outfitted with 384 10GB ports. Not surprisingly, when first introduced in 2014, the Vision 7300 was promoted as having "the most throughput" of any comparable network visibility system with "three times the bandwidth" of competitors' wares.

In the time since the Vision 7300's debut, virtualization has become the "next big thing" of networks and the T & M (and ISS) solutions used to track packets therein. As usual in the wake of any innovation, there is a running debate between makers of conventional boxes and proponents of new breed virtual options over which of the two is faster, more reliable and economical.

Rather than take sides, Ixea has instead chosen to take the lead with virtual NPBs, as well. The Keysight Ixea division's CloudLens Virtual Packet Broker provides similar functionality to its appliance-based counterpart the Vision 7300, but for networks that rely on virtual machines. Both systems receive packets captured by SPAN ports, TAPs and DPI solutions. Both aggregate the packets, filter and de-duplicate traffic, then route duplicated traffic for analysis. As a virtual solution, CloudLens is sold on a licensed basis in BASIC or ADVANCED packages. One caveat: Where processing "muscle" is concerned, a unit of CloudLens is lighter-weight than its hardware counterpart, with a maximum of 2 ports per BASIC package and 12 ports per ADVANCED Package.

Keysight Ixea Packet Capture Module

Expressly designed for its top-of-the-line Vision 7300 Network Packet Broker, Ixea offers its own DPI appliance, the Ixea Packet Capture Module. Based on WireShark, the Ixea PCM captures and decodes packets at line rate from networks up to 40GB. Presence of the Ixea PCM in the product lineup is a reminder that NPBs don't necessarily operate in isolation, and are often found in combination with DPI capabilities. With an assist from the Ixea PCM, the Vision 7300 can open and view full packet content.

On the Test & Monitoring side of the business, the Ixea PCM is most frequently deployed to capture packets and detect anomalies associated with a network issue. As a rules-based engine, the software core of the Ixea PCM can by the same token be programmed to look for specific activities, trends, key words or individuals. Just as critical for surveillance purposes, the device is equipped with a buffer to ensure against packet loss. Captured packets may be stored within the device itself, or on a user's laptop or a remote site. Wireshark then goes to work decoding and analyzing captured data. Activating the Ixea Packet Capture module is a simple "drag and drop" function.

Netcope: Scouting 100GB Ethernet

Before getting in line to purchase ISS solutions capable of intercepting traffic on 100GB Ethernet networks, it is wise to begin with a sanity check. Just how prevalent are 100GB networks? Is the trend real or merely hype?

The short answer is that 100GB is indeed real and growing, driven by insatiable demand for bandwidth. At the same time, network architectures are amorphous. Even as networks evolve toward 100GB and beyond, many smaller carriers may remain perfectly satisfied with 10G or even 1G to meet customer requirements. Law enforcement and government agencies need to be prepared to conduct intercept and monitoring on a wide of array of networks. Speed and volume of data across networks are issues that aren't going away. That includes 100GB monitoring – the metier of Czech-based Netcope Technologies.

World-Class FPGA Adaptors

Netcope claims to be the first company in the world to develop an "FPGA Adaptor" to support, manage and monitor Ethernet traffic at 100GB wire speed. Field Programmable Gate Arrays are integrated circuits that can be configured by the customer after manufacturing. FPGAs are noteworthy for their ability to capture unstructured data – video, test, images, etc. – at significantly greater speeds, lower energy, logic and cost than conventional ASIC chips. FPGAs are commonly referred to as hardware accelerators.

Netcope makes several products that build on the FPGA Adaptor concept. Each houses one or more FPGAs that capture and filter all incoming network traffic at 100GB wire speed with zero packet loss. The data may then be fed to analytics solutions such as deep packet inspection, Flow Monitoring or both.

Filtering at 100GB is an important asset to packet analysis capabilities because it intelligently eliminates high volume apps such as Netflix and Youtube, which now constitute the vast bulk of data streaming on high-speed networks.

Netcope FPGA Adaptors discard these distractions to ensure that analytic engines such as DPI can focus on data relevant to investigations. Without FPGAs in the middle, DPI can bog down and waste precious time on the irrelevant, hampering access to real-time intelligence. Indeed, without such Adapters filtering data in hardware platforms prior to analysis by software, real-time monitoring of 100G networks without packet loss would not be feasible.

Academic Roots

Netcope comes at these capabilities through a distinguished heritage. The company is one of two spin-offs of INVEA-TECH, an enterprise itself inspired and eventually spun off by academic researchers at the University of Brno who helped advance the science of flow monitoring – a solution that gained momentum as data flows on high-speed networks began to overwhelm conventional packet analysis such as DPI.

As INVEA-TECH grew, it became clear that the company's technology had a split focus on addressing the problem of packet analysis in 10GB, 40GB and 100GB network environments. One half of that focus was software: flow monitoring analysis based on representative sampling of specific flows. The other was the use of FPGAs to parse and filter the data *before* it is handed off to CPUs to conduct flow monitoring or DPI.

In 2015 INVEA-TECH decided to split into two separate entities, each dedicated to its specific expertise in managing/tracking streaming data on high-speed networks. The new Flowmon Networks took the software component of packet analysis, and its fraternal twin Netcope Technologies the hardware side.

Both companies continue to work closely together, though independent and free to go their own way. They maintain a close relationship and still co-conduct research with the university. Each contributes its part to solutions used in lawful intercept.

Top Cop Tool – Netcope Session Filter

Netcope offers multiple FPGA-based platforms aka "adaptors" that might be applied to any number of applications: real-time financial trading, cybersecurity or traffic monitoring for signs of criminal and terrorist activities in progress on high-speed networks.

Here we will examine one such solution as used for lawful intercept: the Netcope Session Filter (NSF). The NSF, like other Netcope hardware, uses FPGAs made by Xilinx, typically the Virtex-7.

Whichever adaptor is used, the principle at work is the same: the FPGA board offloads traffic direct from the network to pre-process it before the data is handed off to CPU space dedicated to either DPI or Flow Analysis.

Taking a cue from Flowmon's approach, the NSF views network traffic as being comprised of separate flows numbering in the thousands – each with uniquely identifiable features. Communications in a flow typically shares traits such as commonality of IP addresses, or similarities at the application layer.

Filtering traffic as IP flows in hardware enables the system to single out specific flows of interest to the law enforcement or government agent for further processing and analysis. This is more efficient than sending each packet straight to DPI or other processing. The NSF discards traffic of no interest and focuses on flows most likely to contain evidence.

The user can program NSF to look for flows with specific characteristics based on statistical data common to each type of flow: IP addresses, ports, Level 4 protocol, type of app indicated at Level 7, beginning and ending timestamps, or the number of transferred packets and bytes.

Depending on the requirements of the next phase of analysis – by software – Netcope's filter can forward designated IP flows in whatever format is required for further analysis: headers only, whole packets, or packets shortened to a predetermined length.

Still, even when traffic is qualified and "downsized" on Netcope hardware, it might still prove to be a great deal of volume to transfer into CPU processing. In that event, Netcope's solution is bifurcation, using a single Xilinx FPGA to manage sustained traffic flows, full duplex for 100GB Ethernet.

Measuring Market Demand

According to research by IHS Markit, carrier investment in 100G networks is expected to reach US $13 billion in 2020.

Not surprisingly, demand for 1GB has peaked as carriers look for faster alternatives. However, the largest sector is, and through 2020 shall remain10GB, which will grow to over US $27 billion this year, more than double the expected investment in 100GB networks.

Not everyone is investing in 100GB? Using just the U.S. as an example, hundreds of smaller communications service providers in rural or remote areas in the United States are perfectly satisfied with 1GB and aren't even making the switch to 10GB as yet, let alone 40GB or 100GB. Large network operators in populous nations, on the other hand, most certainly are making the move to 100GB and beyond.

For LEAs that operate in mainstream population centers, owning ISS solutions that can filter target data from 100GB networks is a must-do. As Netcope correctly observes, with the demand for high definition video for entertainment purposes, 100GB is the network standard. Whether to manage the network or protect it from malicious activity, service providers will require real-time DPI. But sheer data volume alone will make real-time DPI via software totally infeasible. The same challenges apply to the monitoring of 100GB by law enforcement.

Which Shall it Be – Netcope Technologies or Flowmon Networks?

Now that we've reviewed both INVEA-TECH offspring – Netcope Technologies and Flowmon Networks – which is best at delivering specified data intelligence off of 100G networks and above? Both can do wire-speed network monitoring and filtering, and use quite similar tools: FPGAs, Flow Monitoring and DPI.

Netcope focuses on FPGA Adaptors, the hardware that directly offloads traffic from any network – 10G, 40G, 100G or 200G – and filters it to eliminate irrelevant streams such as Netflix or Youtube that might consume precious CPU dedicated to DPI analysis. Netcope uses Xilinx FPGAs, but otherwise is tech agnostic on the IP Flow Monitoring or DPI that the customer might wish to use. The Netcope Session Filter for Lawful Intercept comes with filters that can perform either conventional stateless interception on IP networks to determine packet headers at Level 3 and Level 4, or stateful IP Flow monitoring at Level 4 for network flows. Netcope's LI solution provides a wide array of flexibility, but because reprogramming FPGAs requires certain expertise in writing code, it is not for the novice.

Flowmon Networks' INVEA LI uses FPGAs to accelerate packet capture and analysis by its own proprietary Flowmon Probe and IP flow monitoring software, with final analysis conducted by Keysight IXEA DPI. Flowmon probe performs network monitoring and interception at 100G wire speed. INVEA LI also includes a mediation device. It is a classic lawful intercept solution, with the hardware for configuring a court-ordered intercept and formatting collected evidence in the correct protocols. In sum, for surveillance purposes, Flowmon Networks INVEA-LI is the more user friendly.

Netronome: Driving SS8 and Blue Coat

One of the common secrets of the ISS industry is that the technical geniuses behind ISS innovation often act as silent partners and suppliers for big name brands in surveillance. Such is the case with California-based Netronome, which for years has quietly supplied ISS vendors with an array of technologies ranging from classic deep packet inspection with flow monitoring analysis, to multicore processors for hardware acceleration and SSL proxy servers. Clients in the lawful intercept and government intelligence space included SS8 and Blue Coat – now part of Accenture via its acquisition of Symantec, the first company to acquire Blue Coat, back in 2016.

Although largely invisible to the general public, Netronome is well-known among clients for its vision and practical solutions. Netronome holds the distinction of having invented software-defined networks – or rather the central concept – well before the communications industry was even using the term "SDN." Now that the market has caught up, Netronome enjoys the advantage of being the pioneer, and remains head and shoulders above the competition.

To be clear, ISS is not Netronome's mainstream business. Its primary customers are communications service providers and data centers. But Netronome's products do play a central role in solutions offered by ISS vendors. It is fair to say that for selected companies in the ISS space, Netronome technology is *the* core focus.

Inside Netronome

Established in 2003, Netronome is a privately held company specializing in network flow processors that accelerate network, security and content processing at 40GB and 100GB for networks and data centers. As a "fabless" semiconductor company, Netronome's game is design and sales, i.e., they outsource actual manufacture of chips and devices to a third party.

Network Flow Multicore Processors or NFPs, grew out of the need to better manage heterogeneous multiplexed streams of traffic that include both real-time and non-real-time data packets. In order to achieve optimal network performance, networks must first classify these varying types of traffic – streaming video, videoconferences, email, texting, etc. – prior to forwarding to CPU systems for processing.

That task is not so hard at 1GB perhaps, but when networks reach 10GB or more, CPUs suddenly are required to inspect, process and apply security rules to 10s of millions of simultaneous flows. Netronome was first to step forward with a signature NFP platform architecture coupled with Layer 2-7 deep packet inspection. Perfect timing, as it turned out.

Intel – With Netronome Alongside

In the year 2007, technology giant Intel underwent major restructuring. The company's NFP flow unit was found to be "underperforming." As a result, Intel decided to back off that business and turn it over to Netronome, a young startup with a clear vision for the future of network processors. The companies signed a sales, licensing and marketing agreement that put Netronome in charge.

Under this agreement, Netronome began to design what would soon be recognized as the next generation of flow processors, combining Intel's proven IXP28XX network processor technology with Netronome's strengths in flow management and deep packet inspection – all of it powered by 16 "micro-engines." In addition, Netronome and Intel agreed to collaborate on joint technology initiatives with "tight integration between the Intel CPU architecture and Netronome's network flow processors."

The relationship advanced quickly from there. In 2010, Netronome unveiled its NFP–3240 Network Flow Processor designed for x86, a backward-compatible (interoperable with older legacy systems) set of architectures for popular 16-bit Intel microprocessors.

Netronome would continue to evolve its NFPs and introduce other product lines. But from the initial agreement onward, the default NFP for any network with the Intel x86 would be Netronome.

Programmable NFPs, Adapters – and Diverse Software

Netronome is not your father's NFP vendor given to proprietary anything/everything. On the contrary, it proudly describes its appliances as augmented white boxes. Every Netronome device is interoperable with commercial off the shelf (COTS) servers. The ability to flex to any user application, and change on a dime, is key.

Conventional network processors are configured (that is, locked in to an app) at the factory versus being programmable. Processing is generally limited to OSI Layers 2 – 3, and the appliances tend to slow down when faced with heavy assignments such as DPI. Not Netronome. The company's network flow processors are powered by programmable networking cores.

Speed is a major driver. Each core can deliver 2,000 instructions per packet at a rate of 30 million packets per second. A single appliance can readily handle 1GB, 10 GB, 40GB or 100GB processing at Layers 2 – 7 for DPI.

At speeds of 10GB and above with heavy data flows, service providers and law enforcement alike often face a tough choice of performance versus cost. The old school approach: Pile on CPUs and servers.

Netronome has a better way: co-servers called adapters that offload data flows to accelerate processing without adding CPUs. Netronome's SmartNIC "Intelligent Server Adapters" – part of the company's Agilio Networking Platform – are all about doing more with less, and doing it extremely well.

Conventional adapters working in a 16 CPU server format consume the time and power of 12 of those 16 CPUs. Netronome's SmartNIC adapter – just one CPU. That effectively frees up 15 CPUs to take on more processing work at 10GB or 40GB throughput.

When it comes to specific applications, Netronome's software library runs the gamut of possibilities in network flow processing, packet classification and cryptography – or the client may use his or her own custom solution. SS8 and Accenture Blue Coat illustrate the two approaches.

On Duty at SS8

SS8 Intelligo, the company's core product, began life as a lawful intercept solution and evolved for use in detecting data breaches for the enterprise market. Over time SS8 added attractive "bells and whistles" to Intelligo including data retention, analytics, visualization, the ability to reconstruct web sessions of the target, and delivery of real-time alerts to agents based on behavioral analysis of the target and imminent threats presented. This impressive assortment of capabilities lets the agent build a useful historical profile of the target, construct networks of affiliates and act on alerts to intervene and apprehend suspects.

Baseline, all frills aside, Intelligo is a classic DPI solution that begins with the collection and filtering of metadata and communications on targeted suspects. SS8 and Netronome each contribute its part. SS8 brings its own software, which runs on a Netronome NFE-3240 PCIe Network Appliance Adapter. The Adapter uses an x86 platform and is connected to a service provider's network switch to conduct surveillance of the target.

Adapter models are available for 1GB, 10GB, 40GB and 100GB throughput. In use, the adapter acts as a front-end to the x86 platform. Programmed by rules specific to the target and types of communications services used, the combined solution's DPI engine filters out extraneous data and hones in on evidence applicable to an investigation. Like all Netronome Adapters, the NFE-3240 PCIe classifies packets at Levels 2 – 7 and delivers full content – as well as complete details on all the target's communications applications.

Both SS8 and Netronome pride themselves on this latter capability – cracking apps that span the array of heterogeneous traffic from voice calls to text, emails, messaging and streaming video. As Netronome is quick to point out, without the adapter's unique ability to accelerate processing with minimal burden on CPUs, gathering such a complete picture via conventional adapters and servers would not be feasible.

The Real Power Behind Blue Coat

Depending on the source, Blue Coat – acquired by Symantec effective June 2016, then by Accenture in 2020 – is alternately revered or reviled as a pioneer in providing proxy server man-in-the-middle (MITM) attacks on SSL certificates. Netronome plays both a direct and indirect role in these undertakings.

The underlying platform is the Netronome NFE-3240 PCIe Network Appliance Adapter installed on an x86 processing platform. The appliance conducts MITM attacks on Secure Socket Layer-protected packets and decrypts the data into plain text at multi-gigabit speeds.

Blue Coat reportedly acquired the "SSL appliance product line" from Netronome in May 2013, and the solution is indeed now named "The Blue Coat SSL Appliance." However, the underlying hardware – the NFE-3240 PCIe Network Appliance Adapter – is still very much a part of the Netronome product line and is sold for a variety of applications including intrusion detection and prevention, firewalls, Distributed Denial of Service (DDOS), as well as lawful intercept and SSL inspection systems.

Flash Forward to Software Defined Networks

Always the pioneer, Netronome is in the vanguard of innovators that stand to benefit from service providers' adoption of SDNs and Network Function Virtualization (NFV).

In traditional physical infrastructure networks, higher speed by definition introduces new problems for DPI: increased capex to cover the cost of more CPUs required for processing; higher power consumption and its accompanying cost; and in the end a higher total cost of ownership (TCO).

The same problems afflict SDN and its sister development, NFV. Piling on more servers and more CPUs is not the answer. Netronome's Agilio Networking Platform comes to the rescue again, saving x86 CPU cores from being consumed by server-based networking functions.

The Agilio SmartNIC Adapter saves in other ways. Each comes with 96 core processors, all of them programmable, making upgrades easy with no expensive forklift required. The system is "future-proof," able to evolve to new protocols as they are developed, without any changes required to the server.

NETSCOUT: Silent $1 Billion Surveillance Vendor

In the surveillance business it is rare to find a major player that is almost completely invisible. Such is NETSCOUT Systems, a market leader in IP interception and forensics.

A company tracking to reach $US 915 million in revenue in 2020 deserves recognition for all it does, if for no other reason than to better inform prospective customers.

NETSCOUT, founded in 1984 and based in Westford, Massachusetts, is principally known as a maker of computer network performance management products such as its nGeniusONE family of packet flow switches, which analyze traffic to enhance network efficiency.

"Network performance management" is a corporate euphemism for deep packet inspection, the standard technology solution used in managing IP networks – and widely deployed for target monitoring, as well, which is exactly how NETSCOUT applies DPI to the lesser-known side of its business: surveillance for law enforcement, intelligence and other government clients.

NETSCOUT's flagship product in this sphere is nGeniusONE, which can monitor packet flows to intercept and copy metadata and content that transit the Internet.

Although NETSCOUT has long used DPI in its mainstream commercial segment, it was a pair of acquisitions that not only complemented the company's packet-flow strategy, but sped its involvement in the ISS arena.

In September 2011, NETSCOUT unveiled an important acquisition from Fox-IT, a Netherlands cybersecurity and consulting company that also developed IP surveillance products, most notably Fox Replay, a suite of solutions offering interception and forensic capabilities, both real time and after the fact. NETSCOUT understood the value of Fox Replay and acquired the relevant Fox-IT unit and the solution.

One month later, NETSCOUT announced the purchase of Simena, a Herndon, VA-based provider of IP flow monitoring technology used to intercept and create copies of packets, distribute and filter them for lawful intercept purposes. In addition to gaining applicable new technologies, NETSCOUT picked up Simena's global customer base.

Together with a third, comparatively small acquisition of British Telecom spin-off Psytechnics, a CEM analytics shop focused on voice and video quality assurance, NETSCOUT's 2011 spending spree came to $US 47.3 million, according to the company's 2012 10-K filing, the bulk spent on Simena and Fox Replay, and the investment drove solid financial results: market capitalization of $1.76 billion just two years later.

As NETSCOUT described it in a rare mention from its 2012 10-K, Replay Analyst served the "specialized market" of law enforcement, for which it "interprets intercepted data and reconstructs all communications, guaranteeing the ability to "see what the target saw, or hear what the target heard." Combining its own capabilities with those of Fox Replay, NETSCOUT announced the first iteration of nGeniusONE that same year.

NETSCOUT nGeniusONE

nGeniusONE is a real-time streaming and advanced IP monitoring tool. It uses DPI to pinpoint a target's Internet communications and make a mirror image of packets in the protocol format designated by the customer, then routes data to Infinistreams, the company's advanced analytics solution.

nGeniusONE reconstructs IP communications in the exact context and sequence in which the target sees it: a precise view of the same screens, and in chronological order. The law enforcement or other government agency can then slice & dice how the intercepted data is used and by whom: setting up access permission for analysts by individual or group, and what each is permitted to view and work on. For security and convenience, nGeniusONE integrates with the authentication modules of other lawful intercept or intelligence collection solutions the customer may already have in place.

As they work, whether independently or as a unit, analysts can communicate findings to one another, bookmark where they've stopped, and view any remaining intercepted data not yet analyzed so that they can estimate the time required to complete the analysis.

nGeniusONE supports all protocols including IPv6, and can crack packets for VoIP, chat, as well as non-verbal features such as avatars and wallpaper. The product also makes short work of ZIP files and can peel away layers of encryption. Plus the user can not only view, but search the target's communications.

All of the above is delivered at 100GB speeds. While that might at first sound like a massive volume of data to digest, nGeniusONE's focus on processing data per packet versus in huge chunks enables streamlined delivery of intercepted, reconstructed and sequentially tagged target data to Infinistreams, the LEA's collection device, data center, or to laptops of authorized and authenticated analysts.

nGeniusONE appliances must be deployed in distributed fashion across the network in order to capture IP traffic and convert it to intelligence. The device readily integrates with equipment from all major IP network device manufacturers including Cisco, Nokia, Avaya and Dell EMC.

Packet Forensics: DPI for the NSA

To law enforcement and government agencies, deep packet inspection is an essential commodity offered by scores of vendors with little to separate one from another, unless the provider has the cachet of Packet Forensics. Based in Tempe, Arizona, USA, Packet Forensics has built a successful livelihood around features that place it in the top ranks of companies selected by no less a client than the U.S. National Security Agency.

Packet Forensics is not your run-of-the-mill DPI company. What separates Packet Forensics from the herd is the ability to bypass the most sophisticated forms of encryption.

When word leaked in 2012 of NSA contracts with Packet Forensics and Vupen, and in 2013 of the NSA's ability to crack Tor, few linked the two. The agency blazed through massive data loads to find Tor traffic, then singled out HTTP requests between Tor and specific servers. Identifying protocols at Layer 7 of the OSI stack happens to be a function of DPI. Finding and leveraging network points of attack is a job for vulnerability exploitation experts. In essence, the NSA combined the power of Packet Forensics and Vupen to hack the world's best-known anonymous network.

Here, in addition to examining Packet Forensics' more general capabilities in products designed for CALEA compliance by communications service providers, we will also look at features that keep the company in the top ranks of vendors used by U.S. Intelligence Community customers.

Perspective on the DPI Market

While definitions of DPI vary — some argue that it begins at Layer 4 (transport) of the OSI stack, others that it deals primarily with the "payload" of Layers 5 – 7 (session, presentation, application) and still others strictly at Layer 7 applications — general agreement holds that the lion's share of the work is at Layer 7, from which users may harness apps and protocols to see the exact content of the payload, including diverse applications and protocols.

Given these capabilities, the global market for DPI is to grow to some US $19 billion by 2023 as carriers, government agencies and LEAs pile on for reasons that range from improved traffic management, network investment planning, national security and police work. The outlay is small compared to the spend by a US $2.0 trillion telecom industry eager to keep costs (and smaller competitors) under control. Traffic management remains the primary function of DPI, particularly in developed markets with high bandwidth consumption driven by video.

DPI for lawful intercept and intelligence purposes represents a niche of a niche. FOIA requests, for example, show that NSA contracts in 2012 with Packet Forensics and Vupen products totaled US $500,000. For perspective, the largest US wireless operator in 2014 spent US $173 million on LTE network investment – in one city, Houston. Notwithstanding the media furor over the topic in recent years, DPI barely makes a blip on the budgets of government agencies or telecom operators.

However, DPI does hold strategic importance for national security and law enforcement, not only for its use in standard lawful intercept, but where it serves as a gateway for "man-in-the-middle" and "man-on-the-side" attacks on encrypted communications of criminal elements and terrorists.

The Link Between DPI and MITM Attacks

Owing to the relationship between DPI and man-in-the-middle attacks, a brief explanation of MITM is in order. MITM is the use of technologies to come between a target and server to fool each that the other is authentic, and in the process convince the target the he/she is communicating with the intended server. DPI is a process for examining the payload of packets, including encryption protocols such as HTTPS, TLS, SSL, and SSH.

DPI "passively" assists MITM attacks by filtering packets to: (1) find and segment SSL/TLS traffic; (2) set the stage to apply vulnerabilities from libraries of weaknesses such as those sold by Vupen's successor, Zerodium; (3) which in turn provide access to public encryption keys; and (4) present valid, trusted Certificate Authority (CA) permitting unimpeded access to the target's communications and traffic. Once "in," MITM techniques give the user the ability to control a target by manipulating or injecting data.

Back to School on Man-in-the-Middle

The two most prominent gatekeepers of Internet security are the Secure Sockets Layer (SSL) and Transport Layer Security (TLS) protocols.

SSL has been around since 1994 when it was developed by Netscape. TLS followed soon after, under the auspices of the Internet Engineering Task Force (IETF) and is widely characterized as the "successor" to SSL, or its partner technology, as in SSL/TLS. For simplicity's sake, we'll define SSL/TLS as the principal protocols used to secure HTTP traffic, encrypting traffic in Layer 7, the application layer of the OSI stack.

Though available for the past 20-plus years, SSL/TLS was not widely used – except for financial transactions – until late in the first decade of the

21st century. During that relative hiatus, MITM techniques grew up around four schools. An examination of these early day MITM techniques is helpful in providing a good grounding in the basics and shows how the stage was set for today's methods.

No sooner did SSL/TLS catch on in the marketplace than some clever hacker designed a way around it. A client and server use TSL to establish a stateful connection via an interim procedure called the "handshake." Remember that point when we get to "SSL Hijacking" below.

The top four MITM schools of yesteryear:

- **ARP Cache Poisoning.** The oldest, lowest level and most technically simple MITM approach, ARP Cache Poisoning took advantage of a weakness in the Address Resolution Protocol (ARP) between OSI Layer 2 (data link) and Layer 3 (IP) used by hardware to initiate network communications. At work here is the basic IP process of "send and reply." A device sends a signal to locate other hardware with a given IP address, then, on locating the right piece of iron, receives an "all clear/ go ahead and send" reply. In an MITM attack of this nature, the attacker simply sends false signals to a host, which then updates its cache of ARP data – but now with a receiving IP address belonging to the attacker.

- **Session Hijacking.** Whenever a target signs-on to a site with user name & password, he is establishing a stateful connection or session with rules for logging in and logging out. Web interactions with financial institutions and government agencies that hold confidential information are examples of session-oriented connections where the site ensures user credentials to allow access, automatically logs off when no activity occurs after a period of time, or terminates the session on command by the user. Session hijacking works by obtaining key parts of the sign-in procedure in order to take over the ID credentials of either or both parties. In typical cases, the attacker resorts to ARP Cache Poisoning to obtain packets during the log-in procedure, then reveal the IP address of the target and use a proxy to enter that individual's own account.

- **DNS Spoofing.** Domain Name Server (DNS) requests are the lingua franca of the Internet – the ability to type in a url and go to the sought-for site. The transaction is facilitated by assigning a unique DNS or IP address to each site, and a DNS server that completes the transaction and takes you where you want to go. DNS Spoofing, as the name implies, sends the seeker to a faux website operated by the attacker. The attack is made possible by the Internet's routine process of "DNS Recursion," wherein one DNS server checks a request by sending a query to a second DNS server. In a DNS spoof, the attacker might initiate the attack by ARP Cache Poisoning of hardware so it will reroute traffic to a fake host, which can then intercept DNS queries. The attacker can then begin sending spoofed messages which the target will assume stem from a real website.

- **SSL Hijacking.** In 2009 a technique called SSL hijacking emerged, allowing intrusion at the "handshake" between unencrypted and encrypted traffic. The working principle was that traffic initiated on a non-secure connection could be nabbed before being transferred to a secure protocol. In a textbook setting, the attacker intercepted HTTP traffic on a network, looked for HTTPS links and redirects, then mapped the links to look-alike HTTP or HTTPS links. The attacker then could assume the identity of the target, sending Certificate Authority to the web server. To the server the process was invisible because it was still receiving SSL traffic from the target. Just one shortcoming: An astute target might notice that his browser was no longer marking traffic as "HTTPS."

Still, SSL Hijacking marked a big step forward in the evolution of MITM. NSA documents disclosed in 2013 revealed a sophisticated version of MITM attack called "man-on-the-side" taking MITM to a new level via special servers codenamed QUANTUM. With the cooperation of ISPs, QUANTUM servers were located at strategic points on the Internet backbone where they benefited from a speed advantage enabling them to impersonate a legitimate server before the latter could respond. QUANTUM put the hooks of the attack in a target, who assumed he was surfing the bona fide website. From that point, target traffic might be routed to NSA's FOXACID server, described as an "exploit orchestrator." (See Chapter 4: Offensive Cyber)

The identification of targets masked by encryption was a function of DPI. Detailed manipulation of the traffic: the work of classified vulnerabilities held in a library of the NSA's then-named Tailored Access Operations (TAO) or Equation Group. No public information is available on who provided the core technology; however, the 2012 NSA contracts might be a useful indicator.

Since then the NSA and its private sector vendors have moved on. Technology companies from the opposing camp haven't stood still, either. Late in 2014 Google introduced its nogotofail product, designed to test "common SSL certificate verification issues, HTTPS and TLS/SSL library bugs, SSL and STARTTLS stripping issues, cleartext issues, and more." In short, nogotofail is a tool for identifying potential points of vulnerability. It does so by emulating a MITM attack.

Packet Forensics for Lawful Intercept

While most law enforcement agencies (LEA) are unlikely to need the manifold capabilities available to a large government intelligence agency, it is fair to say they benefit from the trickle-down effect of technology advancements.

Packet Forensics DPI products for lawful intercept are CALEA compliant and bear signature traits:

- Integration of hardware and software in one box, eliminating the additional capex typical of solutions that require multiple Intercept Access Points (IAPs) and active mediation devices integrated with network switching equipment.
- Specific, differentiated treatment of voice intercepts and data intercepts, providing optimal intelligence from both – versus providing data intercepts with voice embedded.
- Plug-and-play operation in minutes, and intercepts in seconds, visible on a multi-platform GUI.

Packet Forensics products are offered at price points keyed to user need and budget. One attractive add-on capability for all units is the Packet Forensics cQuery service, which anonymizes queries including reverse "Who is" lookups.

From the "bottom" to the top:

- The affordable **Packet Forensics M1** is a multi-purpose unit equally facile at handling lawful intercept, network security policy and other communications policy management. Flexible enough to accommodate Ethernet, IP and MPLS, the device provides both probe and mediation functionality, records and buffers metadata, is designed to be future-proof for evolving standards and, though comparatively low cost, is 100 percent carrier grade.
- The **Packet Forensics M1S** is the mid-range device, providing similar features with higher throughput. Distinctive features include Ethernet probe, integrated mediation server and accelerated cryptography.
- Top of the line, the **5-Series** has the ability to execute diverse policies, each geared to a different action, e.g., intercepting VoIP calls and forwarding for analysis, mapping RADIUS data with IP addresses or writing pen registers, all while searching the full packet stream. This is a high performance engine, but compact and with low power requirements.

Protei: Russian Deep Content Inspection

Of the top homegrown players in Russian Lawful Intercept, one stands out as a key provider of solutions for the state police: Protei, based in St. Petersburg.

Established in 1997, Protei manufactures a line of equipment that meets the technical requirements laid down in SORM, the set of mandatory capabilities for intercepting wireline, mobile, Internet, social media and other media that transits IP, PSTN, cellular, Wi-Fi and any digital network in Russia, where communications carriers of all stripes are a captive market.

Since the early 2000s, all Russian communications service providers have been required to install and maintain SORM-compliant equipment, and in recent years the rules have been extended to include social media providers, as well.

A glance inside any Russian carrier's network, and those of most members of the Commonwealth of Independent States (CIS – independent nations once part of the USSR) will quickly lead to discovery of the familiar "π" logo of Protei on at least one and likely many items of hardware. Just like the ancient Greek symbol, the company is a mathematical constant where Russian lawful intercept is concerned.

The company's product depth is surprising considering the diminutive size of its development team of roughly 150 IT experts, and the fact that surveillance is just one of its business interests. On a broader front, the company serves mobile and fixed line operators in over 20 countries, with a range of products for emergency first responder and government agencies: a gateway for Intelligent Network Services, SS7 systems for roaming and mobile location, missed call notification, prepaid cards, virtual office and real-time data charging. Customers include service providers throughout Russia, former USSR properties in the CIS, as well as customers in Iraq, Palestine, Jordan and Sudan.

But to the Federal Security Service (FSB), the agency in charge of counter-intelligence, counter terrorism, border and internal security, and criminal investigations within Russia, the primary attraction of Protei is a suite of products that enables real-time tracking of the country's populace of some 150 million souls.

Protei does not work alone. Even in today's modern Cold War environment, Protei, like all other providers of lawful intercept worldwide, partners with free market tech leaders including Nokia Siemens to help improve those companies' market penetration in Russia. Many of the switches and routers used in Russian communications networks are made by Nokia Siemens, Ericsson and Cisco, as well as Nortel devices from days of yore.

Deep Content Inspection

DPI represents an evolution from the first packet inspection technology that surfaced after 1979, when email began to emerge as the "killer app" of the U.S. Defense Advanced Research Projects Agency packet network (DARPANET). Early apps involved packet filtering (examination of IP and TCP headers for firewalls, to reject packet attacks), then evolved to stateful packet inspection (wherein the technology was deployed in firewalls to better understand a packet's source and destination), and finally to modern DPI, which adds analysis of data protocol structures.

DPI as we know it today, available as software or installed as a card in a device, works within the Open Systems Interconnection (OSI) model, which presents the functions of a network as a set of abstractions. There are seven layers of abstraction in the model – 1 through 3 in the "Media Layer" and 4 through 7 in the "Host Layer." Typically, the "OSI Stack" is rendered as follows, with the highest Layer at the top:

The OSI Stack

HOST LAYERS
Layer 7. Application: Network process to application
Layer 6. Presentation: Data representation, encryption & decryption
Layer 5. Session: Managing sessions between applications
Layer 4. Transport: Delivery of packets between network points
MEDIA LAYERS
Layer 3. Network: Addressing & routing delivery of datagrams
Layer 2. Data link: Reliable point-to-point connections
Layer 1. Physical: the physical data connection

DPI analyzes packets from Layers 2 through 7, i.e., from the Data Link to the Application Layer. DCI finds and assesses signatures that cross packet boundaries by keeping track of content across multiple packets. By transitioning away from traditional packet inspection and concentrating on the content and intent of data, DCI provides a comprehensive method of filtering for attacks and malicious content.

Deep Content Inspection

Deep content inspection (DCI), though often referred to as "the next stage of DPI," is in fact a different animal. DCI's own evolution began in the mid-1990s along a separate track.

Like DPI, DCI began as a way to stop spam and malware, using caching proxies that captured and stored IP network traffic. Proxies cached and performed rudimentary inspection of packets before forwarding them to their destination. Also called secure web gateways, proxies were fine for their day but did have their flaws such as limited scalability, routing issues that required network reconfiguration or special routers, and inability to handle any protocol beyond http and ftp.

Next, developers combined proxy-caching with DPI in firewalls for "unified threat management." That temporarily resolved the packet-to-proxy routing hassle, as firewalls were already set up as network chokepoints. But it wasn't cheap or quick. Network operators opted to use proxy-cashing with DPI only when essential, e.g., as a backup.

Finally, in the mid-2000s the first generation of true Deep Content Inspection emerged. Developers created a device that could do the full application Layer 7 inspection, including both the "handshake" and the payload at "wire speed," i.e., the peak rate of transmission at the physical or "wire" level.

End result: object analysis of packet payload for behavioral analysis, "signature" of the sender, seeing any relationship between the packets under examination and those in other, previous sessions, and disassembling the content to see exactly what it said before re-assembly and forwarding to the recipient.

The dividing line between DPI and DCI remains fairly straightforward: a matter of deep versus deeper. DPI is designed for analysis and classification of packets. DCI is content-focused to deliver intelligence about exactly what the sender is saying.

Protei DPI

Protei combines its own deep packet inspection platform with best-in-class solutions – notably the Qosmos ixEngine platform made in France, to provide users all the usual bells and whistles:
- Over 2000 protocol plug-ins, with regular updates as new or revised protocols surface, with the ability to recognize all protocols from Layer 2 of the OSI stack up to and including Layer 7.
- Use of "flow pattern matching," session correlation, plus both heuristic (historic) and statistical analysis.
- Comprehensive metadata extraction, including: volume of traffic per application and per application by target.
- Time aggregation per user or target.

- Target's Web usage at the page level.
- Protocols and apps driving IP traffic flows on wireline and mobile networks.
- Integration with packet processing middleware.
- Ability to filter all packets or just a defined subset.

Users can extract content, disassemble then reconstruct communications so no one notices he's being tracked. And when certain parties disagree online with the state, Protei DPI Platform also provides Web censorship capabilities through content filtering and blocking.

Buyers have three models to choose from:
- Protei DPI 8: 8 Gbps throughput, with 100,000 flows per second (2 million total), for use with a subscriber base of up to 200,000.
- Protei DPI 20: 20 Gbps throughput, with 250,000 flows per second (16 million total) for up to 1 million subscribers.
- Protei DPI 80: 80 Gbps, with 1,000,000 flows per second (50 million total) for 5 million subscribers.

Radisys Wire-Speed Packet Capture

Radisys made headlines in mid-June 2016 when Verizon signed a contract to become the first commercial customer for the Radisys FlowEngine Intelligent Traffic Distribution system, which the former Baby Bell still uses to improve network intelligence and control. Even so, analyst reaction to this major contract win was mixed.

Most gave the deal a thumbs-up and switched their recommendation on Radisys from "hold" (Wall Street-speak for "get rid of it") to "buy." Others expressed concern about Radisys' reputation for flat financial performance over the prior decade. One analyst went so far as to dismiss the Verizon win as a bubble that would be followed by share price crash of 35 – 50 percent. As time would tell, they were far too optimistic.

At the time of its big Verizon win, Radisy traded at US $4 and $5 per share (and as low as US $2.08 in 4Q2014) with revenue of roughly US $215 million in 2016. Those numbers are indeed a far cry from its glory days when the stock hit a high of US $64 (2000) and annual revenue reached US $320 million (2005). At this writing (2020), Radisys shares trade at around $1.50, despite support from tech partners such as Israel's EZChip, a Mellanox Technologies company.

FlowEngine, a wire speed platform for monitoring, is one of several Radisys products applicable to ISS. Others include the Radisys T-100 Ultra for deep packet inspection (DPI) and the Radisys Advanced TCA (Advanced Telecommunications Computing Architecture) that powers real-time, end-to-end radio access network (RAN) monitoring solutions such as Polaris Networks' OmniLocate, a solution used to determine vertical mobile location. (See Polaris in Chapter 5: Mobile Location and Monitoring.)

Inside Radisys

Based in Hillsboro, Oregon, Radisys calls itself a "pure play investment" in telecom and cable, with emphasis on wireless and next generation networks. In that vein, it promotes solutions supporting software defined networks (SDNs) and Network Functions Virtualization (NFV) as the future of the communications and media markets. Radisys casts its work as an accelerator of network performance via software solutions and embedded hardware that leverage network intelligence to improve the efficiency and ROI of carrier infrastructure and enhance speed to market with profitable new product lines.

Radisys' market focus and performance represent a true sea change from the day it first left port. Founded in 1987 by former Intel engineers Dave

Budde and Glenford Myers, the company initially set out as a general purpose computer startup. But with Myers and Budde at the helm, the company switched to telecom just ahead of that sector's 1990 boom years. When the company IPO'd in 1995 it was with US $20 million annual revenue already in the bank. Barely one year later, revenue quadrupled to US $80 million. Then, for whatever reason, Budde left for other ventures. Radisys went through a series of acquisitions that dampened profits, but with Myers as CEO, the company reached over US $200 million in revenue in 2000 – an incredible 10-fold increase from its IPO day just five years earlier.

Then came the Great Telecom Bust of 2000. Overnight, carrier capex for network investment evaporated, the competitive local exchange carrier (CLEC) market collapsed, and many vendors concentrating on network solutions went down with them. Not Radisys, at least not entirely. True, the company's stock entered a whirlpool that sank prices from the US $60 range reached in year 2000 to a low of US $3.90 in 2002, yet revenue continued strong, hitting US $320 million by 2005. But as 2006 opened, the good times ended. Radisys revenue crashed 50 percent, and but for the occasional seesaw up or down, share prices entered a downward spiral that has made it a single-digit stock ever since.

Despite the discouraging financial picture, Radisys continued to raise money for sound strategic investments. In 2011 the company acquired Continuous Computing, whose expertise in DPI, Trillium monitoring software and ATCA hardware would quickly prove important to Radisys' future.

Today the company's largest customers are network hardware manufacturers and integrators – e.g., Nokia Siemens, NEC, Danaher Corporation, Unicom – that find Radisys software systems and embedded products a valuable addition to systems for service providers. Radisys also serves players in the ISS arena, including Verint and military clients, and has marketed direct to customers at ISS World events. Among the core products is Radisys FlowEngine.

Radisys FlowEngine – Fast & Accurate

When Radisys FlowEngine launched in 2011, the company described it as "a deep packet inspection (DPI) software framework" made possible by the just-completed acquisition of Continuous Computing. Radisys has taken Continuous Computing's work further.

FlowEngine is an OpenFlow-compliant data plane software app for network monitoring that drives performance based on the intelligence gathered from captured packet flows. It runs on a Radisys TDE (Traffic

Distribution Engine) platform that interfaces with 10GB, 40GB or 100GB network platforms. The combined Radisys software/hardware platform does classification, forwarding and analytics of high-speed traffic flows.

FlowEngine guarantees 100 percent packet capture, providing "the ability to capture one or many specified flows," all in real-time, an advantage the company says network packet brokers (NPBs) lack. Flows that are tapped can be saved in common format for network forensics or big data analytics.

FlowEngine's OpenFlow compliance is a fresh twist – a major factor in reducing carrier or other users' capex on hardware.

OpenFlow is a protocol first developed at Stanford University and now managed and maintained by the OpenFlow Foundation. What OpenFlow does: allows switches from different vendors – each with its own proprietary interface and scripting languages – to be managed remotely using a single, open protocol. Perhaps most importantly, OpenFlow's efficiency eliminates the need for much of the hardware that dominates networks. In sum, it lets carriers jettison big iron.

Radisys FlowEngine, a network monitoring system that works with OpenFlow, is right in tune with everyone's goal to minimize capex on infrastructure investment.

Why intelligence organizations remain interested: FlowEngine classifies IP flows at wire speed and identifies all traffic. When coupled with a DPI solution, FlowEngine's rapid speed reduces loads on CPU blades, enabling full DPI through all layers of the OSI stack. FlowEngine's wire-speed forwarding capabilities to the DPI engine, and spread of the load across multiple blades, ensure that no packets are lost.

T-Series: The Right Platforms for DPI

With its acquisition of Continuous Computing in 2011, Radisys set itself in motion as a primary provider of first rate platforms for DPI used for both commercial and government purposes. Continuous Computing was among the first companies to move away from proprietary platforms for DPI and embrace open platforms that permit rapid scalability. In particular, the company saw Advanced Telecommunications Computing Architecture (ACTA) as "the perfect fit" for DPI.

Key advantages of ACTA:
- Standardized platform for carrier grade comms/media systems.
- Multiple sources for the platform – not wed to one or just a few vendors.
- Flexibility to accommodate any advance in next generation networks.

- "Five nines" reliability: no single point of failure.
- Easy to service, upgrade and use.

Continuous Computing jumped on the ACTA platform with its FlexTC DPI System in the 2008-2009 time frame. FlexTC was designed for 10GB networks, considered high-speed for that era, and outfitted with the company's Trillium monitoring software for DPI.

To Radisys, Continuous Computing was more than a set of products and software solutions, however – it was a launchpad to a broader portfolio of products for higher margin accounts.

Since acquiring Continous Computing, Radisys has continued to evolve ACTA-based DPI platforms, introducing its state-of-the-art T Series, platforms built on merchant silicon (off-the-shelf chipsets) from a variety of tech partners including Intel, and easily integrated dedicated packet processing systems. The top-of-the-line products are the Radisys T-100 Ultra and T-100 Pro, both with FlowEngine at the core.

Because DPI is a commodity application already explained in abundance in this book, we won't belabor how it works. We will only say that Radisys FlowEngine stands out for its ability to couple scalability, precision and reliability, gathering finite or limitless packet flows at line rate, then work with industry leading deep packet and content inspection capabilities to deliver unparalleled intelligence.

Riverbed and Wireshark Packet Capture

Wireshark is a popular open source DPI solution that often makes an appearance in ISS suites. Wireshark is tightly linked to two enterprises, the founding sponsor CACE Technology and Riverbed, which has made significant improvements to the solution.

Beginning with its 2010 acquisition of CACE – then the principal sponsor of Wireshark – Riverbed has long been recognized as a leader in deep packet inspection for network monitoring purposes that improve customer experience and loyalty. Less well recognized is Wireshark's role as a vital tool in performing man-in-the-middle (MITM) attacks for purposes of monitoring those same customers or specific targets. Riverbed would just as soon play it down, though this side benefit of Wireshark is well recognized by academic experts and a frequent topic at industry events such as ISS World.

With its 2016 buyout of Aternity, Riverbed took the commercial role of network monitoring, and surveillance, too, another important step forward. Aternity's stand-out feature was the ability to monitor targets at the individual user device level. It essentially turned every device, whether physical, virtual or mobile, into what Riverbed called a "self-monitoring platform" providing details on the identity of the user, all the applications they use, a list of all other devices used, and the target's behavior. This set of capabilities has obviously had important implications for intelligence professionals, as well.

Often one sees surveillance technology advances emanating from government-funded programs, then growing into commercial applications. The CIA's In-Q-Tel, for instance, has furthered the commercial success of many companies by funding them to grow for government purposes first. Palantir is the classic example. But Wireshark swims against the current. Initially designed for commercial purposes it is now equally popular for network monitoring of a different type.

Swimming with Wireshark

Wireshark is free, open source packet analyzer software for Unix, Linux and Windows. It does trouble-shooting and monitoring networks "at the microscopic level," per the organization's website, which also hails Wireshark as the "the world's foremost network protocol analyzer." Basically, Wireshark is a manual deep packet inspection (DPI) engine, one of the most commonly used in the world today. It offers sophisticated live capture and display filtering, specialized offline analysis of data screens. Wireshark supports over

1,000 protocols, and can identify and categorize application traffic by IP addresses, ports, protocols or urls.

Other key Wireshark features:
- Captured network data can be browsed via a GUI, or via the TTY-mode TShark utility.
- Rich VoIP analysis.
- Can Read/write in numerous capture file formats: tcpdump (libpcap), Pcap NG, Catapult DCT2000, Cisco Secure IDS iplog, Microsoft Network Monitor, Network General Sniffer® (compressed and uncompressed), Sniffer® Pro, and NetXray®, Network Instruments Observer, NetScreen snoop, Novell LANalyzer, RADCOM WAN/LAN Analyzer, Shomiti/Finisar Surveyor, Tektronix K12xx, Visual Networks Visual UpTime, WildPackets (now Savvius) EtherPeek/TokenPeek/AiroPeek, and many others.
- Captured files compressed with gzip can be decompressed on the fly.
- Live data can be read from Ethernet, IEEE 802.11, PPP/HDLC, ATM, Bluetooth, USB, Token Ring, Frame Relay, FDDI, and others.
- Decryption support for many protocols, including IPsec, ISAKMP, Kerberos, SNMPv3, SSL/TLS, WEP, and WPA/WPA2.
- Coloring rules can be applied to the packet list for quick, intuitive analysis.
- Output can be exported to XML, PostScript®, CSV, or plain text.

Open Source with a Commercial Attached

Wireshark has a curious saga going back to the late 1990s. Originally it was developed by Mr. Gerald Combs and called "Ethereal." Combs worked for a small Internet Service Provider at the time and realized that most of the commercial protocol analysis tools of that era were not designed for Linux or Solaris, the platforms used by his employer. So Combs set out to develop a packet analyzer himself. The end result was Ethereal, and Combs owned the copyright for the program. Unfortunately, however, another company owned the brand name Ethereal. Shortly after switching employers to go to work for CACE Technologies in 2006, Combs changed the name of his DPI solution to Wireshark, a great name that stuck.

Throughout Wireshark's history, Combs has held principal responsibility for the code and issued software updates. As an open source platform, Wireshark has also benefited from some 600 contributing authors, as well as from financial sponsorships. CACE was the main sponsor beginning in 2005, and then Riverbed took over when it acquired CACE in 2010.

Augmenting Wireshark with Aternity

Riverbed has experienced its ups and downs. With the introduction of its Steelhead wide area network (WAN) optimization in 2005, Riverbed rose quickly as the king of that market, serving some 97 percent of the U.S. Fortune 500 enterprise networks. Steelhead was so effective at resolving challenges in WAN performance that it was quickly dubbed "network crack" by managers who simply couldn't get enough. Over the next 10 years Riverbed graduated from a startup to a US $1.0 billion company based almost entirely on the success of Steelhead.

But other technology advances began to have an impact, particularly the move toward software-defined networks (SDNs). While the need for WAN optimization was no less than before, the means of accomplishing it almost overnight evolved beyond what Riverbed offered. By 2014, the company was, to put it bluntly, behind the market. Thus when Riverbed received a US $3.6 billion offer to go private in 2015, management jumped at it. The deal closed late that year.

By mid-year 2016 Riverbed had recovered and was once again on the acquisition trail, first with German vendor Ocela, and then with Aternity. The latter purchase, in particular, held strategic promise both for Riverbed and the ISS community. As a tool that can be used in conjunction with Wireshark to take packet capture down to the individual device level, Aternity provides "a complete understanding of the user."

With Aternity software agents added to its SteelCentral management suite, which incorporates Wireshark, Riverbed enables users to collect data straight from end-user devices, including physical devices, virtual endpoints and smartphones.

An agent can readily determine the name, location and activities of a target, even in today's cloud environment where personal data is distributed across diverse public or private clouds and data centers. In the commercial networking arena, such data helps the service provider track and improve the user experience. For intelligence and law enforcement agencies, the same capabilities ensure an always-on surveillance environment.

Rohde & Schwarz's ipoque DPI

Time was when Rohde & Schwarz's subsidiary ipoque, one of the foremost makers of DPI solutions, was fairly open about the use of its technology for lawful intercept and national security. Then came the Edward Snowden controversy of 2013. From that point on, where surveillance was concerned, one might say that ipoque has gone opaque.

Nevertheless, the company remains a strong player in DPI for law enforcement and intelligence agencies. For that matter, US $2.3 billion parent company Rohde & Schwarz uses ipoque DPI in a test and measurement solution for smartphones, making it equally facile for other, more clandestine purposes than the T & M function for which the solution is primarily promoted.

ipoque's stature in the DPI trade, and the fact that the company's products can manage most common protocols and applications, as well as encrypted and suspicious ones, make the company a serious player in the ISS space.

The Download on ipoque

ipoque launched in Leipzig, Germany in 2005 to provide tools for bandwidth management and network monitoring. The company provided solutions – including both hardware and software – designed to help carriers tame growing volumes of peer-to-peer (P2P) traffic. The solution's focus on the application layer quickly proved its merit over older approaches in accurately classifying traffic. ipoque's skills at sorting all types of protocols and apps including VoIP and streaming media won kudos, as well, capturing the attention of law enforcement and national security interests.

Before long, ipoque was a regular at surveillance industry trade events, where co-founder Hendrik Schulze held forth on TCP/IP, DPI and decoding basics for eager audiences of law enforcement and intelligence agents. As simple but comprehensive explanations of the various stages of IP interception, from probe hardware to traffic filtering, http decoding, analysis and packet reconstruction, Schulze's tutorials were among the most accessible and popular sessions at such events. ipoque proved equally in vogue with commercial clientele on the operator side, raising its profile among suitors interested in the young company's capabilities.

Rohde & Schwarz stepped in to buy ipoque in May 2011. As a prominent, decades-old leader in test and measurement equipment for radio and microwave, Rohde & Schwarz saw ipoque as a stepping stone to grow from its core RF monitoring and location business into Internet traffic monitoring.

Under Rohde & Schwarz's wing, ipoque has evolved the DPI portfolio and grown to 200+ deployments in 60 countries.

Net Sensor OEM Probe and PACE 2 DPI Software

The workhorse hardware of ipoque's DPI system is the R & S Net Sensor OEM Probe, a passive device that can be deployed at the edge of the network for lawful intercept and mass interception. The Net Sensor comes equipped with a fast packet processing library that provides visibility into plain and encrypted network traffic.

The Net Sensor ipoque DPI engine is PACE 2.0, software that classifies protocols and applications including HTTP, DNS, Skype, SSH and FTP, as well as others of the encrypted and obfuscated variety.

As with all DPI products, output is driven by rules programmed into the system on a per-client basis. Protocol-specific filtering might be keyed to the target's IP address, name, or any of 25,000 keywords that the user sets up in rules for the DPI engine.

Net Sensor handles both simple and aggregated "class" events. In the former instance, the probe generates data on a single event per protocol. In "class" cases, it aggregates multiple protocol events that occur during an app session into a single structured event. Class events are valuable because they provide greater semantic understanding so that data streams from the application are automatically correlated. End result: less time and bandwidth consumed during post-processing and analysis.

In a lawful intercept scenario, the solution's metadata extraction provides a complete picture of both call information and content. From an IMAP data capture, for example, the user can see the sender, receiver and full message. In Facebook and Twitter: profile visits, group activities, followers, follower lists and messages. In Webmail: sender, receiver and attachment. And in WhatsApp: mobile number and user nickname. Behavioral and statistical analytics point to trends, patterns and target sentiment.

PACE 2.0 never goes out of date. Already well-versed in literally thousands of protocols and applications, the software library is continuously updated.

Rohde & Schwarz Ipoque DPI Benefits

In its market positioning, R & S stresses its traffic management and cyber defense capabilities, citing Tier 1 mobile operators, cable MSOs and institutions of higher learning that use Net Sensor and PACE 2.0 for these purposes. In that vein, the company likes to point out that its solution:

- Detects 95 percent of all IP traffic "in a reliable manner."
- Needs only 1 to 3 packets to classify most common protocols, hence speeding processing time.
- Comes with a flexible interface that readily accommodates a user's protocols and applications.
- Costs a fraction of competitors' signature plug-ins.
- Ensures the lowest memory usage, and allocates zero memory during run time, reducing processing costs.
- Is always "application and user behavior aware."

Although Germany is ambivalent to private surveillance companies – FinFisher, Utimaco and PLATH base their headquarters there, after all – Rohde & Schwarz has clearly positioned the company to steer clear of controversy or public backlash. But as a DPI vendor with offices in 70 nations, the company is a regular at conferences including ISS World and Milipol.

Savvius Network Forensics

Deep packet inspection is among the most-widely used yet misunderstand technology applications in the tech world and surveillance communities today. Ask "got DPI?" and many vendors and customers automatically pipe up, "you bet – all the way through Layer 7 of the OSI stack!" and assume the job of intel/evidence collection is done.

The fact remains that the measure of DPI is or always should be: (1) not just how fast but how far it can go into Layer 7; and, (2) whether the platform goes beyond metadata or app classification to reveal packet content, meaning and intent for evidence and intelligence. If the user loses sight of this mission he may be driving his investigation at half-speed down a country road, miles off-track.

These are rules well understood by Savvius (formerly WildPackets), a leader in intelligent network forensics, based in Walnut Creek, California, USA. Though primarily geared to the commercial market for cyber threat detection, Savvius network forensics products also hold top value for law enforcement and intelligence officers. For this audience, Savvius network forensics is the road to evidence that culminates in arrests, or even preempts acts of crime and terrorism. The muscle under the hood: analytics powered by rocket-speed yet intensely accurate packet capture and DPI.

The story of Savvius is a legacy of commitment to network optimization and security for the enterprise, with development of forensics tools for the surveillance community, however discretely.

Thumbnail History

The original WildPackets launched in 1990 with a product that would point the way to the company's future horizon: a packet analyzer called EtherPeek for Apple computers. In 1997 that product was expanded to Windows. Then came AiroPeek for Wi-Fi networks in 2001. Spanning the last decade of the 20th century and more, the company's "Act I" was long indeed. But the second Act definitely proved worth waiting for.

The breakthrough came in 2003: WildPacket's OmniEngine Distributed Capture Engine. OmniEngine was a "flag in the ground" for DPI at a time when many still considered packet sniffing state-of-the-art. From then on, WildPackets lived up to its name, growing quickly as bandwidth demand surged and it dawned on ISPs that they needed to start thinking about network optimization. DPI took off and WildPackets began creating buzz and demand for "network forensics," deep analysis of network usage and trends to improve optimization and security.

At WildPackets, application of the same technical advances to law enforcement's needs occurred in tandem, just more quietly. WildPackets entered the field with its Lawful Intercept Software Package (LISP) and began to make the rounds of surveillance trade shows such as ISS World.

A LISP that Cops Heard Loud and Clear

LISP was expressly designed for communications service providers and law enforcement agencies to support CALEA in the U.S., ETSI in Europe and MITA (Modernization of Investigative Techniques Act) in Canada.

From the outset, WildPackets made it clear that LISP was not to be confused with the company's commercial products. They created LISP as an adjunct platform for lawful intercept, providing capabilities not offered with the company's standard network management security portfolio. While the software for both segments did share a common feature – "DPI plug-ins" to capture and classify packet data – LISP was special, using additional customized plug-ins for LEAs only.

LISP was obviously no off-the-shelf platform. Plug-ins empowered the user to capture, record, decrypt, analyze and in real-time see Web, LAN, Wi-Fi, social media, IM, email and other content – the full payload of intercepted data – "in the clear," as though sitting in the chair of the target himself.

Plug-ins that came with WildPackets LISP covered:
- **VOIP:** Media Capture Analysis Modules that grabbed and saved the target's VoIP calls.
- **IM:** Instant Messenger Analysis Modules showing comms via an IM service.
- **SMT, POP3 and Web-based Email:** Email Capture Analysis Modules that reconstructed email and attachments from any email client. Plus full capture for Pen Register Analysis.
- **File Transfer Protocol:** From a list of target names, LISP's FTP Session Monitor could capture FTP comms.
- **Transmission Control Protocol:** Remote TCP Dump Adapter accessed the target's device via SSH, and copied content.
- **Dynamic Host Configuration Protocol Targeting:** Captured and recorded IP traffic between selected targets.
- **Remote Authentication Dial-In User Service:** Captured and recorded RADIUS traffic.
- **Virtual LAN:** Tag Filter for single or multiple VLANs.
- **Metadata Only:** Pen Register capture, recording and analysis.

Around 2012-2013, LISP fell off the map of publicly promoted Savvius solutions, but not off the market. The product is still available through companies such as network Performance Channel in Germany and Austria, GSA Mart in the U.S. and others, and despite the company name change to Savvius, always as WildPackets LISP. Today the words "lawful intercept" rarely cross the lips of Savvius in any public forum. Whether LISP was spun off or simply went undercover at Savvius is a deep dark secret. But Sept. 2014 was the last time WildPackets referenced DPI for law enforcement in its blog.

In public, Savvius has moved on, but by no means away from network forensics for investigations. Today's Savvius products represent an extensive and extensible framework that dramatically increases surveillance capabilities and leaves little doubt as to the most productive way to leverage DPI – whomever the user. Market results: Savvius products are sold in more than 60 countries.

Smart Partnering

The Savvius family of Network Forensics solutions is built around the company's Omnipliance hardware and OmniPeek software. Two key Savvius partners – Procera Networks and Napatech – have a good deal to do with how Savvius products perform. For that matter, Procera and Napatech commonly join forces to provide complementary solutions for many clients. Savvius is one.

Before we dive in on how Savvius leverages Procera and Napatech, a brief elaboration on WildPackets' partnering backstory is in order.

Recall that WildPackets started out with its own DPI plug-ins, a strategy they built on for years. But after the turn of the century the company began to look for ways to expedite processing. Instead of building faster packet capture solutions on their own, why not shop it out? In 2003, around the time WildPackets debuted OmniEngine, a group of Danish math wizards launched a startup called Napatech. The Copenhagen company had an attractive line of products called "accelerators" (FPGAs) that looked like a great option for ramping up packet capture.

WildPackets made the leap, integrating its hardware with Napatech accelerators to speed packet capture for clients on both the commercial and surveillance sides of the business.

Partnering worked so well on that one front that WildPackets decided to try a similar approach with DPI, too. Forensics, after all, was the company's core business. Why not integrate with best-in-class DPI vendors, too? Which they did. Over the next decade, WildPackets proceeded to "vendor hop" between DPI providers. In March, 2014, Ixia was named the company's "network visibility" partner.

But life moves fast in the DPI lane. In May 2015, just days after changing its name from WildPackets to Savvius, the company announced its choice of Procera Networks for DPI, and at the same time affirmed its commitment to using Napatech accelerators to boost packet capture performance. Because the old WildPackets crew and Procera both had years of experience with Napatech, it was a commonsense move. And there Savvius' ongoing evolution stands, for the moment. Lift the hood of Savvius products today and you will find Procera DPI and Napatech accelerators.

DPI with Forensics on the Tail End

From the network monitoring standpoint, the action begins with Napatech.

Napatech specializes in packet capture accelerators that grab packets from networks at speeds as fast as today's networks operate. The company's NT100E3-1-PTP, handles packet capture with zero packet loss up to 100GB Ethernet.

Napatech also provides a 200GB unit, essentially a two-port version of the NT100E3-1-PTP, that is ideal when upstream and downstream 100GB feeds are tapped in separate feeds. Data is combined via intelligent hardware connection before delivery to a cache, ensuring optimal control of the load without overloading CPU cores.

Napatech devices can capture data from any type of traffic, known or unknown, including Port 80 (the most commonly used for receiving HTTP), UDP (user datagram protocol), TLS (transport layer security), TCP (transmission control protocol), and GPRS for 2G and 3G radio.

The data flows up to the Procera NAVL DPI Engine which cracks all 7 layers of the OSI stack, through and including the application layer, identifying apps from IM to file sharing, social media, Web queries, mobile, VoIP and others. NAVL identifies and classifies patterns, conversations, behaviors, flow association and future flow to create a library of apps by type. In that regard, NAVL surpasses the capabilities of classic DPI by providing Deep Content Inspection (DCI), and with Napatech accelerators on the front end of the intercept, it operates in real-time.

DCI goes beyond the traditional header/footer inspection of DPI. DCI examines the entire packet, reassembling, decompressing and decoding it into an application-level object called a MIME (Multipurpose Internet Mail Extension) object. A MIME is an Internet standard that applies email format to all other forms of communications on the Web, each with its own classification signature – whether text or non-text such as audio, images or video. By converting diverse forms of traffic into objects based on a common standard, DCI can thus single out packets by precise type for reliable identification and classification. Then it can use key words and other factors to search for the context and patterns, or anomalies to a pattern. One might say that DCI takes DPI Layer 7 to the next level.

Savvius Network Forensics: The Basics

Given all that Napatech and Procera can deliver, what added value can today's Savvius Network Forensics provide? With these two partners' platforms embedded in its suite, quite a bit, actually.

So DCI successfully decodes identifies and classifies hundreds of types of protocols and apps in use – FTP, Skype, email, social media, etc., and any trends, patterns or intent that surface therefrom. What of it? To the FBI sitting in Quantico, that level of intercept may take an investigation only so far.

Savvius can help in several ways: Omnipliance hardware housing a network recorder that captures data for days or weeks; storage & search functions; OmniPeek software for analytics; and excellent graphics to display data findings.

Omnipliance network recorders can work either of two ways: (1) providing continuous and comprehensive capture of the full load of network traffic, with the ability to quickly identify any suspicious activity that pops on the screen courtesy of Procera NAVL; and (2) close refinement via an "ad hoc" capture of specific and related types of traffic keyed to targets, events, patterns or other factors.

Depending on client budget and needs, Savvius can serve up the right hardware match. There are three basic models, from low-end to high-end:

- **Omnipliance CX:** An affordable model for smaller clients – or big ones with multiple remote offices – the CX provides network forensics and analysis without maxing out the client's capex. Supports 1GB and 10GB networks as the "capture engine" for OmniPeek analytics, and provides storage up to 16 terabytes.
- **Omnipliance MX:** Provides more storage, up to 32 terabytes, and more bandwidth – supports an additional four 1GB or 10G cards. But still affordable. Same capture engine as the CX, and with OmniPeek built in. Commonly used in network operations centers (NOCs), Wide Area Network (WAN) links and data centers.
- **Omnipliance TL:** The TL is Savvius' high performance recorder. It is designed for use with any 10GB or 40GB network and provides storage

capacity up to 128 terabytes. Like the MX, the PL can accommodate add-on 10GB cards.

For those that need to monitor targets while on the road, the Savvius Omnipliance Portable is an excellent tool for capturing traffic from Wi-Fi networks via the Savvius OmniWi-Fi flash drive. Just plug it into the Omnipliance Portable and all systems are go. OmniWi-Fi – which harkens back to the company's first product, AiroPeek for Wi-Fi introduced in 2001 – shows that great companies never stray too far from their roots. They build on them.

One can draw similar links between certain Omnipliance features and WildPackets LISP: complete data traffic collection of any communications across any network. It works for weeks or days, with zero packet loss and offers extensive storage. Another feature perhaps carried over, this time from LISP to the Savvius OmniPeek platform: rich data analysis.

OmniPeek Analyzer Software

Whether used as part of an Omnipliance unit or in standalone portable mode, OmniPeek performs the usual jobs of traffic monitoring, management and optimization regardless of the type of communications over the full network or a network segment. Users can drill down for searches on any type of data and see the results in a handsome GUI to perform Level 7 app ID and classification and reveal real-time call data records. But it is its special talents that make OmniPeek stand out:

- **Decryption.** OmniPeek decrypts common protocols such as WEP, WPA/PSK, WPA2 Personal – as well as SSL packets – to gain context.
- **Real-time or Take Your Time.** Can perform analytics on a network, segment or multiple segments either "in the moment," after-the-fact on archived data – or combine the two for comparisons, trends or anomalies that might point to a threat.
- **Location.** OmniPeek applies analytics to packets by conversation pair, to locate events and targets on the network.
- **Video and Voice Analysis.** Not just metadata CDRs, but full playback of captured voice and video communications, with analytics applied for context, intent and possible "futures" implied.
- **Change Filters "On the Fly."** Remember: DPI and DCI results derive from the rules programmed into a filter. With OmniPeek, users can click to choose diverse filters that select other relevant packets, potentially changing the picture an investigation.
- **Real-Time Alerts.** Every second, OmniPeek checks pre-specified data flows for upticks and downticks. Users choose the "level of severity"

for various risks, in advance. Any change that crosses the line triggers an alert.
- **Customized Reporting.** Analytics results are compiled in a report, with frequency of reporting determined by the user. Reports are logged and saved for future review against fresh reports.

Silicom Denmark: VoLTE Interception

In December 2015, Danish hardware accelerator wizard Fiberblaze made headlines when it was plucked from the market by Silicom Ltd. of Israel. It was a good match. Silicom was an established maker of diverse network hardware and software ranging from Ethernet adaptors to smart bypass switches and adapters for deep packet inspection and packet filtering. Fiberblaze had an equally impressive roster of network processing "accelerators" for data traffic-heavy apps that include lawful intercept.

The former Fiberblaze, renamed Silicom Denmark, has proved a wise investment for the Israeli vendor. Any company with technologies that support the explosive surge of broadband, data centers and the cloud, as Fiberblaze did, is well-positioned to capitalize on market demand across a number of sectors. All that blistering computing and network speed comes with new challenges in performance optimization, quality of service, shifts in customer trends – and data "visibility." All of these factors played a role in Silicom's decision to snap up the Danish company.

In the "NIC" of Time

Fiberblaze, founded in Copenhagen, Denmark in 2011, was a relative newcomer to the network management and surveillance fields, but one that showed rapid growth from the get-go. The company offered a wide variety of products leveraging Field Programmable Gate Array (FPGA) technology. One line was its suite of FPGA-based Network Interface Controller (NIC) devices for telecom, datacom and lawful intercept.

Right from the start, the company joined the global partnership programs of the two FPGA industry titans: Xilinx and Altera, the latter now a property of Intel. Each company's products were at the time considered state-of-the-art in the FPGA realm. Fiberblaze built on the FPGA platforms of both companies, giving customers a choice of NIC devices that used either platform. As systems "accelerators," Fiberblaze NIC devices helped government and law enforcement agencies attack the challenges of monitoring high-speed networks and services including VoLTE.

What Made LTE Great – and Tough

The advantages of LTE are well-documented: faster data rates than its predecessors; reduced latency and delay; flexibility to operate in a wide variety of spectrum; use of MIMO (multiple-input, multiple-output transmit

& receive antennas) to multiply the capacity of individual radio links; flat architecture of the Evolved Packet Core (EPC); lower capex and opex than preceding mobile protocols such as 2G and 3G; and, overall simplicity. When it comes to lawful intercept and other forms of surveillance, however, simplicity definitely takes a hit from LTE and Voice over LTE.

For starters, VoLTE call intercepts are generally implemented at a session border controller (SBC), a network interface device that manages signaling and communications streams to set up, enable, then end the call. Per the GSM Association, the rules for intercepting roaming VoLTE calls add another twist, with "the mandatory point of intercept" being an access session border controller, a separate type of network interface device. Once the intercept solution is deployed, what follows is a new world of complexity.

Among the challenges to commonly used lawful intercept solutions:

- **Traffic speed:** Higher speeds on 100G networks and the impact of this speed on data visibility.
- **Technology coexistence:** Must intercept not only LTE and VoLTE, but UMTS, and to the extent that is still provided, GSM.
- **Protocol proliferation:** Same problem. The solution must accommodate different protocols for each mobile technology.
- **Device proliferation:** Different devices operating at different layers of the network.

Of these challenges, the foremost to law enforcement are traffic speed and data visibility. Many LEAs, including even large federal organizations, are simply not set up to deal with monitoring 10GB networks, let alone 40GB or 100GB.

Unless one is willing and able to match the NSA's budget for computing power at Fort Meade, the cost of filtering and processing the huge volumes of traffic on 100GB networks is simply out of the question. Storage was another issue. To be sure, one can build DPI systems that capture everything, but packet level inspection and long term storage of the data can lead to significant financial outlays.

Fiberblaze addressed these issues with an FPGA platform.

What FPGAs Do

A "field programmable" gate array or FPGA is exactly what the name implies: a sophisticated circuit board that can be programmed in the field (or more often at the user's desk) to virtually any purpose. The programmable parts are logic blocks that implement complex digital circuitry, and logic gates that handle complex computations at high speed.

FPGAs are popular for several reasons. They offer the flexibility to configure the device for a variety of applications. Other pluses: low cost compared to ASIC, and the ability to allow re-reprogramming of one set of logic blocks and gates while others do their job. Finally, as a hardware accelerator, the FPGA can readily assume responsibility for specific computing functions within a system, sharing the computation speeds output.

As deployed in an LTE network, Silicom Denmark NIC devices based on FPGA technology can ramp up delivery and assure the quality of VoLTE and other LTE services on networks running at 100GB and capture all the data.

The first such NIC, the 100GB xR4 Capture Solution with 10/40/100GB capture functionality, debuted at the Interop trade show in March 2014. The xR4 processed data at the network interfaces in real-time at Layer 2 (data link) and Layer 3 (network).

This NIC could also use the XR4 to capture Layers 4 – 7, including the application layer, providing the utmost in data visibility of the utmost variety.

NetFlow and IPFIX

As explained in earlier chapters, DPI examines each packet, stripping the header and perusing the payload. Flow monitoring monitors packets passing any given point of observation in the network, assessing only the header. To be recognized as a distinct packet flow, the set must share common properties – called a key – such as packet header fields indicating the same destination, selected packet characteristics such as number of MPLS labels, and packet treatment. Because it examines headers only, flow monitoring yields significant improvements in performance.

For purposes of optimizing network management, flow monitoring is fine. The CSP can identify flows by type of packet, origination and destination, and arrive at serviceable conclusions about network usage and performance, all with minimal impact on storage. The downside is that information goes missing when headers-only are examined. What's needed to fill the gap: application-aware flow monitoring that combines the speed of flow monitoring with the precision and accuracy of DPI.

That's what Silicom Denmark means when it references its ability to meet "the challenge of full VoLTE visibility" on 100GB networks: FPGA NIC cards coupled with application-aware flow processing.

Silicom: Stalwart of "The Valley"

Among its many products – server adapters, server to appliance converters, intelligent bypass switches. encryption accelerators, external taps and 10

GB – 40GB Ethernet Time Stampers to name just a few – Silicom also makes one with complete DPI functionality, the NetLogic Packet Processor Server Adaptor, with Layer 7-capable DPI module.

The NetLogic Packet Processor is fully programmable, and the system's multi-core XLP processor gets a major assist from 5 separate networking acceleration engines that can offload and share processing tasks.

Like its adopted counterpart Fiberblaze, Silicom is tightly focused on the networking, data center and cloud markets, where most of the data acceleration occurs and the need for data visibility is thus all the more urgent. Silicom's acquisition of Fiberblaze, like Intel's move on Altera, was affirmation that these markets are where the money's at. It was a strategic, well thought-through decision.

That's no surprise when you consider the longevity of Silicom's management team. Ruling the roost from Silicom HQ on the northern end of Israel's "Silicon Wadi": Avi Eizenman, co-founder in 1987 and Chairman today; Zohar Zisapel, Director and also co-founder; and Shaike Orbach, President since 2001. Solid management with a shared vision speaks volumes about any company and the products it stands by. It is the kind of stability that government agencies and law enforcement bank on.

Sovereign Intelligence and Sixgill

With the Deep Web and Dark Web comprising some 90 percent of the Internet, the trick of finding the right data to monitor and analyze for "actionable intelligence" is half the challenge. Equally daunting: automating the process so that alerts are triggered by specific filters without the time and expense of relying on agents in seats. To achieve both missions, a pair of vendors, Sovereign Intelligence of the U.S. and Israel's Sixgill, have embraced artificial intelligence and deep learning to power results from next generation Dark Web search engines.

These two companies offer similar technology solutions for government and enterprise customers, and each has found an eager market on both sides of the marketplace. Not surprisingly, they came into being in nations with a strong reputation for developing and understanding analytics. For that matter, the executive teams are comprised of individuals with extensive backgrounds in government intelligence, top secret clearances, and hands-on experience at catching targets in the act of committing criminal or terrorist activities.

Inside Sovereign Intelligence

Sovereign Intelligence founder and CEO Mark Johnson appears to have been groomed for the job of surveillance from the start of his career. As a young lawyer in Washington, D.C., he was in the right place at the right time when 9/11 happened: interning at the White House and the US Department of Justice. Like many public-spirited individuals of that era, he took action, signing on as a Special Agent for the Naval Criminal Investigative Service. His forte: counterintelligence (CI). With training in CI by the Joint Counterintelligence Training Academy, Defense Intelligence Agency and the CIA he rose quickly to the position of case officer, then senior staff officer, producing innovative CI solutions and conducting CI operations against U.S. foes throughout Asia. His next stop: Baghdad, where he played a role in shutting down the largest vehicle-borne IED operation of that war.

One lesson that stuck with him from these experiences was the lack of synergy between national security and law enforcement agencies. All gathered significant amounts of data, but the lack of synthesis created gaps. The rapid growth of unstructured data – text, video and images – magnified the problem, spurred by the information bomb called social media. The era of big data had arrived, but the tools for making sense of it, learning trends, discerning anomalies and key influencers remained rudimentary. Market leaders such as IBM were in hurry-up mode to catch up on analytics and did so by acquiring niche players such as SPSS and Netezza for real-time predictive analytics (RTPA) capabilities. But their solutions were confined to the known or "surface" Web. Johnson sensed that data scientists were not delving deep enough.

Sovereign Intelligence set to work developing solutions with the ability to scour petabytes of data from the Deep and Dark Web, plus OSINT such as social media. The end result was two products, SI Reconnaissance and SI Cognition, Dark Web search engines that gather data from unstructured, uncatalogued sources outside the surface Web.

With these two products, data is funneled to databanks for storage and subsequent analysis. The evidence collected is made available to users from a single platform. Specific data points can include URLs, pseudonyms, email addresses, IP addresses, social media posts, geolocation and phone numbers of targets and affiliates. SI Cognition can also undertake social engineering techniques to lure targets to a honey pot, then determine where they plan to strike next.

What sets Soveriegn Intelligence apart is the company's specialization in deep learning and artificial intelligence (AI) capabilities that automate the process of investigating Dark web threats, and learn while they go. In so doing, the company draws on the expertise of a key acquisition.

Enter Stage Right - Sensai

Concurrent to Sovereign Intelligence's founding in 2013, a small group of Silicon Valley-based AI experts launched a company to develop deep learning solutions that would help companies qualify and accelerate business intelligence from unstructured textual data. The company: Sensai. At the time, Sensai CEO Jonas Lamis candidly acknowledged that they did not know which industry segments would adopt their solution. However, that problem resolved itself once word got out about what Sensai could do. Companies in the financial, technology, insurance and technology verticals showed interest, and by 2015 Sensai was bringing clients on board. The most important application across all sectors: determining patterns of risk and fraud concealed in a company's big data set.

Lamis and his peers recognized that risk management was the company's primary selling point, and in the Spring of 2016 began to seek a round of new investment to fund further product development in that area. Shortly afterward, Sensai's CEO met Sovereign Intelligence leader, Mark Johnson.

Sovereign Intelligence was in the market for an AI solution that could automate the process of pinging their petabyte-size Dark Web database, and save on the time and cost of using human analysts. Sensai wanted to apply its deep learning skills in the risk management field. Sovereign fit the bill. In addition, owing to Johnson's well-established connections, Sovereign already had strong ties with the Intelligence Community, as well as large commercial clients.

In sum, they were a perfect match. Sovereign Intelligence announced in August 2016 that it would acquire Sensai and the deal closed in October 2016. At the time, as Lamis put it, "Sensai's technology easily digests Sovereign's proprietary data sets, and fits like a glove for the risk analytics processes that Sovereign provides to its customers."

Of note, Sovereign was successfully identifying threats from Dark Web sources for financial, retail, pharmaceutical and government agency clients almost from Day 1 in 2013. But the move to embrace AI came with the acquisition of Sensai.

Meantime, a company based in Israel had been in the same game for months.

Sixgill: Taking Large Bytes Out of the Dark Web

Formally launched in June 2016 after its informal founding in 2014 and two years of intense R & D, Sixgill entered the market as a specialist in automated Dark Web search and analysis, backed by US $5.0 million in seed money. The company's DARK-i solution is a specialized search engine that scours the Dark Web for risk factors, and like competitor Sovereign Intelligence, deploys special AI algorithms to automate and refine the process.

As to the source of the corporate name, the "six-gill" is a rare shark that resides at a depth of 6,000-feet below the ocean's surface during daylight hours. The six-gill is a predator known for for its stealth, speed of attack and bite size.

DARK-i tracks Dark Web activities and actors silently and autonomously, using AI and deep learning to direct the hunt. Closed, open and hybrid Dark web forums are the solution's prey. Detection and monitoring are automatically conducted by Sixgill's proprietary Hidden Service Locators (HSL) in the DARK-i software. The system quickly detects threats and sends red alerts before risk posed by hidden threats turns real and to catastrophe. DARK-i specifies identity of the targets, and through deep learning predicts where and how they will strike their target.

Threat actors often communicate in truncated fashion to conceal their intentions. However, this challenge is not an obstacle for Sixgill DARK-i. The Dark Web search engine can piece together bits of transactions to understand hidden intent and plans in high-risk scenarios. To ensure that the captured intelligence is reliable, the system automatically prioritizes what is important to the user and filters out extraneous data that might cloud decision-making.

Another attractive feature of Sixgill is flexibility. Clients can purchase DARK-i as a cloud-based solution, thus avoiding the lengthy, complex deployment and cost of special hardware.

VSS Monitoring: Network Packet Brokers

Network Packet Brokers (NPBs) and TAPs are systems designed to provide visibility into the network. Both have grown in popularity with the rise of software-defined networks, virtualized networks, the cloud, and the attendant increase in the volume, bandwidth, complexity and variety of network traffic, factors which can limit network insights when the user relies on older network visibility systems designed for legacy networks.

The most commonly promoted applications for NPBs and TAPs are on the commercial end of the business, e.g., for network performance monitoring and security applications. However, NPBs and TAPs, by virtue of their ability to capture both targeted IP flows and specific identifiers of suspicious traffic, also play an important role in lawful intercept and government intelligence.

But even as the market for NPBs and TAPs continues its boom cycle, there is a growing debate among network management operators, security teams and other core users including law enforcement and government intelligence, over the purported drawbacks of each. Proponents of TAPs contend that NPBs are too complex and expensive. NPB advocates counter that TAPs are the opposite – too simple and functionally narrow to effectively monitor software defined and virtualized networks. Adding to the confusion, some vendors offer both types. Which is best, NPBs or TAPs, and what are the criteria for making an informed decision on the best investment?

Using VSS Monitoring as our guinea pig, let us find out. VSS Monitoring is a highly respected hardware vendor in this space, offers both NPBs and TAPs, sells its wares on its own and through partners such as India's Fastech, and thus is an established global presence in the field. Let us say at the outset that VSS Monitoring's approach of offering both types – the commonsense decision to provide choices that meet specific needs – prevails.

Inside VSS Monitoring

VSS Monitoring has an intriguing history for a Silicon Valley operation. The company was founded in 2003 in Sunnyvale, CA, USA, and immediately set its sights on providing software and hardware geared toward monitoring software-defined networks used by large enterprises and database centers and the then nascent market for cloud networks. Through its pioneering work in addressing the challenges of virtualization, VSS Monitoring quickly set the pace in this arena and was out ahead of the competition. The unique aspect: In its first seven years, the management opted to go it on "guts" alone, turning away venture capital investment. That turned out to be a smart move.

With flocks of eager customers beating down the doors, there was no need to give up a part of the business in the early stages. So VSS Monitoring waited for the opportune moment, when larger competitors began to muscle in, before taking action. The perfect time emerged on August 24, 2010 when VSS Monitoring moved on its first round of venture financing, from Battery Ventures, totaling US $20 million. Battery's assessment said it all:

"This is a unicorn company. How often do you find a company that's this old and shipping product in the middle of Silicon Valley with no VC investment?"

Already a bright star in its field, VSS Monitoring immediately became a red hot target for larger companies seeking to capitalize on demand for all things SDN, virtual and cloud-based. Just two years later in 2012, Danaher Corporation, a global giant in the data collection business for science, technology, medical and manufacturing enterprises, snapped up VSS Monitoring for an undisclosed sum. The key incentive for Danaher: VSS Monitoring's superlative ability to extract data from high-speed networks.

But even that eventful corporate union was comparatively short-lived. Fate had different long term plans for VSS Monitoring. Just three years later in July 2015, the unit changed hands again when Danaher spun off its entire communications business to Netscout. Curiously, VSS Monitoring scarcely gained a mention during the transaction and was buried deep down not only in the news that resulted, but in the very announcement issued by Danaher and Netscout. It was all very quiet.

Today VSS Monitoring marches on, still highly successful as a quasi-independent division of Netscout, and remains in Sunnyvale. VSS Monitoring holds as prominent a position as ever in the fields of NPBs and TAPs. Do any web search of either type of technology and the company's brand name almost invariably appears at or near the top of the search engine results page. The words "VSS Monitoring," "NPBs," "TAPs" – and particularly VSS and NPBs, are deemed virtually synonymous. At the same time, while VSS Monitoring goes both ways and offers lines of NPBs and TAPs, the company clearly gravitates toward ultrafast packet brokering. Why is that?

In contrast to TAPS, NPBs can scale to monitor entire networks while at the same time drilling down to provide visibility and analysis of the link layer of IP traffic, seeing and identifying any packet on the network, i.e., full packet capture with accurate time and port stamping and an assist from DPI to open and view full packet content.

Moreover, state-of-the-art NPBs provide "link state monitoring" so that if one NPB device fails, traffic is automatically routed to a functioning NPB device, thus ensuring full redundancy. The device is built to both load-balance and discriminate so that traffic is at once split to multiple ports while packets in a targeted session are routed to a single monitoring tool for accurate traffic and packet inspection. In sum, given its capabilities, speed, intelligence and redundancy, the NPB is perfect for network forensics and evidence gathering when connected to probes, mediation devices and other intercept hardware.

The VSS Monitoring Approach

Always forward-thinking, VSS Monitoring has taken NPBs to the next level: true virtualization. The company's top-of-the-line vBroker and Optimizer suites monitor the network, providing all the aforementioned functionality, but now at arm's length. New "virtual hardware" operates completely separate from the network, using VSS Monitoring's vMesh platform to provide total separation from a communications service provider's physical network. The advantages: speed, zero interference with the network itself, outstanding performance regardless of exponential increases in bandwidth, take-up at any speed, irrespective of burgeoning customer headcount, and – because the "units" are virtual – total freedom from location-dependent restrictions.

There are multiple VSS Monitoring NPBs beginning with the Optimizer class for outstanding work at the 1G/10G level, all the way up to the company's creme-de-la-creme unit, the VB6000 for high-speed networks. It's not an "either/or" or "start at the top" choice. VSS Monitoring helps the customer pick the model that is right for the job, and makes it easy to step up as needs change.

The Optimizer 2400 is a good place to start for users with moderate network performance management requirements. Basically the Optimizer is for 1G networks on the cusp of needing support for 10G. Alternately, one might consider the company's baseline v16x8 unit (16 fixed ports plus 8 "SFP" or smaller form-factorable ports for fiber connectivity), designed for networks still in the megabit range. Like all VSS Monitoring NPBs, the Optimizer duo does not skimp or hark back to the past: It is fully compatible with the vMesh platform, with all the advantages that attend to the needs of network virtualization.

In ascending order, VSS Monitoring's next three products, called vBrokers – the VB120, VB220 and VB 420 – do more, handling from 1G through 40G network loads at up to 640 Gbps throughput. Standard features: hardware-based filtering at Layers 2 – 7 of the OSI stack; flow-conscious load balancing; protocol stripping; and remote or local systems management, and finally, vMesh. Note that this level of VSS Monitoring NPBs is often as far as resellers go. For example, Fastech's line of VSS Monitoring devices sold from India stops at the 40G vBroker.

The premium solution in VSS Monitoring's lineup is the VB6000, for what the company immodestly but correctly terms "ultimate scalability": support for any network from 1G to 100G at 6 terabytes of throughput and supporting 600 ports. With the ability to scale up or down in a flash, the VB6000 sets the bar in the marketplace. And, like all VSS Monitoring products, the VB6000 is fully interoperable with the company's other NPB products so that the customer can build on – without necessarily having to dispense with – prior investment in another of the vendor's units.

Given the dynamic capability, technical superiority and array of pricing options in VSS Monitoring's NPB portfolio, one might naturally assume that the company is done with TAPs. But once again, flexibility comes into play. VSS Monitoring offers its own TAP units, too, also in 1G/10G, 40G or 100G options. What's the difference and why go with an NPB versus a TAP? Answer: vMesh. The NPBs have it, the TAPs do not. Does it matter?

IT market leaders are a good judge of such matters. When IBM was in the hunt for the best IP traffic capture & monitoring solution to couple with its virtual switches, Big Blue picked VSS Monitoring. Ditto for Cisco and ditto for VMware in their respective realms. The final deciders are customers themselves: Some 80 percent of the world's CSPs opt for VSS Monitoring. The solution of choice is generally the NPB.

NPBs are at present the best option for network operators and by extension for law enforcement and government agencies that require rapid, total and intensely refined intelligence. The next logical step in the market's progression may prove to be network switches and routers with baseline NPB support built-in, the better to accommodate leading edge NPBs themselves.

CHAPTER 3

Analytics

Overview

Today terms such as "advanced Analytics," "big data" and "data science" are used with abandon. To appreciate their meaning and value it is worthwhile to delve into their origins: the evolution of computing, the rise of "structured" data and the explosion of "unstructured" data fueled by the personal computer, the Web, universal mobility, streaming data and video. All of these factors have contributed to what we call big data, sets of data so large and complex that they surpass the capabilities of conventional computing.

Big data actually began more than a century ago. At the time it was known as storage by paper documents in file cabinets which accumulated by the ton at government agencies and corporate enterprises. Like early digital data, data on paper, too, defied analytics.

From Punch Cards to Big Iron

The first "computer" to address the vast corridors of plain paper storage was the "punch card" calculator of 1929, which used paper cards dotted with holes in predefined positions representing specific elements of data. Initially confined to simple calculations, the punch card computer evolved into more complex processing machines, but still relied on paper cards for storage, processing and memory. Punch cards would own computing for years to come.

Although invented by a separate company, this semiautomatic processing was quickly taken over by IBM, which dominated the field until a better alternative arrived: the first card-free digital processor, the IBM 701, introduced in 1952. Also known as the Defense Calculator – which gives a hint as to the forces behind its development – the IBM 701 launched the era of computers with internal electronic memories. Within a year, memory switched to tape on reels. In 1956 IBM formed its "Big Iron" division to manufacture what soon became known as the mainframe computer.

By the early 1960s, the business end of enterprise and defense IT were in the hands of numerous mainframes that used large scale computer architectures to process bulk data. Burroughs, UNIVAC, Control Data, RCA and three other companies entered the field, but because IBM commanded 90 percent market share its competitors were often dismissed as the "seven dwarves."

Structured vs. Unstructured Data

As data sets grew more "bulky," engineers sought new ways to improve data management and utilization. Among the key breakthroughs was the development of the "relational database" by IBM in 1970. The importance of the relational database: It provided a simple way to organize data in formally-recognized tables that could be accessed, analyzed and manipulated without having to rewrite the tables that held the data. Properly formatted, such data was termed "structured," that is, cast in predesignated models that defined how data would be recorded, stored, processed and accessed. Though dated in many ways and declining in popularity today, relational databases for structured data still have their advocates.

With structured data, decisions are made on the various fields to be used for storage of each data type – alphabetic, numeric, currency, name, date, address, etc. – plus their *relations*, and any restrictions on types of data to be retained. Structured data is typically stored in a relational database such as SQL (Structured Query Language), which became a formal standard when adopted by the American National Standards Institute in 1986.

SQL databases were, and still are, relatively simple to use within certain parameters. Limitations include the high cost of storage, memory and processing. Also, any data that does not fit a predefined category may "fall out." Such irregular data would eventually become a problem.

In the early 1980s, along came the personal computer, word processing and then in 1989 the World Wide Web, soon to be packed with images, photos and video. None of these new types of data fit the standard SQL models. Being "unstructured," they had no home in the relational database.

Unfortunately for those fond of SQL, the services that hinged on these rebellious packets were also highly popular. By the late 1990s, it was estimated that between 80 – 90 percent of useable data originated in unstructured form, that is, outside the management capabilities of a standard SQL database.

SQL Gets a Sequel

IBM and others had been aware of the problem as early as the 1960s, working on early versions of what would later become the "Not Only SQL" or NoSQL database. The NoSQL database introduced the concept of accessing data via "associative" modeling outside the limitations of SQL's tabular modeling.

NoSQL's arrival was timely for another reason: the bandwidth boom.

NoSQL could "horizontally scale" to other nodes outside the system to either increase or decrease the amount of computing power required for complex tasks. Today the common term for this capability is "distributed computing," the practice of spreading massive parallel processing requirements across a near-infinite array of servers in the network.

NoSQL is not a perfect solution to the challenge of unstructured data, nor in every way a "besting" of SQL. NoSQL offers availability and speed, sometimes at the sacrifice of consistency, although new developments have reduced correction times to milli-seconds. SQL is still better at simple transactions such as one-function needs – bank transactions, transfer of files within a single database, etc.

But NoSQL paved the way for innovative advances such as data mining, Natural Language Processing and text analytics systems that reveal patterns in data. Bulk metadata collection and analysis is one example. Voice tagging used in voice biometrics systems is another.

With the advent of NoSQL, distributed computing and databases, relational databases are quickly fading into the past as standalone solutions. Simply put, NoSQL scales, SQL does not. Prior to the agency's 100 percent transition to secured cloud service, the CIA used NoSQL, as did Google and other organizations with massive data management requirements.

As big data soared to multi-zettabyte and today's yottabyte levels, NoSQL and similar database systems required an enabler to facilitate ever-faster distributed processing.

Distributing Computing – How it Works

Enter the Apache Software Foundation, a U.S.-based non-profit organization representing a community of developers dedicated to finding new and better ways of accelerating big data management. Apache's most famous contribution is Hadoop, a "software ecosystem" that organizes massive parallel processing across multiple servers.

Hadoop itself is not distributed processing, but rather, a set of interfaces that facilitate computing clusters. In action, Hadoop enables creation of a "data lake" of information relevant to specific queries. Other Apache solutions such as Apache Spark build on Hadoop, further accelerating data flow and computing, and accepting content from any data source including traditional SQL databases.

All Apache programs are free and downloadable from their website. From the analytics standpoint, whether for ISS or other needs, the action begins with paid programs that use special algorithms to "score" findings from data, that is, narrow the data field to the most relevant items in real-time – and see what potential futures they point to. These "real-time predictive" capabilities cross the line from traditional to Advanced Analytics.

Real-time Predictive Analytics

To the data scientist, classic data analysis is "heuristic" or historic, drawing conclusions from past incidents. Real-time analytics goes an important step further by collecting data "in the moment," or alternately, at whatever improved time parameter is set by the user. To understand the distinction, consider that definitions of real-time can vary and "real-time" can be a relative term. For example, in a system previously set to collect data monthly or weekly, the transition to "real-time" might mean collecting data by the day or the hour. However, real-time is generally considered as being within seconds, or increasingly, milliseconds, to gather and analyze data from thousands of sources. What keeps the process real-time (and processing costs under control): As a new data item is collected, the oldest is jettisoned.

Predictive analytics operates in similar fashion, aggregating and scoring both heuristic and real-time data to produce "actionable intelligence" based on trends and patterns that point to the most likely set of future actions or events.

Once again, IBM holds the lead. A pair of products – IBM Netezza and IBM SPSS – are, respectively, among the best known and most commonly used software applications for real-time and predictive analytics or RTPA.

What if the findings of Advanced Analytics are themselves so big that they defy comprehension or fail to provide complete understanding? Here visualization tools come into play, providing 2D or 3D graphics that clearly illustrate what numbers alone might not say: the scope of a major data event, even hidden connections that were previously unrecognized. One of the best in that game is Palantir, a company whose visualization products have proven their value in bringing Advanced Analytics to life for users across a range of fields: financial services, first responders, military and intelligence organizations.

Advanced Analytics and ISS

With the influx of communications and Web data covered by surveillance technologies, Advanced Analytics play a vital role in the job of evidence and intelligence gathering. Analytics reduces this maze of information to useable granules that can reveal previously undetected patterns and networks. Examples: the ability to build on social media analysis to identify co-conspirators not even *on* social media networks, or to locate and visualize the hidden source behind distributed denial of service (DDoS) attacks.

The end game of analytics is deeper understanding of meaning and intent: what the data reveals about the identity, relationships, behaviors, areas of activity, timeframes and imminent threats posed by targets known and unknown. The arsenal of analytics now available to address these issues is huge. Among the most popular and best known:

- Real-time Predictive Analytics (RTPA).
- Semantic Analytics.
- Link Analysis.
- Visualization.
- Data Retention with Purpose-Built Search.
- Deep Web Analytics.
- "Cognitive" computing, one of many forms of AI.
- Cryptanalysis.
- Dark Web Analytics.

There is no "one best tool" in the universe of analytics. Each system provides a set of capabilities that contribute to a holistic picture of the target and the threat: RTPA for current insights and future likelihoods; Deep Web Monitoring for lost or forgotten threads of OSINT insights that contribute context; semantics for precise understanding of what is spoken or written; link analysis to build connections; visualization to graphically represent mission critical data and reveal hidden parties; comprehensive storage of structured and unstructured data that is continuously added in a real-time collection environment; and purpose-built navigation of the data sources.

Among the most significant developments in advanced analytics: systems that learn, make decisions based on variations in what is observed, single out potential risk, and thereby come as close to thinking as a machine possibly can. In development for nearly 50 years, cognitive computing, machine learning and the countless other schools of artificial intelligence are far from perfect. But some, such as Elon Musk of Tesla and SpaceX, believe that we are close to seeing the first true thinking machines. Musk is a major investor in AI.

Of course, analytics can only be applied when data is "in the clear," and with the growing popularity of encryption, the mission of cryptanalysis has taken on increased urgency. The challenge to most users is that excellent encryption systems such as PGE (Pretty Good Encryption) can be difficult to master. Tor is simpler to use, but as the NSA and academic researchers have proven, Tor is by no means a 100 percent guarantee of anonymity. For that matter, neither is PGP.

Guard Against the Illusion of "Impeccable Data"

One further point to remember: Not withstanding all of today's fanfare over real-time, predictive and prescriptive analytics, some 90 percent of useful intelligence flows from analysis of heuristic behavior – in other words, from history. To be sure, RTPA is important. But one should never laud real-time data at the expense of heuristic data.

One note on the character of data itself. Some misguided "data scientists" like to boast of their capability in determining "impeccable" or perfect data. This so-called goal emerged concurrent to the rise of big data, when alleged experts faulted systems supposedly plagued by "dirty data." For the record, there is no such thing as impeccable or perfect data – all data is "dirty" to a degree. It is through the analysis of vast amounts of always-dirty data that one arrives at the patterns, trends and relationships essential to the work of Intelligence.

Through massive parallel processing – and increasingly, the use of general-purpose computing on graphics processing units (GPGPUs) essential to current forms of AI such as deep learning and neural networks – we are gaining on the greatest challenge of analyzing big data. Not the volume of data, and not even encryption, but the limitless *variety* of data forms that now confront government intelligence agents, analysts and data scientists.

Big Data Analytics Top to Bottom

While the architecture that supports big data and real-time predictive analytics is complex and constantly changing, in the ISS arena it may for the sake of simplicity be addressed in five layers:

- **Layer 1: Data Ingestion and Storage.** The first layer is Apache Hadoop or similar tools that have emerged for processing very large sets of data including structured, unstructured, batch and streaming data, and for storing such sets in a secure repository commonly known as the "data lake." Hadoop works with NDM (network data management), a widely distributed query processing system that leverages special algorithms to look across and query any type of network and capture the data sought. Increasingly, the cloud is the new data lake.
- **Layer 2: Data Preparation.** At Layer 2 another Apache program (formerly MapReduce, now Spark) combs the data via univariate statistical analysis that winnows down the data mound by single variables such as sex, age, nationality, or frequent travel to a known hotspot of the target's activity.
- **Layer 3: Heuristic and Real-time Predictive Analytics.** At Layer 3 the system applies descriptive statistical analysis that assesses trends based on past activities, then couples these findings with live streaming data on the target's plans and actions in that moment. Predictive analytics applies models that forecast the most likely future outcomes. Real-time and predictive analytics are, respectively, the domains of tools such as IBM SPSS and IBM Netezza, and similar platforms by SAP and others. Bivariate and multivariate statistical analysis provide further refinements.
- **Layer 4: Interpretation.** Within Layer 4, real-time predictive analytics are integrated with representational state transfer (RESTful) and service oriented architecture (SOA). RESTful is a representation of the Web architecture that controls the interaction of components, connectors and data elements to ensure the correct interpretation of data. SOA provides application functionality of the data to specific services, independent of vendor, product or technology.
- **Layer 5: Dashboard Visualization.** Different analysts within an intelligence group might be responsible for varying aspects of a potential target's action or location. Dashboard visualization lets each user see the data customized to his or specialty. As high interest data surfaces on the screen, the agent can pose ad hoc queries to find new or related answers. The dashboard can also be programmed to deliver alerts.

Because situations change in real-time, an agent can then examine or change Key Performance Indicators (KPIs). At the highest level, the intelligence leader can see the big picture unfolding through his own customized dashboard, augmented by visualization that depicts relationships and activities associated with other members of the target's network.

Apache Spark: Not Just Faster – More Refined

In 2014, big data analytics received a major boost from a development by the Apache Software Foundation (ASF), "Apache Spark," or simply "Spark," an open source cluster computing system that integrates with Hadoop, speeding the process of refining intelligence into actionable insights. In the early days of Hadoop, data intake and queries were handled by another Apache program called MapReduce, which used parallel computing to create large but still manageable big data sets. The problem with parallel computing: Separately designed data sets might not "talk" to each other or coordinate on the answer to a query. That lack of cohesion posed the risk of losing critical data. MapReduce was also comparatively slow. Spark fixed both problems.

Using resilient distributed datasets (RDDs), Spark can interactively query up to two terabytes of data in under one second. Spark is the perfect fit for streaming analytics, which assesses smaller amounts of data as they enter the "lake" and enables the user to perform search queries on the fly. Spark works up to 100 times faster than MapReduce, and offers the added advantage of being easy to use. An agent can launch Spark quickly through a user interface on the back end of the dashboard. Spark moves beyond the capabilities of MapReduce to speed target-essential data ingestion into a data lake. In the view of leading analysts, Spark has crushed Hadoop. Fast, flexible and development-friendly, Spark is now the leading platform for large-scale SQL, batch processing, stream processing and machine learning.

Integrating Voice Biometrics, Mobile Location and DPI

Public discussions of big data analytics and ISS often stray into the controversy over the NSA's now largely abrogated practice of collecting bulk metadata in the form of call data records (CDRs) and IP data records (IPDRs). Bulk metadata collection, formerly authorized under Section 214 of the Patriot Act, provided a mechanism to understand links of hops between calls and callers, and to make intelligent hunches on actions. That is not the same as real-time predictive analytics, which far exceeds the capabilities of bulk metadata collection.

There are no hunches or guesswork in RTPA. Analysts are not confined to making assumptions about the future based solely on a record of the past. Every type of data can be assessed and leveraged instantly. An RTPA system can be integrated with other data sources gleaned from tools such as voice biometrics, mobile location, and deep packet inspection to obtain the fullest picture possible of the target. In both the commercial and ISS realms, for example, the Qosmos DPI solution ixEngine is often coupled with real-time predictive analytics to provide detailed insights on the trending activities of customers. Although Qosmos has in recent years distanced itself from ISS, ixEngine has in the past been an integral part of surveillance systems, and notwithstanding company denials, still is used in Russian and other authoritarian nation state intelligence gathering systems.

Onward to AI

Big data analytics is a process of constantly refining data to anticipate the most likely moves of a target and recommend "next best actions" for the agent. Queries rarely conclude with one recommendation, but rather, with a set. The decision on which action to take may be automated or manual. In the latter case, deciding on which course to follow can consume minutes or hours, undermining the very purpose real-time predictive analytics.

However, even automated systems don't necessarily come to a "next best action" recommendation as fast as the human mind does. The mind needn't plow through billions of bits to reach a conclusion. It is more akin to flash memory, which can seize the precise data it needs to make a rapid decision.

Human intelligence is distributed and at the edge, making it easier and quicker to reach decisions. Only when big data architecture adds a sixth architecture level – a Decision Layer that operates in flash mode – will advanced analytics truly compete with intellect and perhaps reach the far shores of Artificial General Intelligence.

Artificial Intelligence in ISS

According to some futurists, Artificial Intelligence (AI), the ability of computers to think exactly like humans and replicate nearly every complex process of the mind, is fast approaching. The advent of AI has its fervent advocates and equally adamant detractors. Some, like Google Director of Engineering Ray Kurzweil argue that human-level AI will find the cure for cancer and reverse global warming. At the opposite end of the scale, Tesla, SpaceX and Paypal entrepreneur Elon Musk has famously warned that those pushing the development of AI are "summoning the demon" – a world where thinking machines pose an existential threat to mankind.

Who is right? Perhaps the contrarian of the lot. Among the most salient insights are those provided by David Gelernter, Yale professor, scientist and a father of parallel processing: the ability of computers to perform multiple complex transactions across oceans of data, and the core technology behind another major development, big data analytics.

In his book, *The Tides of Mind: Uncovering the Spectrum of Consciousness*, Gelernter observes that the processes of computing and thought are vastly different, and that many modern intellects have become so focused on the human potential of digital devices that they have almost completely abandoned study of the infinitely more intricate and complex mind itself, which by comparison is scarcely understood. Most have forgotten that the mind is inextricably tied to the physical body, the emotions, the senses, dreams and the unconscious, all of which influence "high level" analytic thought, some of which represent types of submerged thinking – and none of which can be replicated by a computer, at least not yet, however sophisticated.

Machines, Gelernter says, miss out on many critical elements of thought and thus will always be an incomplete representation of the mind. Moreover, every human mind is unique: You can install a standard software program into a machine and it will perform consistently from one digital device to the next, but not so with human beings, each of whom might think and react differently.

Surprisingly, despite acclamation of AI ushering in a new age, the science behind it is not really new and has, in fact, experienced a prolonged gestation going back decades. To be sure, various forms of AI are present in manufacturing, bioscience and other prominent fields including ISS. But for all the hype over programs that beat chess masters or tromp champions on TV quiz shows, AI remains the perennial teenage prodigy, full of promise and excitement, but short on the tremendous expectations generated by its propentents.

In ISS, so much skepticism has been leveled at AI that some of the most prominent companies in the field long avoided it. Lining up against these doubters was SRI, one of the pioneers in AI (of a kind) and still a proponent, offering a suite of surveillance solutions that supposedly rely on thinking machines. It is a prospect that alarms many.

Are we rapidly approaching the day when machines will form the basis of a surveillance state where everyone is under continuous examination, or are such fears overwrought?

The best answer is to consider the long-running saga of AI, and the position of leaders on both sides of the debate: the camps that support the futuristic vision of a world run by AI, represented by SRI, and the pragmatic folk who dismissed AI in favor of big data analytics and "intelligence augmentation" (IA), represented until recently by Palantir.

Big Data Analytics vs. AI

Big data is the offshoot of the vast array of transactions and records fueled by the emergence of huge data sets and storage capabilities, in large part propelled by rapid advances in communications technologies and services, including those stimulated by near universal mobility and Web access, and the millions of apps that hang thereon.

As previously stated, big data analytics developed in the 1990s as an approach for managing and making productive use of what was fast becoming out-of-control amounts of data. Much credit goes to the non-profit Apache Software Foundation, which developed Hadoop, an open source platform for collecting data relevant to a query into a giant data pool, and Apache Spark, a platform for high-speed data processing. Spark is now the heart and soul of many big data analytics systems today. Commonly available tools that add real-time and predictive analytics capabilities include the popular IBM SPSS and Netezza software.

The essence of big data analytics is to apply massive parallel processing capabilities to data stored in computer clusters. Analytics is by definition rules-based. In other words, a system will search databases, collate data and provide findings, make recommendations – or automatically act on the same – based on the query and the predefined rules for doing so. The same holds true for the field of predictive analytics, which helps users see and assess the most likely set of future events, based on statistical analysis of heuristic (historic) data. Real-time predictive analytics incorporates live streaming data for an "in the moment" assessment. Most analytics solutions also provide what is called "treatment optimization" – a set of recommendations aka "next-best actions" to follow in order to capitalize on the findings of an analytics query.

On the surface, big data analytics and AI would appear to have much in common. Both are outgrowths of advanced computing capabilities. Both vie to reduce or eliminate human error. Both rely on statistical analysis to create models, then test and measure their performance. Both continue to evolve and improve.

Beyond these similarities, big data analytics and AI are separate disciplines.

Big data analytics is a real-world capability that simply deals with the data at hand. AI's ambitions are infinitely more aggressive: to emulate the thought processes of the human mind.

Big data analytics is commercially available and proven. Though the younger of the two fields, there is little contention over methodology or best practices.

AI is a work in progress. It has evolved along many separate paths for some 60 years, with numerous speed bumps, arguments and opposing principles along the way. Today there is no monolith of AI poised to take over from humans. One reason: There are nearly as many schools of thought and as much acrimonious debate among the proponents of each as there are people involved in the field.

As an analogy, think of big data analytics as fully baked. In AI, the chefs can't even agree on the proper ingredients.

AI Timeline

The mission of artificial intelligence research is broad and deep, not just "AI," but the all-embracing concept of artificial *general* intelligence (AGI): the ability of computers to think like humans, drawing on a wide array of knowledge – or to reach out and augment that knowledge, reach a conclusion to a question or challenge an assumption.

In the 1940s, the first efforts at AI were seen in the development of cybernetics systems that used electronics to perform rudimentary acts of intelligence similar to those in lower vertebrates. Cybernetics were the first to adopt the term "neural networks" – systems that attempted to emulate human thinking.

Not long after, cybernetics was overshadowed by the Symbolic school, which endeavored to reduce intelligence to symbols that could be manipulated.

Symbolism had a short life. Arguing that symbolic representation of the mind was fine for simple applications, a quartet of opposing schools

– cognitive, logical, anti-logical and knowledge-based – followed, each with its own angle and group of supporting academicians.

Among the most interesting was the "anti-logic" or "scruffy" school, which relied on commonsense knowledge bases of all the information that people – or specialists – are expected to know. "Scruffy" AI made ad hoc decisions to tackle very complex problems based on general knowledge.

In the mid-1970s, experiments in knowledge-based AI led to the development of "expert systems," the first form of AI to win broad commercial acceptance. Touted as the first AI to "replicate the human mind," expert systems comprised three virtual components: (1) a knowledge base of facts and data relevant to a challenge; (2) rules-based logic; and (3) an inference engine that could apply the rules to the data to draw new conclusions.

But the initial euphoria over expert systems quickly came and went, too. They were excellent at solving one problem at a time, but less effective at switching gears to a new one. Owing to this limitation, enthusiasm over expert systems flamed out by the mid-1980s, but not for long. Expert systems resurfaced in later years and today are the backbone of one-off apps such as automated online customer service.

In the 1990s came one of the more revolutionary developments in AI: the application of statistical methodology, an important move because it meant results could be measured and confirmed. The ability to "prove" results mathematically was widely viewed as a smack in the face to the "scruffies," who had refuted logic as too narrow to embrace human intelligence. (Note that the timing roughly corresponds to the emergence of big data, where statistical modeling is also the central basis of analytics.) From that time on, statisticians could claim the loftier title of "data scientists," and in truth, math did transform AI into a science.

Today's AI takes an all-embracing approach toward the best of what has been produced in the field. AI systems now are seen as multidimensional, integrating the various schools in an "intelligent agent" architecture that applies the methodology most likely to produce a result, given the challenge. In one instance, for example, Symbolic AI might prove most effective. Another might call for the logical school, and a third for the anti-logic or "scruffy" approach. Coming full circle, cybernetics and neural networks have also regained their early luster and are frequently applied in certain situations. Machine learning, when used to orchestrate "playbook" cybersecurity, quickly identifies the type of threat in progress and automatically applies the most appropriate and effective response to defeat it – faster than a human analyst could hope to do. Israel's Fifth Dimension applies this very approach with its deep learning-driven cybersecurity solution.

When programmed to act independently, the AI system chosen for a specific project acts as an "Autonomous Intelligent Agent" that monitors, analyzes and acts rationally to achieve a goal. But there is one catch: Actions are still rules-based. The AI solution that delivers on the expectation of artificial *general* intelligence or AGI – aka "strong AI," with the ability to fully replicate the functions of human consciousness – is still over the horizon, though just how far is open to debate. Which is undoubtedly a source of endless frustration for one of its main proponents, SRI International.

SRI International

Based in Menlo Park, California, where it was founded in 1946 under the trustees of Stanford University, SRI (Stanford Research Institute) International is a non-profit organization that provides research and development activities – and licenses its R & D – to a wide variety of clients and industries. The Institute split off from Stanford in 1970 and has operated as SRI International since 1977.

Over the years SRI has contributed its technical wizardry to projects as varied as the first computer "mouse," the invention of inkjet printing, the first studies of air pollution, and helping the Disney Corporation settle on the best location for Disneyland. Those colorful items aside, SRI International's financial meat and potatoes are dished up by the defense and intelligence communities. Per company documents, some 64 percent of SRI's income derives from R & D performed for defense.

Like any R & D work, SRI's defense program has had its share of out-and-out fiascos, such as the group's initiative in the area of psychic surveillance techniques, a program that lasted nearly a decade with little to show for the effort. But SRI can point to triumphs, too.

Among the high profile AI-related programs initiated for defense purposes was CALO (Cognitive Assistant that Learns and Organizes), involving 300 researchers from universities and commercial organizations in what SRI once described as "the largest artificial intelligence project ever launched." Derived from the Latin for "soldier's servant," CALO ran for 5 years from 2003-2008, producing many innovations, and an important spinoff – Siri, Inc., later acquired by Apple as its personal digital assistant for iOS devices.

SRI's work in the AI arena began in 1957, with early projects at its Applied Physics Laboratory on self-organizing systems and learning machines. But the Institute's full commitment to AI came a decade later with the 1966 founding of the Artificial Intelligence Center (AIC), which remains the center of the organization's efforts in AI to this day.

Success followed quickly with the announcement of "Shakey the Robot," a mobile system with modest capability to perceive and act on its environment, planning and following routes. At the time, SRI was still under the wing of Stanford University, where another brilliant mind, Dr. Edward Feigenbaum introduced the first "expert system," described above. Widely hailed as "the father of expert systems," Dr. Feigenbaum is recognized for this and other work that, for all practical purposes, got AI off the launchpad.

SRI Angler

SRI's play in the ISS arena is a web-based collaborative tool called "Angler" that enables users in multiple locations to confer, cluster and rank intelligence with the goal of creating an objective assessment, free of bias due to prior experience or other factors. Users can interact, providing their input anonymously so that the contribution of "outliers" can be equal weight and provide fresh insights to the group.

A facilitator controls the session and makes the decision on when to terminate the input or brainstorming part of the session and move to prioritizing the assembled ideas via clustering.

With Angler's clustering, each participant in a brainstorming session can sort the ideas of others into color-coded groups. Angler assesses alignment between the contributions of the users, and stores each idea for future reference.

In the final stage, called "Consensus and Ranking," Angler uses an "agglomerative" algorithm to show the consensus of the group, while also allowing each participant to view his or her own input in contrast to the consensus. Looking across the segmented clusters, the group can then rank those with the highest priority.

From what you can see here, Angler is less about machines thinking like humans and more like its opposite – giving humans the power to think like a computer, making sound decisions based on multiple inputs and weighing each to assess its true value.

To determine the next best action, agents can use a second SRI Web-based software tool, the Structured Evidential Argumentation System (SEAS), originally developed for the Defense Advanced Research Projects Agency (DARPA). SEAS uses best-in-class analytics to develop the most appropriate scenario, based on the evidence, to pursue opportunities or fend off potential threats.

Once again, as with Angler, the work is collaborative. SEAS recaps the reasoning behind the group's decision based on the available evidence, determines what may be missing from the evidence pool to reach the best decision, then explores other possible courses of action either recommended by a participant or pulled from a comparable scenario. Provided in simple visual format, SEAS helps ensure rapid, accurate decision-making, accelerating response times.

SRI cites the national security applications of SEAS as follows:
- Assessment of nation-state stability.
- Assessment of terrorist motive and threat.
- Information for mission preparation.
- Counter-intelligence assessment.
- Assessment of infrastructure vulnerability.

Without a doubt, the goal of AI – to replace error-prone humans with infallible computing powers that never tire or take their eye off the ball – is perfectly sound. But the SRI tools described here are in fact little more than an automated spreadsheet that, if anything, elevates the role and influence of users over the outcome of the system. SRI International's AIC possibly has classified AI solutions that do more, but if SEAS and Angler are how they choose to showcase their AI work in defense, it's odd, to say the least.

Palantir

Palantir, well-recognized as a supplier of advanced analytics solutions to the CIA and other government agencies, for many years completely disregarded AI. Palantir's rationale was quite simple: Advancements in AI are simply too slow and unreliable for the surveillance needs of the day, whether the targets are criminals, terrorists or nation states. As Palantir Director Shyam Sankar once put it, the company early-on recognized the limitations of relying purely on AI, and went a different route:

"The team leveraged the fundamental insight that computers alone (i.e., machine learning/AI) could not defeat an adaptive adversary like the Russian mob. The only thing that did work was empowering human analysts with powerful tools to quickly explore data in conceptual ways."

Palantir's approach was to support IA (Intelligence Augmentation), not AI. As Sankar once famously said, "The idea is not to take a toaster and make data from *Star Trek*, but to take a human and make her more capable." Following that line of thinking, Palantir developed its first product, Palantir Government, not by retiring to a laboratory but by making multiple cross-country trips to meet with intelligence community analysts over a period of three years.

Taking a cue from PayPal, Palantir Government initially focused on structured data, i.e., data residing in a fixed field for ready call-up by relational databases. The only problem with that: the vast majority of data spewed out by systems and communications networks is unstructured, i.e., lacks a pre-defined data model: true of text, images, video and audio. Intelligence analysts politely pointed out the omission, and Palantir stepped up, finishing with a government solution that can input, manage, analyze and provide recommendations on any type of data, whether structured or unstructured.

End result: a system very much like real-time predictive analytics. In sum, while Palantir used some AI capabilities at a low level, their core focus was analytics and visualization.

Since that time, Shyam Sankar and Palantir have done an about-face on AI, embracing specific schools such as machine learning to accelerate delivery of critical data to users including warfighters and intelligence agencies.

Google, which has long used AI to power its search engine and highly lucrative ad business, in January 2014 paid $US 400 million for British AI startup DeepMind. Not long after, two prominent names in technology together invested $US 40 million in Vicarious, an AI startup proposing to build a replicant of the human cortex, the part of the brain that handles language processing and spatial reasoning. One was Mark Zuckerberg of Facebook. The other: Elon Musk.

Google continues to fund research in AI, and long ago open-sourced its machine learning library Tensor. John Giannandrea, then Google's head of machine learning, in May 2016 hailed what he termed an "AI Spring" marked by advancements in the use of machine learning for speech recognition and image recognition. Google also unveiled Magenta, the first AI solution able to write music without human assistance. But Giannandrea was quick to add that the kind of super machine intelligence necessary for the dystopian machine-run world feared by some is "decades away."

Yale's David Gelernter has put his finger on what AI is missing – the complex stream of mental processes at all levels essential to creative thinking and problem-solving that are unique to the mind and will likely never be recreated by AI in their fullness. Even aspects of the cognitive process deemed damning and negative by some scientific quarters can play an important role not open to machines. As Gelernter puts it: "We can't have artificial intelligence until a computer can hallucinate."

Axilent ACE Profiler

Information on Axilent founder and CEO Loren Davie is as spare as data on the company itself: scattered career tidbits on Linkedin; tweets that reveal an array of personal interests ranging from music to Tor, the Dark Web, and an affinity for Satoshi Nakamoto, inventor of the Bitcoin protocol. Davie's self-profile is as puzzling as that of Axilent's website, which reveals a great deal about the company's focus on the use of APIs to accurately profile any individual and predict his movement, thoughts, location and trending activities – but offers nary a syllable about the company itself.

Who or what is Axilent? Is it the intellectual power behind "ACE PROFILER," a company that showcases its prowess on the ability to identify any person in some 200 nations simply on the basis of a telephone number? Granted, Axilent has its own "ACE Profiler," and the odds of two companies having the same-name product in a similar space would seem more than coincidence. Yet any comparison of the two comes up short for the obviously ISS-focused of the two. Axilent's Adaptive Context Engine (ACE) is far more sophisticated than any reverse lookup service based on a phone number. Axilent ACE makes a comprehensive match of people and content, based not only on the content they are drawn to but, more importantly, on the data they themselves generate and its context. The resulting profile is as unique to the individual as any biometric measurement and can not only pinpoint said targets but easily predict their next steps. As remarkably, Axilent's product may be used to profile a near-limitless headcount of targets.

Is Axilent in the ISS business? If not, they easily could be. To the public, Axilent is in the business of helping companies modify content to suit an individual's unique interests, with the goal of boosting sales. What makes Axilent's work so pertinent to ISS: However cleverly a target customer masks his identity, Axilent's ACE Profiler knows exactly who he is, where he is, what he's thinking, his proclivity to act – and when.

Part real-time predictive analytics (RTPA) and 100 percent intelligent automation, Axilent ACE Profiler creates a high resolution "image," or rather, series of images from every conceivable angle, that make the target's passage through a communications network unmistakable, thus completely trackable. Indeed, the terrorist or hostile nation state target's only hope of escaping such detection – and perfect characterization – is to opt out of the Web altogether and deny themselves access to any/all information or communications derived therefrom.

That is hard to imagine given that Web addiction, albeit encrypted, is at the core of their very beings. They are perfect subjects for Axilent solutions.

Axilent's pioneering work is highly relevant to the ISS field for being among the best and purest applications of RTPA and profile modeling anywhere.

Axilent at Work

Axilent is software-as-a-service (SaaS). It is comprised of several baseline components: the Adaptive Content Engine (ACE); Axilent APIs designed per respective market focus; and the ACE Profiler with modeling and analytics. Here we will review what Axilent does in simplified "1, 2, 3" step-fashion. In actual operation, the system is a continuum with neither beginning nor end. New data is constantly fed in, processed and added to a profile, always with the possibility of altering outcomes.

ACE makes relevant matches with content and people to establish what the company calls "Real Life Relevancy," the ability to generate content based on proven interests of the individual, the same as would happen when a client meets with a prospect and listens carefully in order to understand exactly what makes the potential customer "tick" and apt to make a purchase. The process almost perfectly matches the approach of sound intelligence gathering. It is a two-way street paved with the prospect's (or target's) demonstrated interests and behaviors.

In Axilent's world, ACE monitors a target's interaction with the client's website, content management system or mobile app, capturing the individual's behavioral input. What types of content did he or she gravitate to, from where, at what time of day, and on what kind of device? Where did he enter and exit, what was the duration of the visit, what patterns emerged, and what is the "saliency" of all this data, so that the gathered intelligence may be leveraged to ensure that this target always sees what he wants, when, where and how he wants it?

Here ACE Profiler takes over. To begin with, the user chooses the appropriate API from Axilent's library for a given project. Then ACE Profiler goes to work.

The Profiler is configured to create "Profiling Models" of the types of individual targets to be put under examination. The plural of "model" is important. There may several or multiple types of individuals that could prove to be of interest. Axilent lets the user create as many as needed.

A Profiling Model is ground zero for identifying and tracking potential targets. Each model has a designated "Deployment Target" that integrates with the API for direction. The Deployment Target is set in auto mode to collect data in real-time on individuals that fit the model.

The killer app, however, is the Profiling Model's predictive capability. The user can set the system to score a Profiling Model as stable, moderate or volatile based on data input, for use in determining predictive outcomes, in other words, to establish the urgency of any response required.

Individuals categorized within each Profiling Model have their own unique profiles with full identity information: name, address, location, historic preferences, sentiment, recent actions, and so on. Each profile may also be subjected to analysis within the Profiler, for example, to correlate the individual target's personal data with events that contribute a detailed chronology of interactions. Further analytics within Profiler can tag the individual as being aligned with known behavioral segments in the Model, score the target by "value," and show all "outcomes."

Outcomes are significant as indicators of imminent actions. ACE Profiler actually has a "Predict" button that instantly pulls together all the data, then shows likely outcomes in order of priority, much like a Monte Carlo Simulation engine. Finally, the system recommends next best actions depending on the outcome.

It's classic RTPA. Well, almost. Actually it's better. Reason: Axilent knows the answer to the following question:

Q: What is the absolute worst type of RTPA solution?

A: The type that either permits human error – or goes to the opposite extreme and embraces AI as a form of "thinking."

Why Intelligent Automation is Central to Near Perfect Modeling

One of the key problems in current thinking on big data and its offspring RTPA is lack of a middle ground.

At one extreme a data scientist might develop analytics solutions that give the user too much latitude in choosing algorithms (whether he understands what he is doing or not), or in meddling with outcomes and next best actions, i.e, second-guessing the analytics and introducing human error.

One of the worst examples of human intervention with dreadful results is CIDNE, the intel system widely used by the U.S. Department of Defense. There were and still are many issues with CIDNE. The most onerous is the system's feature permitting analysts to intervene and discard data they consider irrelevant. As a direct result of this freedom to manually intervene, CIDNE analysts routinely step in and dismiss up to half the data collected. The outcome is rubbish to soldiers desperately in need of accurate intelligence on what lies ahead.

The other extreme, turning into a disciple of deep learning, neural networks, cognitive computing or other terms for unquestioning faith in artificial intelligence can also lead the user down the wrong path. Machines can "learn," in a manner of speaking, but they do not "think." As renowned machine learning expert Michael Jordan noted once, there is at present no corollary between the human brain and a computer and may not be for decades.

One common sense middle ground approach is Intelligent Automation: knowledge and application of the correct algorithms to use in a given exercise, and the strength of character to let the system do its work with minimal manual intervention.

Creators of outstanding RTPA solutions don't expect users to be data scientists themselves. The vendors do the work based on deep understanding of the algorithms and techniques requisite to success, while building-in flexibility that provides choices depending on the case. They may offer the user the ability to set the system on automatic or to manually intervene, but in most cases they recommend going "on auto" unless the user is a data scientist himself. Finally, when the system has proven results to show, they advise against re-interpreting outcomes or ad libbing best actions. The rule of thumb: Trust the system – it's proven.

A fact commonly overlooked in the ISS community is that solutions used for surveillance often have their mirror image in the enterprise marketplace. As often as not, the origins of today's "spyware" can be traced to commercial applications. It is true of deep packet inspection (DPI), Field Programmable Gate Arrays (FPGAs), Flow Monitoring (from Cisco) and real-time predictive analytics (RTPA). It will be true of Network Functions Virtualization, too, when NFV finds its way into the ISS fold.

Axilent is a pioneer of Intelligent Automation, and they continue to advance the field. Listen and learn.

ComWorth SwiftWing SIRIUS

If line rate packet capture is the pipe dream of every law enforcement and government agency, then Japan's ComWorth would appear to be making it real through SwiftWing SIRIUS, a comprehensive portfolio of fast packet capture solutions esteemed by communications service providers, data centers and government agencies alike. Most appealing about ComWorth, perhaps, is the versatility of its product line offering rack-mounted, portable and "mini portable" versions that can collect vast amounts of data "on the fly" anywhere.

Until fairly recently, ComWorth positioned itself as a one-trick pony that specialized in data capture and storage only, and left the job of analytics to others. The company has rectified the situation by introducing a Portable Analytical Intelligence System that complements its line rate capture products.

Inside ComWorth

Launched in Ota, Japan, in 1965, ComWorth – originally named Sankei Trading Co. – began as a manufacturer of diverse electrical equipment. In the early 1970s the company expanded to become a distributor of products by other companies, and it retained this dual-focus model as it turned to specialization in network equipment in 1993, then gradually to monitoring. The company was officially renamed ComWorth in 2001, and in 2005 launched the SwiftWing SIRIUS line rate packet capture suite, its current flagship brand. Expansion followed and ComWorth today maintains offices in Singapore and Kriftel, Germany near Frankfurt. In addition to its own SwiftWing hardware, ComWorth sells some 60 products by 20 different vendors including Keysight, Silicom Denmark, PacketStorm and XENA Networks. SwiftWing SIRIUS products are also re-sold by a network of vendors such as India's Comint.

ComWorth was been something of a lone wolf in the industry, for years insisting that its focus is fast packet capture and nothing else. SwiftWing SIRIUS's quals in this niche are respectable and impressive. The rack-mountable and portable units capture up to 100 Gbps of data to a hard disk, and provide access to stored data up to 1,500 terabytes. Each model is outfitted with SwiftWing's Smart Hardware Filter Engine, which the user can program to eliminate extraneous data such as broadcast video, and to focus on refined searches. Users can comb data by a variety of identifiers including source and destination IP addresses, ports, metadata, and view historical statistics on the data. Notwithstanding all these capabilities, which certainly resemble classic DPI, ComWorth until very recently emphasized that SwiftWing SIRIUS is not an analytics solution. The company wanted to be positioned as a pure play, line rate packet capture company. Unlike competitors that add analytics and other bells and whistles to differentiate their product, ComWorth wanted to stand out for doing one thing supremely well.

In marketing literature, ComWorth rhetorically asks why any customer would consider such an "incomplete" appliance that only captures data, particularly when competing products offer an all-in-one approach that includes monitoring, capture, and analysis.

In answer, ComWorth argues that "all-in-one" solutions that bundle analytics bog down when network traffic accelerates. They contend that the more complete the solution, the slower the rate at which it can capture data for analysis. They conclude that as networks become faster and more complex, such multi-purpose systems will inevitably drop packets. Therefore, in their view, it makes more sense for customers to look to vendors with complementary "shared competencies" that can form a more proficient hybrid: SwiftWing SIRIUS for fast packet capture and a host of others for analytics. And why not? SIRIUS appliances, after all, capture data on high-speed networks without a single packet lost.

The World Before Streaming Analytics

ComWorth's world view dominated in the general marketplace for years. The basic idea: Buy one system for collection and another for analytics – but try blending them and network monitoring would grind to a halt. Through mid-2014, at least, analytics was largely done on data at rest, as in SwiftWing SIRIUS storage. The idea of streaming data analytics still seemed futuristic, or as IBM analyst James Kobelius dubbed it, an "outlier in discussions about big data architectures." But change comes quickly in this marketplace. Just one year later Kobelius was writing about how streaming data *and* streaming analytics have become the new reality of this field.

The sudden advent of streaming analytics is a reminder of how, for years, data experts essentially divorced the concepts of high-speed data capture and real-time analysis. One particular revealing source is this 2010 post by WildPackets (now Savvius) titled "The Truth About Line-Rate in Network Analysis":

There is a common misconception lingering in the networking world. Even though several protocol analysis and troubleshooting solutions claim "line-rate" analysis, the actual network throughput that can be effectively analyzed varies greatly and is highly dependent on a number of factors. One of the most important factors is whether or not the analysis is expected to be real-time, or if all analysis will be performed post-capture. Real-time analysis is extremely demanding, so much so that "real-time" and "line-rate" should not even be used in the same sentence.

At the time, WildPackets would have been in total accord with ComWorth.

In fact, the WildPackets author could be describing ComWorth's line-rate packet capture suite:

The only condition under which line-rate can even be considered is when data is being collected for post-capture analysis, often referred to as forensics analysis. In this scenario all network packets are written directly to disk. The most capable network analysis and troubleshooting solutions available today have a data capture rate of somewhere in the range of 4 – 6 Gbps. Even though these solutions claim 10G "line-rate" captures on fully utilized, half-duplex 10G links, they begin to lose packets if pushed beyond 4 – 6 Gbps, obviously far short of a fully utilized rate of 10Gbps. A solution that could capture at that high rate and not drop any packets would be beyond the state-of-the-art! The fact is that today's networks are actually faster than the available network analysis and troubleshooting solutions, resulting in greatly diminished network visibility.

True enough in the "olden times" of data collection and analytics, but no longer.

ComWorth's Dive into Analytics

After many years as a pure play packet capture hardware appliance vendor, ComWorth ventured into analytics, and with a law enforcement focus, another first for the company. As it often does, the company is talking a big game with its "Lawful Interception" product, described as a "total solution, designed to intercept mass data, capture, analyze and extract data to find actionable information." We don't see very much new in this beyond a spin on the company's established capabilities.

ComWorth has always specialized in streaming packet capture. Now they've simply added streaming analytics to the mix. Considering the company's established record in data storage, ComWorth also touts the solution's compliance with the European Union's data retention directive.

ComWorth has a long record of serving government and military agencies in Japan and throughout the AsiaPac region. By pegging its Lawful Interception and Analytics solution as an add-on to SwiftWing SIRIUS solutions already embedded with existing customers, ComWorth has seen steady growth with LEA and government agencies.

Fifth Dimension Crime Prevention

As pre-crime solutions go, Fifth Dimension is a major step above the usual run-of-the-mill malware, lawful intercept, packet capture and analytics ware. Its key differentiator: deep learning.

Fifth Dimension has documented case studies of preventing planned incidents of crime and terrorism. As a result, the Israeli company was among the first ISS vendors called upon by former French President Francois Hollande's government following the horrific 2016 terrorist attack in Nice. In retrospect it is not clear whether putting France in a long term state of emergency led to the decline in Jihadist terrorist attacks, or other factors were in play.

Fast Out of the Gate

Fifth Dimension is one of two companies co-founded by CEO Guy Caspi that serve the ISS and cybersecurity fields. Remarkably, they were launched within a year of one another. Fifth Dimension, which opened its doors in 2014, caters primarily to law enforcement, intelligence agency and military clientele. The second firm, Deep Instinct, focuses strictly on cyber security. The companies share distinct characteristics. Both apply deep learning, a branch of AI, in ways that radically accelerate the processing of intelligence. Neither stops at identification of a threat. The mission of both is to preempt the threat.

By virtue of its more public nature, Deep Instinct is the far better known of the two Caspi ventures. You can Google any number of videos on Youtube that showcase Mr. Caspi holding forth on deep learning and why it differs so profoundly in its approach to cyber threats. What he says about deep learning is applicable to Fifth Dimension, as well. To put it succinctly, deep learning makes conventional real-time predictive analytics look as dated as something out of a 1920s silent movie.

Inside Fifth Dimension

Fifth Dimension briefly made headlines in 2013 when the company named a new chairman: retired Lt. General Benny Gantz, former head of the Israeli Defense Forces (IDF). The fact that a young venture such as Fifth Dimension could attract such a renowned military figure to its top position speaks volumes about the importance of its work.

Guy Caspi, principal founder, is the company's public face. With advanced degrees in mathematics and machine learning from leading Israeli and U.S.

universities, including Harvard, and some 15 years with an elite intelligence unit of the Israeli Armed Forces as well as government intelligence, Caspi brings both academic prowess and direct experience to the job. Rounding out his military/government resume with commercial experience, he was an executive with the former Comverse (now Verint) before leaving to found Tamares Telecom, a network infrastructure company specializing in undersea cable deployment. Other members of the executive staff and operations team reflect similar backgrounds – a balance of military and ISS experience with an intense focus on advanced mathematical applications.

When Caspi set out to found Fifth Dimension, big data and real-time predictive analytics were all the rage in the IT and communications arena. Palantir was making waves with visualization. However, AI and machine learning remained something of a stepchild in the intelligence and ISS arenas. As a mathematician, Caspi saw this as an anomaly. Google, arguably among the greatest companies on the planet, had scored impressive gains with machine learning, most notably with its deep learning projects. Why not apply deep learning in high profile areas where they are desperately needed, national security and cyber security?

What Sets the Deep Learning Approach Apart

Fifth Dimension claims to be the first company to successfully apply deep learning in the intelligence sphere. The vehicle for this accomplishment is a platform that automatically learns.

Fifth Dimension gets a little carried away by describing the platform as "tens of layers of artificial neurons connected to hundreds of millions of synapses," evoking physical imagery in a purely virtual environment. Putting such poetic license aside, the platform is a suite of algorithms that are accelerated with the use of graphic processing units. GPU offloading is a common practice in computing complex scientific, engineering and mathematical applications, leading to performance gains of 20 -30 percent in tested benchmarks. GPUs play a major role in the world of Guy Caspi.

In an interview on Deep Instinct's cyber security solutions, Caspi notes, "What makes us unique is that it is very difficult to implant deep learning software on GPU." The end result is a cyber security solution that is 99 percent effective at threat detection and prevention – and according to Caspi, 100 times faster than competitors' products. Fifth Dimension is its twin on the ISS side. As described by the company:

The [Fifth Dimension] platform constantly accumulates data, analyzes masses of entity activities, cross-correlates insights and identifies new links. The platform's "brain" performs mapping of entity patterns, anomaly detection and risk ranking. As a result, the system provides the human analyst with prioritized, meaningful intelligence on time.

Speed makes a difference in both spheres. As Caspi has observed of the cyber arena, the sheer mass of new Zero Day exploits and advanced persistent threats has overwhelmed the old strategies for combating breaches. Systems need to think and act in real-time. Furthermore, they need to be distributed at every level, including users' laptops and mobile devices. With deep learning embedded, every part of the corpus is the brain – one that never stops thinking, learning, or getting smarter.

When Caspi and crew were casting about for a company name in 2013-2014, they settled on "Fifth Dimension" for a reason. The termed was originally coined a generation ago to describe the battlefield where the wars of the future would be fought – the Internet. While the term "Fifth Domain" has long since replaced Caspi's naming brainchild, at least in U.S. military quarters, it remains as relevant to intelligence, military and law enforcement agencies as ever. Customers who opt for the company's deep learning approach will be well armed with a solution that immediately raises IQs on intelligence gathering, decision-making and preemption.

IBM i2 Safer Planet

Since the turn of the century, IBM has invested billions of dollars on nearly 200 acquisitions. Among the more pertinent for surveillance purposes are its buyouts of SPSS, Netezza, i2 and CyberTap, undertaken with the goal of making the company the global leader in network analytics. Together with the company's array of in-house platforms developed by a team of 8,000 software engineers, the end result today is the IBM i2 Safer Planet Intelligence Analysis suite. At first blush, IBM's portfolio seems a formidable array of new analytics capabilities for enterprises, government agencies and law enforcement. But a closer look reveals that IBM's product suite is showing its age to the point of losing interest among government, law enforcement and military customers.

The hidden story behind IBM's acquisition binge is low investment in R & D: US $5.99 billion or 7.8 percent of 2019's $77 billion revenue as the company continues its downward spiral from $99.9 billion in 1999. Tech sector peers such as Amazon invest more than triple that amount in R&D every year – US $23 billion per annum, 13 percent of revenue. The difference underscores the rationale behind IBM's ongoing acquisition spree. Contrary to public perception, the company relies on acquiring innovation because it contributes little to technology advancement on its own.

To be sure, acquired products such as SPSS and Netezza have won broad acceptance in the marketplace and sit at the heart of many commercially available real time predictive analytics solutions. Even so, given the longevity of some IBM products, both acquired and developed in-house, it is reasonable to ask:

- Are Safer Planet products best-in-class, or have they survived merely by virtue of carrying the IBM brand?
- Are "advances" touted for IBM i2 Safer Planet Intelligence Analysis simply a cosmetic makeover of old products?

The world at large – including the analyst community – has accepted the "Safer Planet" marketing spiel on faith without asking any of the tough questions.

Framework or Patchwork?

IBM's buying spree for analytics and visualization systems began in 2005 but took off in earnest in late 2009 when the company laid out US $1.2 billion for Chicago-based SPSS Inc., a provider of predictive analytics solutions used, among other things, for crime prevention, i.e., detecting patterns that could be leveraged to foresee and prevent crimes, all in "real time." The following year, IBM purchased Netezza Corporation, maker of a data warehousing appliance that mines big data, performs real-time analytics, and integrates with SPSS for predictive modeling.

Then in 2011 IBM bought the UK's i2, a specialist in systems for law enforcement. With i2, IBM gained two key properties: (1) COPLINK, a database platform that lets law enforcement search for and share criminal data; and (2) Analyst's Notebook, which provides visualization of the data so that LEAs can see patterns and links. Along the way, IBM also picked up two other companies, Guardium and Initiate.

The acquisitions seemed complementary: Netezza for creation, storage and mining of big data, and SPSS for predictive modeling, both working in real time. Then Guardium for secure encryption of Hadoop or Netezza, and Initiate for electronic records matching, integration and sharing. And finally, a specialized data platform, i2 COPLINK, just for LEAs, and the means to visualize the data via Analyst's Notebook.

But problems arose when IBM added its own products to the mix in the company's 2012 surveillance product portfolio, dubbed "Threat Prediction and Prevention." The variety of systems for mining, analyzing and visualizing what IBM called an "ocean of data" involved a veritable tidal wave of products itself, some under the InfoSphere brand, others carrying the name of acquired products, and at least one pulled in from IBM's Watson family of AI solutions. Listing these diverse solutions by functional category illustrates the unintended duplication that occurred:

Real Time & Predictive Analytics
- IBM SPSS: predictive analytics.
- IBM ILOG: decision management (predictive analytics).
- IBM Netezza: real-time analytics.
- InfoSphere Streams: real-time analytics for data in motion.

Database Management
- IBM Netezza: data management.
- InfoSphere BigInsights: Hadoop-based data management.
- IBM Initiate: shared digital records management.
- IBM DB2 Anonymous Resolution: database management software with controls authorizing access to confidential data.

Visualization
- IBM i2: data visualization.
- IBM Cognos: data visualization.
- IBM ILOG: data visualization.

Other Solutions
- IBM Watson Content Analytics: semantic analysis of content for "actionable intelligence" based on intent/sentiment of unstructured data. ["cognitive computing," or artificial intelligence.]
- InfoSphere Identity Insight: for determining the target's hidden ID and links, with alerts for imminent threats.
- InfoSphere Global Name Recognition: "sophisticated" multi-lingual name matching by written, oral and cultural data.
- InfoSphere Guardium: monitoring and encryption security for Hadoop and Hadoop data clusters.

In all fairness, there were slight differences between some similar looking products that justified their inclusion in the "detection/prevention" framework. For example, depending on their requirements, one client might prefer Netezza data management and another the Hadoop-driven version offered in InfoSphere BigInsights.

However, on another front ILOG's visualization was clearly redundant, leading IBM to spin off that piece of ILOG in 2014. IBM Cognos visualization is alive and well today, but primarily in the enterprise space where it began. InfoSphere Identity Insight remains an important tool for investigators, but InfoSphere Global Name Recognition, more useful to enterprises and LEAs, has fallen off the list. Add up the hits and misses and the 2012 "threat detection and prevention" platform was as much patchwork as framework and remains so to this day.

Inside IBM i2 Safer Planet

With IBM i2 Safer Planet, IBM has clearly made an effort to bolt down loose ends, combining and integrating the capabilities into a "new and improved" four-part portfolio: IBM i2 Analyst's Notebook; IBM i2 Analysis; IBM i2 Enterprise Insight Analysis; and i2 COPLINK. However, upon close examination it becomes apparent that the company has never bothered to rationalize or upgrade the suite by jettisoning duplicative and outdated applications.

i2 Analyst's Notebook is visualization software. The product is now in its fourth decade. The new Safer Planet version lets the user move through big data to find the pieces relevant to an investigation – target ID and networks of affiliates, patterns, timelines and imminent threats – from SIGINT, COMINT and OSINT sources, then cap it off with visualizations of the data. IBM notes that the correlating data from these sources can be accomplished in one click versus the six or seven clicks required in prior versions, resulting in a 60 percent improvement in speed. Much of this technology is inherited from the original 2012 threat detection prevention portfolio, thus not really new, at all. The key improvement is the click reduction, indicative of improved integration between legacy products.

i2 Analysis is touted as an "intelligence analysis environment" that integrates with the user's existing infrastructure, culls data from disparate sources, establishes a hierarchy of authorization for access, and lets analysts collaborate on findings and create visualizations. In a word, it's Netezza with DB2 Anonymous Resolution, Initiate and Analyst's Notebook visualization.

i2 Enterprise Insight Analysis offers the identical capabilities in two versions: "Core" for smaller data sets and "Advanced" for big data. This enterprise version comes with a pair of added features: 3-D data and geospatial coordination of targets.

In assessing a product it is always important to consider customer experience. One noteworthy user of Analyst's Notebook stands out: the U.S. Army. In 2009, two years before IBM acquired it, i2 won a US $9.6 million contract to integrate Analyst's Network with the Army's beleaguered DCGS-A, a vast intelligence system meant to deliver real time tactical insights on enemy strength, position and other data to U.S. ground troops. Although most of the kinks have long since been ironed out, for years DCGS-A was condemned as a complete boondoggle by troopers, analysts and the U.S. Congress.

COPLINK, the last product inherited from IBM's acquisition of i2, continues to provide a serviceable shared database of criminal files and records for LEAs, with photo matching capabilities to help identify potential targets for virtual lineups, and pre-crime capabilities that point to likely trends and timelines of criminal activities.

Neo4j and Linkurious Visualization

More than a decade after the official dawn of big data analytics, there is general consensus that raw data findings alone are but the starting point for the true end game of analytics: visualization that simplifies understanding of trends, coming events and perhaps most importantly – relationships. In the ISS sphere, companies such as Palantir and IBM have made a mint providing visualization software that graphically depicts key findings of data analytics. Good as they are, both are dwarfed by the world's most popular graph database solution Neo4j, developed by Neo Technology in 2007 and today used by leading enterprises including eBay, Walmart and Pitney Bowes to understand and meet customer needs. Neo4j, like Palantir and IBM i2, visualizes data faster, more accurately and economically than conventional database tools such as SQL and NoSQL. At ISS World events, Neo4j consultant GraphAware has demonstrated how the Neo4j database may be used to "untangle the criminal web." GraphAware makes tools that speed the process. However, Neo4j is still complex. Using it is a job for data scientists. While large contractors have the math wizards on hand to serve government clients, it's far less less likely to find such lofty types at the typical law enforcement agency.

For that reason, startups such as Linkurious have emerged with tools specifically designed to make graph database searching and queries simple to use. Here we'll begin with an in-depth look at the value of Neo4j, then consider how Linkurious steps in to make visualization easy.

Neo4j: Accelerating Contextual Intelligence

Neo4j provides a fresh approach to database management systems (DBMS) and relational database management systems (RDMBS). Conventional DBMS and RDMBS present and store data in language and numerics. The end result of queries and analytics: numbers and text, for example, on types and numbers of cyber or terrorist threats presented in columns and tables.

The graph database has a different and expanded mission: to show the relationships between data findings. To do so, the graph database uses "graph structures" – lines, edges and arcs – to represent and store data for semantic queries, i.e., queries that reveal context and associations. Each image represents specific items of data stored as a graph, not as text and not numerically, but as "nodes" (circles) connected by "edges" (lines).

The relationships in a graph database are inherently linked together – nodes to edges to other nodes, etc.

As a result, semantic queries on a specific topic can instantly produce a hierarchical structure of graphs in a single image. Color-coded for different features, the graphs reveal not just bare data but *the relationships* between what might otherwise seem disparate findings.

The graph database is light-years removed from the traditional relational database. In a SQL database, for example, the user must tap the "join" function, which then works through the process of combining columns and tables based on common values in each. It is a slow and cumbersome process. In contrast, the data in a graph database is already linked in a "nodes and edges" diagram, and is technically more streamlined.

In a commercial setting, a graph database might show a retail store how female shoppers, age 25 to 30, prefer specific merchandise, have a predilection for complementary purchases on or around set days and times, are more likely to spend when the price is at or below a set level, or take advantage of loyalty discounts. Neo4j might also reveal friends in common, skills in common, whether targets live close or work together, and the degree of personal closeness or separation.

In the ISS arena, the graph database might be used to show the relationships between criminal or terrorist targets of a certain age, education, livelihood, social media preference, communications with others, and similarity in profile to individuals not previously known to law enforcement, but definitely linked.

Raising the Bar on Data Value

The business case around Neo4j is "building sustainable competitive advantage" via the ability to fathom, leverage and profit from customer relationships. In other words, Neo4j boosts the value of data by pulling more insights from it than is possible with conventional RDBM systems. Because the graph database is faster and more economical, costs drop, as well. For all these reasons, to date some 25 percent of enterprises have migrated to graph databases.

Users can get a boost from GraphAware, which provides consultancy and six different software products that can help ease the way into Neo4j. In addition, the company provides training in Cypher, Neo4j's query language, and how to use Neo4j with PHP, the common scripting language for HTML. However, there are simpler solutions.

France's Linkurious

As graph databases began to gain momentum, Paris-based Jean Villedieu saw an opportunity. While the market was flocking to Neo4j, many enthusiasts found they were hitting a wall.

They needed a PhD in Data Science to enter the world of relationship visualization made possible by Neo Technologies and GraphAware.

Villedieu described the problem in vivid terms:

"The problem with graph databases is you don't have tools that make it easy for end users to extract information. In relational databases there are tools like Tableau and Qlikview, and all the business intelligence solutions that do a great job of making sense of relational data. But there's nothing for graph data. You only have tools that are designed for scientists."

With four partners, Villedieu in 2013 created a new venture to solve this problem – Linkurious. With the company's eponymous software solution, users can access visualizations from a graph database such as Neo4j simply by keyword search. Results are displayed in "nodes and edges," the same as a developer or data scientist would see after applying their skills at Neo4j directly or with help from GraphAware.

Linkurious is "Everyman's" entry point to brilliant, insightful visuals that shed new light on investigations and intelligence activities. Linkurious works with popular graph databases including Titan, Datastax Enterprise Graph, Franz Allegrograph and Neo4J. No data scientists or developers required.

MemSQL: Escape from Hadoop Batch Mode

In the Fall of 2014, startup MemSQL was selected for funding by In-Q-Tel, the CIA's venture capital fund set up for the benefit of the U.S. Intelligence Community. MemSQL, often lumped with analytics companies but best described as a "fast data" player, had already accumulated some $45 million in VC funding, so money alone isn't what made news. What did: the fact that a deal with In-Q-Tel is the ISS world's equivalent of an Oscar or Pulitzer. A check from IQT is the precursor of bigger things to come.

Of course, the winner must agree to give the CIA first rights to the technology, which may mean modifying a company's initial vision and direction, but for most fundees, making such changes is easy work. It's a relationship that often leads to healthy buyouts on an accelerated schedule – or to industry status as a giant in the company's field. As evidence consider Palantir, one of the early beneficiaries of In-Q-Tel funding. Palantir, although it began as a company strictly focused on intelligence, today has expanded its market to include multiple government agencies and industries. Credit for this success story goes to In-Q-Tel.

Genesis of In-Q-Tel

In 1998, amid that era's tech boom, science and technology thought leadership within the Intelligence Community came to the realization that Silicon Valley far surpassed in-house government initiatives in fueling innovation. Dr. Ruth David, a former CIA Deputy Director for Science and Technology, and her Deputy, Joanne Isham, conceived the idea of partnerships with the tech industry. In a way, this thinking followed the model of Department of Defense relationships with contractors such as Raytheon, Lockheed Martin, and Northrop Grumman. However, the CIA decided on a very different approach: The CIA would focus on startups through a new government entity that served the role of venture capitalist and incubator.

In February 1999, this new enterprise was formed as a non-profit corporation, and tasked with finding brilliant companies whose work could be brought to bear on enhancing intelligence in the service of U.S. national security. One year later the group was named In-Q-Tel, ostensibly after "Q" of James Bond fame. Since that time, In-Q-Tel has funded more than 300 startups, a number of which have subsequently been acquired by tech heavyweights such as Google and Oracle.

Like many of the startups it backs, In-Q-Tel has experienced rapid growth and expansion of its scope. What began as a program benefiting the

CIA now supports other members of the U.S. Intelligence Community such as the NSA and National Geospatial Intelligence Agency (NGA). In-Q-Tel has also widened its geographic focus beyond Silicon Valley to include tech centers such as Boston. There is a definite "trickle down" benefit to mass markets as technology advances funded for intelligence applications spill over into commercially popular innovations. Touch screens and Google Maps, for example, began in companies backed by In-Q-Tel.

Today In-Q-Tel gravitates toward startups in information/communications and physical/biological technologies. The former, which spans companies that excel in advanced analytic tools, communications/computing platforms and geospatial intelligence, was a clear fit for MemSQL.

Where RAM Supercedes Disc

Analytics vendors provide systems in four principle fields: Diagnostic (heuristic), Descriptive (real-time), Predictive and Prescriptive (rules and recommendations). MemSQL differs from the herd by attacking issues that affect the performance of each type of system: volume and velocity – the tremendous flow of data at ever-increasing speeds. MemSQL's driver is the understanding that to be of value, analytics must be instantly accessible, based on the ability to extract data. MemSQL is able to achieve this capability in three ways: (1) by ramping up the speed of big data lakes such as Hadoop; or (2) by working with Apache Spark on top of Hadoop; or (3) by replacing Hadoop altogether.

MemSQL co-founder and CEO Eric Frankiel describes the foundation of MemSQL as a decision to "scale out" big data databases as opposed to doing it the traditional way of "scaling up." Scaling out permits a user to add multiple commodity machines for storage and processing in tandem (distributed computing), thus scaling "out" to handle volume and velocity.

Typically, big data systems have huge amounts of data (online transaction processing or OLTP) flowing in and creating a bottleneck in the database. As data volume increases, the ability to scale the database becomes an issue. At the same time, more online analytical processing (OLAP), which permits the user to easily extract and analyze data, is coming out of the system. MemSQL begins by eliminating the bottleneck via additional databases, allowing more OLTP to flow in, and more OLAP to flow out for extraction and analysis.

MemSQL's design is based on a simple, two-tiered approach: aggregators and leaf node tiers. Leaf nodes are responsible for storage and computing. The aggregators hold systems intelligence and are responsible for metadata storage.

Because the MemSQL system uses a "share nothing" architecture, the leaf nodes holding the data are not connected or aware of each other. During a transaction, an application connects to a single aggregator, which in turn accesses metadata relevant to the user inquiry. Applications see only one database connection, meaning no time is lost dealing with data location or movement, typical "time eaters" within distributed computing systems.

The core differentiator of MemSQL is its approach to storage.

Most database systems use the classic memory pyramid of RAM, Solid State Drive (SSD) and hard disc, with the bulk of storage handled at the hard disc level. What sets MemSQL apart is an in-memory approach that dramatically expands the use of RAM deeper into the memory pyramid to store petabytes upon petabytes of data. As Frankiel states, MemSQL has made "RAM the new disc," in the process energizing data warehouses, powering operational analytics in real-time, and accelerating transactional applications by eliminating the database bottleneck.

RAM is more expensive than disc. But in MemSQL's view, the trade-off for vastly improved performance is well worth the expense. As data storage costs decline, RAM costs will, as well.

To those who understand big data, MemSQL's platform is a huge step up from Hadoop, which operates in batch mode and is known all too well for its latency. That is a critical shortcoming. Neither intelligence agencies nor major commercial enterprises can tolerate delay. At Citibank, for example, a 100 millisecond of delay in a major financial transaction can drive a loss of US $1.0 million.

Stepping Up to the IC's Metadata Challenges

In addition to In-Q-Tel, MemSQL is backed by Accel Partners, Khosla Ventures, First Round Capital and Data Collective. The company is making, not burning dollars. MemSQL has won significant business from high profile accounts such as Comcast, Zynga, Ziff Davis and "hundreds of other companies," according to company sources. MemSQL has kept pace with the market with releases such as its MemSQL Loader, which enables the user to transfer data from Hadoop to MemSQL's proprietary in-memory SQL database for faster analytics performance. Today, the company's approach to database management is touted as the world's fastest.

NORSI-TRANS Vitok-3X, Vitok-Cluster

Russia's NORSI-TRANS is remarkable among ISS vendors for the depth of its product line – some 40 solutions spanning analytics, standalone high-speed packet monitoring, lawful intercept (for fixed line, mobile and IP), mass data collection/retention, traffic filtering for 10G/40G/100G networks, Wi-Fi interception, and even network management.

This is the full package, whether the client is a law enforcement or government agency, and not just for Russia and the CIS states. NORSI-TRANS solutions are, of course, SORM compliant. (SORM, translated to English, stands for "System for Operative Investigative Activities" – Russia's technical specifications requirements for access by all law enforcement and intelligence agencies to networks of every variety.) As of 2015, the company is also a member of the European Telecommunications Standards Institute (ETSI), playing an active role in the standards developed for EU communications service providers and companies in other nations that adhere to ETSI.

Notwithstanding skepticism from those who question the wisdom of trusting surveillance solutions from a country at odds with the West, NORSI-TRANS has earned its place at ETSI and at ISS World. The company's solutions should be taken seriously and respected as important contributions to the field of evidence and intelligence gathering. A client may purchase and use NORSI-TRANS products in full confidence of their compliance with ETSI standards.

The problem with the NORSI-TRANS suite: It remains confined to the world of "boxes." The company doesn't appear to have committed to the virtualization trend that is rapidly overtaking networks and all apps that reside therein, including lawful intercept and other ISS systems.

One such box is the Vitok-3-X, promoted by NORSI-TRANS as a "universal" platform for advanced big data analytics. While that claim may have been an apt description during the early glory days of big data, does it still hold true?

Inside the Vitok-3X

The NORSI-TRANS Vitok-3X is in many ways a classic big data analytics solution, leveraging parallel processing to sift through massive amounts of data, then filter and score findings. The solution acts identically to other big data systems, collecting both structured and unstructured data, storing it in a data lake for initial processing, then scoring it for specific criteria that uncover intelligence. It also provides visualization that makes it easy to understand trends, see individual targets and their networks, and glean "actionable intelligence" that informs "next best actions."

The Vitok-3X's metier is metadata such as CSP billing records that provide indications on specific communications by and between targets, as well as their geolocation in each instance. Users may also modify the engine to receive data from sources that cover financial transactions, credit, insurance and criminal records.

Not only structured databases such as Oracle and SQL, but also unstructured data such as video, text, images and graphics are readily handled.

Like all NORSI-TRANS solutions, the Vitok-3X may be integrated with other NORSI-TRANS modules such as lawful intercept solutions that collect not only metadata, but specific detailed content, as well.

At the tail-end, NORSI-TRANS Vitok-3X can translate findings from conventional table format into 2D or 3D graphics that show temporary, circular, network and hierarchical relationships among targets.

Otherwise, the solution is a bit old school. Not to imply that there is anything faulty with the NORSI-TRANS Vitok-3X, but rather that the industry has moved on and big data analytics per se no longer carries the same "buzz" as it did in years past. Some of the evidence: Use of Hadoop, the famous engine created by the Apache Software Foundation for collecting and storing big data, declined significantly in 2016, as did MapReduce, which became less relevant to data scientists than newer tools such as Spark. Analytics never stands still.

Caught in a "Box"

The emphasis today must be not only on capturing but analyzing data in motion, developing tool kits that learn from the data, creating algorithms that assist forecasting, extending search beyond the surface Web to hidden sources – and beyond human interactions to the world of interconnected devices. This vision of "today" is already happening. Is NORSI-TRANS there yet? Not yet, at least not where the Vitok-3X is concerned.

What we find missing:
- **Predictive Analytics.** As stated earlier, heuristic data analytics that reveals trends is of highest value to law enforcement and intelligence agencies. But there are occasions when they need the ability to move beyond heuristics to predictive analytics solutions that provide guidance on where and how targets are most likely to strike next.
- **Machine Learning and Cognitive Computing.** Artificial Intelligence (AI) and its related fields of Machine Learning and Cognitive

Computing are transformative technologies, and will soon be ubiquitously integrated into network software. Following suit with ISS solutions is a must, not an option. As with the cloud, the drivers of AI will be higher performance and greater cost efficiency. Using AI "in the wild" of investigations will enable systems to improve performance on their own.

- **Dark Web and Deep Web Search Capabilities.** Conventional data analytics confined to the surface web may be serviceable for some commercial purposes, but sophisticated criminals and terrorists don't operate in that realm to any great extent, at least not when they are conducting business. Fathoming the Dark web requires special search engines designed for the job.
- **The Internet of Things.** To the Intelligence Community, certain aspects of the IoT are ripe for surveillance – a fact that is giving privacy advocates and even some CSP executives sleepless nights. Conceivably, the Vitok-X3 has the flexibility to adapt to leveraging the innate security vulnerabilities of IoT devices, but to date NORSI-TRANS has remained silent on the topic.

NORSI-TRANS Vitok-CLUSTER: NoSQL Data Retention

When it debuted in 2017 at an ISS World Europe conference, the NORSI-TRANS Vitok-CLUSTER left many wondering. Promoted by the Russian vendor as a NoSQL database that provides real-time access to streaming information, with Lambda architecture to facilitate batch processing, clustered servers for multiple access *plus* visualization, the Vitok-CLUSTER seemed too good to be true.

NORSI-TRANS Vitok-CLUSTER is, in fact, all it claims to be: a data retention solution that downloads and stores in NoSQL, provides real-time access to and visualizes intelligence gleaned from data. As with any big data solution, "real-time" is a moving target and a matter of compromise over what is retained and analyzed, and what is jettisoned. The NORSI-TRANS Vitok-CLUSTER solution is the real thing, not just hype.

Vitok-CLUSTER as a NoSQL Database

First, a re-cap. NoSQL means "Not Only SQL" and refers to a "newer" type of database that emerged at the beginning of the Web's heyday to accommodate the wide variety of unstructured data – text, images, video – that did not fit into the simple categories used for conventional SQL or "structured query language" relational databases.

SQL, a relational database, was and still is loved by financial institutions for the simplicity of data categorization it provides, e.g., customer name & account number. However, when marketing, government agency and other types began to hammer SQL databases with sophisticated queries, the end result was a massive gum-up: millions upon millions of queries that didn't fit the SQL format and created huge delays in accessing data, sometimes bringing entire database systems to a standstill.

The initial "cure" for this problem was to install more servers or to ramp up to mainframes, a solution that didn't last long once chief financial officers saw the subsequent spike in hardware spending. Plus, CTOs disliked the notion of analyst types meddling with operational data, so they decided to separate their data from analytics – in data warehouses.

But the data held therein quickly grew in volumes never anticipated, and in varieties that defied the relational database formatting of SQL. Users plying a SQL system with sophisticated queries were often stymied by "locks" designed to safeguard SQL data. The locks were in place for good reason. One unforeseen consequence of having SQL slammed by analyst queries: the queries required changes in the data, a thing SQL was not designed to tolerate. As users continued to pile on with queries, they created a traffic jam, leading to 'SYSTEM NOT AVAILABLE" pop-ups that fueled frustration.

The needed solution was a database that was "not only SQL," that could handle columnar loads of structured data as SQL does, but do much more: use key values to inspect large volumes of unstructured data, or process streaming data at very low latency – or both functions. Ideally, such a system should "cluster" servers, enabling multiple user to access the same data simultaneously. It should also have the ability to simplify the significance of the data by querying it in a visual database such as NEO4j. That describes NORSI-TRANS Vitok-CLUSTER in spades.

As a NoSQL Data Retention solution, Vitok-CLUSTER goes the extra mile. It can accept structured and unstructured data from any interception device, and upload the content/metadata at speeds ranging from 1Gbps to 100Gbps depending on the type of network connection.

The solution name itself underscores an important feature of this product. "Clustering" capability enables multiple servers to interact as a single database, or to connect the users to various instances in the memory of different servers holding the files storing data. Clustering automatically load balances, so that users are connected to a server with the lightest load for quick results. This feature also provides a "failsafe" in the event one or more servers go down. Data is still accessible.

Vitok-CLUSTER also offers the option of either "shared-disk" or database sharing architecture. In the former instance, data is centralized and shared via different servers, while with database sharing no single server is fully independent of the master. While using one "virtual server" to manage data across distributed data caching is technically feasible, the Vitok-CLUSTER isn't there yet, for security reasons. Considering the cost advantage of virtualization, we foresee this capability as inevitable.

One intriguing and attractive feature of the Vitok-CLUSTER is its use of Lambda architecture, which applies both stream and batch processing to the capture and storage of big data. The two methods work together. When the user enters a query, it accesses and processes the data simultaneously via batch and stream processing. Analysis of each data capsule is combined to provide a complete answer. By sharing the load two ways, Lambda architecture bids to balance latency and throughput. The input data is retained unchanged. One downside: Critics rightly assert that maintaining code for the same data in "two complex distributed systems" can be a painful process.

One plus of Vitok-CLUSTER is the ability to store data in a graph database, a crucial and much-desired capability. Vitok-CLUSTER graphics aren't quite at Palantir level, but sufficient for most uses. Final point: The interested customer can use Vitok-CLUSTER anywhere ETSI standards are accepted. NORSI-TRANS is not confined to SORM compliance.

Ockham Solutions: Mobile Forensics

Launched in 2005 by a former law officer and his IT partner, Ockham Solutions serves some 4,000 law enforcement clients in France and has expanded operations to other parts of Europe including the United Kingdom and The Netherlands. The company's Mercure4 solution applies Ockham's analytics tools to financial and criminal records, GEOINT, HUMINT, OSINT or SIGINT, as well as data gleaned from forensic examination of target devices. Highly versatile, Mercure4 works with both structured and unstructured data, and with any type of communications including voice, text, IP or social media.

Mercure4 is COTS (commercial off-the-shelf software) compatible with any computer system. It can be used in remote mode on a laptop or within a networked environment by multiple users. Being a COTS product makes Mercure4 simple, reliable and easy to build upon as technology evolves. It requires no special or costly systems integration. Once downloaded, the software is ready to use. Flexible Mercure4 also may be integrated with complementary investigative tools such as IBM i2 Safer Planet for link analysis and visualization.

Given the current emphasis on preventing acts of terrorism and crime – particularly in the tense emergency environment of Western Europe – it must be stated up front that Mercure4 is not a pre-crime solution. Rather, Mercure4 helps law enforcement pull together the bits of information, evidence and relationships from multiple sources, leading to arrests and convictions after the fact.

It is telling that the company takes its name from medieval philosopher, William of Ockham, best known for "Occam's Razor," a principle of logic summed up as *Entia non sunt multiplicanda praeter necessitatem* (Translation: "More things should not be used than are necessary.") Typically the "razor" principle is interpreted as meaning that the simplest explanation of a problem is the correct one. Ockham Solutions follows the same common sense approach, applying advanced technologies to complex challenges with the goal of finding the most logical answer. And they work overtime to ensure that its solution is user-friendly, or as they put it:

"Our priority has always been to provide tools accessible to the widest audience while meeting concrete field operational needs."

Mercure4, while sophisticated, is meant for use by law officers, not tech geeks.

Inside Mercure4

The principal focus of Mercure4 is analysis of telecommunications traffic, and specifically wireless. When the company launched in the mid-2000s, wireline TDM (time division multiplexing) dominated voice telephony and wireless was still largely a backup accounting for roughly 16 percent of network traffic. Introduction of the first iPhone in January 2007 turned the tide in favor of wireless. In 2020 there are 5.8 billion wireless subscribers in the world, according to GSMA.

Recognizing that crimes and acts of terrorism are often preceded or accompanied by phone activity, Ockham Solutions' Mercure4 makes it possible to input and analyze millions of records in a system that can help identify individuals in relation to the timing and location of an event. The information resources might include: call data records (CDRs) from the network operator; logs of mobile base transceiver stations (BTS) responsible for handling traffic and signaling between mobile phones and network switching subsystems; cell site references and locations; and user identities.

One of the chief challenges of handling such data is the wide diversity of formatting found. Each mobile operator might have its own system. To get around this issue Ockham Solutions has developed its own standard data import system using a logic engine that can "tune" to and automatically sort and input data into a single uniform system regardless of the number of mobile operators. Mercure4 also is programmed to recognize international dialing prefixes to ensure accurate input and analysis of phone data from different countries.

When law enforcement takes possession of mobile devices used by a criminal or terrorist, Mercure4 can also input data obtained from forensic extractions of memory and SIM card data such as texting, phone directories and call logs. Popular mobile forensics products driving data to Mercure4 include MSAB (Micro Systemation of Sweden) product XRY, and Cellebrite UFED Touch.

Built-in queries vastly simplify the job of analysis. Law enforcement officers can search for calls between two numbers and common numbers in phone bills, detect BTS relay stations common to itemized bills for surveillance/reconnaissance, look for chains of calls in specific zones, or search by IMSI and IMEI numbers of devices. Mercure4 lets the user build priority lists of numbers for further analysis or comparison, search by date relevant to specific events, and match one or more targets geographically to an event. Synchronized behavior of two or more individuals involved in an action becomes quickly apparent, as do "entourages" or networks of perpetrators.

Mercure4 helps identify all targets by any network of co-conspirators even when some members may be using false identities.

Users benefit from total immersion in three-day training sessions hosted by Ockham Solutions, and upon graduation can master any Mercure4 analytics task in a few clicks.

Moroever, Mercure4 is designed so that multiple law enforcement agency officers may use the product, each taking a specific area of responsibility in an investigation that contributes to the key findings that emerge from the data. Mercure4 makes the data easy to understand via graphs, charts or detailed visualizations created specifically for police work. For example, the user can view call density and heat maps, cell site profiling and mapping, common numbers charts, call location geographic displays, and billings matched to GEOINT.

Smaller Vendors Step Up

Considering the ongoing spate of terrorist attacks, it is reasonable to ask whether France is indeed safer. The short answer: France is as safe as a host of smaller ISS vendors such as Ockham Solutions can make it. However, France continues to come up short on the type of national intelligence apparatus found in the United States or the United Kingdom. As discussed in Chapter 1, France's own effort at such an initiative is embodied by the country's Plateforme Nationale des interceptions judiciaires (PNij), a national platform first proposed in 2006 for collecting, analyzing and sharing critical intelligence. Administration of the PNij was awarded to Thales as prime contractor with a host of subs in support, but due to internal government bickering and numerous glitches by Thales took more than a decade to launch – and still has problems.

France's Court of Auditors, charged with investigating national intelligence infrastructure, in 2014 determined that the country had spent more than 1.0 billion euros on judicial interception since 2006, and could have saved nearly half that amount – 480 million euros – if the PNij had been operational. Because responsibility for lawful intercept is shared by the General Secretariat for Defence and National Security (SGDSN) and the Ministry of Justice, there are often conflicts in decision-making that impact negotiations with ISS vendors. Even requisitions for ISS services and solutions are hard to keep track of. More remarkable still: the final call for proposals from ISS vendors in 2014 made no mention whatsoever of 4G.

In June 2019, French officials announced plans for new tenders for a complete "re-internalization" of all computer servers within the Ministry of

Justice. Reason: regular systems crashes that left the PNij and all evidence therein inaccessible. Estimated cost: $110 million euros. The driver: fears that 5G's arrival will be a nightmare for law enforcement and government intelligence, leaving them completely in the dark.

The real nightmare is the government's inability to effectively manage ISS resources. Until France can get what it paid for the PNij, Ockham's Razor applies. The simplest solution to a problem is always the best answer: Ockham.

Sqrll: Cyber Analytics with Machine Learning

By some estimates, between 40 – 50 percent of U.S. surveillance and cyber companies are founded and led by former employees of the National Security Agency and other branches of the IC. Founded by NSA veterans Adam Fuchs and Oren Falkowitz, Sqrll fits the mold. Found in 2012, Sqrll operated out of Cambridge, MA for seven years, offering analytics that revealed vulnerabilities prone to cyberattacks. Amazon acquired the company in 2019 to beef up cybersecurity for its cloud service.

Sqrll derived its name from the way most large enterprises attempt to secure sensitive data, by "squirreling" it away in places that seem safe, but really aren't. Sqrll had a better alternative: Accumulo, a solution developed for the NSA in 2008 and subsequently handed over to the Apache Software Foundation.

Accumulo is a NoSQL real-time storage and analytics system, inspired by Google's proprietary Big Table data storage system and still used for key Google apps such as website indexing. Like Big Table, Accumulo uses massive parallel processing to store and analyze structured data. Unlike Google's system, Accumulo also accommodates unstructured data.

Accumulo provides security features that are valuable not only for an intelligence agency, but for any enterprise. One key feature is "cell-level" security access, providing the ability to set strict access controls on every bit of data, dictating who is authorized to access it and how. Considering that Accumulo scales to 10s of petabytes of data, securing each bit is impressive.

Another feature that sets Accumulo apart from the batch-processing mode of Hadoop: analytics of data-in-motion, providing real-time actionable insights that can be presented in graphs or eye-catching visualizations.

Accumulo was the backbone of Sqrll Enterprise and later named "Threat Hunter," a real-time analytics product with granular access for security, and scalability to handle up to 100 million data inputs per second. Refinements by Sqrll for its enterprise customers in the government, communications, healthcare and financial fields included simple out-of-the-box operability, enabling analysts to quickly begin seeing the links in data that point to cyber threats.

From Right Questions to Anomaly Detection

Data analytics vendors tend to present their solutions as providing "all the answers" to marketing, service or cybersecurity needs. As Sqrll's team understood, such an approach put the cart before the horse. The key to deriving value from big data is asking the right questions, i.e., knowing what queries to present in order to derive the most useful answers.

Sqrll helped the user establish a process for building a reliable linked data model based on entities, relationships and features (ERF) that provide context for and define both. The methodology for setting up ERF is to create queries that pertain to common activities, e.g., "Who is logging in?" "How often?" "How is this user identified?" "Where does he fit in the organization?" "What files does he have access to?" "From what location?" "From what type of device?" "How is the device identified?" And so on.

Next up: selecting relevant data – i.e., data that will provide a meaningful response to the predefined queries – from all the raw data generated on a network, then integrating or "mapping" relevant data to the linked data model. By establishing a model responsive to the right questions, the user can create behavioral models wherein pattern matching and pattern discovery are the prelude to revealing security anomalies.

Because linked data analysis embraces the relationships and context associated with an entity, Sqrll's approach was highly accurate at singling out true anomalies without triggering false alarms. Simple illustration: The entity whose number of log-ins significantly exceed the norm for that individual, or an entity that attempts to access data in cells from which it is barred by policy. Having identified the anomaly, the user can then execute the "cyber hunt" in full color graphics to visualize the attack, its perpetrators and scope.

If other leads emerge, an analyst could return to the raw data stored in its original format for forensic analysis that closes the loop.

The security of each cell can be set on a "need to know" basis, commonly known as Zero Trust Architecture, and hidden from those not allowed to see it. If the user lacks access to a cell, that means he doesn't even know it exists.

Performance improvements derive from a virtuous cycle: Create the model; simulate and experiment; analyze results; modify code as needed; refine the model and/or use the output to create better code and new models.

The Uses (And Limits) of Machine Learning

Cybersecurity vendors such as DB Networks tout machine learning (ML) in tandem with behavioral analysis. It's the usual argument: Humans tire and make mistakes whereas machines never do. Did Sqrll buy into machine learning? Yes, machine learning was among the assets they deployed – but always with an eye to the ongoing debate over AI.

Critics fault ML for being "nondeterministic." In plain English, in a nondeterministic system the same data entered twice may come to different conclusions owing to random factors. Thus when changes occur at the tail end of analysis, the reasons may not always be explainable. Sqrll straddled the argument, contending on the one hand that the deterministic model is the optimal route: data-in should preferably produce the same result unless or until the model itself changes.

Sqrll conceded circumstances where the scale of complexity calls for ML. However, the end result must always be reliable. With its direct involvement in the development of Accumulo, expertise in exposing anomalies, and a product line designed to cater to federal and enterprise needs, Sqrll offered the security of Zero Trust Architecture and the speed of machine learning to deliver reliable intelligence discretely and at speed. Today, Zero Trust is the cyber model used by major customers including the US Intelligence Community and the intelligence branches of the DoD.

Verint: OMNIX Intelligence Fusion Center

With Verint's late 2019 announcement that the company would broken into two enterprises in 2021, respectively focused on cybersecurity and customer experience management, analysts rightly pondered the future of its intelligence division. The company had for years promoted its OMNIX Intelligence Fusion Center as the ultimate tool for integrating government intelligence. Whether OMNIX survives the split remains to be seen.

Like many companies, Verint has long played on both sides of the line – enterprise and cyber intelligence providing essentially the same technology-based solutions to customers in each sector. A big data analytics solution designed to help a large commercial entity do customer experience management that improves satisfaction, loyalty and margins can easily be turned to perform "Terrorist Tracking Management" that follows the sentiments, activities, affiliates and likely next actions of ISIS-inspired fanatics or other criminals.

Verint arrived at its own big data analytics solution through a mix of acquisitions and partnerships. The most significant was a strategic partnership with IBM wherein Verint actively engaged in selling a key element of Big Blue's Blue's Infosphere portfolio. Granted, Verint's own moves such as the 2014 acquisition of KANA added some muscle on the big data analytics front in the form of social listening, text analytics, and customer experience management (CEM).

But Verint was already an experienced hand with IBM's big data analytics wares – actually a finalist in IBM's annual "Throwdown" competition that same year in the *IBM Insights 2014* event in Las Vegas. Verint scored high for its solution using IBM Engagement Analytics, a real-time predictive analytics (RTPA) tool. While Verint's nomination was keyed to the enterprise side, the same solution was applicable in the realm of surveillance.

Inside Verint Intelligence Fusion Center

Verint markets its OMNIX Intelligence Fusion Center (IFC) as a "single point intelligence environment" that houses and analyzes data from multiple sources and reviews the data afresh based on whatever cycle is selected – by the week, day, hour or minute – "intuitively" decides how best to present data to the end user, and also sends alerts on anomalies. Data sources can include police field reports, criminal records, vehicle info, personal info, phone detail, call detail records (CDRs), "Web content," as well as mobile location data provided by Verint SkyLock, an SS7-based tool that measures the distance from a target's mobile device to the nearest base station.

Broken down, that is a very extensive list, with some overlap. For example, personal info can span name, age, ID, gender, date & place of birth, nationality, passport number, physical characteristics, known places visited including work/home/social scenes, criminal records, suspicious events, whether an individual is related to other targets and known criminals or part of an investigation. Web content might include data from lawful intercept such as specific content and related metadata, as well as web surfing, social media posts and tweets, and other forms of OSINT.

When Verint tackles big data, they do it in a big way. OMNIX IFC can handle structured or unstructured data (recall that all text and video – the bulk of data – is "unstructured"). The system supports 100 database and 500 file formats, whose ingestion can be controlled via policy set by the user, mapped by type, sent to the appropriate retention component for storage and access, and scheduled for incremental updates.

How OMNIX IFC works: The user can put that data to work in three stages of search: Specific, Structured and Group.

At the outset, an agent might look for a specific type of criminal or terrorist activity – bombings, shootings, kidnappings, drug transactions, etc. – in a given geozone, using a predetermined data set such as Twitter posts with a set keyword, call detail records (CDRs) with a certain 3-digit prefix – or both. Next, the Intelligence Fusion Center might spit out thousands of facets, i.e., instances that link to other factors. The agent can then decide to focus on CDRs, see the option to choose calls or texts and opt for the latter, then narrow the focus to a city instead of a region. OMNIX IFC shows a more refined set of results, visualized by table or map.

The agent moves to the second phase, a structured search that maps findings of the specific search against all available types of data. Up pops the most likely target, complete with all personal details, his or her image, where they routinely travel as disclosed by SS7, any anomalies in travel patterns, and related public records – weapons purchases, court appearances – corroborated by every atom of communications he ever transmitted or received, and stored in the IFC database.

Finally, OMNIX IFC may be used to conduct a group search to determine the target's network of affiliates, be they family or partners in crime. Visualization of each key connection is shown by color-coded links for each type of relationship, with frequency of communications indicated by the variable density of lines connecting the target to potential partners.

The visual can be viewed in either radial or hierarchical format, a common feature in all such systems. The agent can then expand the view of the target to reveal any new relationships, add suspects to the list of those who bear watching, and initiate monitoring of any or all individuals revealed by OMNIX IFC. In other words, based on the findings OMNIX reveals in the data, the agent can tune monitoring solutions to produce *more* data that hones in on the target.

Of note, the target cannot hide. OMNIX is adept at discovering fake names and any accounts they set up and adds them to the mix of targets to be observed further.

The end game of Verint's OMNIX is the ability to produce a "fused intelligence entity" that represents a potential or imminent threat. Security is rock solid from the get-go. Strict cyber measures guard against network penetration and trigger real-time alerts on any attempted breach. Similar backup measures continuously test both physical and logical aspects of OMNIX for full integrity across all external interfaces.

In sum, OMNIX is a fine solution, for what it does. Arguably, any flaw in OMNIX lies in lack of innovative spirit by its developers. A look around the big data analytics marketplace shows many companies with similar baseline capabilities. Then you find a MemSQL that makes "RAM the new disc," accelerating transaction speeds. Or a Recorded Future, which adds a vital dimension to real-time threat analysis by blending GEOINT with OSINT. Either could leave OMNIX Intelligence Fusion Center in the dust.

CHAPTER 4

Offensive Cyber

Overview

To the general public, offensive cyber solutions are intrusive tools with the ability to capture device traffic and content, "keyboard sniff," view web surfing, view and modify content uploads or downloads, take charge of a device's microphone, camera or video camera, and send messages on behalf of the target.

Less well-known, but arguably far more dangerous, are offensive tools planted in critical infrastructure IT systems – digital time bombs that lay dormant until activated by the attacker. STUXNET, jointly developed by the U.S. and Israel, and deployed in Iran to make that nation's nuclear plants self-destruct, is the most famous example. It is generally agreed that Russia, China and North Korea have deployed similar, equally aggressive cyber weapons in Western nations' critical infrastructure sectors.

Among the most sinister are those that target space assets. Ground control and communications systems, transport for network traffic and data, and satellites – many launched in pre-Internet days and still operational – are easy prey for offensive cyber tools that overpower "command and control" (C2) and other satellite systems. End result: subtle but lethal alterations in the performance of intelligence collection, weapons systems, GPS, and atomic clocks used for accurate timing and other functions essential to national security. To understand the gravity of such attacks, imagine fleets of naval vessels lost at sea due to dysfunctional GPS, or major trading centers gone haywire when automated financial transactions that ordinarily take milli-seconds are mysteriously delayed.

At ground level, more conventional offensive cyber solutions may be installed directly via USB into the target's PC or laptop, or remotely through the subterfuge of phishing, emails, advertisements, "rogue" (fake) websites, social media and social engineering that gull the target into clicking on a link that uploads the intrusion.

Also growing in popularity: "drive-by" attacks via rogue websites engineered to the target's known "likes" or via DNS hijacking, i.e., reconfiguring the target's domain name search (DNS) capabilities to route him or her to a website that plants malware. Drive-by attacks are similar in nature to "packet injection," a hacking technique commercially introduced by Gamma Group International around 2007 and possibly inspired by the NSA. Packet injection is a sophisticated form of intrusion that injects packets into backbone networks or specific devices, redirecting targets to a rogue server which then launches a man-in-the-middle (MITM) attack. The technique is identical the NSA's famous QUANTUMINSERT program, which followed the same process and redirected the target to the agency's rogue FOXACID website for a MITM attack and malware plant.

Offensive cyber often leverages Zero Days, i.e., system vulnerabilities that are not published or known. For years, Vupen was a key provider of military grade Zero Day exploits for Western intelligence agencies. Vupen officially closed shop in May 2015 and was replaced by Zerodium under the same management team, led by legendary hacker, Chaouki Bekrar, who by strange coincidence was born near NSA's offensive cyber HQ at Fort Meade, Maryland. Zerodium now offers bounties to researchers for premium Zero Days, but professes that it no longer produces them.

Providers of communications devices often unwittingly open the door to Zero Day exploits and malware by failing to implement patches for weaknesses. Even Apple, renowned for its device security, has its lapses. A much publicized vulnerability in Apple's Thunderbolt USB port went uncorrected for many months until January, 2015. More recently, in January 2018, Apple confirmed that nearly every device the company makes is prey to the Specter vulnerability – for which there is no known fix. Offensive cyber that works against Apple's "closed garden" approach to cyber defenses and strong encryption remains the Number 1 item on Zerodium's and other bounty hunters' lists – and earns the highest bounties.

Offensive cyber technologies that invade and control a target's device have been available since the dawn of the Internet, and are commonly used by law enforcement, intelligence and government agencies worldwide.

Despite efforts to curb it, offensive cyber continues to grow, spurring new market entrants from the Middle East and Asia. Contrary to public perception, in the U.S. the vast majority of such work stems from government services contractors, notably Raytheon, Northrop Grumman and other companies that specialize in computer network operations (CNO), the industry's trade name for this field.

Among the most highly-paid and sought-after government contractor employees are experts who know how to find vulnerabilities, then devise, deploy, operate and manage offensive cyber solutions. The market for these individuals is so tight that government agencies by and large cannot compete for their services salary-wise. Hence, even in the lofty realm of the U.S. Intelligence Community's most prominent agencies, the really sensitive work is performed by members of the private sector.

Such individuals hold security clearances and work at "sensitive compartmented information facilities" (SCIFs) or even more classified settings either on-site at U.S. government agencies or in secured private offices whose location is kept secret and off-limits. In a SCIF it is forbidden to carry a smartphone, bring a personal computer or USB drive, use private email, texting or any other medium of communication with the outside world. In a word, the SCIF is a walled-off, air-gapped facility. These extreme security measures are designed not only to prevent external attacks, but as importantly to preempt "insider threats," i.e., the next Edward Snowden.

The hack of the Office of Personnel Management (OPM) in 2015 led to an ongoing shortage of security-cleared offensive cyber experts, at one point creating an estimated backlog of over 700,000 unfilled security-cleared positions among government contractors. While that number has fallen considerably in 2020, it is still common to see contractors win large awards, then find themselves unable to find sufficient numbers of qualified, security-cleared experts to do the work.

The obvious solution to this critical manpower shortage would be to change the focus from human talent to an "outcome-based" approach, i.e., automation via AI solutions such as machine learning, deep learning, cognitive computing and neural networks. That essential change may still be years away. The majority of U.S. defense and intelligence agencies continue to issue RFPs that call for X number of warm bodies – full-time employees or FTEs – to do the work of offensive cyber. Contractors follow suit and give government customers what they want: "butts in seats."

One potential outcome of the COVID-19 pandemic will most certainly be the realization that outcomes-based solutions leveraging AI are the only rational alternative when those seats cannot be filled.

Cracking Cryptocurrency

Law enforcement investigators tend to look upon Blockchain – the peer-to-peer encrypted record of transactions used by Bitcoin – with a mix of horror and frustration. Blockchain's system of anonymizing transactions makes it difficult to detect who is using Bitcoin, whether for illicit purposes such as drug dealing, money laundering, murder for hire, funding terrorism or purely innocent purposes. How can the investigator find a felon or terrorist from an oblique Bitcoin "signature" consisting of random numbers, letters and punctuation marks and protected by public key cryptography?

To this day, Bitcoin remains the bogeyman of digital or cryptocurrency, certainly the best known and at the same time the one most scrutinized by law enforcement. Often overlooked is that Bitcoin is just one of many cryptocurrencies, in fact of some 700+ clones. The circulation of any one of the larger competitors dwarfs Bitcoin's mere 16 million units of digital currency. Examples of the largest and best-known, with number of units shown in parenthesis:

- Siacoin (23 billion)
- Mintcoin (24 billion)
- Ripple (37 billion)
- Dogecoin (108 billion)
- Bytecoin (182 billion)

All share common Blockchain characteristics: peer-to-peer decentralized transactions and record-keeping; anonymity of the user; and "proof of work" reliability conducted by "miners" that use massive parallel processing to validate transactions. If they are leery of using Bitcoin owing to the scrutiny it has drawn it has drawn from law enforcement, there are a variety of alternatives to turn to.

But there is an important upside for investigators: Whatever the digital currency chosen by offenders, all operate in a similar Blockchain sandbox.

Understanding Blockchain Public Key Cryptography

What exactly is Blockchain and how does it work?

In Blockchain, a block contains the digital record of a transaction that is "hashed" with the label and value of the transaction, and encoded in what is called a "hash tree" or "Merkle tree" of interconnected Child nodes (data structures assigned a value). Each block contains the hash data of the block that preceded it, forming a linkage that creates a chain wherein new blocks are authenticated by their predecessor. The common hash makes it possible to authenticate each block in the chain all the way back to the original block. The result is that data in a blockchain cannot be manipulated or changed in any way. Thus, the value inherent in a blockchain-secured transaction is inviolable and secure, as is the user's identity – or so the theory goes.

As a decentralized public ledger of online transactions, Blockchain provides security by performing transactions and recording the data on a multitude of computers spread worldwide. By confirming each transaction where a unit of digital money was spent, Blockchain eliminates the risk of duplicating data or bitcoins themselves.

Anonymity is provided by public key cryptography, which uses two sets of keys: public keys open to all, and private keys known only to the user. The public key encrypts a message or transaction, but only the holder of the private key can decrypt it.

Public key cryptography is generally considered secure – and anonymous – because of the difficulty of calculating the private key simply from knowledge of the public key.

Current techniques of cryptocurrency analysis focus on "de-anonymizing" transactions and finding anomalies – preferably both.

Specialists in Blockchain analysis have successfully applied machine learning and deep learning to reveal the identities of those engaging in illegal activities using Blockchain. In so doing, data science thought leaders are using methods that detect anomalous trends and transactions and de-anonymize perpetrators of criminal acts conducted with digital currency.

On the Lookout for Anomalies

Many cryptocurrency clients use digital "on-ramps" and "off-ramps" such as exchanges or wallet services to access their accounts. In those instances, an account for a digital currency such as Bitcoin is still very much like a bank account. It is dedicated to one user. If an investigator can map the account to a user, or alternately map specific transactions to one account, it becomes feasible to learn the IP address and personal identification of this target.

Targets also leave bread crumbs when they convert cryptocurrency to legal tender backed by a government (fiat currency). True, users may have recourse to opening more than one account, but in most cases targets limit themselves to just a few. In any event, investigators can learn a great deal by analyzing transactions that connect multiple addresses, and by scouring IP addresses used in transactions.

Two other features of cryptocurrency transactions work in law enforcement's favor. First, because Blockchain is a public record, law enforcement and government agencies do not require a warrant or court order to pursue and obtain the data, although there are times when they might choose to do so. Second, because Blockchain is an international platform and the currencies that use it are global, the "coinage" issued is truly global in scope. Thus, pursuing a cross-border investigation does not require the cooperation and permission of different national authorities under a mutual legal assistance treaty (MLAT).

To avoid detection, clever cryptocurrency owners often resort to "tumbling" or "mixing" services developed on the Dark Web for the express purpose of laundering money. Mixing services break the connection between a Bitcoin address sending currency and an address the Bitcoins are sent to.

But there are problems for criminals. The supposedly "clean" Bitcoins obtained from the mixer may, in fact, already be dirty and on a list closely watched by law enforcement. And in the end, Blockchain still remains a permanent record. Like Tor, Bitcoin and other digital currencies are already on law enforcement's extensive data monitoring radar, and law enforcement is patient. Once one transaction is cracked, the identity of the criminal element involved will over time be tracked right back to the beginning.

When criminals get greedy and try to move a large quantity of Bitcoins through a mixing service, that is another anomalous red flag that investigators immediately latch onto.

"De-anonymization"

One of the innate qualities of cryptocurrency such as Bitcoin is its built-in inefficiency and decentralization. If data flows were coordinated in equal measure then it would indeed be impossible to single out one transaction or set of transactions from another. In the real world, however, the structure of a digital currency is never top-down. Data relays are anomalous.

A single computer may conduct one digital currency transaction at random intervals, with one IP address. Alternately, the same user may do multiple transactions, again all from the same IP address. In so doing, the user inadvertently sets up a trail of Bitcoin transactions that map to a specific device identifier and ultimately, personal ID.

Even multiple addresses may be clustered to recognize one entity. A method pioneered by the husband/wife team of Philip and Diana Koshy at the University of Pennsylvania applies an "algorithmic approach for mapping Bitcoin addresses to the IPs that own them."

The Koshys' creative technique involves six phases:

1. Pruning of transactions to remove "noise."
2. Monitoring of relay patterns to hypothesize an IP owner for each transaction.
3. Creating granular Bitcoin pairings by breaking down transactions into individual IP addresses.
4. Computing statistical metrics for the pairings.
5. Identifying pairings that tag owner relationships.
6. And, finally, eliminating pairings that fall below pre-defined thresholds.

The Koshys succeeded in mapping IP addresses to over 1,000 Bitcoin addresses.

Blockchain Alliance's Role in Helping Law Enforcement

Much of the impetus for helping law enforcement crack cryptocurrency comes from the digital currency community itself.

The Chamber of Digital Commerce and Coin Center in 2015 established a new affiliate organization, the Blockchain Alliance. Chief among the group's missions is to promote the members' work to law enforcement agencies and to educate and provide assistance in catching individuals engaged in money laundering and other illicit activities via cryptocurrency.

The Alliance quickly attracted 25 of the largest cryptocurrency analytics firms, and nearly that many law enforcement agencies. Right out of the gate, the Blockchain Alliance had working relationships with major US law enforcement and government agencies including the Department of Justice, the FBI, Homeland Security, Secret Service, Marshals Service, Immigration and Customs Enforcement, and the Commodity Futures Trading Commission.

The Alliance then moved to acquire working relationships with large metropolitan LEAs, as well as international crime-fighting organizations such as Europol and Interpol. The Blockchain Alliance selected Steptoe & Johnson as legal consultant and named a prominent S & J partner head of the new non-profit private/public forum: Jason Weinstein, former deputy assistant attorney general in the Department of Justice in charge of cybercrime.

Not all members of the cryptocurrency movement look with favor upon the Blockchain Alliance. Bitcoin Foundation Executive Director Bruce Fenton once criticized the Alliance's work as a threat to privacy and inferred that the Blockchain Alliance was essentially creating "back doors" into cryptocurrency that would compromise the integrity and security of Blockchain.

DNS Hijacking: The Art and Science

DNS hijacking is a form of cyberattack that involves using malware to infect a target's device, gain access to its TCP/IP data, change these settings, then re-route the target to a fake duplicate of a bona fide site to either capture personal information, track the target, or both.

As a stock in trade not only of black hats but also "ethical" malware companies and law enforcement agencies, DNS hijacking surfaces in association with malware such as FinFisher, The Hacking Team (now part Memento Labs), STUXNET or CIPAV (Computer and Internet Protocol Address Verifier). One of the better known examples is the FBI's use of CIPAV to lure a bombing suspect to a rogue version of an *The Associated Press* article, from which the agency quickly identified the suspect's IP address, MAC address and location, and put him under arrest. More innocuous forms of DNS hijacking are commercial versions that simply redirect an individual to advertising, or are used to collect data for marketing purposes.

DNS stands for "Domain Name System" or "Domain Name Server." It is a protocol that enables the Web user to reach any site simply by keying in [or speaking] its name. The name is automatically translated to the desired site destination's IP address, and there you arrive without having to remember the IP address itself. Every end user device and server in the network has its own unique IP address which plays a critical role in all Internet communications applications, from simple Web surfing to email, SMS and social media. Thus when the user keys in a site name he or she wishes to reach – be it *The New York Times* or Google apps for gmail – DNS takes over and automatically converts the words to code for the sought-for site's IP address.

The process of establishing a connection between an end user device's IP address and that of another destination is called DNS Resolution. The user could enter the IP address itself with the same result, but why bother? DNS Resolution makes connecting simple.

The problem – or opportunity, depending on one's point of view and need – is that it is comparatively easy for an intruder to hijack the DNS process and put it to work for other purposes, as the FBI, other government agencies and malware companies routinely do.

In the case of the FBI's fake *AP* story, the perpetrator actually turned out to be a high school student who hacked into the school's email system by bombarding it with spam, then left hand-written bomb threat notes at his alma mater. As a result, the school went through forced evacuations every day for two weeks. Stymied on how to find the perpetrator, local police turned to the FBI for help.

In this instance, the young "bomber" was posting threats to what he assumed was an anonymous myspace.com social media account. The means used to explain how the FBI penetrated that myspace.com account has never been revealed, though as you will learn in subsequent sub-chapters, the feds have virtually unlimited hacking power and capabilities – all of it "legal," albeit by circuitous means that stretch the law. Thus, in the end, the agency had a fairly decent short list of bombing suspects. Their next step was to create a false *AP* article on the bomb scare, slyly praising the lad's tech savvy by observing that the perpetrator had not yet been identified. The FBI then emailed him the false *AP* article with a Trojan embedded. When the boy clicked on the link, the Trojan provided access to his computer and CIPAV took over, quickly determining the lad's IP address, MAC address, web surfing habits, email record, etc., and he was taken into custody.

The story of the fake *AP* article and FBI CIPAV emerged months later in a roundabout way, through a court hearing on another case. Initially a good deal of confusion accompanied the leak, including errant accounts of the FBI creating an actual fake website mimicking *The Seattle Post Intelligencer (PI)*. The PI's editorial staff reacted with moral indignation over having their logo and look absconded, and were soon joined by the Electronic Frontier Foundation (EFF), whose chief technologist roundly criticized the use of malware and DNS hijacking as an outrage to the public trust. However, in this case the only item "faked" was a single news story, not an entire newspaper. In other words, given the targeted mechanism of confining the attack to one individual by email, there was no way that others – that is, "innocents" – could have been deluded and had their computing and communications devices comprised.

True DNS hijacking is a good deal more sophisticated. The attacker hijacks the target device's IP address by using a program such as CIPAV, watches what websites the target likes to visit, mimics one, then routes the target's traffic to the rogue server of the faked website and installs a Trojan on the suspect's device.

In recent years tech giants such as Apple and Google have made a big play in end-to-end encryption for texting. Google, for example, in 2015 announced its first end-to-end encrypted text service for Android users, called "Allo." But in the end, Allo encryption was merely an option, not an automatic, always-on feature. Users of Android devices had to turn it on. If they did not, then everything they keyed-in was open to interception. Google has improved since then. Apple's iMessage service, on the other hand, still has gaps – messages are only encrypted when sent between iMessage users. Messages sent to an Android device appear as text.

Use of any encryption tends to raise red flags with law enforcement and intelligence agencies. Investigators can tell immediately when an individual is defaulting to end-to-end encryption, a fact that may inadvertently put innocents on a list of potential suspects. Such is the fate of Tor users, too. As discussed in other analyses herein, the NSA's TAO aka The Equation Group has found ways around and into the Onion Router.

How Targets Catch Themselves

Not that it takes a great deal of effort to identify anyone trying to hide. One fact well-known to investigators and agents is that targets involuntarily turn themselves in by making simple mistakes. Bad guys' overarching career ambitions are crime and terrorism, after all, not IT or coding. One of the most common mistakes: being too careful by combining various anonymity tools and muddling their attempt to remain secret in the process.

For example, using a virtual private network (VPN) is a very popular method of ensuring communications privacy and preventing attackers from seeing one's IP address. What targets typically forget, however, is that VPN is not really impervious to investigators, either. When a target pays for his or her VPN service, that payment mechanism is the weakest link to his personal identity. Financial institutions, after all, insist on real names and credit card numbers. As revealed in the prior sub-chapter, even Bitcoin payments can be converted to real, identifiable money in the end, and tracked right to their owners.

Combining Tor with other tools can also compromise the best efforts at safeguarding one's identity. In illustration, use of a VPN always establishes a single node of entry and exit – which can be tracked. As to more "hops" on multiple VPN servers, as often as not, in the end the user is merely creating more points of potential attack. Plus the connection between the user's device and the VPN node is, as often as not, unencrypted. Plus VPNs can be troublesome for the uninitiated to deploy. If a VPN connection is broken or needs to reboot, then the target's communications are "in the clear" during that interim. Many VPN users simply forget to turn it on, or to check to ensure that it is operating. Finally, as with Tor, when the target uses a VPN it is evident to his service provider that he is doing so. Under a court order, the Internet Service Provider is required to cooperate with law enforcement.

Targets with the cash and expertise to do so may always set up their own VPN servers and thus avoid the "money trail" created by a commercial relationship with the vendor. But watch out: encrypted communications may still be visible at the Tor exit node.

Criminals and terrorists with nominal knowledge of computing often make the most obvious blunders when it comes to leaving a digital trail. One such example of such oversight stems from use of Microsoft Word. Every Word document contains almost as much information as an old-fashioned business card: user names plus "folder paths" to the target's device. Target-specific metadata is easily extracted.

Need Help? Don't Contact the FBI

Do you think you've covered all the bases and are now impervious to DNS hijacking? Don't be so sure. If you experience DNS Resolution or the dreaded "404 Error Page," the root cause may be a problem with your device, host or the sought-after website. These common nuisances can also result from DNS hijacking.

Help is hard to find. True, the FBI's website showcases a page with links to free services that ostensibly help detect DNS hijacking. But good luck trying to use either of the two English language versions provided. One leads to a 404 Error Page, and the other has been out of service for years.

Man-in-the-Middle Attacks

The man-in-the-middle (MITM) attack is a common form of hacking that allows hackers or government agencies to eavesdrop on communications by acting like one or both of the participants. Legacy MITM methods that involve compromising trusted certificates, while still used, are child's play compared to current state of the art MITM. For a look at the latter we shall consider the work of Blue Coat, a pioneer in advanced MITM.

Prior to its acquisition by Symantec in 2016, Blue Coat was for years among the main IT equipment manufacturers that play in both the commercial enterprise and government surveillance realms. The company's core product was ProxySG, still an important commercially available MITM solution despite Symantec's serial acquisition by Broadcom in September 2019 and then by Accenture in January 2020. In the enterprise market, ProxySG gives companies greater visibility into SSL traffic. Among its chief talents: telling SSL from non-SSL traffic, spotting HTTPS traffic – a layering of SSL over Hypertext Transfer Protocol for added security – and pegging a website by identifying its SSL server certificate hostname.

For commercial users, the benefit is an encrypted private tunnel for secure communications. For government agencies, the prize is the ability to capture targeted content and metadata via man-in-the-middle attack.

ProxySG may also be used to selectively intercept and decrypt SSL traffic – in other words, to break in without being noticed, which is exactly how the intelligence community uses the device. By some counts, the ProxySG is among the most popular solutions in the world for intelligence-gathering activities.

At its most basic level, ProxySG acts as a web cache – a device for temporarily storing web documents that pass through any network even when the device itself is not deployed directly in the physical path between client and server. ProxySG achieves this end via servers that are programmed to redirect specified packets to the ProxySG for storage, decryption and analysis.

While still in business, Blue Coat received an important assist from one of the world's biggest and best known manufacturers of servers: Cisco.

Cisco is the author of the Web Cache Communications Protocol (WCCP), which among other purposes may be used to program routers and switches to transparently redirect network traffic from its intended destination to a ProxySG. With ProxySG connected to the specified server by a "virtually inline" connection, the government or law enforcement agent may intercept any target.

Herein resides an important difference between the "explicit" proxies used for commercial purposes, and "transparent" proxies commonly used by law enforcement and intelligence.

In commercial applications, the ProxySG does "explicit" proxies for the purpose of helping companies improve the efficiency, performance and prioritization of network traffic. The client browser is configured to use a proxy server, and the browser knows that all requests will go through a proxy. The browser is given the IP address and port number of the ProxySG. Or companies may use a Proxy Auto-Configuration (PAC) file to configure the browser to download the proxy settings from a Web server. When a user makes a request, the browser connects to the proxy server and complies with the request.

When a transparent proxy is enabled, the targeted client does not know that his traffic is being processed by a proxy rather than the originating server. The agent's server is configured to intercept traffic for a specified port, or for all IP addresses on that port. A WCCP-configured router then redirects selected traffic to the appliance depending on policies the user creates for the redirection device.

What if the target or those he communicates with aren't using Cisco routers? Blue Coat, its successive owners and similar vendors offer options in that event. Their proxy device can take traffic from a Layer-4 switch, or via its own software bridge.

Consider how the proxy play might work in a surveillance scenario. The "client," in this case perhaps a foreign military officer, wants to let his superior know the status of an upcoming field operation. He sends an encrypted message, thinking it's safe from being discovered. What really happens: Instead of his message going straight to his commander, it is mirrored and sent via WCCP-enabled virtual connection to an intelligence agency that has one or more Blue Coat ProxySGs installed.

Based on rules established at the outset of a surveillance operation, traffic from the target goes straight to the designated cache engine, which then decides on the action to take. In this instance, Blue Coat Encrypted Tap, which can be built in to ProxySG, deciphers the SSL-protected message and quickly forwards it to the monitoring intelligence agent for analysis and forensics. In the same moment the message is forwarded to the target's superior officer. When the latter replies, the process is repeated, this time going the other way: data is cached for de-encryption and analysis, then the commander's response is forwarded to his subordinate.

Throughout this process, the client server never knows that it is interacting with a ProxySG. This capability makes the ProxySG ideal for surveillance. Other features:

- **Scalability and Re-Routing.** If the user so chooses, he may have target traffic distributed automatically to up to 32 ProxySG devices. If for whatever reason a designated ProxySG goes down, the target traffic automatically ships to other ProxySGs in the user's group. Result: no data loss, whatever the traffic volume.
- **Flexibility.** The user decides what communications to intercept and where to send it. He can redirect traffic, filter it to capture only specific traffic, or choose the precise ProxySG to send it to.
- **Safe from Intrusion.** WCCP devices are password-protected to bar intrusion from outsiders. In addition, users can configure specific access control lists on the router so that traffic is sent only to designated ProxySG appliances.
- **Failsafe.** What if all ProxySG devices were to go down or become unavailable? That's not likely, but let's say it did happen. The target would still have no clue he's being monitored. The WCCP router would simply send his communications to the originally-designated destination until ProxySG traffic is restored.

Mobile Hacking via SS7 and Diameter Protocols

An historical factoid often lost on Millennials in the U.S. is that from 1913 to 1982 when it was broken up by order of the U.S. Court of the District of Columbia, AT&T and its local Bell operating companies held virtually total control of all local communications services in the United States. Any effort to compete with these monopolies was tightly restricted by the government. When the U.S. communications industry underwent massive deregulation via the Telecommunications Act of 1996 (TA96), a key goal was to open the local telecom market to competition.

In the U.S. in the space of a few years following enactment of TA96, a local telecom market once confined to fewer than a dozen "competitive access providers" (CAPs) sprang into a vibrant bazaar of nearly 1,200 competitive local exchange carriers (CLECs). One of the chief barriers torn down to make competition possible was the prior restriction on access to Signaling System 7 (SS7), a protocol developed by Bell Laboratories in the mid-1970s to accelerate call set-up for and between carriers handling calls. Previously confined to monopoly control, SS7 became available everywhere – un-gated and with completely uninhibited access.

The CLECs came and went quickly, brought low by foolish management, dot.com-like spending habits, and legal maneuvering by Baby Bells such as Verizon at the regulatory level. But wide open access to SS7 and its successors SIGTRAN for IP and the Diameter protocol for advanced mobile networks live on today in the even bigger role of speeding calls and data connections across and between thousands of local, national and global networks.

Somewhere, the ghosts of CLECs are laughing. They know that the tools meant to create vibrant competitive markets everywhere are not only un-gated but largely unprotected, in essence wide open to cyberattacks that leave all networks – including those operated by the giants who killed competition – vulnerable to hacking.

A Weakness Neglected for Decades

SS7, an "out of channel" signaling protocol for setting up and taking down calls on the public switched telephone network (PSTN), was introduced in 1975 and quickly embraced by the International Telecommunications Union (ITU) as a global standard. It was deployed in an environment of trust between network operators worldwide, which for the next 20 years would remain a finite and closed community. As a result, while the vulnerabilities were present from the outset, security was not a top priority. Even after the expansion of the marketplace to include hundreds of new providers, the security weakness of SS7 and SIGTRAN rarely surfaced.

More than three decades would pass from the launch of SS7 to a signature moment in 2008 at a Black Hat conference. Karsten Nohl and Jakob Lell from German security firm SR Labs demonstrated the ease of penetrating mobile networks via SS7 vulnerabilities. But the story came, went and was soon forgotten.

Then in 2014 security researcher Tobias Engle demonstrated mobile network hacking via SS7 at a 2015 meeting of the Chaos Computer Conference. After capturing brief attention, that revelation blew over, too. Most graphic of all, in mid-April 2016 SR Labs resurfaced to demonstrate hacking into a U.S. Congressman's mobile phone on "60 Minutes." The popular TV show succeeded in getting policymakers' attention.

In June 2016 the U.S. Federal Communications Commission asked "Working Group 10" of the Communications, Security, Reliability and Interoperability Council (CSRIC) to delve into the growing problem of attacks on networks via SS7, SIGTRAN and Diameter. CSRIC is an advisory body of industry and government thought leaders who recommend steps toward government policy. CSRIC Working Group 10 was comprised of representatives from the top U.S. wireless and cable operators, the Alliance for Telecommunications Industry Standards (ATIS), IT companies and the U.S. Department of Homeland Security. The task put to Working Group 10: "to assess existing and potential threats and current defensive mechanisms... and security challenges present in SS7 and other communications protocols."

Working Group 10's report is still an eye-opener – not so much for what it revealed as what it proved had been neglected for decades.

The CSRIC Working Group 10 Report

The CSRIC report that emerged from their research acknowledged multiple factors as responsible for a geometric expansion of the "attack surface" available to hackers. Key causes included competition, the mandatory opening of SS7 to multiple carriers, reliance on the old "trust ecosystem" and the expansion of SS7 from wireline to wireless and to "new and novel" uses such as SMS. Within the SS7 stack, CSRIC drew attention to the "Mobile Application Part" (MAP), a protocol that was developed to support wireless. Together, SS7 and its MAP component at the time coordinated global roaming of users on some 800 mobile networks.

Through its research, Working Group 10 discovered that MAP was the favorite target of attackers for initiating exploits. From there, the hacker could gain access to the Home Location Register, the principal database used by mobile network operators to store subscriber information. They found it easy to hack any number of accounts and penetrate the various apps used on any mobile device using on the network.

The problem did not end with SS7. As stated in the report:

The fact that access to networks is possible through any number of complicit or cooperative operators means that all telecommunications protocols used to interconnect networks are potentially at risk. This includes SIP interconnects, as well as Diameter.

Diameter today is the industry standard protocol used in LTE networks to exchange authentication, authorization and accounting (AAA) information. To date, no hard evidence exists that Diameter has been exploited. However, in laboratory work, researchers have used the Diameter protocol to undertake attacks very similar to those known to have been executed against SS7 – and at the same time proven the existence of other vulnerabilities in Diameter that support new and different exploits.

CSRIC Working Group 10 faulted poor network security. In contrast to the marketing hype by Verizon, AT&T, Sprint, T-Mobile and their kin in other nations outside the U.S., mobile networks were revealed as festering sores of vulnerabilities. Mobile data communications were not protected by end-to-end encryption. Only the air interface between the mobile device and a radio tower was encrypted, while the call itself was transmitted "in the clear." As users roamed cell-to-cell or network to network, communications were secure to a degree when handed off between trusted nodes. But even if even a single node was compromised, any call transiting that node was also vulnerable.

As a solution, CSRIC recommended cooperative information sharing and increased investment in end-to-end encryption measures.

But because CSRIC is not a rules-making body, new policy on safeguarding mobile protocol remains to be seen. For hackers, it is still open season on SS7, SIGTRAN, Diameter and many other communications protocols spawned in an era of trust when the notion of attackers – whether criminal or government – was way over the horizon.

NetFlow Versus Tor: Conquering Anonymity

Tor (The Onion Router) enjoys legendary status as a service that encrypts network communications end-to-end, is completely impenetrable to surveillance and thus safeguards the identity of users, be they ordinary individuals, journalists, freedom fighters or less savory characters. However, researchers have demonstrated that with commonly available tools it is not only feasible but simple to "slice the onion" and de-anonymize targets on both ends of the hook-up.

The hero or villain of this story, depending on your point of view, is Cisco NetFlow, which has proven 100 percent effective at cracking Tor in the lab, and 81 percent effective in field tests.

ABCs of Tor and Tor Interception

Tor is a proxy-based, low latency anonymization network that effectively masks the identity of the user, his or her content and the communications destination. Tor clients obtain anonymity by establishing a circuit that routes SMS and email through a random, ever-rotating set of globally-distributed servers from the start point (entry node) through to the terminating point (exit node). Each server involved in this secretive digital relay race acts as a proxy that masks the target's IP address, hence the term "proxy server."

The onion analogy refers to how Tor messages are encrypted and decrypted. To begin an anonymous message the user first picks a set of proxy servers from Tor's directory. These nodes will use the Tor algorithm to create a circuit. The servers then create the cryptologic keys that will protect the message. When a message is sent it is packaged in cells of a fixed length – each cell being a 512-byte packet – and encrypted in different layers using the keys. Now the onion process begins. As the message transits the Tor network, each proxy server strips away one layer of encryption before forwarding the message to the next node. At the exit node, the message is released to the recipient "in clear" via TCP connection.

Note that as a low latency service, Tor sends each 512-byte cell via inter-arrival delivery, the arithmetic mean of packet distribution at separate times. In theory, this delay would make the system vulnerable to attacks and penetration. In practice, however, the hand-off of communications between multiple proxy servers – each constituting an autonomous system (AS) – means that any such surveillance is impractical for an attacker save those with limitless resources, e.g., the U.S. National Security Agency and its counterparts in other nations.

Another method is to attempt interception at major aggregation points. Attempts to monitor Tor by placing passive monitoring systems at key Internet exchange (IX) points have met with some success, but again, such systems rely on significant infrastructure commitment to capture traffic at 100 GB speeds and up.

Looking for another way into Tor, the U.S. Federal Bureau of Investigation in 2012 contracted with The Hacking Team for remote intrusion that could infect Tor users' devices via classic phishing. After an outlay of US $775,000 to The Hacking Team, the FBI reported the results as "interesting," but no prosecutions resulted therefrom, and the agency subsequently terminated its agreement with The Hacking Team.

Other attempts to hack Tor rely on exploiting vulnerabilities in software modules or on simple mistakes made by the target, e.g., errors during setup or failure to follow the Tor manual to the nth degree. It should be noted, though, that when any vulnerabilities crop up, Tor does an excellent job of patching holes and notifying users. In addition, government and military agents fully realize that the odds are increasing against the likelihood of mistakes by sophisticated users. For example, during its brief reign in the Middle East, ISIS operated a 24 X 7 help desk on encryption for novice terrorists.

For law enforcement and national agencies with small budgets, Tor long proved an insurmountable obstacle. But no longer. Now available are "new" methods of interception that use IP flow monitoring to lift the curtain on target identity and the destination of intercepted traffic. We put "new" in quotes because the technology used – Cisco NetFlow – has been available for years. Tor cracking is simply a new app for this proven work horse.

NetFlow and IP Flow Monitoring

As a monitoring tool, NetFlow can distinguish keys that include:
- Flow start and end times.
- SNMP (Simple Network Management Protocol) source and destination IP addresses.
- Port numbers.
- Protocols.
- Number of packets and bytes per each packet.
- TCP flags in the TCP header.
- MPLS Labels.
- Packet treatment.

As it works, NetFlow time stamps the packet flows it captures as either "active" or "inactive" and continually updates its records. Recently recorded data is marked "active" while older data streams that have not been renewed are marked "inactive" and sent to data retention. System administrators have the flexibility to set different time parameters for flushing the data so that NetFlow flow capture is always current and reflects real-time exchanges for higher accuracy of analysis.

What if an agent could use NetFlow on his own attack server to deliberately insert a marker or "perturbation" into traffic? If that were possible, then the agent could begin tracking specific flows.

As an analogy, think of how wildlife conservationists shoot a wolf with temporary anesthesia, then collar it with an RF transmitter that transmits its location for the rest of its life. Only with NetFlow monitoring, the packet flow would be attacked and tagged with a perturbation while the beast is still on the move at full speed, all completely unnoticeably. Applied to Tor, this technique would allow the user to identify the origin and destination of the packet flow, and apply mathematical analytics to follow it through proxy servers to the end node, and "de-anonymize" both sender and recipient.

The introduction of such flags into packet flows could then be leveraged one of two ways. In one method, the user might be a nation state actively attacking the entry and exit nodes of a Tor-based anonymous communication via mathematical analysis of perturbations that track the packet flow. A second way – less cost and infrastructure intensive – would be to use similar markers and the same mathematical techniques to track the passage of data streams into and out of Tor's own network of proxy servers.

Academicians Weigh In

Led by Dr. Sambuddho Chakravarty at Columbia University and Dr. Marko V. Barbera at the Sapienza Universita di Roma, a team of computer scientists set out to test this second methodology. The experiment: To test results when using Cisco NetFlow to monitor the traffic entering and exiting a "victim node" in Tor, then apply statistical analysis to determine the traffic destination.

The computer scientists conducted their experiment in two phases: one in the lab, the second in a live Tor environment.

The team's overall strategy was to apply statistical correlation to entry and exit traffic in a Tor network server. How to overcome the problem of sorting through and identifying one traffic flow from potentially tens or hundreds of thousands? Simple: introduce an easily recognizable flag that makes the flow stand out. Planting the flag was the issue.

In order to create a sustained period of time for observation purposes, the academicians decided that file downloads or videos would be the ideal enticement. In this event, the academicians, operating from their own "attack server," would force the proxy server to send the target highly desirable downloadable files, for example, pirated copies of movies in current release. Because video consumes large amounts of bandwidth, the packet flow would be readily identifiable.

In the academic team's model, the attack or "perturbation" would occur at the exit node from a Tor server. While the target downloaded his movie or other large file, the attack server would infect the victim server's outgoing TCP connection with multiple repeatable traffic patterns that together would be as unique and recognizable as a fingerprint. One such pattern, for example, might be a rapid switch between bandwidth values, and another might be a switch between several predesignated bandwidth values set by the attacker.

In theory, this attack model worked perfectly. The victim proxy server would be made to cooperate with the attack server and introduce a large download at the exit node, where it would be easy to inject fingerprinted traffic patterns, halt the injection after sufficient data was gathered, and then collect flow records for analysis. But in the lab the computer scientists encountered a problem: variance in the time alignment of IP flow pattern arrival at the victim proxy server, with some entering at 5 second intervals, others at 15 seconds, and some with a separation of up to one minute. Remember: Tor operates on latency, with cells sent and arriving at different intervals. Timeouts and network loads can also play a factor in cell arrival times.

Here the scientists were helped by Tor. In correlating the bytes sent from proxy servers to the exit node, the computer scientists quickly recognized that packet flows with the highest coefficient were always associated with the victim client. This led to revelation of a previously unrecognized attribute of Tor routers: Data cells with multiple time lag intervals often lead Tor to create one large, easily identifiable large interval. From this finding, the team was able to approximate different time arrivals and peg targeted IP flows. It worked in the lab with 100 percent accuracy, delivering infected traffic to the attacker's server for analysis. Next in line was testing the model on Tor itself.

For the live test, the computer scientists used victim clients in Texas (US), Leuven (Belgium) and Corfu (Greece). Each communicated through Tor routers to an attack server with NetFlow installed and under control by the testers in Spain. Under examination: actual Tor traffic with hundreds of Tor circuits operating simultaneously. In each case, targets downloaded a large file, circuits were infected and fingerprinted at the TCP connection upon exiting, and data was gathered by NetFlow from both the server and the exit node, at which point "flagged" IP flows were mapped and could be tracked straight to the destination. Overall, the team identified 90 measurements – 30 per each victim – that located the victim's precise location, tracking from the final exit node to the destination target itself, with 81.4 percent accuracy.

The Upside and Downside of Cracking Tor

As in the world of hacking, monitoring Tor remains a race between law enforcement and government agencies on one side, and sophisticated groups and individuals who leverage each new technological advance to cover their tracks and deeds. Areas that will require further investigation include: (1) Tor's built-in traffic shaping parameters, which might introduce traffic throttling that reduces the effectiveness of NetFlow and other commercial IP Flow monitoring tools; and (2) Traffic Padding, i.e., the use of "dummy" traffic to mask IP flow patterns including those infected with "fingerprinted" patterns.

For law enforcement and the intelligence community, the work of Drs. Chakravarty and Barbera was a starting point for revealing the identity and location of criminals, terrorists and other undesirables who use Tor to hide in the Dark Web. For individuals who rely on Tor to reveal the dark actions of government, this same accomplishment may prove a major blow to privacy, security and freedom itself. Following this initial research, reports have emerged of nation states investing significant resources to crack Tor, e.g., by monitoring traffic at bona fide Tor exit nodes (Russia) and by operating their own Tor nodes to track traffic back to targeted individuals (US - NSA, UK - GCHQ).

Tor has countered with further efforts to increase security, including bounties awarded to those who share any vulnerabilities discovered.

Tor OONIprobe: Outing Surveillance

The Open Observatory of Network Interference (OONI), an affiliate of the Tor project, has since 2014 offered a mobile app called OONIprobe that lets Android and iOS users test to see if they are blocked from visiting specific websites, and if their device is subject to "equipment interference" by government or law enforcement agencies using surveillance technologies.

The OONIprobe mobile app, available free of charge on the App Store and Google Play, reveals the names of specific surveillance industry vendors involved, and identifies each country where such solutions are deployed. OONI collects data from the tests and publishes the results on a map that shows the extent of tests conducted in each country, and the results proving the presence of censorship and government surveillance.

The declared mission of the OONIprobe app is to "redefine internet freedom for smartphone users all over the world." However, by exposing the presence of surveillance solutions in target devices of terrorists and criminals, the OONIprobe could prove a challenge to national security and public safety interests. Used correctly, OONIprobe in the wrong hands is a free pass to avoid monitoring.

One facet of the OONIprobe is software that conducts a "Web connectivity test" to determine whether websites are blocked and how, whether by DNS tampering, TCP connection, RST/IP blocking or by a transparent HTTP proxy. Specifically, the test does the following: resolve identification, DNS lookup, TCP connect and HTTP GET request, over a control server and also the network used by the target. Per OONI:

If the results from both networks match, then there is no clear sign of network interference; but if the results are different, then the websites that the user is testing are likely censored.

A second test called HTTP Invalid Request Line detects the presence of network components ("middle boxes") which could be responsible for censorship and/or traffic manipulation.

Rather than issue a normal HTTP request, this test emits an invalid HTTP request line that includes an invalid HTTP version number, an invalid field count and a large request method to an echo service set-up to listen on an HTTP port. An echo service simply sends back to the originating source any data it receives, and is a common measurement and debugging tool. Echo services also detect the presence of a "middle box" such as DPI solutions, and proxy devices that conduct man-in-the-middle attacks.

For example, if a Blue Coat ProxySG device is present in the tested network, the invalid HTTP request line will be intercepted, triggering an error message that is shipped back immediately to the OONIprobe app in the target's device. The error message is a dead giveaway of an abnormal presence on the tested line, and that traffic mirroring or manipulation are taking place.

One caveat: OONIprobe can't always specify the exact kind of box or software that the invalid request has hit. But even in those instances, there is no mistaking the fact that some form of outside intrusion is taking place. And as often as not, the OONIprobe can actually identify specific ISS vendors from the content of the error messages returned.

If a problem is discovered, OONIprobe offers ways to resolve it, such as helping the target change a DNS address. Alternately, the savvy target can simply ditch that smartphone and obtain another one. Throwaway phones are the favorites, not only among terrorists and criminals, but the privacy-dedicated, too.

What if the line is clean and no middle box is revealed? Then the echo service returns the invalid HTTP request without change, exactly as it was issued, and the connection is deemed clean.

The bad news for ISS vendors is that OONIprobe works, not just for ordinary citizens or journalists trying to safeguard their privacy, but for malevolent individuals, as well.

OONIprobe's HTTP Invalid Request Line test has uncovered the use of web censorship and surveillance solutions in China, Finland, France, Greece, Iran, Italy, Moldava, Myanmar, The Netherlands, Portugal, Spain, Saudi Arabia, South Korea, Sudan, Switzerland, Turkey, Uganda and the United Kingdom. Upon receiving evidence of monitoring in a specific country by any entity conducting surveillance or censorship of smartphones on a mobile network, OONI posts the result to the global map on its website.

OONI is certainly no newcomer to helping individuals check for censorship and surveillance. The organization has been in action since formed by the Tor project in 2012. What's different, and from the ISS vendor's standpoint alarming, is the simplicity and ease-of-use of deploying the new OONIprobe app.

Prior to launching OONIprobe, OONI required users to download its software package with a Command-Line Interface, a legacy era tool that interacts with a software program by issuing a specific set of commands via successive lines of code. While developers are generally familiar with the use a Command-Line Interface, average users far prefer graphical user interfaces. The OONIprobe meets that need head-on, offering a GUI interface that is simple to work with. The OONIprobe app is as easy to install as any other app available on the App Store or Google Play, enabling anybody to test a smartphone for its security in a matter of minutes.

The one downside for users: Just as government agencies can easily monitor who is using Tor, they might as easily detect the presence and use of OONIprobe. Merely downloading Tor is a red flag to investigators. Unless users opt out of having their test results go live on the OONI website, they will leave traces. OONI itself acknowledges:

Published data will include your approximate location, the network (ASN) you are connecting from, and the time when you ran OONIprobe. Other identifying information, such as your IP address, is not deliberately collected, but may be included in HTTP headers or other metadata. The full page content downloaded by OONIprobe may include such information if, for example, a website includes tracking codes or custom content based on your network location. Identifying information could potentially aid third parties in detecting you as an OONIprobe user.

China's Move on Quantum Cryptology

China has hacked Anthem, Bank of America, top U.S. military contractors, the passports of millions of Marriott Hotel guests, and even the White House – to name just a few targets – but formidable as their cyber capability may be, it could pale alongside one of their other feats: the first long distance fiber network protected by quantum cryptology.

Quantum cryptology has long been touted as the "next big thing" in encryption, supplanting mathematics with physics. Why it's different: cryptologic platforms are based on photonic keys that cannot be penetrated without immediately altering a message in a way that makes the detection obvious and reveals the attack.

Note that this doesn't mean quantum systems are invulnerable – merely that no attack goes unnoticed. Or so the theory goes.

Quantum cryptology has been used for limited commercial purposes since at least 2008, but with drawbacks. Qubits – the workhorse carrying encrypted load – must be managed at temperatures near absolute zero (-273 Celsius) and typically travel only a limited distance before dispersing. However, recent advances make it possible to use quantum cryptology over long distances, and China has been hard at work on such an ultra-encrypted network connecting Beijing and Shanghai since at least 2016. Small wonder, according to our sources, that the USA's National Security Agency is currently investing hundreds of millions of dollars on quantum computing, cryptography and cryptology. The agency needs to catch up with China, big time.

This development would appear to signal China's leap ahead of the U.S. in the two countries' quiet but still tensely fought war via cyperspying. That is, unless the U.S. succeeds at cracking quantum encryption.

Public-Key and Secret-Key Cryptology – Soon to be Outmoded?

For many years, encryption was been dominated by two forms of cryptology, i.e., coding to safeguard plain text from intrusion: public-key and secret-key. Public-Key Cryptology (PKC) uses two keys: one that everybody knows, and a second key that is known only to the recipient. The sender himself does not know the secret key.

In Secret-Key Cryptology (SKC), just one key is used and it is known to both sender and recipient. The secret key is embedded in the cipher (message) sent. An attacker can see the cipher, but without the key the cipher cannot be opened.

For its time, given the billions of mathematical combinations possible for a key in PKC, even billions of computers operating at billions of transactions per second would require one trillion years to hit on the correct key to decode a cipher. The problem now – at least for those hoping for impenetrable encryption – is the coming generation of computing power that is far more advanced and could crack PKC easily: quantum computers operate on an infinitely small (and fast) level.

SKC has issues, as well. The only perfect way to ensure the security of a secret key is for two parties to meet in person and decide on that key. Any attempt to transmit an agreed-upon key must also be encrypted, otherwise a third party can and will listen in and obtain the secret key. For parties communicating via SKC from remote locations, this issue is called the "key distribution problem."

The Download on Quantum Cryptology

At the core of quantum physics work in encryption is the qubit, the tiniest measure of light. Qubits are constantly in motion, spinning in all directions at once. Light in this unpredictable state is unpolarized, i.e., without direction. This unpredictability is the basis of quantum physics and of the Heisenberg Uncertainty Principle, which states that one cannot know an object's velocity and principle simultaneously. By extension, it is impossible to measure a qubit without affecting its behavior.

The same principle can be turned to advantage in using qubits for encryption.

Cryptographers schooled in quantum physics today use light emitting diodes (LEDs) to create a stream of unpolarized qubits, one at a time. With special filters, they can then polarize each qubit to take a specific state or spin, such as vertical or horizontal. Qubits that emerge from the filter will be in the state defined by the filter.

When sending a message, the user assigns a binary code to each qubit, and applies random filters, recording each qubit's assigned polarity. Thus, each qubit has a distinct polar signature identifiable through a filter known only to the sender. For the purpose of this explanation we'll keep things simple and assume that only two filters are used.

When the signal arrives, the recipient applies his filters to each qubit in the encrypted message received. What follows next is an unencrypted exchange between both parties, in which the recipient matches filters to each qubit and the sender replies "correct" or "incorrect." Because the recipient isn't revealing the sequence of filters applied to qubits or his measurements, it is impossible for an eavesdropper to rebuild the order of the qubits.

At that point, knowing the correct filter per qubit, the recipient can decipher the message by applying the correct filter to the appropriate qubit and translating the binary code.

Why Quantum Cryptology Could Be Bad News for Cryptanalysts

For years the NSA and China's various schools of hackers, from Axiom to the 20 or more military Advanced Persistent Threat (APT) units, have played a back and forth tit-for-tat game of cyber espionage. For obvious reasons of security, the NSA is very close-lipped about its own activities. China's, on the other hand, are the subject of intense investigation by respected organizations such as Mandiant and others. In addition, because Chinese cyberattack groups attack commercial as well as defense and national security targets, their assaults, once identified, are widely publicized.

Largely government-sanctioned and in many instances conducted under the direct authority of China's military, the APT arsenal is not going away anytime soon. However, once China implements quantum cryptology, some analysts fear that the challenge to Western cryptanalysts could rise by an order of magnitude.

The essence of classic decryption is to passively intercept a message and then work to decode it without the knowledge of the communicating parties. Such interception is accomplished via mechanisms ranging from traditional wiretaps to today's more sophisticated malware and man-in-the-middle attacks.

Quantum cryptology is the first initiative to successfully thwart passive interception. Remember the Heisenberg Principle, its meaning relative to measuring a qubit and the potential impact of such measurements on the qubit's behavior.

After being polarized by a specific filter, a qubit cannot be measured again except by the identical type of filter. So if a qubit with a horizontal spin is measured by a spy or hacker using a vertical filter, one of two things happens: Either the qubit won't pass through the disparate filter, or its behavior changes and it adopts the identical spin as the filter – vertical. The qubit's original polarization is lost, as is any information linked to its original spin.

In a passive intercept, the attacker will capture the qubits, measure them through his own set of filters, then use an LED to re-send them along to their intended destination. The problem, for the hacker, is that merely by measuring the qubits he has inevitably changed their polarity. So when the qubits reach their destination they either act differently or cannot pass through the recipient's filter – considered a dead giveaway of an attack.

Most frustrating for the hacker: By having to guess at the filters and likely using the wrong ones at least half the time, he is destroying the binary code message attached to the qubit. Is that bad news for Western hackers? Maybe. But as with so many things in this field, new attack methods seem to arise as quickly as those designed for defense.

Nothing is Un-hackable

In practice, it is generally unwise to claim that any new technology is foolproof or bombproof. Even much-lauded quantum cryptography has its vulnerabilities.

It has been known for some years that a hacker can blind a qubit detector with a strong pulse, making it incapable of seeing inbound qubits. That could present an opportunity to hack qubits the user is not aware are arriving.

Researchers at the Massachusetts Institute of Technology (MIT) have taken advantage of a phenomenon known as "photon entanglement," wherein a single qubit has two interdependent states. By measuring one state they were able to accurately forecast the other state and decode a photon cipher without alerting the receiving photon detector that it been hacked.

So much for the Heisenberg Unpredictability Principle.

Then, too, recall the current common practice using LEDs to create and encode one qubit at a time. There is a possibility that the device may create a second qubit with the identical secret information, which could be intercepted by an adversary without the users' knowledge.

Another headache could be false alarms. Under certain conditions users of quantum cryptography might think they're being hacked when, in fact, nothing adverse occurred. A photon detector might have failed to register the arrival of a qubit, raising concerns that the system is under attack.

Dark Mail vs. Offensive Cyber

A former General Counsel for the National Security Agency once commented that if an agent could obtain sufficient metadata on a target, then "content" became almost irrelevant. Metadata alone would provide sufficient indication of the target's whereabouts, identity and the same factors for whomever he was in communications with.

Loss of the ability to track targets by bulk metadata collection explains why the NSA was sufficiently concerned about passage of 2015's USA Freedom Act to request a six-month extension of the controversial Section 215 of The Patriot Act – and why the agency fought to retain bulk metadata collection under Section 702 of the Foreign Intelligence Surveillance Act (FISA), which allows the capability when applied to foreign entities suspected of engagement in terrorist activities, but as often as not collects "incidental data" on millions of harmless individuals.

However, it has been suggested that other work underway by a small but highly skilled group of cryptologists may do away with any and all access to metadata via an emerging app. The app is a form of email, made possible by an emerging protocol with the ability to provide end-to-end encryption. If broadly adopted, this ostensibly un-hackable form of email would literally make it impossible to ascertain one jot of metadata about who is communicating with whom.

The group's work is being closely monitored by major providers of commercial email systems including Google gmail and Microsoft Outlook, and if successful will become a key addition to the companies' arsenal of methods for protecting user privacy – potentially making it difficult to impossible for any agency, however sophisticated, to crack the code.

The app is called Dark Mail or Dark Internet Mail Environment (DIME), a revolutionary approach to email that holds the promise of making metadata interception impossible.

Dark Mail is the offshoot of the Dark Mail Technical Alliance (DMTA), a group founded in late 2013 by Ladar Levison, Phil Zimmerman and other pioneers of earlier secret email services and modes of encryption.

Levison is perhaps best known as the founder of Lavabit, an encrypted email service that he voluntarily shut down rather than comply with FBI requests to provide encryption keys for a single customer. Zimmerman is the renowned encryption expert who gave the world Pretty Good Encryption (PGP), arguably one of the best systems of its type ever developed.

DMTA is small, comprised mainly of similar-minded tech experts, and relies entirely on contributions for its survival and advancement of the Dark

Mail cause. However, the group has hope of a "bright," if that is the right word, future for Dark Mail.

One might well ask, "Why bother with Dark Mail, at all, if an end user can simply log on to Tor or take advantage of PGP?" We've already answered the first part of that question: Merely being on Tor is a red flag that makes the user suspect in the eyes of government agencies, and there are proven methods of hacking The Onion Router.

As to PGP, it is an excellent system, to be sure, but is so complex that it has had virtually zero uptake in the decade-plus since its development. But an email system with encryption already embedded? That would be a different creature altogether.

The Evolution of Security

In the early days of online communications, "security" had a very different meaning and intent: safe and reliable military and government communications in the event of nuclear war. With that goal in mind, the Defense Advanced Research Projects Agency (DARPA) took on a secret and very special project: ARPANET, the forerunner of today's Internet.

Recognizing that the public switched telephone network (PSTN) – and the private military lines therein – would be primary targets in the event of war, DARPA devised a better alternative, a system based on multiple routers that could safely transport email and FTP transfers. Every message was broken into multiple packets that followed separate paths for final reassembly at their destination. At every stop, servers stripped packet headers to determine sender and recipient, then reassembled the data and sent it on its way.

The next challenge: protecting the data network from penetration or sabotage by foreign foes. In the mid-1990s, the U.S. Naval Research Laboratory came up with the solution, The Onion Router. Tor works by encrypting communications at the application layer and transmitting it to random servers, which "peel back" one layer of encryption at a time to reveal only the IP address of the next router as the message is passed along to the exit server.

A public version of Tor was released in the Fall of 2002 and quickly gained credence as a safe way to surf or communicate on the Web without observation. Tor today is little different. With Tor, each of some 7,000 servers is donated by multiple participants globally. Every server's IP address is masked so that it may not be identified. And finally, each server sees only the address of the next server in the chain, not the identification of the sender or recipient. As Tor puts it:

Tor helps to reduce the risks of both simple and sophisticated traffic analysis by distributing your transactions over several places on the Internet, so no single point can link you to your destination. The idea is similar to using a twisty, hard-to-follow route in order to throw off somebody who is tailing you – and then periodically erasing your footprints. Instead of taking a direct route from source to destination, data packets on the Tor network take a random pathway through several relays that cover your tracks so no observer at any single point can tell where the data came from or where it's going.

To create a private network pathway with Tor, the user's software or client incrementally builds a circuit of encrypted connections through relays on the network. The circuit is extended one hop at a time, and each relay along the way knows only which relay gave it data and which relay it is giving data to. No individual relay ever knows the complete path that a data packet has taken. The client negotiates a separate set of encryption keys for each hop along the circuit to ensure that each hop can't trace these connections as they pass through.

The Dark Side

Conceptually, Dark Mail or DIME is similar to Tor. A message or other Web transaction is encrypted at the outset. Once the message enters Dark Mail it traverses a series of routers, each of which can see only the destination server address and then forwards the message along to the next router in the chain. Just as with Tor, at each stage a layer of encryption is removed, but only enough for the hand-off.

One attractive feature of Dark Mail is the system's use of "signets" that are unique to each user and known only to him and to the party he communicates with. Signets are designed to simplify the ordinarily complex process of encryption and key management. A personal profile of the user might be stored in the signet, but only he and those he communicates with can break the seal.

In one iteration of Dark Mail, the system is managed by a master key management system. Each transition point has two sets of keys – one private and one public – to handle decryption of metadata relevant to reaching the exit node, then encrypting it again before it is sent on. The parent entity involved, be it corporate or government, in turn serves as the repository of keys for each of its servers, and email addresses for each of its users.

DMTA likens the process to DNS Resolution, the system by which Internet users type in a name of the website they wish to find and have DNS automatically translate the word or words into the destination's IP address. In the case of Dark Mail, DMTA's own servers perform the functions of Mail

Transfer Agent and Mail Delivery Agent. What if one party's server does not support Dark Mail? Not a problem. Dark Mail can create keys on the spot and automatically encrypt the user's and recipient's email so that they are at that point communicating securely. Both parties' Mail Transfer Agents, now fully encrypted, are uploaded into their servers' DNS capability. This "trustful" mode lets an organization or individual use Dark Mail for specific instances and gain the confidence to upgrade and use Dark Mail full-time.

"Trustful" mode is the lowest of three layers of secure communications planned by DMTA. The problem with "trustful" is self-evident: trusting the keys and the encryption to one master organization, which could itself be hacked. DMTA prefers to recommend its other two levels, "Cautious" and "Paranoid," wherein communications are encrypted on the user's own device and only the owner – never another organization – has access to the keys.

Will the Curtain Go Up on Dark Mail?

Already several years in the making, Dark Mail still remains a work in progress as of early 2020. DMTA has issued a test version called Magma which interested parties may explore. DMTA also offers technical specifications for Dark Mail on its website to generate interest and input. The chief obstacle is lack of endorsement by any major service provider, customer, or standards-setting body. Support by the Internet Engineering Task Force (IETF) would be a major boost, and the group is certainly familiar with Dark Mail and the work of DMTA. However, to date the IETF has not issued any formal word on its attitude on Dark Mail. Similar inaction is evident among major service providers, equipment manufacturers, potential customers and Internet groups. The dominant attitude appears to be "wait and see" or "let somebody else go first."

The rationale for such lethargy may be "cautious" and "paranoid" itself. It would seem a rule of thumb in the security realm that any system touted as being invulnerable is begging for a beating. If you can crack Tor, as has been proven most possible, then why not hack Dark Mail, as well?

Darktrace: AI Versus Offensive Cyber

One of the more intriguing outcomes of the black hat arts is the evolution of cyber breaches that use cognitive computing techniques with the ability to learn about a potential victim's network or personal interests and to make modifications that will leverage a vulnerability. Such "smart attacks" offer a major advantage over the majority of antivirus and cybersecurity systems which act based on known signatures. Basically, the latter are helpless if the signature evolves to work around a firewall.

Of course, from the standpoint of ISS ethical malware, such a weapon would be ideal. One company addressing one and possibly both sides is Darktrace of the UK. Darktrace is a unique presence in the cyber marketplace in several regards.

Inside Darktrace

The company, founded in 2013, provides a cyber suite that uses AI to continuously scan a client's network, learn the "subject" and implement essential safeguards that improve security on an ongoing basis.

Darktrace was launched by mathematicians at Cambridge who could draw on their skills with algorithms to assist clients. But the idea for the company came from members of the British Intelligence community who approached Cambridge for help.

This is an interesting twist. In the US and UK it is common practice for private contractors to develop solutions that provide the core of both defensive and offensive cyber used by government. But when the inspiration for an advanced technology solution comes *from* a government agency, that is something of a rare event.

Finally – not that sex matters – but it is worth mentioning all the same: Darktrace co-founders Nicole Eagan and Poppy Gustafsson are women, an anomalous and frankly welcome change of pace in the male dominated ISS community. The team quickly drew other experts, including former high-ranking officials with the UK's MI5, and key cyber directors from the NSA.

Darktrace's combination of cyber and math skills quickly drew the attention of venture capitalist Mike Lynch, who dropped an estimated £1.3m into Darktrace in September 2013, not coincidentally the date that the company rolled out its core platform.

Learning from the Human Immune System

Unlike the static viruses, worms and malware of yesteryear, today's most serious cyber threats mimic actual viruses from the native world. They evolve and take new forms. Just as human vaccines are a "hit and miss" approach to preventing disease, essentially guesswork on which strain of virus might prove most prevalent and lethal, the dominant approach of most cyber defenses is to attempt walling out threats based on known signatures. Oftentimes, the human immune system, which itself shows the power to mutate and adapt, is a far better defense against disease. Darktrace adopts this principle to its cyber model, the "Enterprise Immune System."

The basis of the Enterprise Immune System is a complete understanding of all networks, computers and end user devices – and every crumb of data that transits these vehicles – throughout an enterprise client. The starting point is "Darkflow," a big data tool that uses real-time predictive analytics plus heuristic data search to capture and analyze all transactions past and present. Darktrace calls the process "Real-Time Total Network Immersion." Any anomalies discovered are further assessed via algorithms that examine and prioritize them for risk.

Next up, Darktrace hones in on any point of weakness found in the network or legacy cybersecurity system. It might be a previously undiscovered Zero Day, a flaw in command code, an unprotected end user, or vulnerability in any one of a myriad of IoT-connected devices – among the favorite targets of black hat attacks used for systems penetration, data theft, DDOS, ransomware and similar ends.

Similar to many big data analytics solutions, the Enterprise Immune System comes equipped with 3D visualization. Dubbed Threat Visualizer, the system provides what Darktrace calls "topological network projection technology," a pregnant phrase that underscores the use of mathematical formulas to provide graphic insights – not a static view, but change in motion, i.e., threats as they're happening.

To those familiar with the ways that real-time predictive analytics is applied in the commercial arena to better understand and serve customers, or in the ISS world to identify and track targets, there is nothing all that new about Darktrace's use of the identical methodologies in the cyber defense sector. Darktrace follows a similar approach, but focused inside on the client's own network versus outside on the world at large.

Invariably, the weakest link in all such systems is the human analyst on the tail end, who is presented with a variety of Monte Carlo Simulation engine-type forecasts and given the opportunity to choose, that is meddle, with the best course of action.

Here, Darktrace's true differentiator comes into play: machine learning. The company's Antigena solution is AI's version of the human immune system, identifying the threat, mutating to defeat it, and self-defending the network.

A Problem for Ethical Malware?

As cybersecurity experts universally acknowledge, cybersecurity is a game of one-up-manship where new threats continuously emerge by the thousands to challenge traditional defenses. Thanks to machine learning, Darktrace would appear to offer a significant advantage through its solutions' ability to morph with the threat.

As unlawful target groups advance their own sophistication via adoption of AI, the question arises whether conventional ethical malware sold by ISS vendors will have "the wits" to bypass cyber immune systems used by terrorists or organized crime. To retain the loyalty of law enforcement and government agency clients, ISS vendors are racing to embrace machine learning, as well as other forms of AI. Darktrace is well ahead of most.

DARPA Memex Dark Web Intelligence

The teaming of companies such as Diffeo, Uncharted Software and Hyperion Gray in the DARPA Memex partnership program from 2015-2017 spurred major advances in alternative search technologies, making it possible to range all over the Deep and Dark Web via open source software.

Without question, the types of information now available through Memex-inspired search capabilities far exceed what is possible with conventional solutions such as Google and Bing. Even so, skeptics question whether the Deep and Dark Web deliver intelligence worth the effort and investment. We believe the answer is "yes – absolutely." Where disagreement arises it is often due to a misunderstanding of what is "deep" and what is "dark."

Deep Web Versus Dark Web

"Deep Web" pertains to content, websites or older forms of Web communications such as GIFs that are not indexed or are indexed with low page rankings by conventional commercial search engines led by Google, Bing and Yahoo. Deep Web content may be as simple and innocuous as academic treatises, scientific papers, registration-required web forums and dynamically-created pages like Gmail accounts, or news, text, video and images that were not deemed worthy of saving for future days when some might wish to view and retrieve them.

However, arbitrary decisions either by the content owners' authors or by modern search engines does not necessarily negate the value of Deep Web material. Nor does it mean that Deep Web carries any evil connotation, or that this content has been deliberately hidden. Deep Web simply pertains to information that was overlooked and omitted from commercial search engine capabilities, whether by intent or accident. By some estimates, and these vary widely, the Deep Web comprises some 96 percent of the Internet.

"Dark Web" refers to any material on the web that is unindexed and deliberately hidden from view by encryption. It is a tiny subset of the Deep Web, responsible for less than .01 of the Internet. The most popular vehicle of the Dark Web is Tor (The Onion Router), a separate browser that encrypts and then randomly bounces communications through a network of relays, separating identification and routing. Users typically download and access Tor via a virtual private network (VPN) connection that assigns a different IP address for every log-on, thus creating an added layer of invisibility. Tor and similar Dark Web browsers are popular with journalists and activists, as well as criminals and terrorists.

Tor is not perfect. Relying on an elaborate network of volunteer servers worldwide, Tor can be maddeningly slow. It is not "friendly" to high-bandwidth apps such as videos or any streaming data, which bog down en route. Nor, as discussed, is Tor 100 percent reliable at concealing identities. Users often make simple mistakes that make them easy to find. In addition, there are proven ways of cracking Tor, beginning with the relative ease with which law enforcement and government agencies can quickly determine who downloads and uses the service.

The key difference between information on the The Deep Web and the Dark Web is "low" versus deliberate "no" indexing. Man-on-the-street average users might understandably confuse the two. More surprising is that public sources which should know better continue to use the terms interchangeably.

To its credit, BrightPlanet, one of the earliest companies to enter the market offering services for both niches, has long made a religion of distinguishing Deep Web from Dark Web. While the company has evolved to a dual-purpose firm supporting commercial enterprises, its ISS roots, unfortunately, tend to make BrightPlanet a "voice crying in the wilderness" on the matter.

Reassessing Web Search – Where Google Went Wrong

The fact that the Deep Web exists, at all, may in large part be blamed on Google. By focusing on and providing commercial advantage to Web content that follows the search engine giant's rules of engagement for indexing, Mountain View deliberately draws a line between what is easy to find and what falls through the cracks and becomes obscure. By definition, searchable content always has an involved owner and master. All else is orphaned. Rewarding those entities who follow the indexing rules, and creating opportunities for tag-along advertising sales, is Google's whole business model.

Google is rightly credited with innovating the concept of centralized indexing that today powers some 70 percent of global web searching. Microsoft's Bing and Internet antique Yahoo share 7.0 percent, and the rest is divvied up among various small players. One caveat to these stats: They do not cover Baidu, China's #1 search engine with dominant market share in that nation. Google has gone back and forth on doing business in China, leaving at one point over disagreement with adhering to Chinese surveillance laws the company considers invasive, then announcing in late 2018 that it would return. Google does maintain a nominal presence in China, but in the months and years the Mountain View, CA company waffled, Baidu took command of web search and ranking in that nation.

Going back in time to Google's emergence in the U.S. in 1998 one finds a very different story – the classic tale of innovators who watch patiently while the pioneers in a new marketplace get arrows in their backs.

Google learned from and leveraged its mistakes.

In a comparatively short time in the early 1990s, the World Wide Web had accelerated from a finite planet to a sizable universe. In 1994, the early search engine "World Wide Web Worm" boasted that it had succeeded in indexing 110,000 web pages and received 1,500 queries per day. By 1997, the top search engines claimed to have indexed 100 million web documents. Altavista alone claimed to process 20 million queries per day.

In an academic paper presented at Stanford University that year, young Sergey Brin and Larry Page postulated that by the year 2000 search engines would be slammed by "hundreds of millions" of queries per day, creating epic problems in scalability and quality of search results. Brin and Page felt strongly that the quality of search results was sloppy. Too often, they claimed, the searcher ended up with "junk results that often wash out any results that a user is interested in." Their solution was Google, a scalable search engine backed by massive processing power to keep pace with crawling, indexing and storing pages, and the use of hypertextual information – notably link structure and anchor text to serve as filters providing a higher quality and relevance of the search results. The end result was "PageRank" – top notch, highly focused search results based on the quality ranking of each web page. PageRank, they believed, would serve as "an objective measure. . .that corresponds well with people's subjective idea of importance."

And so it did. The overlooked side effect of Google's approach: With time, large swaths of data fell by the wayside. By serving the subjective interests of searchers, Google became sole arbiter of what was "objective." This populist approach to search engine design quickly became tyrannical by jettisoning poorly-indexed pages. Material with low PageRank filtered out into what we now call the Deep Web. Today, Deep Web content and pages are virtually impossible to find using conventional search. Ditto for Dark Web pages with deliberately low PageRank.

As it happens, law enforcement and government intelligence agencies often take a different view on what counts as important on the World Wide Web. The centralized, filtering PageRank or Google and Bing won't take them there.

Even more problematic: Dark Web activities on Tor, Freenet and I2P (Invisible Internet Project) are unindexed and invisible, thus a perfect haven for criminal activity.

DARPA Memex to the Rescue

In February 2014, DARPA (Defense Advanced Research Projects Agency) decided to help fill this gap, issuing an RFP for multiple proposals by companies that could deliver innovative solutions that would provide better insight into the Deep Web, and even the ability to crack the Dark Web.

DARPA had for years been pursuing the challenge on its own through a program dubbed "Memex," to identify human trafficking operations. Now DARPA wanted to go further by encouraging the involvement of private industry.

The inspiration for Memex came from a Dr. Vannevar Bush of MIT, who during World War II headed U.S. military R & D for the Office of Scientific Research and Development responsible for the Manhattan Project and other weapons with far-reaching strategic impact. Following the war, Dr. Bush published a popular article, "How We May Think," calling for new approaches to problem-solving that went beyond linear reasoning to emulate human creative thought, but at greater velocity and with limitless storage and access capabilities. Initially called "At Random," the concept grew into a new name "Memex" (memory/index) widely credited for inspiring the first hypertext sessions.

In borrowing the name some 70 years later, DARPA paid tribute to and sought the same "outside the box" thinking to provide new methods of search for the Deep Web and Dark Web. Dissatisfied with Google's and others' limited "one size fits all" approach to search, DARPA told applicants to move beyond reliance on centralized indexing.

DARPA's biggest problems with conventional search:

- Limited scope and richness of indexed content, which might not include relevant components of the Deep Web such as temporary pages or pages behind forms.
- An "impoverished index," missing shared content across pages, normalized content, automatic annotations or content aggregation.
- Basic search interfaces that render every session an independent solo act with no collaboration.
- No history beyond the search term.
- Requirement of exact text input for search.
- Standardized interaction with top-line web content, requiring one-by-one manual queries that invariably return the same results – from the same indexed content.

What they sought:
- Domain-specific indexing and search capabilities.
- Mechanisms for content delivery, information extraction and retrieval.
- User collaboration.
- Ability to address distributed aggregation.
- New ways to analyze web content.

DARPA open-sourced key components of Memex so that others could build on the platform to provide law enforcement-specific applications. What evolved was collaboration by think tanks, universities, and private companies in the ISS field. Among accomplishments that stand out:

- **Easy-to-Use Interfaces:** With Uncharted Software in the lead, private and academic resources combined their efforts to provide state-of-the-art front interfaces for law enforcement that simultaneously process and can display vast amounts of information from the Deep and Dark Web.
- **Web Crawlers That "Think" Like People:** Bringing Dr. Bush's concept of human "at random" thinking to life in a machine environment, web crawlers by Hyperion Gray sidestep password-protected authentication to enter Dark websites. They can also complete forms to determine whatever evil might lurk therein – human trafficking, drug deals, murders for hire and more – without alerting the host site. Another tool by Hyperion makes it possible to do something impossible with Google: find matches on limitless numbers of seemingly unrelated pages without benefit of hyperlinks.
- **Freedom from PageRank:** Diffeo produced a tool called Dossier Stack that can search the Deep Web with precision, filtering out non-essential results by exclusion, e.g., "Search for ISIS in Turkey, not Iraq, not Syria." As the search proceeds, Dossier Stack's algorithms learn and apply new filters to refine results. Arbitrarily uniform page ranking, typical of Google, never enters the picture. The user exercises full control to look for and find what he wants. Diffeo evolved from its principals' involvement in setting up a partnership for Memex and the National Institute of Standards and Technology (NIST) called the TREC KBA (Text Retrieval Conference Knowledge Base Acceleration) project. TREC KBA continues to contribute to the field of Deep Web penetration. Among its many signature achievements: development of "Dynamic Domain" – the ability to conduct collaborative searches for multiple domains of interest, with "multiple runs of user and search engine interactions." Dynamic Domain learns from what it finds and adjusts en route to improve the field of findings.

Other partners contributed real-time web crawling and ways to lower the cost of parallel processing. NASA's Jet Propulsion Laboratory (JPL) built applications on open source Memex that enable streaming analysis of text, images and video, then rank findings by relevance to create a picture that would never come into focus using conventional search tools.

Endgame Zero Day Exploits

Years have passed since US Marine Corps veteran Nate Fick took the helm as CEO at Endgame, shifting the company's focus toward providing vulnerability intelligence and away from the company's once-vaunted Zero Day exploit prowess. The question is whether Endgame would entirely abandon such a valuable asset as knowledge of offensive cyber.

At the heart of this conflicted public vs. private stance is the ongoing battle over the "ethics" of offensive vs. defensive cybersecurity. Privacy advocates and other critics decry the practice of using exploits at all, even when the "victim" is a nation state or criminal hacker using the self-same technologies to attack others. Proponents of Zero Day exploits say it's madness not to retaliate in kind.

Before public exposure of its business led to Mr. Fick's about face, Endgame was clearly in the latter camp. The switch proved a winner with financial backers, pulling in a fresh infusion of US $23.0 million the moment Fick took charge. But without some involvement in Zero Day exploits, Endgame might be little different from your run-of-the-mill penetration testing outfit. They are, in fact, far more advanced and sophisticated.

In the Beginning: All-Out Offensive Cyber

When famed cyber expert Chris Rouland launched Endgame in 2008, there was no doubt about the company's mission: providing the means of preemptive and/or retaliatory cyberattacks for government agencies. Specifically, Endgame provided not only Zero Day exploits for the assault, but a blueprint of the target's exact systems, devices and locations via a software tool called Bonesaw, which according to reports, became a vital weapon of the NSA, CIA, Cyber Command in the U.S., as well as for the UK's MI6.

All was smooth sailing for Endgame and Rouland until 2011, when Anonymous hacked into company files of a partner, releasing documents exposing Bonesaw and other details of Endgame's business. Rouland promptly took down Endgame's website, gave orders never to use the company name in a press release and for all practical purposes led the company into a black hole. All to no avail, however, as there was further damage to come in the form of charges that Endgame had been involved in a hack of Wikileaks.

By late 2012, Rouland was out the door, replaced by Fick, who immediately set in motion a one-hundred-eighty degree turn on Endgame's market focus. Within a year, he was doing the formerly unthinkable – speaking to press. In a memorable interview with *Forbes*, Fick made no bones about his feelings on Bonesaw and Zero Days. "The exploit business," he said, "is a crummy business to be in."

Not everyone shares his feelings.

Gone But Not Forgotten

At a Sept. 2013 Carnegie Council on Ethics in International Affairs, Rouland hammered the lack of deterrence capabilities in cyber that leaves enterprises wide open to attack. Harking back to the days of Cold War deterrence, he said, "I do think eventually we need to enable corporations in this country to be able to fight back," by which he meant go on the offense.

At the time, Rouland acknowledged that the U.S. Computer Fraud and Abuse Act disallows such private hack-backs. A point no one raised: Hackbacks are completely legal when done in partnership with law enforcement, as occurred in 2013 when Microsoft joined forces with the FBI to launch an attack on Citadel "malware-as-a-service."

Citadel Trojan malware, likely the product of Russian or Ukrainian hackers, emerged in early 2012 as an offshoot of an earlier virus, Zeus. Citadel was used to steal IDs and empty bank accounts via botnets. Users of Microsoft products were among the prime targets.

With court approval Microsoft's cybercrime unit, accompanied by FBI agents, broke down the doors at Citadel's U.S. hosting operations. Captured forensic evidence pointed straight to the IDs of infected devices, enabling Microsoft to send alerts to users.

Although Microsoft's well-orchestrated campaign put an end to that incident, the hackers got away scot-free. No thought was given to fighting back or punishing the transgressors. The rationale for going only so far: concerns that retaliation or any form of brinksmanship might lead to even worse cyberattacks than Citadel.

Big Data + Zero Day Intel for Robust Vulnerability Intelligence

Endgame's public suite of services offers "vulnerability intelligence" informed by analysis of a client's cyber defenses, and a wide array of other client data, matched against Endgame's own extensive research on cyber threats.

Endgame's approach is characterized as big data analytics based on parallel processing to reduce massive amounts of data to manageable blocks. Beyond that, Endgame uses real-time predictive analytics (RTPA) to query and score "data in motion" driven by evolving changes in the world of cyber threats. Actionable insights alert the client to potential weaknesses in their cybersecurity system, staff, vendors and other openings that might be leveraged by intruders.

Note: this is more than just penetration testing or "pentest." Vulnerability assessment involves close examination of business processes, data sources, BYOD by employees, hardware, network infrastructure, firewalls, intrusion detection and prevention systems, VPNs, encryption, policies, and typical weak points such as remote offices.

If there is a hitch in this picture it may be the challenge of educating the customer. The marketplace is rife with firms that claim to offer both pen testing and vulnerability assessment but are in fact weak on the latter count.

The factor that makes Endgame stand out is the one it strives hardest to distance itself from. Continued involvement in offensive cyber is crucial to the research Endgame banks on when serving both government and enterprise clients, and a key differentiator. As has often been said, the best defense is a strong offense – or at least deep familiarity with the latter.

FBI Offensive Cyber Powers

Does the U.S. Federal Bureau of Investigation undertake investigations using advanced technologies that cross the line and exceed the agency's legal mandate? Strictly speaking the answer is: No. The FBI always acts within the law. However, the extent of powers available to the FBI via Department of Justice regulations and Executive Order might surprise many.

The FBI's use of offensive cyber to investigate and apprehend cyber criminals and terrorists falls under rules outlined in two federal tomes, one very definitive memo and a key Executive Order, the latter dating back to 1995 and President Bill Clinton:

- **Tome #1:** The FBI Domestic Investigations and Operations Guide (DIOG) Sections 2.8.7, 2.8.9 and notably Section 2.10, (U) *Use of Classified Investigative Technologies*. The DIOG is a bible of rules first implemented in December 2008, updated in 2011 and stretching 248 pages, of which Section 2.10 on *Classified Investigative Technologies* comprises a single seven-line paragraph.
- **Tome #2:** Section 14, Attorney General's Guidelines for Domestic FBI Operations (AGG-Dom), Part II and Part VI.B. The AGG-Dom, issued on Sept. 29, 2008 by then-Attorney General Michael Mukasey, is a 46-page set of "guidelines" on FBI responsibilities for "federal crimes, threats to national security, and foreign intelligence." Of note, the AGG-Dom also details the FBI's authority to act as "an intelligence agency."
- **The Memo:** This is a Jan. 22, 2002 memorandum from then-Deputy Attorney General Larry D. Thompson, titled, "Procedures for the Use of Classified Investigative Technologies in Criminal Cases."
- **Executive Order 12958:** authorizing the President and his designees to declare any investigative technology "Top Secret" and off-limits to the public.

Section 2.8.9 of the DIOG is essentially a disclaimer leaving FBI's field of play wide open. It says that while the AGG-Dom applies to "domestic investigative activities" it does not "limit other authorized activities by the FBI." It goes on to say, "The authority for such other activities may be derived from the authority of the Attorney General as provided in federal statutes, guidelines and Executive Orders."

In other words, carte blanche. Section 2.10 of the DIOG discusses inappropriate use of classified technologies. It is interesting primarily for its reference to the FBI organization that holds authority for these technologies: the Operational Technology Division (OTD).

Then we come to AGG-Dom, which sets the stage for the DIOG. Basically, the AGG-Dom is an abbreviated version of the DIOG filled with legal generalities. The one exception: reference to the OTD's authority under the "Procedures for the Use of Classified Investigative Technologies in Criminal Cases."

These procedures are found in former Deputy Attorney General Larry D. Thompson's 2002 memo outlining the breadth of classified investigative technologies that may be deployed by the FBI, the Drug Enforcement Administration (DEA), the U.S. Marshall's Service or the U.S. Immigration and Naturalization Service (INS).

Per the Deputy AG of the George W. Bush era, "Classified Investigative Technology" defines "any hardware, software or investigative technology" that "is designed to intercept or acquire information of evidentiary value pursuant to Executive Order 12958 of April 17, 1995, as amended, or other successor Executive Order. . ."

Under Executive Order 12958 the President, agency heads and "officials designated by the President" hold the authority to declare any investigative technology as classified, Top Secret.

According to the above rules, federal law enforcement has Platinum Carte Blanche to use virtually any surveillance technology.

Who's on First at the FBI – Defensive or Offensive Cyber?

The essence of successful federal investigations is cooperation among all parties involved. The key to personal *advancement* in government is evidence of accomplishment. Within the FBI's OTD one finds two groups at odds with one another over job assignments that clash, with inevitable negative outcomes for personal advancement and cooperation in investigations. Those OTD units are, respectively, Offensive Cyber and Defensive Cyber.

The mission of Offensive Cyber is the development of offensive cyber tools and techniques that can be planted in targets' systems or devices to leverage a system's weaknesses. The mission of Defensive Cyber is to develop protections that guard against such offensive tools, and to create patches that fix systems weaknesses. According to inside sources, the two groups do not get along very well, and for good reason: What possible incentive is there for an expert in offensive cyber to cooperate with counterparts in the defensive cyber group, whose personal advancement hinges on undoing the work of the former? As a result, the two groups tend to operate at a distance. Defensive cyber staff, who like to look and act like heroes, receive the most public attention and plaudits.

However, all within the DoJ and its various law enforcement agencies are fully aware of the critical value of offensive cyber. When intra-agency cooperation comes up short, what is their fallback? Ostensibly, it is outsourcing, to literally hundreds of contractors.

It is well-known that in the wake of major U.S. cyberattacks such as those on SONY and ANTHEM, the FBI routinely claims to have called in vendors to assist with forensics. Less well-known: the FBI itself possesses massive parallel processing powers to conduct cyber investigations. Contrary to claims by former Director James Comey, the hack of the San Bernardino terrorist's iPhone in 2015 was not done by an outside vendor. According to FBI officials speaking on condition of anonymity, the San Bernardino iPhone hack was actually an inside job executed internally via custom hardware attack. Next let's look at other tricks the FBI has in its bag of cyber investigative techniques.

FBI Network Investigative Techniques

Reports on FBI Network Investigative Techniques (NITs) typically skirt two key issues: What is the legal authority for lawful black hat activities; and once a warrant is obtained, what are the specific technologies involved? As it turns out, both the legal and technical aspects of NITs constitute a gray realm that blurs the black and white simplicity of lawful intercept under the Communications Assistance for Law Enforcement Act (CALEA). NITs are serious capabilities that take agents places CALEA can't touch, in some instances right to the heart of Tor to capture evidence on targets.

Just two points to bear in mind.

First, there is no specific U.S. surveillance law that authorizes use of offensive cyber by a federal law enforcement agency. When the FBI or the U.S. Department of Justice (DoJ) references obtaining a warrant to use NITs, the legal instrument used is outside the bounds of surveillance law. Instead, agents rely on a little-known procedural rule that extends the arm of the law far beyond the reach of Title III, CALEA, ECPA, or the Patriot Act.

Secondly, NITs are sophisticated intrusion techniques that may involve malware as well man-in-the-middle attacks and Zero Day exploits – the very cyber tactics most commonly associated with and used by criminal, terrorist or nation state intelligence agencies.

In both instances, NITS are the farthest thing from "nits" in law enforcement surveillance capabilities. They are among the most potent tools used by federal law enforcement today. The question: What legal "cover" do federal LEAs have to obtain warrants for using NITs?

Section 41(b) – Lawful Malware's Version of Area 51

U.S. federal surveillance law is spread across a patchwork of laws that have evolved over the last 50 years.

Of note, while current laws such as ECPA (which extends law enforcement agency wiretap privileges beyond telephony to include transmitted electronic data) and the SCA (which requires ISPs to disclose stored communications) might be applied to NITs, there is no law on the books that expressly addresses the granting of federal warrants for use of offensive cyber. So when the FBI or DoJ refers to warrants for NITs, what is the applicable law? This gray area falls under jurisdiction of an obscure section of the U.S. Federal Rules of Criminal Procedure, put into effect by the Supreme Court by authority of the Rules Enabling Act of 1934. Under this law, Congress granted the judiciary the power to determine the rules of civil and criminal procedures.

The relevant passage in the Procedures is Rule 41(b).

Rule 41(b) of the Federal Rules of Criminal Procedure authorizes magistrate judges to issue search warrants for electronic storage data, a code that is broadly applied to enable seizure of data by means including black hat technologies. To cite the Rule, a warrant may be issued for any of the following:

(1) Evidence of a crime.

(2) Contraband, fruits of crime, or other items illegally possessed.

(3) Property designed for use, intended for use, or used in committing a crime or designed for use in a crime.

(4) a person to be arrested or a person who is unlawfully restrained.

Federal law enforcement agencies have used this legal mechanism to obtain warrants to use offensive cyber and other intrusive surveillance technologies since at least the late 1990s.

One limitation of the code is that it confines the geographic scope of the warrant to the magistrate's own jurisdiction. However, a 2017 amendment to Rule 41(b) – approved by the US Supreme Court – allows warrants for search and seizure of electronic data by any means in any locale, including internationally, with or without the permission of affected countries. The revised language authorizes "a warrant to use remote access to search electronic storage media and seize or copy electronically stored information within or without that district. It also applies to virtual realms where data has been "concealed by electronic means."

In simple terms, federal law enforcement may execute search and seizure warrants under authority of procedural rules that derive from an Act on rules – not from a law on surveillance. And the warrant may be used anywhere in any circumstances, including the investigation of targets using anonymous services such as Tor.

From the evidence revealed in court records, the FBI does not appear to have waited for approval of the proposed revision of Rule 41(b).

Hacking the Dark Web

Beginning in 2011, Dutch police authorities set out to capture rings of child pornographers selling their horrific wares in The Netherlands via Tor. The mechanism for catching the felons was a special web browser specifically designed to collect Tor addresses for websites hosted on the anonymous service. After capturing numerous Tor addresses, Dutch police carefully studied each and singled out sites dedicated to child pornographers. The challenge then was to determine the physical location of the sites, no simple task since Tor

IP addresses are generated algorithmically and masked by multiple layers of the "onion" router system. However, one target made the error of failing to password protect his administrative address, which led to discovery of his real IP address. The perpetrator turned out to be located not in The Netherlands but in Nebraska, USA. Dutch agents forwarded this intelligence to the FBI.

FBI authorities here took an innovative twist. Rather than arrest the target, they instead approached the U.S. District Court of Nebraska for a search warrant based on use of network investigative techniques. The mission fanned out. The FBI wanted not just the porn dealer but his customers. NITs would prove critical to succeeding on both counts.

Officially speaking, the warrant authorized the FBI to do three things: (1) to determine the target server's IP address; (2) to obtain a "unique session identifier," i.e., evidence of the illicit content exchanged; and (3) find the type, version and architecture of the server's operating system. In reality, of course, the FBI already knew all three items courtesy of their peers in Holland. The NITs used empowered federal agents to do far more. They compromised the server, but rather than shut it down deliberately kept it running to track users, leading to arrests and charges not only against the owner but his customers.

Never mentioned in official documentation: To compromise the server, agents used a Zero Day exploit to penetrate and take control of the offending site without the operator's knowledge. Once under the FBI's control, the corrupted server planted malware on the device of every website visitor, making it possible to track them and identify their unique IP addresses.

Buoyed by its success, the FBI has since gone on to use similar NITs to lure and capture the identify of thousands of users of other child porn sites. One example: an Oregon case that netted IDs of more than 1,300 such individuals, with federal charges brought against 137. The singular aspect of FBI Network Investigative Techniques is that none has involved a warrant obtained under the auspices of a U.S. surveillance statute, or referenced the use of offensive cyber tools and skills. The case law used by the FBI to obtain the warrant: Rule 41(b) of the Federal Rules of Criminal Procedure.

FinFisher FinSpy Rootkit Infection

In reviews of how FinFisher FinSpy, The Hacking Team's RCS and similar "ethical malware" programs work, the discussion generally begins with the statement, "They plant malware on a device to take control" and proceeds with a list of all the capabilities that derive therefrom. All fine and well, but what are the steps leading up to the malware plant that make it successful? And how do such solutions overcome anti-virus software, sandboxes, etc. to do their job invisibly without drawing any attention?

Here, without going into an encyclopedic ramble on coding, we will endeavor to explain in laymen's terms the process behind successful malware attacks, using FinFisher FinSpy as an example. In essence, remote control ethical malware operates on three principles: (1) gaining trust of the target; (2) fooling the target's device; or (3) both together. In each case the user is leveraging vulnerabilities, either of human nature or systems design.

First, a brief general overview of malware.

Of Human Cyber Bondage

With all the public alarm over malware and botnets, most individuals – and particularly targets in the criminal or terrorist community – are acutely aware of the dangers of malware, Trojans, botnets and viruses. Even the term "Zero Day," a vulnerability in an operating system, which for years was arcane terminology confined to the likes of VUPEN and Zerodium, has become common parlance. At the most basic level, people know to avoid clicking on emails from complete strangers or to view "out of the blue" notices from entities posing as a bank, the Internal Revenue Service, or doctor's office. But there are still many who fall prey to such common techniques of planting malware, variously dubbed phishing or "social engineering." It is all too easy to gain their trust, and from there deploy malware that takes over a device.

More insidious are "drive-by" malware attacks that are increasingly common and take place more from negligence than outright foolishness. Here a favorite, trusted browser or news website might be rigged with exploit code that attacks when the victim visits the home page. The malicious code then begins searching for vulnerabilities often found in apps the device owner has failed to update. "App hogs," individuals who download multiple apps they then quickly forget about, are favorite targets.

Malicious code may be "environmentally aware," taking advantage of runtime differences between software and hardware. It may continuously alter its signature to end-run malware blacklists used by signature-based antivirus software. It may evade detection by running only at selected dates and times, or after specific events such as rebooting, and remain dormant at other times. Another type of malware changes its name to hide behind trusted features, for example replacing Application Program Interfaces (APIs) with hashed values that avoid parsing, then communicating with a port on the hacked device to encrypt the malware and the data being stolen. Similarly, the code might create executable malicious files that masquerade as legitimate ones.

Malware may also attack via "shellcode," a line of code that provides a base for directly accessing the target device's operating system, then issues commands to take control. Such attacks are called "rootkit infections" because they provide root access to the operating system and replace factory-installed administrative procedures with ones that serve the interests of the attacker, yet are concealed so that operation seems normal to the target.

As mentioned, drive-by or other types of malware attacks can be hosted on a trusted site serving as a victim. However, in the case of ethical malware, the user is typically acting from behind proxy servers that create mirror images of the real website page to upload malicious code. The proxy servers, in turn, report to a Master Server that remains hidden from view.

Many of these techniques will sound familiar as we enter the world of FinFisher FinSpy.

Inside FinSpy

Customers of FinSpy use a Master FinFisher Server and FinProxy servers to host and deploy the malware solution. The Master Server is configured to link to all proxy servers, listing the ports and the external IP address of each. Proxy servers are set up for listening and uploading the FinSpy module. The Master Server, which acts as monitoring center for intercepts, is configured for the ports to be monitored, certificates used and logging files. The Master Server also includes directories of all ports, certificate paths, logging files, locational files and destination paths.

Most importantly, the Master Server holds a "TargetModules" directory listing executable files that bind specific target data to the FinSpy Trojan via rootkit infection. Targeted files might be of any type, including Word, pdf, image, video or audio. Executable files are "hollow" and attach to target files so that both run – first the Trojan, then the legitimate file.

To prevent being caught by and stuck in a secure zone "sandbox," FinSpy executes anti-sandboxing code to ensure the rootkit can be "dropped." Using a process called Structured Exception Handling (SEH), the anti-sandboxing code creates several random bytes for purposes of distracting the device's

"Exception Dispatcher," which immediately launches an exception code in the operating system. The Dispatcher tries three times then quits, assuming the bytes are a non-working app. At that point, the way is open around the sandbox. The FinSpy "dropper" unleashes a rootkit infection to commandeer the operating system.

All very quietly, mind you. FinSpy hides in the target device's "temporary" folder and decrypts its spyware tools packed in from the resource library. Then it generates a legitimate system function, "hollows it" and injects a Dynamic Link Library (DLL) tool unleashing capabilities to perform all the wonders of ethical malware.

One other key feature of the rootkit: It modifies the targeted device's "Master Boot Record," that portion of a hard disk or diskette that tells the operating system to load (and when) into main storage or RAM. In so doing, FinSpy gains control of the device's "Process Monitoring" in charge of real-time process and threat activity on the system, Reason: to lower the malware's level of activity below antivirus software's ability to register and report it as an activity. When the event goes into memory it is under a false name that is almost impossible to detect.

Captured data is encrypted in 256-bit AES and stored in the very directory the original rootkit was packed into. The data is then routed to a FinProxy server and on to the Master Server for analysis. End result: the ability to hack into a target's remote file access, keyword logging, activate his still and video cameras, record his voice live by his own microphone, lift his passwords, access his communications in real-time – voice, SMS, email, social media or video chats – and geolocate him for arrest, or "special handling" by Special Ops.

The FinFisher Portfolio

FinFisher may be installed either on-site via use of a USB device, or remotely via Trojan. Specific solutions include:

FinFisher USB Suite. The FinFisher USB Suite is a set of two USB Dongles, two bootable CDs and the FinFisher HQ – a Graphical User Interface (GUI) – for analysis of retrieved data. The FinFisher USB Suite is engineered for use by any agent, informant, or other individual with government authority to gain access to a target computer. Simply insert the USB into the target computer and extract its entire contents including user names and passwords, e-mails, files and other critical system and network information.

FinFisher Remote Hacking Kit. When physical access to a target computer isn't possible the FinFisher Remote Hacking Kit provides all the tools used by professional hackers to remotely gain access to target computers. The kit includes a notebook running the FinTrack operating system, wireless equipment, and a USB to hold default password lists and rainbow tables. The FinFisher Remote Hacking Kit can be used for internal security assessment as well as intelligence gathering targeting public servers or personal devices.

The FinFisher Suite

The essence of FinFisher, and the reason buyers flock to the company's wares, remains the ability to exercise total command and control of targeted devices. The core capabilities FinFisher provides:

FinSpy: a cutting-edge, professional rootkit infection that enables remote access and monitoring of target devices. Basic functionality includes Skype Monitoring, Chat Logging, Keystroke Recording, accessing printed and deleted files. The Trojan is completely hidden and all its communications are covert.

FinFly: a transparent HTTP proxy that can modify files while they are being downloaded. There are two versions of this software; the FinFly-Lite and the FinFly-ISP. FinFly-Lite can be used within a local network to append FinSpy or a custom Trojan to executables downloaded by a target computer. FinFly-ISP can be integrated into an Internet provider's network to infect multiple targets or individual computers.

FinCrack: a high-speed super cluster for cracking passwords and hashes. It supports password recovery of any device, system, network or document including password-protected files.

FinWi-Fi KeySpy: a device for remotely sniffing keystrokes of commercial wireless keyboards that are within Wi-Fi device range. Agents can also remotely control the wireless keyboard and thus the target device, too.

FinBluez: a product that enables agencies to do various advanced attacks against Bluetooth-enabled devices such as mobile phones, headsets and computers. For example, FinBluez is able to record the audio stream between a headset and a mobile phone or utilize common Bluetooth headsets as audio bugs.

FinSpy, the Trojan, is the heart of FinFisher. Customers that purchase FinSpy have a comprehensive product that includes:

- *FinSpy Client:* The user interface providing access to the target's system to gather information or control (reconfigure or remotely delete) the FinSpy Target.

- **FinSpy Server:** Central server where all infected clients connect and publish their availability and basic system information. The server is also contacted by the FinSpy Client to obtain the infected target list.
- **FinSpy Target:** The package used for the infection and installed on the target system.
- **FinSpy USB Dongle:** A USB dongle that contains software to deactivate all running Anti-Virus/Anti-Spyware software and install the FinSpy Target component.
- **FinSpy Antidote:** Software that can detect and remove any countermeasures to FinSpy Target that might prevent the installation.
- **FinSpy Proxy:** Forwards connections between FinSpy Target and FinSpy Server that can be used to have multiple active public IP addresses, thus minimizing detection by researchers.

FinFisher also offers specialized FinTraining courses on Basic Hacking, Hacking VoIP, Hacking Mobile Systems and Basic Cryptography.

Gamma Group: FinFisher's Parent

Gamma Group and The Hacking Team (acquired by Memento in late 2019) are without question among the best known providers of network injection solutions, surveillance industry jargon for inserting bogus reconfiguration commands that take control of routers, switches and intelligent hubs, then crash or take over targets' communications devices. Yet while their products are similar in many ways, the companies and their histories are completely unlike. One is an open book – the other a dark hole on the bookshelf.

As surveillance companies in this niche go, The Hacking Team was comparatively out in the open until being acquired. The founders/owners, developers, locations and start date are a matter of public record. Hacking Team's founder and CEO often held forth in public interviews that left an indelible impression. To top it off, The Hacking Team's former public face, marketing VP Eric Rabe, was once a PR person for Bell Atlantic (now part of Verizon) and well-known in the Washington, D.C. and New York metro areas. Rabe left The Hacking Team in March 2018 to take a similar job with another malware vendor, Grey Heron, which is equally open about its business.

Malware companies with their own spin doctors? As happens from time to time, some of the things one comes across in the surveillance sector are so richly offbeat there is no way one could make them up.

The Gamma Group, in contrast, is engulfed in mystery, even when it comes to the most basic "who-when-where-what" information about the company. Not surprisingly, the company's intense desire for secrecy has had the opposite effect, generating a cascade of rumors. In the guesswork that follows by many analysts, facts blend with myth, become indistinguishable, and raise more questions than answers. A few examples:

- **Start Date:** Gamma Group claims a company launch date of the "1990s." While that may be true, did they begin offering FinFisher FinSpy or other malware at that date, or much later in time?
- **Corporate Shell Game:** Which corporate entity really makes and develops the company's FinSpy network injection suite? Is it purported parent company Gamma Group, or its subsidiary Gamma International Ltd in the UK, or subsidiary Gamma International GmbH, or FinFisher GMBH which replaced Gamma International GmbH and opened in Germany in late 2013, or the bafflingly named marketing arm, "Lench IT Solutions PLC"?
- **Location:** Where is the malware unit based – in the UK, Germany, or somewhere in the British Virgin Islands, where Gamma operates from a shell company?

And then there is the truly curious material:
- **Transition to Network Injection:** Around the same time that the U.S. National Security Agency (NSA) switched from classic spear-phishing to QUANTUMINSERT/FOXACID network injection hacking, Gamma followed suit with FinSpy. Mere coincidence?
- **Munich's Centrality:** Gamma's German HQ in Munich and its environs are also home to former and still occasional partners DreamLabs, Desoma and Elaman. Elaman is alternately described as an independent firm that was once Gamma's largest retailer. Or was Elaman simply an offshoot of Gamma – or vice versa? For a time, Gamma International GmbH and Elaman shared the same address and phone number in Munich.
- **Der Bundesnachrichtendienst (BND):** Do Gamma International GmbH's roots go deep into Germany's foreign intelligence service, the BND? Is there any connection between the former Gamma International offices in Munich and the nearby Bad Aibling military station – used for years by U.S. Army Intelligence and the NSA before being turned over to Germany's BND?

Let's dig into these riddles.

Gestation of Gamma Group

Gamma Group and it predecessor companies have always centered around two individuals: William Louthean Nelson, a clandestine elderly gentleman with diverse business interests in defense, security and various odds and ends unrelated to either field; and his son, Louthean John Alexander Nelson, who is principally responsible for the company's diversification into the government surveillance arena.

"Louthean," obviously a family name, stands out for its uniqueness and macabre sound. While the name's origins are as mysterious as Gamma itself, the motto on the family crest certainly fits: *Non dormit qui custodit*, Latin for "the sentinel never sleeps." Watchfulness would seem to be in the Louthean Nelsons' DNA, carrying over to the present day business.

William Louthean Nelson's business activities track to the year 1991, when he turned 58 and unveiled a company called Compass Military Limited Services, which specialized in military vehicles. Compass Military occupied the elder Mr. Nelson for most of the decade. Then in 1999 the family took a different turn, introducing an IT service company, ComputePlus, beginning a spree of company launches and ventures into surveillance.

The following year, 2000, saw the debut of the "Gamma" brand with Gamma TSE (Technical Surveillance Equipment Ltd.), a carryover of Compass

Military adding military audio-video surveillance systems to the line of military vehicles, and G2 Systems (operational and technical training), followed by Gamma Cyan Ltd. in Beirut (Middle East sales) in 2003.

One anomaly: the launch of a waste management company bearing the Gamma name. Finally, in 2007, the Louthians launched Gamma Group itself in a modern industrial park in Andover, UK. At the same time, two new subsidiaries were formed: Gamma International UK, also in Andover, UK, and Gamma International GmbH in Munich. Of note, the younger Nelson surfaced for the first time, now as a director of both subsidiaries.

2007 was a signature year. Most of the older corporate entities – including Gamma Cyan Limited, ComputePlus and Gamma Waste Management – were either dissolved or made inactive. Gamma TSE would continue, still in the military vehicle and mobile surveillance business at least through 2011. But by 2007 it is clear that Gamma Group, Gamma International UK and Gamma International GmbH had become the main corporate players, all with a unique focus: intrusive surveillance technology, i.e., offensive cyber.

FinFisher rapidly emerged as a suite of products with the ability to remotely penetrate and take control of devices via classic methods of intrusion such as spear-phishing. However, here the company's distinctive ties to Germany entered the picture. While produced by the Louthean clan, FinFisher was for nearly five years marketed by partner Elaman, an established surveillance systems company based in Munich. That arrangement lasted past the official launch of Gamma Group itself, until the latter part of the decade when Gamma Group took full charge of marketing and distribution. But in the earlier years, Gamma and Elaman did indeed share an address and telephone number in Munich, and for the best of reasons: Elaman served as the principal marketing arm for Gamma, worldwide.

Why the Germanic focus? Credit the rising sun of the family, Louthean John Alexander Nelson. From the start, the younger Nelson was the prime mover behind the company's shift into malware. Beginning in 1989, he served as a director of PKI, a provider of electronic surveillance systems headquartered in Germany and led by founder Peter Klüver (hence the "PK" in PKI). It was through his relationship with Klüver and PKI that the younger Nelson found his true calling. In fact, Gamma's investment in intrusive malware began in the early 2000s, concurrent to introduction of the Gamma brand name, but well before there was a Gamma Group headquartered in Andover, UK. Gamma's and the Nelsons' longstanding relationships with PKI and Elaman, and later Desoma – and for that matter with Germany itself – would have a lasting influence.

The Dark Star of Gamma Group

By 2007, lawful malware had earned an established track record as a tool for successfully penetrating and taking over targeted devices. However, compared to today, the methods of device infection were comparatively crude, relying on classic phishing and similar techniques that required the target's own engagement, e.g., through opening an email or downloading a document or file to commence the infection. But lawful malware was about to undergo a strategic change.

Gamma Group's opening day in 2007 coincided with the arrival of German national Martin J. Muench. Schooled as a network security specialist, Mr. Muench had for several years turned his talents to another sphere: malware. He was the perfect candidate to help run the new company.

Mr. Muench's significance to the operation was underscored by the Nelsons' decision to grant him 15 percent ownership of Gamma International GmbH, a holding that he retains to this day. Concurrent to parent company Gamma Group's opening in Andover, UK, Mr. Muench came on board as managing director of Gamma International GmbH in Munich, reporting up to the younger Nelson, and later was named a director of Gamma Group itself.

"Lench IT" as a holding company or other entity came into being with Muench's arrival at Gamma. The origins of the name have puzzled many an analyst. One hypothesis, which seems as plausible as any: Lench is simply a contraction of Louthean and Muench.

From the outset, Muench had different designs that would change and improve ethical malware forever. In this endeavor, Muench had critical assistance from an American partner, CloudShield, an established military contractor of the U.S. Department of Defense (DoD), and by connection, of the U.S. National Security Agency.

As a reminder, since many forget the fact, we will note here that the NSA is a military intelligence organization that falls under the DoD. NSA's work in the mid- to late 2000s on new approaches to intrusive surveillance is relevant to commercial companies' endeavors in the same area – notably The Hacking Team and Gamma Group – and merits a brief digression.

NSA QUANTUMINSERT

The concept of network injection as a popular form of government hacking first became public knowledge in June 2013 with the Edward Snowden revelations. Among the more intriguing disclosures were twin NSA programs called QUANTUMINSERT and FOXACID, both developed by the NSA's formerly-named Tailored Access Operations (TAO) division – now known as The Equation Group.

Staffed by an estimated 1,000+ software and hardware engineers, military and intelligence analysts, designers and computer hackers – the great majority of them government contractors – TAO or The Equation Group is one the largest groups within NSA, with primary headquarters at Fort Meade, Maryland, USA.

QUANTUMINSERT was one of TAO's core accomplishments, and a flagship service used for many years by the NSA. It was a government hack that injected packets into backbone networks or specific devices, redirecting targets to a special FOXACID server which proceeded to conduct a man-in-the-middle (MITM) attack compromising the target's system or device.

QUANTUMINSERT got a lift from two additional programs known as TURMOIL and TURBINE. TURMOIL comprised a set of high-speed processors that connected directly to Internet backbones. TURBINE facilitated real-time packet injection via FOXACID servers.

Prior to the development of QUANTUMINSERT, the NSA performed FOXACID MITM attacks the old-fashioned way: by luring targets to perform actions such as downloads, essentially spam, via classic spear-phishing.

Over time, however, savvy targets became increasingly cautious about accessing any received content that looked suspicious or emanated from a very realistic-looking duplicate of an authentic source. QUANTUMINSERT resolved this issue by initiating infection of the target's network or device when he or she surfed any of scores of common sites in HTTP.

At this writing, the exact date line of QUANTUMINSERT's arrival remains known only to the NSA. However, it is possible to work backwards toward a date by logging an important fix made to QUANTUMINSERT early in 2011. Prior to then, QUANTUMINSERT hacks often found themselves in a race to beat legitimate servers with a response. The solution to this problem, prototyped in 2011, involved the use of virtual machines situated closer to the target. That date would obviously place QUANTUMINSERT's deployment back to at least 2010 or earlier, to a time when Gamma Group was readying its own network injection solution within FinSpy, with help from CloudShield.

Network Injection in Under Four Months?

CloudShield Technologies, originally a San Diego, California, USA-based cybersecurity specializing in "deep packet processing," had its share of ups and downs. During its first decade, from 2000 – 2009, the company was a successful defense contractor offering its CS-2000 high speed processor for network threat detection. One bright but largely unheralded moment came in its last year of operation as an independent company.

In 2009-2010, CloudShield struck up a critical relationship with Gamma Group. Beginning in September 2009, CloudShield engineers met and worked closely with none other than Gamma International GmbH managing director Martin J. Muench.

In less than four months, CloudShield's people turned a new page in their company's story by adapting the CS-2000 to go beyond threat detection monitoring into fresh territory: as a high-speed platform for FinFisher. As significantly, CloudShield worked hand-in-hand with Muench to transform the original Gamma product by switching to network injection – the self-same tool used in the NSA's QUANTUMINSERT.

The relationship between the two companies was short-lived, as behind the scenes CloudShield was also at work "engineering" it acquisition by SAIC. That deal was announced on January 14, 2010, bringing the Gamma/CloudShield alliance to an abrupt end. The CS-2000 outfitted with FinFisher never came to market, but FinFisher with network injection most certainly did. Moreover, the CloudShield CS-2000 didn't vanish entirely from Gamma's world; as a complement to Gamma's malware, Gamma's 2010-2011 partner Desoma used the CS-2000 to power its own mass packet capture (DPI) and filtering product, DAISY, for government clients.

Following its acquisition by SAIC, CloudShield became lost in the works as its new owner commenced an extended break-up into two companies, a new SAIC and Leidos. When the split came in 2014, CloudShield went with Leidos, which was to concentrate on the defense industry with primary clients including the DoD and NSA, as well as the health sector. Leidos, widely expected to soar under new management focusing on these core markets, instead fell off a cliff revenue-wise in its first year. Leidos' M.O. also went downhill, led by a management attitude of "what can we divest to exit this financial mess?" Within a year, by February 2015, CloudShield was sold off to LookingGlass Cyber Solutions.

In retrospect, the contribution of CloudShield in large part made Gamma what they are today. It is fair to say that the Gamma/FinFisher suite owes a huge debt to American engineering, and arguably one other key aspect of the long arm of Uncle Sam.

Until 2013, NSA's QUANTUMINSERT was completely unknown not only to the general public, but even to some of the cyber industry's most prominent figures. Renowned cryptologist Bruce Schneier in October of that year took credit for outing the story, ignoring the irony that QUANTUMINSERT had operated right under his prestigious nose for half a decade before its so-called "discovery."

Given the highly covert nature of NSA's use of network injection in QUANTUMINSERT, the question naturally arises: How did CloudShield and Gamma Group learn about it? Or did CloudShield actually have a role in the creation of the NSA's primary network injection tool? Government agencies are not laboratories or innovation centers, after all. As noted many times in this encyclopedia, upwards of 80 percent of "staff" in the U.S. Intelligence Community are government contractors with security clearances. Contractors, in turn, generally innovate very little on their own. Most are systems integrators of others' technology genius.

Another answer to this riddle might lie closer to home: within the close-knit world of German intelligence, and specifically the military SIGINT community in and around Munich. A few miles east of the city sits the Bad Aibling Station (BAS), run exclusively by the NSA for years until turned over to Bundesnachrichtendienst (BND) – Germany's foreign intelligence agency – in the early 2000s, then ostensibly closed in 2004.

At its peak of operation, Bad Aibling housed more than 1,000 staff. Under the NSA and with direct support of the BND, it was the third largest site outside the U.S. and UK of ECHELON, the global SIGINT network of "The Five Eyes" – the U.S. UK, Canada, Australia and New Zealand. Agents at Bad Aibling focused on intercepting mass communications including wireless, wireline, satellite and Internet. Although main operations have long since been transferred to a site in northern Germany, the NSA reportedly still maintains a strong presence at Bad Aibling with an assist from the BND intelligence network.

Might word of NSA's QUANTUMINSERT program have filtered through NSA or BND channels to Munich's extensive surveillance industry community? As in the U.S., the ties between surveillance vendors, many of whom are led by individuals with military intel backgrounds, and the Intelligence Community, are just as strong and tight in Germany.

Exactly how, where and under whose auspices network injection began will remain a mystery. To conclude that Gamma Group and CloudShield worked independently and developed network injection solutions identical to QUANTUMINSERT in four months stretches the bounds of luck and coincidence.

The Hacking Team

Until acquired by Switzerland's InTheCyber in April 2019 and blended into a new company called Memento Labs, The Hacking Team – Italy's notorious offensive cyber vendor – was a train wreck virtually doomed to financial ruin. It remains to be seen whether the land of cuckoo clocks can get Hacking Team ticking again, and on par with Israel's very capable NSO Group, as promised. But as we wait to see what happens, here is a close look at what the Swiss have gotten themselves into.

Founded in 2003 by David Vincenzetti as an IT consulting firm, The Hacking Team gradually moved toward "ethical hacking." Unfortunately, Mr. Vincenzetti's company was eventually exposed for being less than ethical itself.

The company's flagship product "RCS" was sold to governments only, supposedly empowering law enforcement agencies to remotely track suspects. To that end, The Hacking Team openly described its product as "spyware, a Trojan Horse, a bug, a monitoring tool, an attack tool, a tool for taking control of end points."

What they claimed to monitor and log:
- Any action performed on a PC, whether Windows or Mac architecture: web browsing, keystrokes in any unicode language, printed documents, chat, email, instant messaging, remote audio spy, and Skype voice conversations.
- Any action on a smartphone including iOS, Windows, Blackberry and Symbian: call history, address book, calendar, email and SMS messages. It also intercepted call signal/location info and voice calls and provided remote audio spy function.

Encryption? Not a problem, supposedly, per Hacking Team's marketing spiel. Ostensibly, RCS sailed right through it. The company promoted RCS as being completely invisible to the target. Neither antivirus, anti-spyware nor anti-key-logger programs could sniff it out. It supposedly provided the power to search and view data on a hard disc, conduct remote execution of commands, and modify content on the target's device. It could also be configured to send data triggered by specific events, e.g., only when the target's screen saver was active, or be configured to remove itself from the target's device at a time or situation programmed by the user.

The Hacking Team used what they called an "attack cookbook" that provided diverse anonymous attack scenario analyses, flexible to the user's need – for example, working "on the fly" to infect the device of a user who is a moving target on Skype. Net net, The Hacking Team's RCS promised "centralized management of unlimited heterogenous targets."

At its peak, The Hacking Team claimed to do a thriving business in more than 30 countries.

Begging for Attention

Of the various ethical hackers, The Hacking Team experienced more problems than most, some owing to its comparatively late arrival to the market, others to its dependence on outside vendors for critical tools. Having a notoriously acid-tongued and publicity-hungry founder and CEO did not help. Time and again Mr. Vincenzetti appeared in international press ranging from *Business Insider* to *The New York Times,* usually in hostile articles. It's been said that if a man can't be praised, he'll settle for being talked about – words that certainly fit Mr. Vincenzetti, who pursued fanfare even when it made him look a satanic evildoer.

In 2015 an unknown hacker who went by the name "PhineasFisher," broke into The Hacking Team's servers and company records, disclosing confidential records such as executive emails, including the various codes used in Hacking Team products.

The company soon found itself under investigation by Italian regulatory authorities who dealt the company a crushing blow. For more than a year, The Hacking Team was banned from exporting RCS or any other ethical malware product. The Milanese vendor cleverly (and illegally) side-stepped the ban by working through offshore resellers for a time. Then it was dealt an even more lethal blow by a Russian tech analyst.

Independent reverse engineer Vlad Tsyrklevich revealed that The Hacking Team had been a tad sloppy in testing the Zero Day exploits it either made or purchased prior to selling them to buyers in the law enforcement and intelligence community.

According to Mr. Tsyrklevich, a big contributor to The Hacking Team's early Zero Day challenges stemmed from its comparatively late entry to the market. By the time The Hacking Team decided to shift from IT consulting to malware, they were well behind established players such as Gamma Group International (GGI). Another drag on getting out the gate quickly: The Hacking Team was short on in-house talent. As a result, the company had to go shopping for Zero Day exploit suppliers.

A big problem from the get-go was that premier Zero Day outfitters of the day were already "in tight" with Gamma Group International (GGI) and other leading ethical malware companies. Too often, the low-cost exploits The Hacking Team purchased were hacking's equivalent of table scraps. Some weren't even true Zero Day exploits. But eager to get started, The Hacking

Team went forward with inferior solutions anyway, The inevitable result: Some of what it sold to government customers just didn't work well, or at all.

Despite what Tsyrklevich described as a "continuing stream of negative reports about the use of Hacking Team's software" well into 2012, the company's market presence somehow grew. But by 2013, Mr. Vincenzetti had had enough, and publicly acknowledged his company's shortcomings in obtaining decent Zero Days and in-house talent. As a result, the Hacking Team shifted that same year to new suppliers including Netragard, Vitaliy Toropov, Vulnerabilities Brokerage International, Rosario Valotta, and in 2014, Qavar Security.

Even so, The Hacking Team lost customers, including two high profile accounts in the U.S.: the Federal Bureau of Investigation and the Drug Enforcement Administration. In mid-2015, the DEA said in Congressional hearings that it had "recently" cancelled its contract for The Hacking Team's core RCS product because it was seldom used, and they had experienced "technical difficulties" on the handful of occasions – just 17 times including one offshore application – when DEA did put it to work. At the same time the FBI cancelled its own agreement with The Hacking Team, without citing reasons.

Insight on how The Hacking Team handled technical glitches is revealed in one of those company-internal emails hacked and exposed to the public. The email recounts a malfunction in booting up the RCS malware infection during a demo and how, instead of coming clean with the customer, The Hacking Team covered up the problem:

"UEFI [Unified Extensible Firmware Interface] infection: the "UEFI part" worked good and the BIOS [basic input/output system] got infected (as far as we could see), but during the first boot after the infection the OS got stuck and we had to shut the system off and then on again. After that, we couldn't see any agent synchronizing/running, so we solved just running a silent installer while Serge was distracting the customer."

"Distracting" the customer from a product malfunction seems a peculiar way to build trust and confidence. Equally telling is the date on that email: not back in 2014 when The Hacking Team's CEO set new marching orders for the company, but January 30, 2015 when the company had supposedly resolved its issues.

Can the new Swiss owner transform The Hacking Team into a successful, reliable surveillance vendor? Time will tell, but bad memories live long.

NSA Equation Group and TAO

In August 2016 the Equation Group, aka the Tailored Access Operations (TAO) division of the NSA, attracted unwanted headlines with reports that anonymous attackers had hacked into the group and made off with critical intelligence plus high-end malware. The purported hackers, known as "The Shadow Brokers," made their success known by publishing 300 megabytes of unencrypted data including samples of specific Zero Day exploits that could zip right through security measures of Juniper, Cisco, Fortinet and TopSec firewalls to plant malware. None of of the vulnerabilities enabling hacks had previously been revealed to the target companies.

In all, this was the most serious breach of U.S. national security since the Edward Snowden incident of 2013. It was also the first time the Equation Group made major news in over a year. Initial concerns that one of the world's foremost engines of advanced persistent threats (APTs) had itself been hacked ultimately proved wrong. In the end, the evildoer turned out to be just another private contractor who had simply walked off with the files and stored them for months in his home and vehicle.

The real problem was figuring out whether the thief had ruined years of work and set back the U.S. Intelligence Community. Some, including former TAO employees, jumped to the conclusion that the Shadow Brokers had snatched and broken the "keys to the kingdom." Few are convinced that this is the case, or that much has changed as far as the Equation Group is concerned.

When first outed by Russia's Kaspersky Labs in February 2015, the NSA's ultra-secret hacking group was like a floating island in a distant sea – remote, previously unknown and completely uncharted. Lacking an official name for the group, Kaspersky simply made one up, the "Equation Group." The APTs attributed to this organization – EquationDrug, DoubleFantasy, TripleFantasy, Grayfish, Fanny or EquationLaser – were simply nameless code. As the doctors who birthed them for public view, Kaspersky invented names for each, on the fly.

Certain lines of code resembled the STUXNET worm jointly developed by the U.S. and Israel and used successfully against Iran in 2010, but beyond that alleged lineage, no one knew for certain what this "group" was, who they were aligned with, where they operated, or even when they began deploying what have been rightly termed some of the most sophisticated APTs ever created. Everything surmised about this body of unidentified and thus far unidentifiable body of experts, including the set of coined names for their work, was inference and speculation, a tribute to the Equation Group's talent for secrecy.

The Shadow Brokers raised the curtain by releasing samples of the stolen malware code, and actual names of APTs. But given that the exploits were at least three years old at the time, and the speed with which offensive cyber evolves, it is doubtful whether much damage was done by the Shadow Brokers' breach.

If we can assume anything about the Equation Group in its current state, it is that the APTs whose breadth and power stunned the security community in February 2015 and resurfaced in 2016 have come and gone as the group moves on to break new ground. EquationDrug, Grayfish, Fanny, etc. have already been analyzed to the nth degree. More salient is consideration of how these earlier efforts serve as a benchmark for even more advanced developments that have taken their place, and what this signifies for the future of APTs.

Attacking Air-Gapped Systems

STUXNET was unique at the time for its focus on "programmable logic controllers" (PLCs) that automate electromechanical processes, and for its ability to attack "air-gapped" computer systems that are walled off from the Internet and other external networks. The method used was interdiction, i.e., infection by USB device that was physically intercepted and planted with the worm. According to Kaspersky, the STUXNET worm used Zero Day exploits similar to a pair developed by the Equation Group in 2008 for Fanny, hence the assumption that the two are connected.

Fanny in operation was basic Zero Day exploit work. When plugged into an air-gapped network, the UBS mapped the network and relayed it to the attacker when the gullible user reinserted the stick into an Internet-connected device. Lastly, when the user plugged the USB back into the air-gapped network, Fanny knew the command structure of the PLC and easily took control.

The Equation Group by no means holds a monopoly on this form of attack. In roughly the same time frame, a USB found in the parking lot of a U.S. military base in the Middle East successfully hacked into U.S. Department of Defense networks.

In 2014, the former Symantec (now owned by Accenture) reported that a similar worm called "Dragonfly," originating in Russia or Eastern Europe, had compromised the computer infrastructure of more than 1,000 electric power companies in North America and Europe, but in this case using common social engineering techniques versus a USB.

Reprogramming Hard Drives

Perhaps the most singular achievement of the Equation Group is the reprogramming capability of two platforms, named EquationDrug and Grayfish by Kaspersky. These platforms override firmware, embedded systems that control computers, laptops, smartphones and other devices.

Firmware is designed to be "non-volatile," that is, always available and unchanged whether a system is turned on or off, and after an operating system is re-booted or reinstalled. EquationDrug and Grayfish, however, had the ability to embed a custom payload that re-programmed firmware and retained persistent control of storage and commands in the hard disk drive. The attack was completely invisible and permanent. The moment an infected OS booted up, EquationDrug or Grayfish took charge.

The platforms were engineered to operate on an array of systems for hardware by companies such as Samsung, Hitachi, Seagate, Toshiba and Western Digital.

Apple wasn't immune, either. While the majority of Equation Group malware pertained to Windows systems, researchers discovered malware callbacks for MAC OS X-equipped Apple devices that connected to the Web through Mozilla, as well as malicious forums using server PHP open scripting language to trap iPhone users and perform the same firmware reprogramming.

Considering what the Equation Group could do with reprogramming hardware, the ongoing debate over "back doors" between law enforcement agencies and tech companies such as Apple would appear to be long resolved in favor of the former. Through the work of the Equation Group, back doors are already endemic.

Disappearing Act

Within days of any story on the Equation Group surfacing, news on the group typically goes dark again for months. One possible factor behind the Equation Group being "forgotten" so quickly is the proliferation of APTs. Examples include a pair of backdoors discovered in Juniper Networks' NetScreen firewalls. Weaknesses were exploited in the DualEC random number generator used for encryption, allowing hackers to passively decrypt VPN traffic and take advantage of a hand-coded password that grants administrative authority in SSH and Telnet.

By some accounts, the NSA (or rather, its private contractors) engineered back doors in the algorithm, giving intruders a free pass for an undetermined period of time. Security experts – and Juniper, for that matter – had known of the vulnerability since at least 2007, but did little about it. A patch issued by Juniper in December 2015 missed a line of code, leaving users in standby mode for a real fix. NetScreen is an older firewall product, but that only exacerbates the problem since it was widely used.

Remember that neither the NSA nor any other member of the U.S. Intelligence Committee do much "engineering" on their own. For the most part they rely on government services contractors to provide expertise in high-end offensive cyber skills.

One recognized player in this space is Raytheon. Another is a far smaller government services contractor whose name is the perfect cover for its operations, as the unobservant man might assume it is a merely a day labor personnel vendor. It is just that, in a very special way. This enterprise wields top-flight offensive cyber manpower whose tech prowess renders it a giant in that realm, highly valued by the CIA, NSA, NRO, NGA, and all intelligence branches of the U.S. Department of Defense.

In the offensive cyber arena, this secretive company and Raytheon lead the market in supporting offensive cyber used by the U.S. Intelligence Community. In fact, though Raytheon is many times our nameless vendor's size, both revenue- and headcount-wise, the two companies earn roughly the same amount for offensive cyber used by leading IC agencies. Both play a direct role in foreign intelligence operations.

NSA/GCHQ Hack of Gemalto SIM Card Keys

It was the biggest hack of SIM card keys and mobile networks ever: The joint NSA/GCHQ breach of the Gemalto SIM card empire, cracking the security of billions of mobile phones and scores of mobile networks worldwide. Bypassing encryption, the allied U.S. and UK intelligence agencies succeeded in collecting the keys of each phone, gaining the ability to passively monitor any mobile device equipped with a Gemalto SIM card. When news of the multi-nation breach broke, the world's largest SIM card maker was oblivious to the hacking of its products, internal emails, confidential computer network – not to mention billions of SIM card keys – that had persisted for five years without leaving a trace.

February 19, 2015 dawned grim for Gemalto, the vaunted provider of SIM cards to many of the world's largest operators, including AT&T, Sprint, T-Mobile, China Unicom and Japan's NTT – in all some 450 mobile network operators in 85 countries. At the time, Gemalto was selling more than two billion SIM cards per year, effectively owning the market while it sold under the corporate banner, "Security to be Free." Within 24 hours of the breaking crisis, the company found itself "free" of US $500 million in market valuation, shattered by loss of shareholder confidence.

Gemalto tried to play down the incident as old news and no longer damaging. And, in fact, the secret GCHQ documents that leaked to the public dated all the way back to 2010-11. That said, the SIM cards produced by Gemalto during the entire period from 2010 to 2015 were still compromised, and the company was clueless about the breach until the news broke on that dark day in February 2015. Net net, a company specializing in cybersecurity was hacked for 60 months without ever realizing it.

Perhaps the biggest takeaway is that the techniques used to hack Gemalto are still very much alive and, if anything, smarter than before. While the methods of deployment may vary, the methods used to penetrate Gemalto bear close similarities to extant techniques.

How exactly did the NSA and GCHQ pull this off, and where do we see similar approaches in use today?

Non-Malware Endpoint Exploits

As widely publicized at the time, GCHQ achieved the coup de grâce of hacks, collecting the encryption keys of every phone equipped with a Gemalto SIM card, and gaining the ability to conduct massive passive surveillance of each and every device without the knowledge of the user, the relevant mobile network operator, or Gemalto itself.

Given the rise of smartphones from 2007 to the present day, capturing a device's encryption key is an essential task for government intelligence agencies. Otherwise, agents face the grueling job of deciphering the intensely sophisticated algorithms used in 4G LTE and 5G protocols. Such decryption is certainly feasible, but it involves great time and effort just for one device. Deciphering the encryption keys of every phone on the planet would be akin to Mission Impossible.

GCHQ was smarter than that. The agency went around not through encryption, leveraging vulnerable endpoints for "non-malware" attacks. Endpoint security is a methodology for safeguarding networks that are accessed routinely by remote devices such as employee laptops and mobile devices. Particularly in today's BYOD work environment, every such device endpoint represents a potential doorway for intruders to enter. Hence, endpoint security is a hot button for cyber vendors and smart corporate and government clients. As often as not, though, endpoints are poorly protected.

One notable weakness: Both legacy and even machine-learning anti-virus programs used for endpoint security are typically designed to detect only malware attacks. Non-malware attacks sail right by. When the attacker's aim is to exfiltrate data, non-malware vehicles are perfect for the job. They enter, collect data and remain unnoticed indefinitely.

As illustrated by notes on the CIA's offensive cyber programs [Wikileaks Vault 7 - March 2017], such endpoints are simple to conquer. In Gemalto's case, the weak link proved to be the company's manufacturing and distribution pipelines. It was far easier and less time consuming to hack these vulnerable points and steal Gemalto SIM card keys by the billions than to go the decryption route. All that was required was breaking into and taking control of Gemalto's network so that keys could be lifted and copied before shipment to mobile operators. Gemalto operated multiple plants around the world. A single daily shipment alone might lead to a "take" of half a million SIM card keys.

Again, the GCHQ attacks on Gemalto were malware-free. Malware invariably leaves a footprint. Bypassing encryption and entering a network via its endpoints does not, which is why the attack on Gemalto went completely unnoticed until secret GCHQ documents on the topic were released to the public.

The endpoints revealed at Gemalto: communications "in the clear" between the engineering staffs of Gemalto, network operators and SIM card personalization companies.

The technology vehicle for this hack was likely supplied by the NSA. Reports indicate that GCHQ used the NSA's X-KEYSTORE, either to stalk Gemalto engineers' communications, or to simply pore over and extract data on Gemalto already collected by the NSA and then penetrate the company's servers to take control of communications. From there, X-KEYSCORE "scored" the individuals tracked to determine which ones had direct control over the current Gemalto SIM card key stash.

The same methodology was applied to finding and mining the communications of targets at mobile network operators and card personalization companies.

In the first instance, SIM card keys were "lifted" from the carriers themselves. How that is possible: A copy of each SIM card key is always provided to and stored by mobile operators so that they may identify, serve and bill for usage of each mobile device connected to the network. GCHQ also hacked the networks of SIM card personalization companies that burn the unique data set into each card, including its encryption key, integrated circuit card identifier (ICCID), international mobile subscriber identity (IMSI) number and the authentication code used by operators to confirm identity of a mobile phone logging on to the network.

How hard was it to hack in to Gemalto? All too easy, as it turned out. Gemalto, network operators and card personalization companies were all guilty of transferring keys "in the clear" by email, or by FTP files protected by simple encryption or none, at all.

End result: massive passive surveillance made easy, without decryption, and without the slightest risk of being discovered by targets or carriers. In a short time, GCHQ had established an automated system for harvesting SIM card keys from any of three sources: Gemalto-produced SIM cards, network operators and personalization companies.

GCHQ's work with leveraging vulnerable end points was by no means a first – the NSA pioneered and perfected the technique months before.

Nor is it the last.

A CIA Favorite Point of Attack – Endpoints

From evidence in the March 7, 2017 Wikileaks Vault 7 disclosure of intrusive technologies used by the U.S. Central Intelligence Agency, the now classic "endpoint attack" is popular at the CIA, too. One point to bear in mind. Despite claims that Vault 7 exceeded the 2013 NSA leaks by Edward Snowden, the material gleaned on CIA hacking is miniscule by comparison and reveals the agency's primary interest as being individual versus mass surveillance. No surprise there.

The most important "revelation" – long documented as a routine capability of government intelligence agencies including the NSA – is that the CIA can easily bypass tough encryption tools including PGP, VPNs and the Tor browser, and routinely does so. Note our emphasis on the word "bypass." The CIA generally does not bother cracking encryption – they simply go around it and gain access to intelligence by attacking endpoints. Does the CIA have malware? Certainly. But remember that malware always leaves a footprint. By attacking vulnerable, weakly-protected endpoints in a corporate or enemy supply chain, CIA agents can easily access any content without leaving a trace. In other words, they use the exact same capability that GCHQ exercised to obtain the Gemalto SIM card keys.

Such "sharing" of intrusive capabilities is fairly routine amongst members of the U.S. intelligence community and its allies. While the CIA has the capability to develop its own surveillance technologies, the NSA – or rather, its government contractor supply chain – has more mathematical and engineering might to commit to such ventures and is often the fount of innovation for intrusive techniques subsequently borrowed by the CIA, the U.S. Federal Bureau of Investigation, the GCHQ and other members of the "Five Eyes" intelligence alliance.

When they choose to do so, the aligned agencies can and do use malware in order to put the blame on other parties such as Russia and China. It is a simple matter to lift code from known exploits used by those nations and "plant it" where the victim, or a third party cyber company such as Kaspersky, will find it. The real party behind the malware attack walks away with the data, while its enemy counterparts are "caught" and "exposed" by national security journalists.

One amusing footnote to this tale. When The Hacking Team of Italy was hacked in 2015 and suffered the indignity of having its entire code for Remote Control System (RCS) aired in public, the CIA briefly toyed with the idea of inserting RCS code as the fingerprint for agency attacks on targets. But the notion was quickly abandoned. Evidently, posing as a hacked intrusive malware vendor already caught with its pants down seemed a little too obvious and unbelievable a ruse.

NSO Group: Offensive Cyber King

For obvious reasons, most companies that specialize in offensive cyber like to fly below the radar. Yet invariably, providers of intrusive systems are outed to the public sooner or later. Sometimes such exposure comes about purely by accident. Other times it results from the work of equally talented spyware experts working for privacy advocates or interested academics. One classic example: the 2017 uncovering of NSO Group Technologies, an Israeli Zero Day and malware company that has successfully covered its tracks for years. The NSO Group bust was the work of the The Munk School of Global Affairs' Citizen Lab in Toronto.

Munk School/Citizen Lab earned early renown for locating the servers of The Hacking Team, nation by nation. They've been in the business of tracking down ISS companies ever since, often working with privacy advocacy groups.

Munk unmasked NSO Group by meticulously tracing mysterious server attacks on mobile devices. Their crowning moment: linking a United Arab Emirate hack of a dissident's iPhone to malware provided by the Israeli vendor.

Munk's success was a fine piece of detective work. What made the story sensational: the unheard of and seemingly impossible. NSO Group exploited vulnerabilities in Apple's mobile operating system, iOS, until then widely regarded as the IT equivalent of Fort Knox. In the end, Munk's work was as much an exposé of an arrogant tech giant – Apple – that for years led the public on by touting the Olympic strength of its security. In the end, the iPhone and iPad were revealed as not being not quite so "iNfallible."

The Munk School and Citizen Lab were careful to quietly alert Apple before issuing their report on NSO Group so that the iPhone maker could develop patches and alert customers. But the damage to Apple's reputation – and its engineers' egos – was loud and irreversible. The company that launched the supposedly impenetrable iOS 8 operating system in August 2014, then thumbed its nose at former FBI Director James Comey's request for cooperation with investigations, got its comeuppance in the form of a public spanking by an Israeli hacking company.

To NSO Group, discovering the iPhone's vulnerabilities turned out to be child's play. Munk School and Citizen Lab had to work much harder to find NSO Group's own "vulnerabilities" leading to public exposure.

Inside the NSO Group

NSO Group Technologies was founded in 2010 by entrepreneurs Niv Carmi, Omri Lavie and Shalev Hulio who, like many professionals in the

Israeli ISS domain, cut their teeth in Unit 8200, the SIGINT branch of Israel's military intelligence service. Just as impressive, the company's Chairman of the Board until recently was retired General Avigdor Ben-Gal, former head of Israel Aircraft Industries, and before that a legendary hero of the Six Day War of 1967 and Yom Kippur War of 1973.

General Ben-Gal was not always a fan of technology. In 2007 he issued sharp criticism of the Israeli Defense Force's state of battle readiness, and lambasted the IDF's shift to technology as a waste of time. But evidently what he saw emerge from the minds of Omri, Lavie and Hulio changed the irascible general's thinking. When NSO Group geared up in late 2009, Ben-Gal signed on as Chairman of the Board. Adding Ben-Gal to NSO was a good move, blending seasoned leadership with advanced technology to drive results. Eager investors lined up.

Initial investment in 2010 of US $1.8 million led by Genesis Partners set in motion a company that was generating US $40 million by 2013. Along the way, American venture firm Francisco Partners stepped in and purchased NSO Group in 2014 for $120 million. By late 2015 the NSO Group was earning at least US $75 million per year. General Ben-Gal passed away in February 2016, by which time Francisco Partners was rumored to be seeking a buyer for the now $250 million company at a sales price in the range of US $1.0 billion. No buyer surfaced, and ultimately, in February 2019, the founders bought back their company from Francisco Partners for an undisclosed sum.

From the beginning the NSO Group has been based in Herzliya, a suburb of Tel Aviv. Now several hundred employees strong, the company focuses on development. It handles direct sales through its internal units such as WestBridge Technologies in the U.S., as well as channel partners including Israel's Ability – a familiar model that many ethical malware companies follow to keep themselves at arm's length from customers and the risk of public exposure. That said, the company has a track record of major contracts in the Americas and the Middle East, including the United Arab Emirates.

Like other companies that know the Zero Day/malware business cold, the NSO Group leverages its cyber knowledge on the other side of the marketplace, selling cybersecurity solutions. The vehicle is a second company, Kaymera, formed by two of the original NSO Group partners, but run separately. Kaymera develops and sells solutions for "comprehensive, multi-purpose mobile security."

NSO Group's focus is the polar opposite: bringing down mobile devices through an ingenious mixture of malware and social engineering. The core product is called Pegasus.

On the Wings of Malware

That NSO Group would choose the name Pegasus for its malware solution is a poetic touch. The white winged stallion of Greek mythology was born of the god Poseidon and the monster Medusa, thus had the DNA to do both good and evil. Pegasus also helped Greek hero Bellerophon fight and defeat the dreaded Chimera, which sounds a little too much like NSO's fraternal twin and cyber opposite, "Kaymera," to be pure coincidence.

The company has given just as much thought to how Pegasus works to penetrate and take over a target's device. As NSO has put it, the technology is "clever enough," but also relies on human fallibility.

Pegasus comes with two installation vectors: one-click and zero-click. In both cases, Pegasus is able to perform a Zero Day remote jailbreak of an iPhone.

The most common vulnerability exploits have been dubbed the "Trident Chain," and take place in sequence as defined by Mitre Corporation's "Common Vulnerabilities and Exposures" (CVEs):

- **CVE-2016-4657:** Visiting a maliciously crafted website may lead to arbitrary code execution.
- **CVE-2016-4655:** An application may be able to disclose kernel memory.
- **CVE-2016-4656:** An application may be able to execute arbitrary code with kernel privileges.

With one-click vector, the target receives a text message with a link to a malicious website which leverages one or more exploits for the target's mobile web browser, be it Safari, Firefox, Chrome, Google or other search engine. Once the Zero Day exploits do their work, the malware is planted and the target's device essentially becomes a slave to the attacker.

For a zero-click vector, the user sends a link via a WAP Push SL message that causes a phone to automatically open a link in a web browser. The target is infected without even clicking on a link.

In both instances – one-click and zero-click – the target is routed to a Pegasus Installation Server, which quickly looks at a target's mobile device software agent to see which exploit chain will work with the device. If an infection fails, the Pegasus operator can direct the target's web browser to a legitimate site to avoid detection.

Once the infection is accomplished, the attacker can exfiltrate any and all data from the device: Emails, text messages, address books and phone call history, keystroke logging, search history and screen shots.

The user can also:

- Record audio and video streaming from device memory, bypassing Skype cryptography.
- Use the device microphone to pick up conversations in a room occupied by the target.
- Activate mobile device cameras.
- Take charge of GPS system to locate the target.
- Extract Wi-Fi passwords.

To protect the user's identity, data collected from a hijacked device is sent to the user's Pegasus server via a Pegasus Anonymizing Transmission Network (PATN), essentially a VPN. To ensure against discovery, Pegasus goes to extra lengths to see that certain features don't "tip off" the target. For example, Pegasus only allows an attacker to take control of an iPhone or Samsung camera when the device is in idle mode.

Earlier versions of Pegasus also had the ability to self-destruct, if need be to avoid detection.

How Munk Caught NSO Group

It all began when UAE dissident Ahmed Mansoor received what he considered suspicious-looking text messages that offered him "new secrets" about detainees tortured in UAE prisons. All he had to do was click on a link in the SMS text. Having been hacked in previous years by government officials using Gamma Group FinFisher and The Hacking Team's Remote Control System (RCS), Mansoor was naturally cautious. He forwarded screen shots of the dubious texts to contacts at the Munk School of Global Affairs/Public Citizen in Toronto for their examination.

As it happened, Munk School had for months been examining what appeared to be a Mideastern set of servers being used to infiltrate mobile devices. All evidence pointed to a little known Israeli company called NSO Group and its malware, Pegasus.

"Long before Ahmed Mansoor had forwarded us any suspicious links he received, we had mapped out a set of 237 servers and linked this set to NSO Group," Munk reports. The screen captures sent by Mansoor contained links whose domain names just happened to match Munk's list of servers associated with NSO Group's Pegasus.

A payload identified by Munk contained files from a base library for tools used to record chats and messages from apps such as WhatsApp and Viber. In each file, Munk found "hundreds" of strings that included the text "kPegasusProtocol."

Munk matched historical data against the Mansoor attack and found key data that identified NSO Group by address, phone number and email:

Registrant Street:	Medinat Hayehudim 85
Registrant City:	Hertzliya
Registrant State/Province:	central
Registrant Postal Code:	46766
Registrant Country:	IL
Registrant Phone:	972542228649
Registrant Email:	lidorg@nsogroup.com

Coda

In its very occasional public statements, NSO Group notes that its mission is to protect public safety "by providing authorized governments with technology that helps them combat terror and crime." NSO Group adds that it "sells only to authorized governmental agencies, and fully complies with strict export control laws and regulations. Moreover, the company does NOT operate any of its systems; it is strictly a technology company."

That statement could be taken as a swipe at competitors such as the former Hacking Team, which in 2016 ran afoul of Italian regulators who revoked its export licenses for any sales outside of Europe. But more importantly it is a defense of NSO Group's business against critics such as Munk School.

Whether or not one agrees with their position, Munk School deserves credit for its investigative effort, and for an in-depth understanding of the technology in play that ultimately proved every bit as impressive as that of the "target" it pursued. Talk about role reversal.

However, in the end the Canadian think tank may have unwittingly done the Israeli malware firm a priceless service in the form of free global publicity. Zero day vulnerabilities and exploits by definition always have a limited life, and there is no doubt that the ones found by Munk were nearing their sunset. In the wake of such incidents a company like NSO Group often disappears for a time, then resurfaces stronger than before, perhaps with a different name as NSO has done several times in the past.

Ouroboros: Assault on Critical Infrastructure

When first discovered, Ouroboros, a family of malware variants with distinct ties to Russian intelligence, constituted a greater threat to the security of Western nations than Al Qaeda, ISIS, the Taliban, lone wolf and all the other Islamic fanatics combined. Since at least 2011 and perhaps earlier, Ouroboros has been used to successfully penetrate critical industrial and government sectors including energy, defense, aviation, finance and pharmaceuticals, and for that matter, intelligence agencies, too. Even today it could bring any – or all of them – crashing down.

Unlike terrorist attacks, this malware doesn't hit and run. Embedded, it sits and waits – holding the power to single out, cripple or devastate its targets. Such malware is less like a bomb than an elusive threat that hangs over the heads of nation states. It's the launch code that can trigger economic and social devastation preceding total national collapse. It levies ransom in the coinage of fear.

At times compared to STUXNET, the U.S./Israeli malware that for a time disabled Iran's nuclear program but was later exposed when it "went into the wild," Ouroboros hides deep in the systems it compromises, is highly sophisticated, difficult to track and hard to eliminate. Both attach themselves to a target's supervisory control and data acquisition (SCADA) computer systems used to manage critical infrastructure IT. Like its U.S./Israeli cousin, the Russian malware, while not a viral "worm," has the ability to penetrate and spread quickly through its targets.

Why the name "Ouroboros," and its related moniker, "Snake"? Credit goes to the Russian code for their familiarity with the legend of the creature. The myth of the Ouroboros dragon or snake that devours its own tale transcends cultures, appearing first in ancient Egypt and Greece and passing down through the centuries to Nordic lore. As late as the early 20th century, the monster was the subject of science fiction and fantasy, appearing in works such as British author E. R. Eddison's 1926 novel, *The Worm Ouroboros*.

The common thread of these legends and myths is timelessness, eternity, a thing without beginning or end. That is the guiding principle of the malware Ouroboros, too, which is exactly what makes it so dangerous. Where it ends, it begins again anew. Once attacked by Ouroboros, the victim may never be 100 percent rid of it, even after extensive de-bugging.

Inside Ouroboros: Rootkit and Kernel-Centric

First discovered in 2013, Ouroboros presents two very different ways of creating cyber spyware. The first is a classic rootkit infection with two files: a driver and an encrypted virtual file system. The second is an attack on kernel-centric architecture.

A rootkit infection is a particularly insidious type of malware that gains privileged access to the "root" or Administrator account of an operating system. It is designed to hide itself, its actions – and even the penetration itself – thus being able to continue operating behind the scenes indefinitely. Rootkit infections are masters of subverting detection or analysis.

The kernel-centric attack focuses on the operating system kernel. Because the kernel controls input/output of a processor, it is virtually impossible to eliminate an attack here short of deleting the current OS and installing a new one. The kernel-centric design of Ouroboros is often described as unique, giving the attacker unprecedented flexibility and sustainability. Even when nearly all infected hosts are put out of action by the target, Ouroboros only requires a single remaining undetected host – still infected – to carry on the penetration, spread the malware, and exfiltrate sensitive data.

Ouroboros is ready-made for Windows systems, including 32-bit and the more secure 64-bit system – which, ironically, Ouroborus has learned to master by taking over common anti-spyware, essentially controlling the hound created to catch the fox.

Once inside a system, Ouroboros works peer-to-peer, infecting other computers in an industrial or government environment that communicate with each other. Soon, all are controlled by the attacker. All that's required is one device connected to the Internet. From there Ouroboros can spread system-wide, infecting and commanding all devices in the network – even those not connected to the Internet. Captured data is routed back through the target's network to the original victim, and from there over the Internet to the attacker. The authors of Ouroboros know full well that an enterprise's most sensitive and secret data is likely stored in a device not connected to the Web. The malware's peer-to-peer workaround is a clever and highly effective strategy to defeat any such obstacle.

With these capabilities, Ouroboros can steal confidential data at will, including files and network traffic; issue arbitrary commands that cannot be countermanded, e.g., shutting systems down or making them operate too slowly or too quickly; and even allow its owners to modify and mold the malware over time, adding new capabilities. In the view of most experts, such a complex and sophisticated tool could only emanate from an intelligence service with very deep pockets. This level of skill represents a significant investment.

How did something this big and ominous remain undiscovered for so long? One view: Ouroboros hides its activities – and collects secret data – by the use of "hooks," codes that misdirect the OS kernel from performing legitimate functions to those directed by the malware. In this case, the hooks are embedded in the Ouroboros rootkit drive. Using inline patching, the driver alters the system's start-up to redirect to a malware function then immediately return to the legit function. Once the hooks are in, the malware's commands are deemed valid by the victim system. The malware owns and is hidden by the attacked system itself. Functions now hidden include: Ouroboros driver and file system files, rootkit handles, rootkit virtual file systems, and any record of the malware in the attacked system's registry. As a final flourish, a hook completely erases any sign of the rootkit when the OS is shut down, and restores the driver in the event it is deleted.

Similarly, traffic containing stolen secrets is concealed so that it is indistinguishable from ordinary web traffic of the target – communicating within the very process that the victim uses for his own business or intelligence communications. Ouroboros blends with the browser traffic, bypassing firewalls and remaining completely low profile.

How does Ouroboros get in, in the first place? To date, the answer to that question remains a mystery. No one knows whether the infection vector occurs via USB flash drives, spear-phishing, drive-by or social media attacks. It could be any one of them – or all.

Within the cyber community there is even disagreement over whether Ouroboros is the work of Russian independents, the government – or someone else entirely. However, the evidence of Russian language in the coding, the sophistication of the program, and the close similarity to malware behind the most infamous attack on U.S. military intelligence almost certainly point to Vladimir Putin's realm.

Whomever's behind it got their start with an infamous piece of work called Agent.btz.

From Parking Lot to U.S. Military Ops in the Mideast

USBs, or flash drives as they are commonly known, are among the most common tools for spreading malware. In one of the worst-ever malware attacks on a major national intelligence system, the victim actually picked up the malware-riddled USB in the parking lot of a U.S. defense facility. Some accounts say the event occurred in the Middle East, and others cite the Pentagon as the attack vector. Wherever it began, once the target installed the flash drive on his laptop, he helped usher in a new type of malware. Dubbed Agent.btz, it began its dirty work, infecting other computers within the network and before long entering the Valhalla of U.S. Military Intelligence, Central Command.

Months passed as the virus spread throughout U.S. military operations in Afghanistan and Iraq. Not until late in 2008, a year or more after the first breach, did Pentagon's IT and highest brass realize they had a problem. The immediate impact was an executive order banning the use of USBs by all military, a real problem for ground forces who relied on the ease and simplicity of flash drives for carrying tactical plans and other vital data. Longer term, the key impact of Agent.btz was to inspire creation of the US Cyber Command.

Six years after the public humiliation of Agent.btz to America's military, Kaspersky Labs issued a report on a "new" malware attack called Ouroboros. But Kaspersky saw some similarities that reminded it of the prior breach.

Looking back at Agent.btz, it was confirmed that attacks occurred most commonly via UBS. However, that wasn't the only way. Another infection vector came via peer-to-peer connections within a network. When a clean computer attempted to map a drive letter to a shared network resource infected by Agent.atz, it would by default open the damaged device's "autorun.inf file" and follow instructions to load the malware.

As one expert put it, "Once infected, it will do the same with other removable drives connected to it or other computers in the network that attempt to map a drive letter to its shared drive infected with Agent.btz – hence, the replication."

Sound familiar? Agent.btz begins where it ends and goes on forever. Agent.btz was detected 13,832 times in 107 countries across the globe in 2013 alone, after half a decade's IT sector familiarity with the Trojan, and exhaustive campaigns to eliminate it.

Just like Ouroboros. There is no doubt about the connection between Agent.atz and Ouroboros. They bear traces of Russian language. Both focus on stealing data. They use the same file names as Agent.btz – "mswmpdat.tlb", "winview.ocx" and "wmcache.nld" – for log files stored on infected systems. And they use the same "XOR key" or cypher to encrypt their log files.

The question whether Ouroboros is a later development or simply part of the same family of malware – successful in avoiding detection – is the subject of endless debate. BAE Systems Applied Intelligence has argued that Ouroboros, Snake and Agent.btz are all variants of the same malware. BAE Systems further believes that a sample of the malware compiled in January 2006 would indicate that Ouroboros has been around since at least 2005.

Wintego Cracks WhatsApp Encryption

Wintego created an uproar when a leaked product brochure claimed the capability to crack end-to-end encryption (E2EE) used by the popular messaging service, WhatsApp. Security analysts quickly dismissed the claim, asserting that the Signal Protocol used by WhatsApp is impervious to attack unless Wintego CatchApp possessed extraordinary powers never before achieved – a possibility that most found dubious. The doubters were quick to add that the Wintego brochure predated WhatsApp's embrace of Signal and concluded that CatchApp was more likely just a catch-as-catch-can product.

Then out of the blue in March 2017 came Wikileaks' disclosure of CIA documents interpreted by pro-privacy interests as showing that the lauded intelligence agency has the same Signal Protocol defeating capabilities claimed by Wintego. Looking back on the Wintego CatchApp brochure, Wikileaks tried to put the two leaks together. Were both entities onto something big and leveraging some previously unknown vulnerability in Signal that made the content of WhatsApp users easy game for the U.S. Intelligence Community?

Closer examination showed that Julian Assange's secret-disclosing organization had stepped off a WikiCliff with its analysis. The CIA documents made no such claim. However, the intelligence agency's documents did indicate that it is not only feasible, but a routine matter to get around WhatsApp E2EE simply by taking over a device via multi-vector attacks on end points. Analysts grudgingly concurred, with the caveat that the CIA's capability would be too cumbersome for conducting mass surveillance, and thus the vast majority of WhatsApp users – now numbering over 1.5 billion – could rest easy regarding fears of being silently monitored.

That facile bit of rationalization missed the point entirely. The CIA does not endeavor to monitor whole populations' SMS. Their aim is to target finite categories of individuals who fit the profile of severe threats to national security.

The same conclusion on end point attacks might initially be drawn about nation state or law enforcement users of Wintego CatchApp. Except CatchApp does, in fact, do a bit more. Research proves the presence of a "back door" in the Signal protocol. That is where Wintego CatchApp plays.

Wintego at Work

Because we review classic Wintego solutions elsewhere in this book (See Chapter 5: Mobile Location and Monitoring), here we will simply provide a brief re-cap of the relevant fundamentals – Wintego WINT, DEX-Inline and Wintego Data Extractor – before moving on to CatchApp.

Wintego WINT is a backpack unit used for Wi-Fi interception. Its sister product, DEX-Inline, is for intercepting one or more targets on 3G, 4G and fixed networks. WINT leverages four different access points to single out targets on Wi-Fi, plus high-gain antennas for remote targets. Both WINT and DEX-Inline share a common platform, Wintego Data Extractor (DEX), which collects live traffic as well as stored data on the target device.

WINT and DEX-Inline operate by conducting a man-in-the-middle (MITM) attack on targeted devices. Originally this goal was achieved by overcoming Secure Socket Layer (SSL) and Transport Layer Security (TSL) encryption, the common method used by hackers for years.

However, a new challenge for Wi-Fi interception arose when carriers dispensed with SSL and TSL, and moved to the supposedly impenetrable Signal protocol for E2EE. Those networks that didn't embrace the upgrade remained susceptible to downloading MITM attacks.

Conquering Signal's E2EE proved no simple matter. Ordinary methods such as recourse to brute force or even FPGA-facilitated multi-vector PRINCE (PRobability INfinite Chained Elements) algorithm attacks were simply too time-consuming and spotty to achieve results in a real-time intercept during the limited time that a target might be within reach on a specific Wi-Fi network.

In the view of several vaunted security analysts any such attempt to crack Signal E2EE by CatchApp by ordinary means was inconceivable. Per one Johns Hopkins University cyber expert:

"They would have to defeat both the encryption to and from the server and the end-to-end Signal encryption. That does not seem feasible at all, even with a Wi-Fi access point."

If that is true, then what is Wintego CatchApp all about?

For the answer, return to what both Wintego and the CIA actually say in explaining how they deal with Signal. Neither one says that it defeats Signal E2EE. What they do instead: "bypass" encryption and attack vulnerable endpoints. Signal developer Open Whisper Systems, of all sources, says it best:

"This is about getting malware onto phones. None of the exploits are in Signal or break Signal Protocol encryption. The story isn't about Signal or WhatsApp, but to the extent that it is, we see it as confirmation that what we're doing is working. Ubiquitous e2e [end to end] encryption is pushing intelligence agencies from undetectable mass surveillance to expensive, high-risk, targeted attacks."

To the extent that intelligence agencies and other clients seek out Wintego for help via WINT CatchApp, it is also a confirmation of market demand for the Israeli company's services. We should add that where CatchApp is concerned, there may be far more at work behind-the-scenes than many are willing to admit. WhatsApp's approach to using Signal, as it turns out, creates a back door, after all.

Built-in Vulnerability of WhatsApp

WhatsApp succeeds with customers by making the service eminently accessible by all. "Usability" is the byword. That includes scenarios when customers on the receiving end of a WhatsApp message change out devices or SIM cards, a common practice in many parts of the world. But when that happens, a potential vulnerability opens up.

In such instances, the recipient goes offline to reboot or reinstall WhatsApp, a process that generates a new set of encryption keys. In the interim period while the user is offline, certain messages sent to him or her may have been stamped "undelivered." WhatsApp, meantime, has generated that new set of encryption keys for the recipient. This forces the sender's WhatsApp account to re-encrypt *and* re-send the undelivered messages. The process is automatic. No action is required by the sending customer. The sender is not notified by WhatsApp that his message has been re-sent until after the fact. Successful attacks via rootkit infection can be executed during the reboot.

WhatsApp has solid business reasons for sticking with this vulnerable process: They don't want users to complain of not receiving messages when they were offline. Both the sender and receiver of messages in such circumstances are at risk of having their messages intercepted.

The question then is how often such an opportunity might come into play for an interested government agent. Fairly often, as it happens. Not all WhatsApp customers have reason to go to the trouble and extra cost of changing out their mobile devices' SIM cards on a routine basis. Suspects who seek to avoid identification are high on the list of individuals who do.

The point is often made that E2EE is of little value to a target if an attacker or agent simply crashes into his device and takes it over, using solutions that involve a combination of Zero Day exploits and malware, or as in Wintego WINT's case, MITM and malware to leverage vulnerable end points. Yet there is no certainty that the ability to crack a device will by definition provide an entry point to E2EE on WhatsApp.

That said, an app that can take advantage of WhatsApp's own Signal back door opens new horizons. As so often happens in the world of comms service, ease of use is Rule #1 for market leadership, but often introduces weaknesses that invite access by individuals other than friends.

ZERODIUM: Zero Days for CyberSecurity

In the world of Zero Days – undisclosed software or hardware vulnerabilities that leave a system open to attack – few companies are better known than the legendary VUPEN, a company founded by master hacker Chaouki Bekrar and dedicated to finding and providing such weaknesses for Western clientele, with NSA at the head of the list.

Perhaps to avoid any of the negative attention or legal liability that might have ensued after the Edward Snowden leaks, Bekrar shut down VUPEN in July 2014 and launched a new and different type of Zero Day enterprise: ZERODIUM.

ZERODIUM's sole interest is cybersecurity – leveraging new knowledge of potential exploits to help clients foil Zero Day attacks. The medium: ZERODIUM's subscription-only Security Research Feed, or "Z-SRF." The clients for the new service are defense agencies and contractors, financial institutions and other corporations most at risk.

ZERODIUM is:
- NOT in the offensive security business.
- NOT part of VUPEN or an expansion of VUPEN's interests – again, VUPEN is defunct.
- NOT in the business of creating or incenting deliberate creation of exploits.

ZERODIUM is a powerful force in stopping Zero Day threats to national security, economic well-being and public safety.

Given the ramp-up in cyber terrorism and hostile nation activity, the importance of a company with ZERODIUM's credentials is significant indeed.

Are Zero Days Out of Control?

It goes without saying that impact from the dark side of hacking has reached an unprecedented level. It is sufficient to say that the NSA has cited cyberattacks as the greatest threat to U.S. national security.

Not that anybody listens to or learns much from the warnings or experience of expert voices. After every attack, consultants urge adoption of multi-factor authentication. Bureaucrats go through the motions, little is resolved, then comes the next breach and the cycle starts anew.

Cyberattacks often begin with Zero Days, whose number is certainly growing. The number of Zero Days discovered rises every year. We underscore the word "discovered." Those that go undetected are likely as or more numerous. The range of victims is scaling up, as well.

No system, software or device is immune. That is why ZERODIUM is ratcheting up the bounties. They're not interested in buying run-of-the-mill vulnerabilities, just the truly dangerous ones that could crash businesses, critical infrastructure sectors, whole economies – or nations, at the push of a button.

ZERODIUM will accept "high risk vulnerabilities only with fully functional exploits" and strictly for major products. Only the best are eligible for inclusion in Z-SRF and reward via ZERODIUM's highest premiums. If a proposed Zero Day or research on even a partial high value exploits passes ZERODIUM's rigorous evaluation, the company vows to pay bounties that beat those of any software, device or Web company – Google included. Winners are paid in full by wire transfer. Bonuses can enter the picture when the exploit meets specified lifespan requirements.

Researchers who hail from or bank in any nation sanctioned by the U.S. or United Natons are not eligible to participate.

The types of Zero Days that ZERODIUM considers pertain to mobile devices, operating systems, browsers, and readers/players. What they seek first and foremost are Zero Days that satisfy one or more of the following needs, depending on the target: privilege escalation, code execution, disclosure of sensitive data and/or sandbox/bypass execution. Also of interest is any truly unique research. The most desired Zero Days include those for:

- Mobile: Apple iOS 13.1; Android 10.0.
- OSs: Apple Mac OS 10.14 and 10.15 Microsoft Windows 10; and Linux CentOS, Ubuntu, Tails.
- Browsers: Apple Safari OSX and iOS; Google Chrome; Mozilla Firefox; and Microsoft IE.
- Readers/Players: Adobe Flash Player or PDF Reader; Microsoft Office Word and Excel; and VLC Media Player.

While the menu might seem brief, it covers major systems, software and devices most widely used today. As ZERODIUM points out, the focus is on quality, not quantity.

Worth noting: The menu is out in the open. By engaging the public, ZERODIUM is widening the network of potential participants, for profit certainly, but still for "good works" that serve the public interest.

Conspiracy Theory Syndrome

Critics can be quick to judge and eager to jump on board conspiracy theories if doing so serves their own interests. The tendency can be toward left- or right-leaning political bias that immediately discredits the analyst's or author's objectivity.

Casting developers as potential criminals with a penchant for deliberately creating systems weaknesses is definitely a step over the line – from logic to tabloid-level conspiracy theory. As the party responsible for any predefined aspect of a system, software or device, a developer would be the first suspect brought in for questioning if new Zero Days mysteriously cropped up. The money trail would be completely transparent.

Another concern voiced is that a Zero Day purchased could escape into the wild. There is always that possibility, but in this case the risk is remote. Zero Days are a precious commodity. Clients paying the ticket for ZERODIUM's Z-SRF – defense and intelligence contractors and financial institutions – hardly have the incentive to reveal knowledge of an exploit. As a failsafe against human fallibility, access to Zero Days would be limited to a readily identifiable few to whom they are entrusted. To his credit, Mr. Bekrar appears unaffected by critics, open about his plans, and emphatic that ZERODIUM, like VUPEN, will sell exploits only to democracies. That candor puts a major dent in ill-wishers' innuendo. When the key figure eschews secrecy, conspiracy theories fall apart.

CHAPTER 5

Mobile Location and Monitoring

Overview

As of 2020, 5.2 billion people subscribe to mobile services on planet earth. According to GSMA, just one in five subscribers will be on 5G by the year 2025, with the balance using LTE and legacy options. Rest assured, the ISS sector has developed mobile location solutions to find all.

Mobile location solutions enable the user to determine the target's precise or approximate physical whereabouts, either in real-time or historically. The solutions include:
- A "beeper" attached to a suspect's vehicle.
- Continuous signaling ("pinging") of the suspect's cell phone.
- Use of Signaling System 7 (SS7) to find the cell tower nearest to the target in real-time, often used under authority of a court order to track foreign intelligence operatives offshore.
- Use of Session Initiation Protocol (SIP) on VoIP networks.
- Special interfaces for the Diameter protocol that support location-based services on networks supporting LTE (but not 5G, which presents special challenges discussed in our sub-chapter on 5G blind mobile location).
- IMSI catchers that emulate a mobile base station and determine target device location, plus perform a man-in-the-middle attack to intercept communications content.

- RF pattern matching that calculates target device location based on the unique radio signature of each point in a cell network.

Two commonly used mobile location data methods:
- In the U.S. a non-tracking mobile location service under the Communications Assistance for Law Enforcement Act (CALEA). This method is based on heuristic data confined to the target's proximity to the nearest cell tower.
- Use of a target's mobile location records, stored by his or her service provider, to associate the use/day/time/whereabouts of the target's mobile device with the geophysical location of a criminal/terrorist event. Similar to CALEA mobile location data.

Beepers and Continuous Pinging

"Beepers" are radio signaling devices attached to the target's vehicle to transmit his location continuously. "Continuous Pinging" leverages the built-in location-based tracking capabilities of cellular networks (e.g., used in 911 services) to signal the target's device and track its physical whereabouts. Continuous pinging is conducted via court-ordered cooperation with the target's mobile service provider, or may be implemented directly by a law enforcement agency to save time in "exigent" (emergency) circumstances.

Signaling System 7

Signaling System 7 (SS7), a technology recognized in 1980 by the International Telecommunications Union (ITU), came about to provide a more efficient means of managing point-to-point voice calls – using a separate data signal with all call routing information embedded that sped ahead to alert the network on the specific handling required for a call. Because wireline, wireless and data messages may still be routed from their point of origin to their destination via SS7, and some mobile service providers internationally use SS7, it provides a convenient mechanism to track a target's location on those networks, whether in the U.S., or for calls made between the U.S. and another country. Agents can discretely send single or multiple tracking queries to mobile operators through the SS7 network and obtain a target's location with "cell accuracy."

The upside of SS7 for mobile location is that the SS7 network is global. The downside: "cell accuracy" merely determines location of a target to the nearest cell tower, thus its accuracy may be greater in an urban mobile network with multiple cells that are smaller and closer together than in suburban, rural or remote areas with larger cells that cover more territory. Also, the SS7 protocol stack applies strictly to UMTS networks. LTE uses the Diameter protocol on top of TCP/SCTP/IP stacks for signaling.

Mobile Location on LTE Networks

Tracking of LTE device position via network-enabled methods hinges on the type of location-based service selected by the mobile operator. Commonly applied methods include:

- **Cell ID (CID):** This is LTE Diameter's version of SS7 mobile location. Cell ID is accurate only within range of the nearest Evolved Node B (eNB or mini-base station) to the target's LTE device. It's not accurate but it's cheap compared to other methods, hence its popularity with network operators.
- **Enhanced Cell ID (ECID):** Uses added radio measurements to fine tune CID, but the accuracy of mobile location is confined to the distance between the target's device and the nearest base station.
- **Observed Time Difference of Arrival (OTDOA):** OTDOA measures timing of downlink signals received from three or more eNBs to pinpoint the target via triangulation.
- **Assisted Global Positioning System (AGPA):** For GPS-enabled LTE devices, AGPS is the most accurate form of LTE network-enabled mobile location. AGPS enhances the accuracy of an LTE's embedded GPS receiver with supplemental SATCOM data on reference position and time. Caveats: AGPS tends to break down when tracking targets indoors or in high-rise buildings. If the operator adds ECID and OTDOA, mobile tracking inside or in densely populated cities may improve.
- **Uplink Time Difference-of-Arrival (U-TDOA):** The best that carriers have to offer in network-enabled mobile location, U-TDOA determines location based on the time it takes for an uplink signal from an LTE device to reach special receivers in a base station. U-TDOA takes advantage of multilateration, the difference in the distance to two stations at known locations by broadcast signals at known times using multiple measurements.

IMSI Catchers

IMSI (international mobile subscriber identity) catchers, commonly known by the slang term "pineapples" to police and government intelligence agents, are fake mobile base stations that are used to lure and attract the mobile devices of criminal suspects and other targets. IMSI catchers are highly effective, but controversial in that their use does not require cooperation by a mobile operator. IMSI catchers work in two modes: "active," to locate targets on the network and capture communications and content; and "passive," to map all mobile devices in a given mobile network cell or area.

In active mode, the IMSI catcher locates the target device by triangulating the device's signal links to other mobile base stations. It then emits a slightly stronger signal than the bona fide mobile base station nearest the target in order to lure his device to authenticate and connect with the fake base station. Because mobile devices always hunt for the strongest base station signal, they are easy targets for an IMSI catcher and invariably log in.

Once a target's device authenticates with the fake base station, the IMSI catcher launches its MITM attack, decrypts the device and intercepts calls, messages or any content on the device. In markets where legacy phones still prevail, mobile devices are typically outfitted with A.5/1 A.5/2 encryption, and all IMSI catchers come ready to decrypt A.5/1 A.5/2 in real-time. In the case of an LTE network, the IMSI catcher often can use circuit-switched fallback, a technique that signals the device that it is on a GSM network and must revert to 2G. Where UMTS networks are not in service, advanced IMSI catchers use sophisticated techniques discussed later in this chapter to make LTE devices log in with them.

In passive mode, the IMSI catcher does not interfere with the network by acting like a legitimate base station. Instead, it tunes into a mobile base station to receive uplink signals from mobile devices and downlink signals from the base station. The uplink signal is the information being sent by the mobile device, while the downlink consists of replies from the base station. Passive mode also lets the IMSI catcher see all mobile devices connecting to a network base station.

Another technique involves radio signature technology, which is growing in popularity as an effective, often lower-cost mobile location alternative. RF signature mobile location matches the location patterns of the target to the unique location signature of a point in the network.

As 5G grows to dominate networks, mobile location will present new challenges such as "radio localization," the use of many small cell sites outfitted with small antennas to efficiently and economically handle massive bandwidth via multiple paths. Radio localization's reliance on multi-path propagation undermines radio direction finding. The investigator may see multiple signals arriving from a variety of different paths, making it impossible to track an RF signal by its direction of arrival.

In Germany, MEDAV and Rohde & Schwarz have made significant gains in developing "5G Blind Mobile Location" that pinpoints the location of subscribers or even of "throwaway" 5G phones. Keysight is another contender in this space.

U.S. Restrictions on the Use of IMSI Catchers

Law enforcement and government intelligence agencies are increasingly more interested in a target's exact location in real-time versus tapping into his communications. And yet the use of mobile location data in lawful intercept and intelligence gathering remains one of the most contentious points of disagreement in the ongoing surveillance vs. privacy debate. Relevant federal laws, which date to the 1986 Electronic Communications Privacy Act, do not address location privacy. Local court jurisdictions apply a widely varying set of legal standards depending on the type of mobile location technology a law enforcement agency wants to use in a lawful intercept. Some states have passed laws placing onerous restrictions on the use of mobile location technologies for surveillance purposes.

Adding a note of uncertainty on the use of IMSI catchers by federal law enforcement in the U.S., the Department of Justice on September 3, 2015 issued strict guidelines intended to rein in potential abuse by police. All federal law enforcement agencies including the FBI now must prove "probable cause" of a suspect's criminal activity in order to obtain a warrant to use an IMSI catcher. Other changes restrict the capture of content and the retention of data intercepted, which under the new rules now must be destroyed immediately. The rules also apply to joint investigations with state and local law enforcement: the moment federal law enforcement becomes involved, the DoJ's new limits on IMSI catchers apply.

In addition, under U.S. Department of Commerce Bureau of Industry and Security (BIS) rules, American-made IMSI catchers are subject to export rules for restricted countries – although U.S. vendors may apply for exemptions.

Internationally, proposed rules under the Wassanaar Arrangement – an accord but not a formal treaty between 41 nations on arms control measures – would require member countries to place "intrusive" and "surveillance" products such as IMSI catchers and other mobile location tools under weapons-grade export requirements. Until such rules are agreed to, the global marketplace for mobile location remains wide open.

LTE Protocol and Chipset Vulnerability

Despite the market longevity of LTE, an oft overlooked problem for the security-minded are the protocol and device chipset vulnerabilities that render LTE devices an easy target for monitoring, interception, jamming and DoS attacks.

Security weaknesses in Long Term Evolution (LTE) were originally discovered by academic researchers from Aalto University, the University of Helsinki and the Technical University of Berlin.

There is a touch of irony in the Finn's involvement in this research. The world's first commercial LTE service was launched in 2009 by Teliasonera, a joint Swedish/Finnish mobile operator. Teliasonera also initiated the practice of describing LTE as synonymous with 4G service, which at the time was far from true. LTE was initially based on GSM and UMTS technologies and for several years did not begin to meet the speed requirements that define true 4G. But the myth of high-speed 4G LTE persisted for several years until network speeds caught up with ITU standards, and new fairy tales evolved around LTE being supremely secure and impenetrable. To their credit, the Finnish researchers have brought the latter erroneous legend to heel, also.

3G's Mutual Authentication

In classic 2G/GSM networks, the user's mobile device authenticates with mobile base stations or towers, but the towers themselves do not. This common vulnerability enabled early IMSI catchers to easily set up as a "fake base station" that could capture a target's mobile device simply by emitting a slightly stronger RF signal than an actual network base station.

LTE changed the basic security of wireless by introducing mutual authentication, i.e., new signaling protocols that required both the end user device and the network to authenticate one to the other. IMSI catchers for a time gained some respite from this increased security by resorting to circuit-switched fallback (CSFB), returning a signal that the tower was ostensibly on a 2G/GSM network, thus bypassing the mutual authentication rules of LTE. CSFB is still used by some older model IMSI catchers. However, because many nations' largest network operators have set aggressive deadlines for shutting down 2G and providing LTE based on the Evolved Packet Core architecture only, the days of CSFB may be numbered.

As of June 2020, with some 79 carriers preparing to enter the 5G arena, there were 4.7 billion LTE users worldwide. Mutual authentication, mandatory integrity protection, re-use of UMTS Authentication and Key Agreement (AKA), extended key hierarchy, and security algorithms are now integral parts of LTE security.

Even so, vulnerabilities in LTE remain and may for quite some time. As the Finnish academicians point out, there remain basic weaknesses in LTE protocol standards and baseband chipsets that will leave the door open to interception for years to come.

Problems with Temporary Mobile Subscriber Identity (TMSI)

Even when 2G GSM networks were in development, designers made mobile location privacy a top priority. To do so, they developed the Temporary Mobile Subscriber Identity (TMSI), a brief identifier assigned when mobile user equipment (UE) logged on to the network. Thereafter, all signaling messages between a network and UE referred only to the latter's TMSI, and not to the UE's permanent identifiers such as a phone number or IMSI number.

Reliance on use of TMSI continued as mobile network architecture evolved from 2G/GSM (circuit switched) to GPRS/UMTS (combined circuit and packet) to today's Evolved Packet Core (packet only). TMSIs are random and change frequently, typically every three days, or more often when the user moves to a new location. As a result, when introduced, TMSIs were considered a good deterrent to passive monitoring efforts trying to determine target identity.

But there was a catch. An attacker might trigger a paging request to a targeted user with a given phone number. Paging is simply a radio resource controller (RRC) process for "waking up" idle user equipment to indicate there is data to be delivered to it – a downlink, change of system information, emergency notification, or change in access technologies. Paging normally happens as part of a cycle after a mobile device logs in to the nearest mobile base station. The equipment knows to "wake up" at specified intervals to check for data. Alternately, downlinks can be generated independently by the base station – or by an attacker who uses a paging request in order to access a target's TMSI.

Researchers from Aalto University and University of Helsinki discovered that paging requests can be initiated by social media network messaging apps. If a Facebook user receives an instant message from a non-friend, Facebook automatically places the IM in its spam folder, "Other." When an LTE phone has Facebook Messenger installed, such spam automatically generates a paging request and a TMSI which a passive attacker can use to track the target's movements for as long as the TMSI is still valid with the network.

One further issue explored by the Finns: the problem of "failed connection" reports issued by LTE base stations.

LTE access protocols follow reporting procedures that help the network assist in handoffs between cells, recover from connectivity interruptions or troubleshoot problems. For example, a dropped call will trigger the target's device to send a "failure report" to the nearest mobile base station. The report includes metrics on all nearby base stations and their distance from the user equipment as measured by signal strength. Some device reports even reveal the target's precise GPS location. A Denial of Service (DoS) attack that initiates a service disruption, then utilizes IMSI catching to intercept the "failed connection" report from user equipment, can easily facilitate triangulation of a target's location.

How did these and other problems with LTE security evolve? The Finns cite protocol and chipset vulnerability.

Problems with the Radio Resource Control Protocol

Whenever a user turns on his or her smartphone on an LTE network, the device initiates a multi-step process that sets up and manages call connectivity. Call connectivity set-up is handled by a 3GPP protocol for LTE called RRC (Radio Resource Control).

The 10 specific steps under RRC are:
- **MIB, SIB1, SIB2 and other SIBs:** Master Information Block and System Information Blocks – broadcast information messages on system bandwidth and system frame number.
- **PRACH preamble:** Physical Random Access Channel for initiation of random access.
- **RACH response:** Network assigns channel and adjusts timing.
- **RRC Connection Request:** User device requests a Radio Resource Control (RRC) Connection.
- **RRC Connection Setup:** Establishes dedicated and signaling radio bearers – DRBs and SRBs to carry Non-Access Stratum (NAS) data.
- **RRC Connection Setup Complete + NAS:** Non-Access Stratum is a set of protocols in the Evolved Packet System used to convey non-radio signaling between the User Equipment (UE) and the Mobility Management Entity (MME), the signaling node used to manage paging and authentication.
- **Attach request + ESM:** During the attach procedure, the network requests EPS session management (ESM) information if the user's device indicates that it has ESM information that needs to be security

protected. ESM is a control plane function that supports all aspects of signaling for Packet Data Network (PDN) session management.
- **PDN connectivity request:** The user device requests setup of a default Evolved Packet System (EPS) bearer to a packet data network.
- **RRC:** Downlink (DL) info transfer of NAS information.
- **Authentication request:** Once the NAS signaling is established, the network initiates mutual authentication with the user device. The authentication message is sent un-cyphered by the Mobility Management Entity (MME) to the user device.

In layman's terms, in the space of a few seconds the Radio Resource Control (RRC) protocol walks the user equipment through a complex process to deliver an authentication request to the nearest mobile base station. This process takes place over Layer 3 (network) of the OSI stack – between the user equipment and the radio access network.

A simple explanation of RRC from Wikipedia:

The major functions of the RRC protocol include connection establishment and release functions, broadcast of system information, radio bearer establishment, reconfiguration and release, RRC connection mobility procedures, paging notification and release and outer loop power control. By means of the signaling functions the RRC configures the user and control planes according to the network status and allows for Radio Resource Management strategies to be implemented.

Remember that broadcast information messages (MIB, SIB1, etc.) initiate the network log-in procedure. However, at this stage that, as researchers from the University of Helsinki and the Technical University of Berlin point out, this network information is "neither authenticated nor encrypted."

The list of flaws in LTE security goes on. When a smartphone user travels from cell-to-cell, RRC ensures a smooth hand-off by sending measurement report messages to base stations. Once again, however, neither RRC requests – which as discussed earlier take place as paging – nor the messages or reports are authenticated or encrypted.

A final flaw in the RRC protocol occurs with the assignment of IMSIs and TMSIs by the network node. This assignment and accompanying network/broadcast information is also not authenticated or encrypted.

Problems with the EMM Protocol for Tracking Target Mobility

In an LTE network, the EMM protocol (short for EPS Mobility Management) controls tracking of a mobile device within the network. In tandem with EMM, a second protocol called MME (for Mobility Management

Entity) is the main signaling node in the evolved packet core, responsible for initiating paging and authentication of user equipment.

While an ordinary user is on his smartphone, EMM is continuously performing what is called the Tracking Area Update (TAU) procedure. The device emits a TAU request to notify of its position to ensure continued connectivity by the closest and/or most powerful base station. During the TAU request, the MME – responsible for paging and authentication – determines what level of connectivity the device will use. Depending on location, the connectivity might be LTE, GPRS or old-fashioned 2G/GSM. Although a smartphone will always seek out an LTE connection, the network itself makes that determination and may "reject" the request.

Such "reject messages" are not integrity-protected. In other words, they could emanate from any source, including an IMSI catcher or other RF interference equipment.

Another problem with the EMM protocol emanates from the "LTE Attach" procedure.

In order to receive services, the user device must register with the network, a process called network attachment. An EPS bearer ensures always-on IP connectivity for the device as a backup during network attachment. There may be more than one temporary EPS bearer used while the network establishes a dedicated bearer during the attach procedure. While the interim bearers are at work, the attach procedure accesses the identify data – IMSI – of the device. The IMSI is not protected during the attach procedure. As a result, the user/target device information is open to a "bidding down" attack that coerces the user away from the default route optimization scheme to one determined by the attacker.

LTE Chipset Vulnerabilities

In certain instances, smartphones themselves will "leak" their own identity. In one example, the Finnish scholars noted that "TAU reject" messages sent to any number of popular smart devices responded to the network by sending Non-Access Stratum (NAS) reports that included devices' International Mobile Station Equipment Identity (IMEI) numbers. IMEIs are typically printed inside the battery compartment of a smartphone, or can be displayed on the screen. IMEIs are absolute identifiers of a mobile device. While a number of manufacturers responded by patching the leak, some were slow to act.

In addition, it also proved simple to outsmart smartphone chipsets in regard to Radio Link Failure (RLF) reports. When an actual temporary network outage occurs, the network signals all devices in the area, requesting an RLF report. All mobile devices check in and respond with their RLFs, which may include key identifiers on each device. Such network requests are never authenticated.

The RLFs provided by smart devices are not encrypted. Since a Denial of Service (DoS) attack can be masked as an outage, it is simple for an attacker to initiate one, then issue a request for RLFs and capture data on all devices in range. GPS positioning is vulnerable in the same way – an unauthenticated request followed by unencrypted response.

Why 3GPP Neglected Standards Vulnerabilities

Lin Huang of China's Qihoo Technology Ltd. has provided an excellent explanation of the Black Hat presentation by researchers from Aalto University and the University of Helsinki. Ms. Huang focuses on IMSI catcher redirection attacks made possible by 3GPP protocol vulnerabilities. A redirection attack could be used to force targets onto a fake network or a rogue network. Huang reveals that the issues leading up to LTE's protocol weaknesses are not a new problem, at all. They were first recognized in January 2006 when a paper entitled "Security Vulnerabilities in the E-RRC Control Plane" was presented at a 3GPP meeting.

The paper illustrated a forced handover attack in which an "attacker with the ability to generate RRC signaling. . .can initiate a reconfiguration procedure with the UE [user equipment], directing it to a cell network chosen by the attacker." The author concluded that "the ability to force a handover serves to expand any form of attack to UEs on otherwise secure systems," making any poorly secured network "a point of vulnerability not only for itself but for all other networks in its coverage area."

In November 2006, 3GPP addressed these concerns in a two-part ruling that left major vulnerability concerns unaddressed:
- RRC Integrity and ciphering will be started only once during the attach procedure (i.e. after the AKA has been performed) and cannot be deactivated later.
- RRC Integrity and ciphering algorithm can only be changed in the case of the eNodeB handover.

In LTE parlance, "eNodeB" (for Evolved Node B) is the modern mobile base station. 3GPP ruled that security for call set-up could only be changed in the event of "handover." Most often, eNodeB handovers occur in extreme cases. Such incidents might include natural disasters, terrorist attacks or other events that lead cell phone users to overwhelm particular base stations. To ensure load balancing and network availability, busy base stations perform a handover, redirecting inbound calls to other eNodeBs.

Barring such emergency scenarios, 3GPP concluded that "integrity and ciphering" would remain fixed. In retrospect, such a decision might seem short-sighted to privacy advocates, but to law enforcement and government agencies that need ready access to mobile network intelligence, 3GPP's approach has proved an absolute boon.

MegaMIMO 2.0 Wi-Fi Interception

In 2016 the Massachusetts Institute of Technology announced development of MegaMIMO 2.0, a set of signal-processing algorithms that tripled the speed of Wi-Fi and doubled its range. Designed to optimize spectrum utilization of Wi-Fi in congested areas such as stadiums, concerts and other venues, MegaMIMO 2.0 was at the time hailed as a major advance that would deliver performance gains in Wi-Fi transmission/reception without the need for additional routers or access points (APs). Overlooked in the flood of kudos that came after was whether the new algorithms might also advance the job of Wi-Fi interception in venues favored by spies and terrorists: airports, railway stations, bus terminals or any busy street in a major urban center.

Wi-Fi allows wireless devices to connect to a wireless local area network (WLAN) in the 2.4 GHz and 5.0 GHz radio bands. This is unlicensed spectrum, meaning anyone may use it. Private WLANs are generally password protected. Public WLANs are not, thus creating a highly popular access option for smartphone users, to the extent that Samsung and Apple make cheaper smartphones specifically designed for Wi-Fi use only, saving consumers the additional cost and commitment of carrier contracts. Absent a carrier agreement, such phones are made without SIM cards used for network sign-in and authorization.

Cities in the U.S., the UK and India have moved toward implementing metro-wide free Wi-Fi access. The magic word "free" has, of course, boosted the popularity of Wi-Fi, leading to congestion, particularly in the 2.4 GHz band. Simply put, the more who pile onto Wi-Fi, the greater the likelihood of interference, slow speeds and dropped connections. It does not help that baby monitoring devices, portable home phones, Bluetooth and other gadgets use the same sliver of spectrum and further crowd the Wi-Fi channel. In addition, as mobile devices merely pass by free Wi-Fi, the more congestion grows on an already crowded WLAN. Those devices are checking in with the WLAN whether their owners realize it or not.

MIT's Angle of Attack

Researchers at MIT's Computer Science and Artificial Intelligence Lab set out with a basic premise: Merely adding routers and access points (APs) to a WLAN does not necessarily improve connectivity and more often has the opposite effect. As one author of the research, Dr. Ezzeldin Hamed, observed, "You can't solve spectrum crunch by throwing more transmitters at the problem, because they will all still be interfering with one another."

In RF technology, radio waves that perfectly overlap tend to boost the power of each. However, when radio waves overlap unevenly the result is interference.

When they meet "peak-to-valley," both signals cancel out. Multiply this effect hundreds of thousands of times over and the result is a massive outage driven by network crowding.

Among the more spectacular historic examples was the collapse of Wi-Fi connectivity during Steve Jobs' famous launch of the Apple iPhone4 in 2010. Thousands of audience members watched in amusement as Mr. Jobs tried to connect his device to the network while his handset face – displayed on a giant screen – read: "Could not activate data network." Reason: too many devoted fans logging on to the WLAN at the same time.

Dr. Hamed and team set out to resolve this classic Wi-Fi logjam by building on an earlier advance called "joint multi-user beam forming" or JMB, developed by MIT in 2012, using it to build on MIMO (multi-input, multi-output) technology. A MIMO transmitter sends multiple individual packet data streams – and on the same channel – to receivers with fewer antennas.

Typically, a two-antenna access point transmits two packets at the same time to two single receivers. With JMB, MiT showed that it was possible to have multiple APs on the same channel deliver packets to multiple receivers, synchronizing the phase in a distributed manner without creating interference. From this, they learned that it was possible to accurately predict and fix frequency misalignment. However, it required continuous measurement across an infinite number of devices, which itself consumed voluminous amounts of bandwidth.

Fast forward to 2016, when MIT researchers had a better idea: Don't send more signals, simply measure the uplinks of devices already connected to the Wi-Fi access points and from there infer what the downlink channel will be. Tests got underway using a system with four access points and four laptops – the latter wandering about on mobile-connected robots like human users at a convention or conference.

Not only did the system reverse the loss from having multiple access points share the bandwidth, it was 330 times faster than conventional Wi-Fi. That was impressive in itself, but Dr. Hamed and others believed that this initial work only scratched the surface of what MegaMIMO 2.0 could do. The more APs added, the faster MegaMIMO can go. MIT said the technology would soon be "commercialized," and how right they were – not only in the field of Wi-Fi. MegaMIMO 2.0 today is a vital technical underpinning of 5G mobile. Law enforcement and intelligence agencies see promise in what MIT has accomplished in advancing the interception of targets operating on Wi-Fi networks. By improving the speed, quality and reliability of connections, MegaMIMO 2.0 helped ensure the integrity of evidence collected on Wi-Fi networks experiencing congestion.

Stumping IMSI Catchers Via Multi-IMSI Phones

As mobile handset makers and operators move to multi-IMSI SIM cards, some conclude that IMSI catchers by Harris, Rayzone and other vendors will hit a wall. More likely: Mobile location detection by IMSI catcher – or other means – will live on.

Imagine a scenario where the old fixed Subscriber Identification Module (SIM) card is dead & gone and each mobile device now has a multi-IMSI SIM card, or a programmable or embedded one, or perhaps no SIM card at all. That world may be fast approaching thanks to pressure applied by Apple Corp. and Samsung on mobile operators to do away with an authentication method that has passed the quarter century mark. To some in the law enforcement and ISS arenas, that day spells dread for mobile location capabilities and perhaps "The End" for the much-revered IMSI catcher. How will national security and law enforcement agencies track suspects on mobile networks if the latter hide behind devices outfitted with multiple IMSI numbers, or with eSim cards that can be easily reprogrammed?

SIM Card Genesis

SIM cards in their current popular state are "tamper proof" microprocessor cards installed in a mobile device to authenticate the device with a specific mobile operator network and thus permit its use on that network. The geographic range in which the SIM card permits access depends on the location of the user and the class of geo service signed up for, e.g., strictly in-country national or including specific out-of-country international domains. Since their introduction in 1991, SIM cards have "grown" progressively smaller and more sophisticated but retained their popularity with mobile service providers. Reason: Thanks to SIM cards, when a customer buys a mobile communications device, he is essentially locked into the provider's network.

Mobile operators like SIM cards for another obvious reason: additional revenue flows derived from customers that wish to roam internationally with their devices. The traditional options were either to buy or rent a new phone for use in a specific country or countries, or to upgrade an existing contract to include roaming. Roaming fees are expensive. In regions such as Europe where multiple countries cohabit a continent and cross-border travel is common, roaming fees once added up to staggering sums. Another inconvenience early-on was that such arrangements generally required the user to carry multiple SIM cards, and that the phone number of the device would change per SIM card for each country where the user traveled.

From the law enforcement standpoint, SIM cards hold the International Mobile Subscriber Identity (IMSI) number central to the operation of IMSI catchers. The IMSI contains a Mobile Country Code (MCC), Mobile Network Code (MNC) and a Mobile Subscriber Identity Number (MSIN). The most essential of the three, for IMSI catcher and mobile location purposes, is the MSIN, which is the permanent identity of the SIM card.

Also important is the Mobile Station Integrated Services Digital Network (MSISDN) number – the actual phone number assigned to the SIM card. The MSISDN uniquely identifies the SIM card in a GSM, UMTS, LTE or 5G network, though in 5G phones the SIM card is encrypted.

Thus, both the IMSI number and the MSISDN identify a mobile device. The difference? An MSISDN can be changed by the network operator at the request of the user. The SIM card contains its IMSI number, which is unique because of the Mobile Subscriber Identity Number (MSIN) that is "burned in."

As it happens – network operators and law enforcement aside – not everyone is a fan of unique IMSI numbers.

The Uprising Against Conventional SIMs

Questions about the utility of solo SIM cards arose more than a decade ago with the advent of smartphones and the ability of users to access the Web for multiple types of content and services. Previously confined to voice and SMS, the mobile world opened up to a range of practical and entertainment apps ranging from music to video, e-commerce and so on. The inability to skip between providers depending on bandwidth options and local service quality was perceived as a weakness.

As early as 2009, analysts began to call for multi-IMSI mobility that would break the bonds imposed by SIM card-loving carriers, freeing customers to range between accounts with the ease of switching public Wi-Fi access. SIM-swapping was certainly do-able for these purposes, but deemed an inconvenience. "Dual-IMSI" mobile network operators were among the first to experiment with providing customers an alternative to expensive roaming plans. Pioneers claiming "virtual mobile roaming" capability offered "national rate" service when traveling in either the U.S. or the UK, but customers were still plagued with having to use a different phone number in each country. Plus the customer remained captive by one carrier in these arrangements.

One year later, the movement for change gained impetus from Apple Corp., a company always interested in maintaining absolute control of its products and chagrinned at having to relinquish power over SIM cards to network operators. In 2010 Apple tested the first "embedded" or eSimcard, made by Gemalto and installed in an iPhone. As designed, this arrangement allowed installation of operator information only after purchase of an iPhone direct from Apple. The GSM Association (GSMA) weighed in with its support, and European mobile operators began to buy in to the idea of an eSim installed by the mobile device manufacturer, but with a caveat: eSimS were deemed acceptable for M2M or Internet of Things applications, but not for communications devices. Reason: Carriers didn't like the idea of outsiders cutting in on their business. And there matters rested for another four years.

Embedded SIMs for Wearables and M2M

Then in 2014 The Netherlands approved regulations that allowed any company to issue its own SIM cards without being configured to any particular network operator. That same year, Apple introduced the LTE iPad with the Apple SIM, allowing customers to select any carrier and to switch between service providers, at will. Finally, in July 2015 it was revealed that both Apple and Samsung were in discussions with operators including Vodafone, AT&T, Deutsche Telekom and others on development of a GSMA eSim standard for all smart mobile devices. The standard, completed and released in 2016 enabled a consumer device to store more than one operator profile concurrently, although only one operator can be in use at a time.

GSMA describes the eSim as a "Universal Integrated Circuit Card (UICC) or "new generation SIM" with major advantages over the historic SIM card. Chief among these is the ability to support over the air (OTA) provisioning of a specific device's profile to the serving network.

The eSim has been widely touted as the end of the conventional SIM card, and of SIM card swapping while roaming. In the eyes of some analysts, the eSIM with OTA provisioning and the ability to switch between pre-programmed authentication schemes with multiple operators also poses a major challenge to surveillance, specifically monitoring and mobile location by the venerable IMSI catcher.

It that true? That might be the case but for the residual Bell monopoly mentality of US carriers AT&T and Verizon. Though the GSMA standard applied to all mobile devices including smartphones, the Baby Bells routinely challenge the law simply by ignoring it. In April 2018, *The New York Times* reported that the US Federal Communications Commission had undertaken an investigation.

The FCC suspected that nation's largest mobile operators were colluding to prevent proliferation of eSIMs that would enable customers to provision their own smartphones and switch between operators. As of May 2020, the investigation was still in progress.

As a result, where monitoring and mobile location are concerned, the eSIM does not yet pose a threat to law enforcement's old standby, the IMSI number of targets caught by Rayzone Piranhas, and Harris StingRays and Hailstorms. But LEAs could find themselves blindsided by another trend: multi-IMSI SIM cards, a concept first developed to combat high mobile roaming charges.

Multi-IMSI Roaming

Since the day of the first mobile device, international travelers have been up in arms over sky-high rates for mobile roaming. In response, the European Union on Feb. 1, 2017 reached a "provisional agreement" with the EU Parliament capping wholesale roaming charges of mobile operators within the 28 nation member states. The goal was to make the cost of calls while roaming in the EU close to or on par with charges incurred in the home country.

Mobile operators cast this decision as Doomsday for their companies, but to no avail. As of September 2018, Europeans can use their mobile devices anywhere in the EU for a few extra pennies per second, whether calling or texting. The EU Parliament did grant one important exception: Mobile operators that make significant network infrastructure investments to pave the way for 5G will be allowed to charge more for roaming. By and large the EU Parliament's decision was good news for government agencies using IMSI catchers in those markets, as well, as it reduced populist incentives for pushing eSIM or the multi-IMSI SIM card.

The EU is, of course, just one market. Elsewhere, pressure for alternatives continues, and in the absence of an eSim standard for smartphones and tablets, sentiment for multi-IMSI SIM cards will likely gain momentum. This is already a healthy growth market targeting the enterprise, and served by innovative leaders such as Telna, a subsidiary of Canada's KnowRoaming. With Telna's "Multi-IMSI Roaming Hub" product, the SIM card automatically picks an IMSI appropriate to the visited country and "minimizes" roaming charges. For the user, the service is completely hands-off.

As an interesting twist, Telna can also provide a "Multi-MSISDN" software solution that provides multiple local phone numbers per country, all embedded in a single SIM card. Targeted for multi-national corporations, the service lets the user be reached "at local costs from multiple countries." At present the service is available throughout the U.S. and on a more limited basis in Canada, the UK, Australia, Poland and Latvia. Companies in other nations offer similar Telna-like services.

Work-arounds by Conventional Hacking

For the law enforcement agent, trying to identify, let alone track a target on a device with constantly changing IMSI and MSISDN numbers might sound like a nightmare. One saving grace is that the marketplace for multi-IMSI SIM cards and multi-MSISDN numbers remains embryonic. For the present, at least, the case for IMSI catchers being at risk from some sudden rise in multi-IMSI SIM cards seems more theoretical than real world.

Similarly, OTA reprogramming of embedded SIM cards hasn't reached the broader "human market" of devices to any great extent, other than wearables, and is largely confined to M2M.

When and if these developments evolve to the point of being major factors in the mobile market, and by extension to government agencies in need of accurate mobile location, Harris and Rayzone will need solutions at hand to resolve these challenges. If IMSIs and Mobile Subscriber Identity Numbers (MSINs) go dark for investigative purposes, so might the IMSI catcher market. Other tools such as black lists of targeted IMSIs would become useless. Pursuing targets will require greater creativity, as well as attention to other options.

One "upside" for investigators: Despite efforts to improve security, mobile networks remain permeable and prevalent to data leakage. Commonly available exploits used by hackers readily determine a target's IMSI number, MSISDN, IMEI (International Mobile Equipment Identity) number, Access Point Name, customer and account name. Verification might be achieved via reverse look-up in the Home Location Register, the main database of permanent subscriber information maintained by every mobile network operator. Black Hats can be counted on to come up with work arounds.

One factor that would appear to undermine IMSI catchers in the future: on 5G phones, as mentioned, IMSI numbers are encrypted. However, academic hackers from Purdue University and the University of Iowa quickly dispelled that myth by cracking 5G encryption via brute force attack and presented their findings at the Network and Distributed System Security Symposium, held February 24, 2019 in San Diego, California USA. To date, major mobile operators offering 5G in that nation have not responded.

Boeing DRT Box

With all the fanfare over IMSI catchers and the importance of mobile location and interception to government customers, one very important player tends to get overlooked: the Boeing DRT Box or "dirt box," as it is commonly called.

The "DRT Box" gets its name from its original maker, Digital Receiver Technologies (DRT) of Germantown, Maryland. The company was founded in 1992 as Utica, then changed its name to DRT five years later. DRT started out making specialized radio equipment for the military and Intelligence Community, and by the end of its first decade in operation had evolved into a specialist in IMSI catchers and RF frequency monitoring tools.

DRT's expertise attracted the interest of defense contractor Boeing, which acquired the company in 2008. DRT is now a wholly-owned subsidiary of Boeing and continues to make the very same types of products for the same clientele: the U.S. Army, U.S. Navy, Special Forces, the CIA, NSA and other federal agencies such as the U.S. Marshal's Service.

"DRT box" is a generic name that applies to a number of software-defined radio (SDR) products made by the Boeing subsidiary. While Boeing's entries in the IMSI catcher marketplace are similar in some ways to competitors' products, DRT does do a few things differently that make its boxes attractive to their unique clientele. Separately, Boeing/DRT also manufactures products specific to military intelligence, providing advanced devices with the ability to intercept, decipher and analyze cellular, RF and SATCOM enemy signals.

Classic Dirt Boxes

DRT receivers automatically scan multiple RF channels including mobile and can identify all signals being emitting in a targeted zone. They monitor for up to 10,000 "entries" (specified targets), and record the interceptions. In the case of a mobile network intercept, the process involves the usual IMSI catcher drill of base station emulation, deciphering and man-in-the-middle attack. DRT systems are offered with the option of antennas that perform direction finding (DF) for mobile location.

Original standard models include the DRT 1101B, DRT 1183B and DRT 1201C, units noted for providing IMSI catcher capabilities over a wider geographic range than competing products such as the StingRay.

The DRT 1101B was developed as a multi-protocol device, able to accommodate up to four tuners for automatic scanning of channels. The

DRT 1101B in active mode operates by emitting a stronger signal than nearby mobile base stations to lure targets. Once the target authenticates with the fake tower, Boeing's product initiates a man-in-the-middle attack on the target's device. In passive mode, the DRT 1101B monitors all mobile device users within range of the receiver, jettisoning those not of interest.

Next in line was the DRT 1183B, offering much the same functionality, only faster, using a field programmable gate array-based wireless processor – DRT's WPM3 – for wideband digital signals, as well as narrowband streams for further processing.

The DRT 1201C monitors as many as 544 half-duplex channels, comes with software configurability for use with any mobile network protocol, processes multiple RF formats in tandem, and like the 1101B and 1183B automates channel monitoring to quickly pick up targets of interest. Once the signal is captured, the 1201C's DF option goes to work locating the target. The 1201C also leverages the FPGA-enabled speed of the DRT WPM3 wireless processor for both wideband and narrowband signals.

Portable Models

While DRT continues to support these older units, the company's current emphasis is on lighter-weight tactical models such as the DRT 1301C, the newer 1301C+, and the star of the show, the DRT4411B.

The 1301C is the DRT "manpack" version designed for field use in rugged environments. Like its bigger, older brothers, the 10.5-pound 1301C receiver is software configurable to adapt to multiple protocols and to process multiple formats in real-time. Being smaller than its rack-mounted counterparts, the more compact 1301C handles fewer simultaneous channels, just 16 full-duplex and 32 half-duplex. Like all DRT receivers, the manpack version includes direction finding as an option. It has its own fan-operated cooling system.

The newer 1301C+ model offers important upgrades: monitoring of up to 72 channels, and an integrated 8-channel parallel tracking GPS receiver for precise location. Heavier-duty that the 1301C, the 1301C+ model weighs 12 pounds.

The top of the line DRT4411B is a software-defined receiver that is light-weight (2.5 pounds with battery), miniature and easy on power consumption. The 4411B can simultaneously capture traffic from any mobile network including GSM, CDMA, EV-DO, UMTS, TD-SCDMA, LTE-FDD and LTE-TDD.

The 4411B can handle traffic from up to three wideband RF tuners in a frequency coverage range of 2 MHz to 3000 MHz. The system is very fast, using high-performance FPGAs. Direction finding is superb, based on a 50-channel parallel tracking GPS receiver built-in. Intercepted data can be collected on the 4411B's removable storage or streamed to other devices for further analysis. The device is controlled by a Windows XP laptop system using a simple Ethernet interface.

The 4411B's small size makes mobile interception and location tracking both flexible and covert. As DRT puts it, the 4411B is "ideal for small air and ground vehicle applications," and is commonly used in fixed wing surveillance scenarios. Not that it matters since the 4411B is often found on small planes or in military vehicles, but the device is completely quiet, as well – with no fans required for cooling.

When you hear or read about "dirt boxes" used by federal law enforcement flying by overhead, typically the device in question is the DRT4411B.

cellXion LTE IMSI Catchers

Based in Caterham Surrey UK, cellXion is a mid-sized player in the mobile location/interception field, concentrating exclusively on IMSI catchers. The company was founded in 2004 and remains 100 percent privately owned by one individual, Mark Brumpton. The company has two offices in the UK and one US subsidiary, cellXion Networks LLC, owned by Mr. Brumpton's spouse, Debra, who served as a Director through 2014. Another family member, Tobias Brumpton, joined in 2009 as Company Secretary.

One very important non-family member is Anthony Timson, who has served as a Director since 2014, and before that as a key consultant from the company's earliest days. Indeed, Timson might fairly be called the engineering father of cellXion, as he is directly responsible for pioneering the company's first IMSI catcher. With his ascension from consultant to Director, the company's QMTS/QRMS became the central mission.

Inside the QMTS/QRMS

The QMTS (Quad Mode Telemetry System) is described as a "multi-role communications platform" that can "utilise" (access) the wireless data networks of up to four mobile operators. The device acts as a router that "manages packet queues" in order to reveal "full IP connectivity" of targeted PCs. Intercepted wireless data is secured by encrypted virtual private network (VPN) line connectivity in order for operations to remain entirely discrete and unobservable to the target.

Each QMTS is connected to a central server, the QRMS. As a concentrator, the QRMS can serve multiple QMTS devices at diverse locations. The QRMS sets policy, with the ability to single out specific mobile devices and block others.

As a multi-role platform, the QMTS with QMRS central server may be dual-purpose, both for test and measurement of network efficiency, and for directly connecting to target devices.

In the former scenario, the server connects to mobile networks, and from there moves to the QMTS for individual testing purposes. For monitoring, take out the network operator: The CRMS server controls the QMTS which directly identifies and intercepts traffic from specific mobile devices. Each QMTS is outfitted with a GPS 12-channel continuous tracking receiver for accurate mobile location of the device, and of targets via triangulation of QMTS devices.

Like its sophisticated competitors, the Harris Hailstorm and Rayzone Piranha, the QMTS/QRMS can "bite" diverse types of mobile networks: LTE, UMTS, and when needed, 2G via circuit-switched fallback.

The bands monitored vary by market. In the U.S. this includes LTE bands 700/1700 MHz and UMTS/HSPA 850/1700/1900 MHz. In Europe, the QMTS/QMRS duo can monitor LTE bands 800/900/1800/2100/2600 Mhz and UMTS/HSPA 900/2100 MHz.

When on the move, users can attach a cellXion to any adjacent Wi-Fi network, continue to monitor targets, and be assured of secure VPN connectivity to the QRMS central server. The QRMS server may, when requested, be partitioned for use by multiple agencies using their own QMTS devices for separate surveillance activities. Buffering of captured data ensures that any packets dropped during an interception may be retransmitted without data loss.

Not surprisingly cellXion never comes straight out and calls the QMTS/QRMS platform an LTE or any other type of IMSI catcher. But industry insiders and law enforcement customers understand what the product does. With touted benefits of capturing streaming video "and other surveillance traffic," remote monitoring and control of devices, and even vehicle tracking and monitoring, the QMTS/QRMS is what its makers come shy of admitting: an LTE IMSI catcher with UMTS/HSPA interception and mobile location capabilities.

Still available are the company's legacy line of GSM/UMTS IMSI catchers, offered in several formats: the UGX 330 Series (GSM & UMTS case-fitted for vehicle or static location); the UGX Quadra (GSM & UMTS – in a rucksack for field use by agencies or military); and the GX Duo (GSM only – rucksack). cellXion also makes a hand-held Android OS Client for GSM, which lets the agent intercept/mobile locate in close proximity to the target, with data fed back to an IMSI catcher. The Android OS also intercepts UMTS via circuit-switched fallback.

Switching gears, cellXion offers NetSecure, a device for detecting IMSI catchers. The small portable device operates by measuring normal network activity in a specific area – police stations, embassies, military bases – to create a "fingerprint" of the zone. Any abnormal activity is quickly spotted and measured for its potential threat as an IMSI catcher being used by intruders.

IMSI Catcher Patent Lawsuits

For the purpose of full disclosure, a digression on how cellXion moved so quickly into the mobile interception/location ISS field is in order.

Both Mark Brompton and Anthony Timson got their start in the IMSI catcher business as employees of M.M.I. Research Ltd., a British engineering firm launched in 1995 for the express purpose of providing cellular telecommunications services to law enforcement, military, and intelligence agencies. M.M.I. was purchased in 2008 by Cobham, where it operated as a semi-independent unit.

In the early days of the business, M.M.I. Research took credit as the developer of the first IMSI catcher. Unfortunately, so did Germany's Rohde & Schwarz. As a result of this disagreement the two firms entered a lengthy patent battle over the technology. The suit ended in M.M.I.'s favor. Thereafter, the two companies reached an agreement where both became co-owners of the patent.

It was the first of many legal struggles that would ensue over patent ownership of the famed base stations emulator. cellXion was up next.

When Brumpton launched cellXion, the British court decision on M.M.I. Research versus Rohde & Schwarz was already history. M.M.I. Research and Rohde & Schwarz had clear ownership of the patent for IMSI catcher devices.

Nevertheless, Brumpton brought his M.M.I. colleague Timson on-board to develop cellXion's version of the IMSI catcher. The end product was the cellXion "DX918." Datong Electronics was retained as distributor of the DX918 in the UK and U.S.

This arrangement lasted until M.M.I. caught wind of it and filed suit for patent infringement against Brompton, Timson, Datong and both the UK and U.S. branches of cellXion. Rohde & Schwarz, the patent co-owner, also testified. In late February 2009, the Royal High Court of Justice – Chancery Division – Patents Court ruled the DX918 and Timson "jointly liable" for patent infringement.

So M.M.I. Research and Rohde & Schwarz won. Or rather, Cobham and the German company won, because by the time the British court issued its decision, M.M.I. belonged to Cobham.

Similar lawsuits continued. Rohde & Schwarz again found its way to court in 2011 to defend its patent for the IMSI catcher. This time the decision came quickly, and not to the German company's liking. The Courts of Appeal for England and Wales ruled that any patent for the IMSI catcher to be "invalid for obviousness," i.e., not sufficiently inventive to merit a patent, at all.

Harris Corporation, makers of the StingRay, has launched numerous patent infringement suits over the years, but never – as far as we are able to determine – against makers of another IMSI catcher. The first commercial Harris StingRay went into operation for the police department of Palm Beach,

Florida USA in 2001, and was doubtless patent-protected at the time. The StingRay II is definitely patent-protected.

In 2013, Raytheon filed for and received patent approval for its "portable base station emulator," ostensibly for emergency service location of individuals during natural disasters. In plain language, the device is an IMSI catcher.

Why companies continue to file patents for a technology deemed not sufficiently inventive to merit one remains a puzzle. cellXion is way beyond that. By carefully going about its business and maintaining a low profile, the company has built a solid business for itself in the UK and Europe.

Cisco Hyperlocation: Tracking on Wi-Fi

The name "Cisco" automatically evokes a brand that has owned the router space almost since the dawn of the Internet. But routers aren't the only market dominated by the giant hardware/software firm named after its native city, San Francisco. Growing in importance is the market for Wi-Fi equipment such as access points. Here Cisco has a solid lead, too, owning 45 percent market share for equipment used in wireless local area networks (LANs). Aruba/HPE, Extreme Networks and Ruckus Networks take the bulk of what remains. As Cisco grows its footprint in this vibrant niche it is layering expansion with related solutions. Among the most promising is the Cisco Hyperlocation module which, when installed in Wi-Fi networks using Cisco hardware, can locate individual targets to within one meter.

Cisco didn't develop Hyperlocation with ISS or surveillance in mind, but rather, as part of its Internet of Things strategy that vies to make mobility a means of improving the "customer experience," a popular marketing euphemism that translates to "selling more merchandise." To consumers, the drill is perhaps all too familiar: A customer strolls by the cologne counter in an airport and sees a pop-up ad on her smartphone offering a discount on her favorite scent if she makes the purchase then and there in that store. Despite Cisco's public posturing against the use of personal data including location for national security purposes, the company has no qualms whatsoever about helping companies use the same or similar technology to exploit that data for commercial gains. For similar reasons, Hyperlocation is a natural fit for surveillance purposes, as well.

In developing Hyperlocation, Cisco drew inspiration from the "business and personal benefits" of GPS and mobile mapping to consumers in outdoor settings. Cisco's goal was to create something similar for use in the "indoor enterprise space" where Wi-Fi rules: including shopping malls, conference centers, music concerts, sports arenas, airports, train stations and similar commercial and public venues. Other companies preceded Cisco into the interior Wi-Fi mobile location niche, not just for up-sell apps but also to help visitors find their way through hospitals or let airport management measure security line wait times. Like many such pioneering technology solutions, these early ventures in Wi-Fi mobile location were often flawed.

Failures included solutions that located individuals on Wi-Fi, but only to within five or 10 meters. Others failed to refresh data as often as needed, thus showing updates on location long after the target had moved on. Some approaches combined real-time video surveillance with RF technology, which worked fine, but too often at significant cost.

Cisco took its lead from Bluelight Low Energy (BLE) beacons that emit a signal that is picked up by mobile devices. There are some 3,000 BLE beacon solutions in use today. However, initial attempts to use BLE beacons for precise mobile location were at times plagued by inaccuracy. Cisco thought it could improve performance.

Cisco engineers made the decision to begin by leveraging the best of what was already available: widespread Cisco Wi-Fi local area networks (LANs) and the BLE beacon principle – with a new twist. Instead of one BLE beacon per network access point (AP), Cisco decided on five separate BLE beacons for each AP. They were on their way then to what they would call "FastLocate."

What the Commercial Product Means to ISS

Hyperlocation is a combination of hardware, software and higher math. Cisco has meticulously designed the unit to integrate with Cisco Aironet access points in the 3700 and 3800 series.

Each solution is actually two products: the Hyperlocation Module and its antenna. Users simply plug the module into a 3700 or 3800 access point, then plug the antenna into the module.

Each device contributes an important component of intelligence. The module itself contains the BLE modules and "FastLocate," algorithms that continually refresh data from targets' own smartphone BLE beacons in real-time. The antenna provides highly sensitive angle-of-arrival measurements that precisely triangulate target location using received signal strength indication (RSSI). In fact, the antenna is the secret behind Hyperlocation's remarkable accuracy in pinpointing a target's precise location.

The Cisco Hyperlocation solution can pick up signals from BLE beacons throughout a venue, further enhancing accuracy. Because the solution plugs directly into a Cisco access point, no battery is required.

To learn what the target is doing while being tracked, the user can take advantage of analytics capabilities embedded in a complementary software package, Cisco Connected Mobile Experience (CMX). From the public perspective, Cisco promotes the CMX platform for its ability to improve the experience (read "profitability") of consumers in a given venue. But the same or similar analytics programs, when integrated with Hyperlocation Wi-Fi mobile location data, can also reveal intelligence about a target's behavior.

Finally, the price is right. The combined cost of the Hyperlocation module and its antenna comes in at under US $1,000.

Cisco is guarded about marketing Hyperlocation for security or surveillance purposes, though it is a natural fit for such applications. Remember that Cisco did not design its routers for use in the Great Firewall of China, either, but Cisco routers most certainly are deployed therein, and are considered by many to be the backbone of Internet monitoring and censorship in that nation.

The long term appeal of Cisco Hyperlocation in ISS work might appear evident on several fronts: ready availability, compatibility with existing infrastructure, and low cost. Does Cisco's newest brainchild represent a threat to traditional active mobile location solutions such as those by Harris Corporation, Rayzone or FinFisher? No, because Cisco's product lacks their functionality in intercepting, monitoring, and taking control of target devices. Cisco's Hyperlocation product is complementary.

As the world – and network operators themselves – turn increasingly to reliance on Wi-Fi, Cisco has demonstrated impeccable timing. Its Hyperlocation solution leverages existing network access points, works plug-and-play off the shelf, and turns in a sterling performance at tracking targets right down to their most recent footstep, all at minimal investment.

CyberSeal IMSI Catchers and Detectors

At industry conference sessions on mobile interception, it's not often that the name CyberSeal surfaces. As a subsidiary of Israel's Magal S3, CyberSeal is but one component of Magal's broad portfolio of physical and cybersecurity solutions sold to governments and critical infrastructure enterprises. But tucked away inside its giant parent company, CyberSeal is a hidden gem that produces both IMSI catchers and a device to monitor for the very same types of StingRay device.

This complementary approach to products in a company's suite makes sense when you consider CyberSeal's primary focus as a critical addendum to the physical security of sensitive facilities: embassies, military facilities, weapons test grounds and correctional facilities. Magal S3 builds walls, provides the sensors, videocams, and a highly sophisticated tri-monitor video monitoring intelligence center for physical security.

CyberSeal, though far smaller, has an equally if not more significant role: safeguarding the airwaves to protect against mobile-based cyberattacks. Keeping intruders out, prisoners in and all under video surveillance is a fairly straightforward endeavor. Protecting secrets from theft by RF monitoring – not so simple, particularly in an era marked by rampant cyberattacks of all stripes: phishing, botnets, malware and ransomware. How do you know those you're spying on aren't spying right back? With CyberSeal on board, Magal S3 has created a comprehensive solutions suite for both needs.

Although cyber gets slight play from Magal S3 in public, its importance to the company's revenue stream is evinced by their market presence. Based in Israel, Magal is well-positioned to serve client partners in the West. But the company hardly overlooks the other side of the globe. Key branches are located in China, Russia and India. Is facility security important in those nations? Certainly. But with the volume of botnets and armies of hackers all three countries contend with, cybersecurity is vital, particularly in the mobile realm that dominates communications in these countries.

Yttrium – Using IMSI Catchers to Protect Facilities

CyberSeal's pure play mobile interception package is built around three products: the Yttrium IMSI catcher; Vanadium IMSI catcher detector, and Rubidium, a combined mobile/cyber monitoring center presenting a consolidated view of all threats.

Yttrium is a classic IMSI catcher device, capable of working in both active and passive modes. It can seek out specific black-listed IMSIs associated with known threat actors, acting as a "fake base station" that lures suspects via a stronger sign-in signal than adjacent commercial networks.

Alternatively, Yttrium's passive mode can monitor all traffic and single out suspicious entities.

The publicly-advertised purpose of Yttrium is to "stop unauthorized cellular use in restricted areas," hence the focus on sensitive facilities such as military bases, Intelligence Community operational or training sites, as well as critical infrastructure. The primary focus is Wi-Fi client networks, on which Yttrium can monitor all incoming and outgoing traffic and review captured IMSI details.

Suspicious calls are identified on a 24 X 7 basis and compared to a "white list" of devices authorized for use by individuals in the restricted site. Suspect numbers are "black-listed." Most importantly, Yttrium categorizes any unauthorized traffic as a potential threat and blocks it.

Wi-Fi networks offer vital details that Yttrium and other IMSI catchers can scoop up. A common feature of Wi-Fi transmission is inclusion of a device's MAC (media access control) address, a number that is unique to every device. The MAC address of a mobile device is completely unsecured. Moreover, the MAC signal is transmitted regardless of whether the device is in use, idle or turned off, making mobile devices easy targets for gathering this data.

Wi-Fi was long criticized for being susceptible to security breaches due to the vulnerability of the WEP protocol. Newer protocols such as 802.11ac have moved Wi-Fi to the 5G band and accelerated speeds five-fold. When combined with CCM (Counter Mode with CBC-MAC) encryption, the new protocols do improve security. The hang-up: They are not backward-compatible with older chipsets used in some mobile devices. Any such shortfall represents a potential vulnerability in a facility's Wi-Fi security. For CyberSeal, any residual weakness in Wi-Fi security is, of course, a selling point the company can take advantage of in winning new orders for Yttrium and sustaining existing contractual arrangements.

That said, IMSI catchers in general, and this includes Yttrium, face new challenges ahead. The rise of eSims will definitely make Yttrium's job harder. But for the moment, the market for diverse versus a single IMSI per mobile device remains in the early stages.

Vanadium – Turning the Table on IMSI Catchers

In addition to marketing the Yttrium IMSI catcher, CyberSeal crosses to the other side with its Vanadium product for cornering illicit IMSI catchers. Vanadium represents a significant differentiator for the Israeli firm.

The IMSI catcher detector market has by and large remained in a nascent stage since the end of the first decade of the millenium. The Android IMSI Catcher Detector (AIMSICD), an open source project for developing methods to detect and thwart fake base stations, has developed one app for Android devices only, and has seen some commercial outgrowth on its work.

Among the commercially-available solutions, one worth noting is Snoopsnitch, developed by German security researchers Alex Senier, Karsten Nohl, and noted cyber expert, Tobias Engel. Snoopsnitch gathers and analyzes radio data, looking for IMSI catcher interference, user tracking and SS7 attacks. When introduced, Snoopsnitch was applicable only to rooted Android phones using Qualcomm chips, but the team, under the wing of SR Labs, has since moved on to making the app work on other devices. Snoopsnitch is currently an app available on Google Play.

With Vanadium, CyberSeal has taken IMSI catcher detection to a new level: a device by a respected security vendor that is specifically designed for protecting Wi-Fi networks. Vanadium continuously monitors the client's Wi-Fi network and issues an alarm when a threat is recognized.

With the exception of these examples, the question arises how often an illicit IMSI catcher represents a threat to sensitive facilities. Although CyberSeal offers affordable pricing on its IMSI catchers versus the cost of a Harris StingRay, these highly-specialized "pineapple" devices are notoriously expensive, costing well into the six-figure U.S. dollar range.

Another point: the purchase of IMSI catchers is tightly controlled by vendors and – via strict laws and regulations in most countries – rigid licensing requirements. Granted, it is conceivable that dangerous elements might obtain one by theft or disguise, or even develop their own. But the odds of either event are slim, as is the likelihood of smuggling one unnoticed past Border Control facilities in nations such as China and Russia, let alone the U.S. Training in the use of an IMSI catcher would be another obvious obstacle. But, for those who need one, the ability to buy an IMSI catcher detector made by CyberSeal is a plus.

CyberSeal rounds out its Wi-Fi security portfolio with its SIEM (security information and event management) Rubidium monitoring center. Rubidium provides rapid-fire analysis of aggregate data on cyberthreats, and issues alarms as needed – basically a recap of what Yttrium and Vanadium already do.

For combined physical/Wi-Fi security, adding the Magus S3 Fortis 4G GIS 3-screen video surveillance system is a reasonable consideration. Fortis 4G is a state-of-the-art perimeter security solution using real-time video and perimeter sensors. Combining Fortis with CyberSeal Yttrium and Vanadium assures full protection against physical and over-the-air security threats.

Harris StingRay IMSI Catchers

While many companies such as Rayzone, cellXion, CyberSeal and others manufacture IMSI catchers, Harris holds the dominant market position in the U.S.

Harris has long benefited from an exclusive arrangement that lets law enforcement agencies purchase the company's $six figure IMSI catchers with grants from the Department of Homeland Security. Sales are kept secret by non-disclosure agreements (NDAs) between state and local law enforcement agencies and the FBI. When cases come to court over use of IMSI catchers, LEAs are forbidden to name the vendor, an unusual privilege that provides Harris with a degree of cover from public exposure.

DHS grant arrangements tied to NDAs have likely continued even in the wake of the U.S. Department of Justice September 2015 clampdown which, as stated in the chapter introduction, affects only solo federal law enforcement investigations. Ranked as the 11th largest government contractor for products that serve a range of needs including electronic surveillance, Harris continues to dominate the U.S. IMSI catcher market.

Mobile Location Basics

Ostensibly invented by Rohde & Schwarz and patented in 2003, the IMSI catcher was actually born in the laboratories of Harris Corporation. User manuals kept in closest secrecy until leaked in 2016 prove that the Harris Stingray was in use by U.S. police departments at least as early as 2001.

Today, the best-known IMSI catcher is the Harris StingRay, which intercepts mobile communications to and from a target's device. Other options include the Harris Kingfish, Gossamer, Triggerfish, Amberjack, Harpoon, and the Hailstorm.

StingRay is a portable device that mimics a mobile network base station sending out signals. As the target travels through a network cell – the area serviced by a base station – his device assumes that the signal originating from the StingRay is from a mobile service provider's radio tower. In the field, the StingRay can capture and track hundreds of device addresses, make a record of all calls made to and from the area, or focus on one target. Small and portable, the StingRay can be used anywhere, including in a vehicle. While the base unit is limited to signaling, Harris sells software add-ons that let the agent intercept call and message content.

Like StingRay, Kingfish tracks mobile devices, identifies users and connections to other devices, only in a more compact unit that can be concealed in a briefcase.

Gossamer provides similar functionality but in a hand-held model that resembles some of the original mobile handsets of the mid-1980s – clunky by comparison to a smartphone, but easier to tow around and use without notice. One important difference: the Gossamer lets the user initiate attacks on a target, essentially taking his mobile device out of action.

With Triggerfish, we're back to a box design like the StingRay's. Triggerfish also does mobile location – but in a much bigger way. The device can collect signaling data from over 60,000 devices simultaneously. Triggerfish also lets the agent intercept a target's conversations on his cell phone, just like a software-enhanced StingRay.

Amberjack is a directional antenna that looks like a home smoke alarm. Magnets and a tie-down kit secure the antenna to a car's roof, where it sits unobtrusively during an intercept. Used with StingRay, Kingfish and Gossamer, Amberjack's contribution is signal monitoring that further helps the user ascertain the target's location.

Harpoon, a companion device, is a simple amplifier that comes into play when the target moves out of range, or if the agent wants to maintain his distance. Harpoon amplifies the signal to and from the target's device in either event.

Hailstorm provides 4G/LTE and malware capabilities, and replaces earlier versions of StingRay. The suitcase-sized device, deployed in a specialized vehicle, was first used by U.S. troops in Iraq. Hailstorm provides mobile location data and can intercept content.

While the DoJ's embargo on federal use of IMSI catchers for capturing and retaining data certainly impacts government use of the Harris Hailstorm, it would appear to have little or no effect on the company's primary customer base of state and local law enforcement agencies – or on DHS funding of purchases.

Innova GPS Tracking Device

An inside joke of the ISS industry is that Italy has more surveillance companies per square mile than any other nation on earth – upwards of 200 vendors including recognized brands such as AREA, BEA, Endoacustica, ESIM Global, The Hacking Team (now Memento Labs), Innova, iPS, RCS and SioS. The punchline: They're just one company with 200 aliases. Of course, that's not true, although given the similarity in product lines between many, it is easy to understand how one might come to that conclusion.

Based in Trieste, one standout is Innova, which makes the usual lawful intercept hardware but excels in an aligned area: audio monitoring and GPS tracking spy gear for following suspects on the move.

Know Them by What They Pitch

Innova markets itself as a provider of "integrated interception systems for lawful activities and intelligence operations" that provide soup-to-nuts lawful intercept (LI) capabilities including: wireline and mobile monitoring; deciphering of encrypted communications; data analysis and storage; mobile location and tracking; and topnotch audio monitoring – in that order. That's true enough, but we're at pains to point out that at least 25 major vendors provide a similar assortment of LI capabilities (See Chapter 7: Lawful Intercept Multi-Play Vendors). Probes, deep packet inspection, metadata and content interception, storage and analytics are all part of the package in this very competitive global market.

Innova knows that, so while they give fair play to their EGO line of LI capabilities in order to look versatile and "integrated," the company's real skills lie in creating miniature audio monitoring devices that can be planted on targets' vehicles to track their whereabouts 24 X 7, and for the right price, eavesdrop on every conversation in that vehicle.

If you meet or see Innova at trade shows, or review their website and promotional material or talk to a sales representative, the products they're primarily focused on pitching are micro systems for undetectable mobile location and/or monitoring.

MicroIP

Innova's MicroIP is the company's front line market differentiator, a miniature audio surveillance device developed in their own laboratory facility in Trieste. MicroIP offers distinct advantages over competing products. It can be quickly installed in a target's vehicle or living quarters and immediately begin delivering high quality, stereo mode monitoring delivered to the agent via GPRS, UMTS and HSDPA network connectivity.

Unlike most "bugs" in this field, MicroIP interception experiences no signal degradation, leading to perfect recordings of target conversations that constitute solid evidence.

Absolute concealment is another major plus. A target using the typical scanner will not be able to find a concealed MicroIP because the device uses UMTS and HSDPA with spread spectrum technology, unlike traditional GSM-based devices that were easy targets for a scanner. As an additional safeguard, the user can set timers on MicroIP transmissions to deter detection.

A major shortcoming of conventional monitoring devices is signal attenuation that impacts quality of a monitoring session and its recording. When a signal is lost, so is priceless evidence or intelligence. Innova's solution to this problem is to outfit the MicroIP with a sizable internal memory that records conversations within the device itself as a backup to radio transmission. When network connectivity is re-established, MicroIP sends the buffered data on to the user's monitoring center. Innova ensures data security by encrypting the stored evidence with its own in-house developed algorithms. Another outstanding feature is device memory, which can store up to 264 hours of recorded evidence when there is no network capability.

Each MicroIP is equipped with SiRF4 GPS, providing accurate mobile location to within five meters.

GPS Innova

If the agent's first priority is mobile location, a complementary product by Innova fits the bill: GPS Innova, a miniature satellite tracking device that provides real-time location coupled with software capabilities that offer visualization and analytics.

Tiny and magnetized, the GPS device can be placed on a target vehicle quickly and discretely. Moreover, the system can accommodate up to five GPS devices so that the user tracks not only the primary target but also his or her affiliates in relation to the user's own position.

Power consumption is low to help ensure long life, and antenna accuracy is guaranteed even with weak signals. But just to be sure, the GPS Innova borrows an important feature of the MicroIP: huge built-in memory and buffering that stores the location data and also transmits it on demand to the agent.

Also like MicroIP, the GPS Innova devices use SiRF4 GPS for precisely tracking mobile location. GPS Innova can feed mobile location to an Innova

Ego lawful intercept system or other monitoring center, and also offers a mobile monitoring unit.

For all the fanfare over IMSI catchers, there is still room in the market for lower cost devices that perform the same key services – mobile location and interception – without breaking the bank as many "fake base stations" do. Beepers are still popular for those very reasons.

For the majority of surveillance scenarios, the MicroIP is completely serviceable and provides both mobile location and monitoring, the latter at vastly superior quality compared to competing devices. The GPS Innova provides added capabilities for agencies that need to track multiple targets in real-time.

Micro Systemation Mobile Forensics

In the hotly-contested market for mobile forensics – the science of recovering digital evidence from mobile devices to support an investigation – there are many contenders, but just one king: Sweden's Micro Systemation AB.

Founded in 1984 in Stockholm, Sweden, Micro Systemation or MSAB spent its first 19 years refining expertise in advanced data communications and mobility. The company's growth spike began in 2003 with its commitment to mobile forensics. In the ensuing years, MSAB's performance skyrocketed, fueling expansion into 90 countries, with direct sales and support offices in Sweden, the U.S., UK, France, Canada and China – and the aid of 28 local partners on five continents.

Micro Systemation's banner product is XRY, evidence extraction software that runs on the Windows platform to extract data from any type of mobile device. MSAB-made hardware running XRY comes in all shapes and sizes to help investigators glean the evidence they need, whether at HQ, on the road or in the line of fire on a battlefield.

The ubiquity of Windows and versatile array of supporting equipment are a definite boost. But the company's core differentiator, like forensics itself, lies beneath the surface: deep understanding of the prime points of attack.

Of Chipsets and App Versatility

While the count on mobile device types reaches well into the hundreds, makers of the chipsets that run most devices are a finite group. The more popular the chipset, the wider the range of devices where it will be found. As a result, any number of different manufacturers may use the same chipset components. Focusing on this commonality, as Micro Systemation does, can provide an edge.

Chipsets perform multiple functions within a device's circuit board. For the purpose of forensics, there are three key modes to understand: read only memory (ROM); the flash drive containing "bootloaders" (code that delivers instructions on launching the device's operating system); and the mobile OS itself. Bootloaders typically are designed with security features that prevent data loss and intrusion. However, it often happens that segments of a boot chain's security are not activated at the factory.

In boot-up sequence, these functions act in sequence. ROM switches on the flash drive, which activates bootloaders that launch the operating system. Each of these points in the chain can present opportunities for data extraction by forensics.

Micro Systemation XRY attacks at all three points:
- Chipsets where secure modes are not active.
- Bootloaders with not-so-secure debug features that can be cracked by inserting code.
- Operating systems with generic profiles that, following successful penetration of one chipset OS, can be conquered in others, as well.

Of the three, operating systems present the greatest challenge to forensics because a mobile device OS typically contains the most significant security measures. Still, while Micro Systemation concedes that there is no guarantee of conducting successful forensics in all instances, the company has established a strong record of performance at the three levels.

Another strength of XRY: manhandling apps. Today, smartphone owners have the ability to use multiple downloadable apps to perform voice and data communications functions over-the-top versus on the device's resident capabilities. Skype, SMS, and Twitter are the most common, but beyond these mainstays the user has access to a vast array of lesser-known OTT apps that do exactly the same things.

Micro Systemation claims "the best smartphone app support in the mobile forensics market." XRY supports forensics of hundreds of apps, and as the app market evolves, the company keeps on top of it.

Complementary solutions help investigators and intelligence agents find the evidence they need and manage it to produce results. The company's XAMN analytics family assists in finding key evidence from massive data volumes, searching and filtering for data critical to the investigation. The XEC suite enhances managerial control of all data extraction and analytical findings and establishes policies that ensure maximum efficiency of operations.

Legal Controversy

In the mobile forensics sector, intense competition at times moves companies from the marketplace to the courtroom.

In August 2013, Israel's Cellebrite filed suit in U.S. federal court in Alexandria, Virginia, USA against Micro Systemation, then amended its filing in December of that year, broadening the lawsuit. Cellebrite, which entered the mobile forensics business in 2007, initially claimed that MSAB used the Israeli firm's technology for extracting data from Android and Blackberry devices. In subsequent charges, Cellebrite said that MSAB had gone so far as to steal software directly off Cellebrite servers.

MSAB denied all charges, stating that Cellebrite had filed the suit "in a bad faith attempt to injure MSAB and gain a competitive advantage."

MSAB promptly counter-sued. Eight weeks later the case was settled, and the two companies dropped their claims. The court dismissed the case "with prejudice," meaning it reached its decision based on the merits of the case, and the plaintiffs cannot file another lawsuit on the same grounds.

With these legal challenges settled, Micro Systemation is on a roll.

The company today offers multiple hardware formats for any requirement: desktop, laptop, tablet and handheld, plus a ruggedized model for military, add-on platforms that provide advanced analytics, mapping and detailed visualization, and new solutions for extracting logical and physical evidence from drones and vehicles.

Market response remains strong. MSAB continues to expand, with notable gains in Europe and the U.S. The company has announced major wins with the French Ministry of the Interior, the US Federal Bureau of Investigation and other agencies of the U.S. government. In the UK, MSAB's mobile forensics solutions are used by 97 percent of police departments.

Polaris Wireless: Vertical Location

An alternative to the classic IMSI catcher is offered by Polaris Wireless, which provides undetectable mobile location tracking via its OmniLocate and Altus products. These solutions don't do everything an IMSI catcher can, but they do offer one vital differentiator that Harris, Rayzone and other standard bearers in the industry can't touch: the ability to go vertical and provide accurate 3D location of the target by altitude – an obvious mission-critical capability whenever the target moves above ground level.

Polaris Wireless is based in Mountain View, California USA. Launched in 1999, the company has distinguished itself with "wireless location signature" (WLS) technology that would ultimately find a home in the surveillance community.

The basis of WLS is the premise that every coordinate within a mobile network cell has an RF signature unique to that location. OmniLocate leverages signal strength and interference metrics that are used by mobile protocols to determine "handoffs" to different base stations. The system then uses sophisticated algorithms that pattern-match the target's signaling with the most likely RF signature for that position in the network. The system updates the database of potential signatures as frequency modifications and other changes are made in the network.

As with so many success stories, timing has played a role in this company's success. While Polaris was finishing the first cut of its WLS technology in 2001, the U.S. passed laws requiring mobile operators to provide location capabilities for E911 services. Polaris was in the right place at the right time with the right technology. The mobile operator community showed strong interest, as did venture capital fund Draper Fisher Jurvetson (DFJ), which pumped US $4.0 million into the young startup.

After testing in the U.S. and Europe, Polaris Wireless won its first operator contract for E911 mobile location with T-Mobile, in 2003. Other U.S. mobile operators quickly followed suit, and by 2005 Polaris Wireless turned profitable. The company then began to explore kindred uses for WLS. Mobile location for surveillance was an obvious fit. In 2008 Polaris won its first customer for surveillance-oriented mobile location, with an AsiaPac client.

The company never looked back. By 2012 Polaris Wireless had 120 employees, US $20 million in revenue from a growing base of LEA and government agency clients, and regional headquarters in Dubai for EMEA and Bangalore for AsiaPac.

When DFJ decided to cash out in 2013, the VC received a healthy return – a payout of US $40 million in cash, i.e., a multiple of 10 times their initial investment. Polaris financed the deal with a fresh US $10 million infusion from Industry Ventures, plus assumed debt of US $15 million.

What Polaris reaped: 60 percent control of the company and the autonomy to negotiate a bigger sellout down the road when the time comes. But not for awhile yet. Polaris has big ambitions on the international front, showcasing OmniLocate and Altus at ISS World events in offshore venues where the company serves as a principal sponsor.

What Makes Polaris Wireless OmniLocate Special

OmniLocate is a multi-technology mobile location system in a "big iron" framework – the Radisys ATCA flexible blade server. While WLS is the headline technology, the system also provides mobile location technologies including Assisted GPS (A-GPS) and others. Offering different flavors of mobile location enables a hybrid approach that can improve accuracy, or give the user flexibility to leverage the most effective technology for given surveillance conditions.

For example, A-GPS provides a high-level of accuracy by using coordinates from multiple satellites, as well as data from a mobile network. But A-GPS delivers the best results when used in a line-of-sight scenario. It is not ideal in dense urban markets with numerous physical obstacles. In the latter conditions, WLS outperforms. Alternately there are situations where WLS benefits from A-GPS. A hybrid approach using both can deliver superior results.

OmniLocate is equally versatile in the range of networks it supports. It works with any dominant mobile interface (3G, 4G LTE) across all device technologies (A-GPS, Wi-Fi, Bluetooth, Ultra Wideband), and in microcell environments that will become familiar with the broad rollout of 5G.

OmniLocate in Operation

Like an IMSI catcher, OmniLocate can handle individual or multiple target mobile location needs across a network – or networks. The system works quickly, locating the target in under five seconds. Tracking is completely unnoticeable – it does not change or interfere with the target's use of the phone or have any impact on a mobile base station.

OmniLocate can also stretch to handle other valuable functions. The system ramps up to perform mass location, scaling to pinpoint all mobile devices operating within a given mobile network environment. The system can geofence an area, providing alerts when a target enters or exits pre-set boundaries, for example near sensitive government facilities, fortifications or boundaries. It can also tell when the target is "upwardly mobile," that is, active in a high rise building. OmniLocate's vertical location feature can find the target to within eight to 12 meters, with 90 percent accuracy.

Data from OmniLocate mobile location can be fed into an accompanying analytics module to reveal trends, patterns, networks of affiliates, target movements over time, most common location, frequency of meetings with affiliates and where – and the likelihood of repeat behaviors leading up to an incident.

Analytics deliver another benefit. Targets using disposable mobile phones can't escape OmniLocate – at least, not for long. The system tracks devices by phone number, equipment ID and SIM card. Regardless of whether the target discards one, two or all three, his behavior creates a pattern that is trackable by analytics and mapped back to him.

What if the agent feels a need for call content such as voice, SMS, data, web surfing – or surveillance video relevant to the target in the same time/space references as the mobile location data? That is a definite "can do." OmniLocate provides inputs that integrate with lawful intercept and video surveillance feeds so that the agent gains a holistic picture of the targets: who and where they are, what they're planning, and what they look like.

All of the aforementioned data can be transferred to and viewed in an LEA monitoring center – or used on the go. In addition to the OmniLocate platform, Polaris Wireless offers a smartphone app called Altus that lets field agents see and take action on everything visible in the monitoring center.

The Network Connection

OmniLocate and Altus may be used standalone or integrated with other surveillance systems. Just one catch. Whether the user is a service provider purchasing Polaris Wireless products to satisfy legal compliance requirements, or a law enforcement or government agency working firsthand, direct access to mobile networks is a must. OmniLocate calculates the target's location RF signature based on information provided by the network. The RF signature database is updated the same way, by direct connection to the network.

Depending on the nation state involved, cooperation by mobile operators may vary. Law enforcement agencies that use IMSI catchers can skip that step, operating with complete independence from the network.

Does that one caveat have much of an impact on Polaris Wireless's global ambitions? Likely not, considering the company's wins in Sub-Saharan Africa, Asia and the Middle East.

But at home, there may be a different kind of obstacle. The U.S. Department of Homeland Security (DHS) has proven to be a reliable funding resource for local police departments in need of mobile location technology. DHS has an exclusive agreement with Harris Corporation.

Pro-Solve IMSI Catchers

With global market giants such as Harris Corporation and Rayzone to contend with, and tough competition on the home front from specialty vendors including cellXion, how does a comparatively small player like Pro-Solve International not only survive but make a strong "go of it" selling IMSI catchers?

In a crowded field, Pro-Solve distinguishes itself with a trio of assets: decades of hands-on experience in RF test and monitoring equipment, smart partnering with other IT companies that bring their own strengths to the mix, and a wise decision to focus on design – while offshoring equipment manufacture.

Though small in personnel, Pro-Solve has used this model to create the kind of size that matters most: footprint in the marketplace. For Pro-Solve, now in its second decade of operation, it's a winning formula that has led to a loyal client base among British and selected offshore LEAs and government agencies.

Inside Pro-Solve

Based In the west side London suburb of Marlow, Pro-Solve incorporated in November 2006 and made its public debut at the Singapore Global Security Asia exhibition the following March. There the company's principals unveiled an impressive line of products that included IMSI catchers, RF jammers, sensing devices, as well as "first response" solutions using iDEN (Integrated Digital Enhanced Network) and TETRA (terrestrial trunked radio) technologies. Pro-Solve even threw secure SATCOM solutions for remote response teams (Special Ops) into the mix.

Quite a showing for a startup. Obviously, the company didn't accomplish all the aforementioned in four months time. Pro-Solve explains what appears to have been a sudden leap into ISS hyperspace as really being the result of two years' intensive work developing both IMSI catchers and iDEN/TETRA solutions, followed by SATCOM work.

Extensive experience in radio monitoring gave the executive and engineering teams a leg up on accelerating from R & D to commercially available products for their targeted clientele. Key team members hailed from RACAL, formerly known as Aeroflex Burnham Limited, a respected specialist in the manufacture of electronic instruments and appliances for RF testing and measurement, founded in 1951.

During their tenure with RACAL and similar enterprises, the men who would birth Pro-Solve International gained the experience and contacts to create a winning venture. Paul Canning knew design and manufacturing of electronic devices inside-out, had first-hand experience installing them and had built a successful management career opening and running offices in France, Germany, Italy, India and the AsiaPac region. Peter Connell and Mike Kenyon each brought 30 years experience in developing commercial and military wireless T & M hardware and software.

The Pro-Solve IMSI Catcher Suite

Pro-Solve's first bids in the IMSI catcher arena were devices designed respectively for GSM, UMTS and CDMA networks. Roughly on par with competing products by Harris Corporation, Pro-Solve's hardware performed the familiar functions of active mobile interception devices: base station emulation with a stronger signal to lure targets of interest; man-in-the-middle (MITM) attacks to capture identification of the target; real-time interception of voice and data traffic; use of triangulation to accurately determine position; and the ability to either jam or take control of target mobile devices.

Like Harris, Pro-Solve introduced supplemental equipment such as booster antennas – the ProSolve "Zoneshield" – to augment coverage areas. Pro-Solve also answered the call for greater portability, providing a battery-powered version that was lightweight and easy to conceal, for use by agents on the move in the field.

Other key advances began in 2013. Pro-Serve set out to produce and offer sophisticated analytics and data retention modules – SearchNet and SearchBase – providing a new dimension to its mobile location and interception suite: broad-based intelligence gathering coupled with forensics.

All the while, LTE was engulfing the world of mobility. Old school IMSI catchers could no longer rely on the old trick of circuit-switch fallback from 3G to 2G. 4G LTE introduced a new level of complexity.

It is hard to say which IMSI catcher vendor led the charge into LTE. But Pro-Solve certainly kept pace, working steadily throughout 2014 until its LTE IMSI catcher solution was ready for market as the 2015 New Year opened.

Add-on Services

Almost from the start, Pro-Solve has taken a few detours so that the company isn't a "one trick pony." The company's ventures into TETRA and SATCOM are typical of such initiatives and broaden the company's appeal beyond law enforcement and government intelligence to reach military clients.

When a new service involves expertise in an unfamiliar area, they find a good partner. One of Pro-Solve's first such ventures was the addition of a counter-intelligence solution that detected GSM bugs, made possible by a partnership with a leader in counter-intel, OCC Interscan, in 2010. More recently, Pro-Solve has engaged with an AsiaPac company to produce a passive mobile interception solution that can monitor entire networks.

They're also good at finding new applications for their core product. Pro-Solve IMSI catchers are often found in prison settings where they are used to locate illicit mobile devices.

With offices in Hong Kong and India, where production takes place, Pro-Solve has the geographic reach to explore new markets, both for clients and to find the best partners. Pro-Solve may never reach the scale of a company such as Harris, but by their very presence they provide an invaluable service beyond their innovative suite: price competition.

Rafael PowerSpy: Amping-Up Location

In the time since Rafael PowerSpy rocked the ISS world with news of a revolutionary approach to mobile location, this remarkable advance has disappeared from view. The initial report emerged in the most innocuous way, via a 2015 academic paper published by the Stanford University Computer Science Department in coordination with financial backer Rafael Advanced Defense Systems Ltd. of Israel.

Clearly it was not the intention to create waves or even ripples in the surface of the mobile location marketplace. But when first revealed to the public, PowerSpy generated a tsunami of criticism in the privacy media, where it was characterized as a venomous new trick for invading personal privacy. Just as quickly and eerily, PowerSpy vanished from public view, like a phantom. Since March 2015 not a single fresh insight on this venture has surfaced.

In all, the tale of PowerSpy is one of the strangest stories in surveillance history. What exactly is this technology, who's behind it and most importantly, how does it perform compared to traditional tools of the trade?

PowerSpy works by measuring variations in power consumption by smartphone apps as a device (and its owner) move through the network – a distictive approach to determining mobile location. In tests conducted, the technology is 90 percent accurate in locating target devices.

Is PowerSpy, as characterized by WIRED and other media, the next generation of technology designed to "spy on your cell phone?" That may come to pass when a commercial version arrives, but it's not true yet. Notwithstanding all the initial hype, PowerSpy at this time is still not a commercial product. But certain aspects of PowerSpy are so high-value that it is wise to assume that this mobile location technology solution is already valued for select use cases by military and intelligence customers.

The basics:
- **No Malware Required.** The smartphone power sensor, designed to keep track of remaining battery juice in every smartphone, is unprotected. PowerSpy can jump right in to access data on power consumption.
- **Machine Learning, AI, Stanford and SRI International.** PowerSpy uses algorithmic modeling that facilitates computer learning without human intervention, i.e., it leverages artificial intelligence. Stanford has an engineering school dedicated to machine learning and AI. Stanford was also the original intellectual home turf for the entity that would evolve to become SRI, which today specializes in commercial AI for defense and intelligence.

- **Rafael's Involvement.** Rafael was founded as Israel's National R&D Defense Laboratory for weapons and military technology in 1949, then spun off as an incorporated company in 2002. The US $3.0 billion enterprise is the second-ranked defense contractor in Israel, with a gamut of military platforms and solutions produced for the homeland and other nations around the world: cyber intelligence, air defense, air-to-air systems, land and naval systems, air-to-ground systems, aerial surveillance, rocket motors, warheads and space propulsion systems. Rafael partners with large U.S. government service contractors such as Raytheon - most recently (in August 2019) to co-produce the Iron Dome interim cruise missile defense capability for the U.S. Army. The Israeli company spends 10 percent of its annual sales on R & D in areas including electronics, microelectronics and software. When Rafael backs research, it means that there is a serious military or ISS application in the works.

PowerSpy Mobile Location Leverages Smartphone Apps

PowerSpy differs from conventional mobile location technologies by leveraging apps and a sensor that come built-in from the factory. The resident apps that come embedded in smartphones are sufficient for PowerSpy to begin tracking a target's whereabouts via the power used by apps – and measured by the smartphone's own power monitor, the sole sensor in iPhones and similar devices that is completely unprotected from intrusion. From the data gathered on a smartphone's power consumption, agents can draw a highly accurate picture of the target's location.

PowerSpy offers distinct advantages over conventional approaches to mobile location. For starters, it is not subject to the usual hoops associated with lawful intercept: a court order and notification to the mobile service provider.

Because PowerSpy involves no special equipment, it will likely – when and if made commercially available – be significantly cheaper than intrusive methods such as IMSI catchers, "off-air" mobile interception systems that circumvent cooperation with the service provider.

PowerSpy is more accurate than Signaling System 7 (SS7), an early signaling technology developed for wireline networks. SS7 includes a Home Location Register (HLR) that provides location of the network node nearest the target in real-time. But SS7 is generally considered a mediocre form of mobile location because a targeted device might be hundreds of yards from the nearest network node.

Unlike other mobile location technologies, PowerSpy never touches a device's GPS or Wi-Fi systems, and thus requires no special access permission from the manufacturer, operator or device owner. Nor does it use the IMSI catcher approach, emitting a signal that creates a fake base station to capture the device's IMSI number, personal ID, then use an MITM attack to take over the device. All PowerSpy does is latch on to a smartphone's ampere meter to measure power consumption, and from that data track the target's proximity to base stations.

As put by the Stanford/Rafael team:

Suppose that an attacker measures in advance the power profile consumed by a phone as it moves along a set of known routes in a predetermined area such as a city. We show that this enables the attacker to infer the target phone's location over those routes or areas by simply analyzing the target phone's power consumption over a period of time. This can be done with no knowledge of the base stations to which the phone is attached.

The PowerSpy Lab Test

Because a mobile device's power consumption directly correlates to its distance from a tower, the smaller the distance the less power used, and the greater the distance the more power used. As the signal drops, the gain must increase. The same holds true when the device is sending or receiving data, which requires continuous transmission with the base station.

To demonstrate this point, researchers measured signal strengths of a device on a predetermined drive through multiple network cells. The results showed a consistent pattern of high and low strength signals through the course, determined by the device's proximity to a tower.

Next they tried a second series of trials on the same roadways, but this time measuring power consumption. The measurements were not as consistent as those drawn from signal monitoring. The slightest change in the way a mobile device's modem reacted to variations in signal strengths made a difference in power usage. For example, when a signal from a base station became too weak, the mobile device worked overtime to connect, and as a result, power usage by apps increased significantly.

Similarly, stability cratered when the target approached a base station from a different direction. Why? In conventional mobile networks, phones switch to different cells when the signal strength in one base station is surpassed by another. Signal strength can vary depending on geography and obstacles.

Initially, these findings demonstrated that an "attacker" – that is, an agent using PowerSpy – would be forced to use a consistent direction or route of travel by the target's mobile device as a reference for power measurement.

If a target under surveillance via power-consumption-based mobile location took a side trip, he went "off the grid" from PowerSpy's perspective.

But all was not lost for the research team. One positive finding: Different makes and models of smartphones showed the same reaction to signal strength variations while on the same path. The power samples aligned, allowing researchers "to obtain a reference power measurement without using the same phone as the victim's."

There were still issues. Targeted drivers who set out from Point A to Point B might not necessarily move in a straight line, or at the identical speed over the same period of time. How could one create and "score" profiles with so many variables in play? Further complicating matters, some variables might be "latent," i.e., unknown.

Enter Artificial Intelligence (AI) – and Success

To handle these variables, the team used an offshoot of machine learning: Dynamic Time Warping (DTW), an algorithm for measuring the similarities between sequences that vary in time or speed in order to create an "optimal match." DTW is used in speech recognition and voice biometrics to identify a target even when speaking at different speeds or in varying contexts. For PowerSpy, DTW was applied to profile power consumption at different drive times and distances along the same path.

To refine the results, the Stanford/Rafael team then measured power consumption along pieces of the travel route, using a variant, the "Subsequence DTW algorithm."

What about the target who might decide to take a ride in country, drive in circles, or otherwise go out of his or her way to be tracked? The solution was to test and record the power profiles of every conceivable route in an area predetermined to be possible for the target.

Here, the algorithm that came into play is one well known to the voice biometrics world: the Hidden Markov Model (HMM), a dynamic Bayesian system based on the assumption that some states or variables are latent, i.e., unseen and unknown. For PowerSpy, an example of such an HMM might be a route traversed by intersections which, if taken by the target, would introduce multiple potential consequences in direction, distance and power consumption.

As an added assist, Stanford & Rafael introduced the Monte Carlo approximation or engine. The Monte Carlo engine is a software tool that weighs all variables to produce multiple possible futures or outcomes. For PowerSpy, the filter showed researchers all probable states of an unknown factor based on samples that approximated the probability of a target's movement at each point.

Accuracy was improved by removing device "noise" and normalizing the target's location profile against points along a route where power consumption was greatest.

With this model in place – DTW + Subsequence DTW + HMM + Monte Carlo filter + noise removal – the Stanford/Rafael team aimed for 90 percent accuracy for its power-consumption-based mobile location trial.

Using a data set of 43 power consumption profiles and four separate routes of roughly 19 kilometers resulted in a success rate of 93 percent in pinpointing the target. Adding three additional routes and another eight power consumption profiles, the target's mobile location was still tracked with 90.2 percent accuracy.

One problem did occur when tests were conducted in more densely populated areas with higher numbers of mobile base stations. In such instances accuracy fell to 78 percent. Researchers cited higher density of cell coverage and "monotonous" power profiles as the culprits. Given that urban areas were defined as the principal field of operation in test parameters, any such deficiency in PowerSpy clearly needed to be addressed.

Comparing PowerSpy to Conventional Mobile Location

The Stanford/Rafael team maintained that PowerSpy is on par with "fingerprinting" techniques that use pre-recorded radio maps of an area to infer locations via "best matching" techniques. The researchers referenced signal monitoring and RF measurement of path loss as alternatives to their approach, but did not pursue a line item comparison of PowerSpy performance to that of these more conventional mobile location methodologies.

As a result, some reviews of PowerSpy were harsh. Paul Ducklin, at the time a senior advisor for network security company Sophos, panned the experiment:

To summarize their main result. . .they correctly guessed which of the four known routes were driven, from power usage alone, 93% of the time. To be honest, that's not a spectacular outcome, especially when they admit that driving from A to B is considered an entirely distinct route from B to A over the same path.

That characterization, and the likelihood that Stanford/Rafael never anticipated the rush of media criticism that erupted from an academic paper, may explain why PowerSpy went quiet and has yet to resurface. However, in all fairness, it must be said that nearly all major technology innovation stumbles out the gate the first time.

Stanford's long history of advancing machine learning and AI should never be discounted or dismissed. When it reappears one day, refined and perfected, PowerSpy could well take the lead in the race for mobile location leadership – depending on where it used.

The arrival of 5G may prove a significant jolt for solutions such as Rafael PowerSpy. If the technology is confounded in dense urban areas due to the multiplicity of base stations and RF signals generated, it is safe to say that 5G, which relies on even greater number of bases stations and far more powerful signals, will be the ultimate challenge for tracking mobile location strictly by power usage. But in field applications, where such conflicts are sparse, PowerSpy may have already proved to be a desirable solution.

Rayzone Piranha LTE IMSI Catcher

Based in Tel Aviv, Israel, Rayzone Group is actually two companies: Rayzone, founded in 2011, focused on lawful intercept and intelligence products; and TA9, launched in 2014 as a data analytics specialist. Since the TA9 analytics platform is sold by its eponymous parent and Rayzone we'll refer to both simply as Rayzone.

Although a relative newcomer to the industry, Rayzone has a versatile and sophisticated product suite spanning: active and passive interception; IMSI catchers; interception for 3G and 4G LTE, Wi-Fi, satellite and fiber/coax; ethical malware for mobile, desktop and apps; mobile location; deep web harvesting and social media monitoring for OSINT; and the aforementioned TA9 data analytics platform. Credit founder and CEO, Aviram Solomon, who brings 15 years experience at Verint and TTI Telecom, and the contribution of Ron Zilka, a veteran in the cyber and intelligence fields, for Rayzone's quick ramp-up.

Note that Rayzone is careful not to package its diverse portfolio as an "a la carte" assortment of unrelated capabilities. Just the opposite. It's the wisdom of Solomon to underscore the importance of each module in creating a full-fledged matrix of intelligence for the end user. The company makes a fair point in saying that the matrix – combining LTE for mobile location, IMSI catcher capabilities for in-the-moment tracking of the target, mobile Trojans for taking over a device, social media monitoring and OSINT for trending on threats, and analytics/visualization for deeper understanding of target context and networked peers – provides the clearest possible picture of a dangerous scenario unfolding.

At issue is whether Rayzone's new LTE functionality in an IMSI catcher adds all that much. For the answer, we'll first examine what makes LTE such a tough target for classic IMSI catchers.

Complexity of IMSI Catching in LTE

Intercepting mobile target content from LTE via IMSI catcher is difficult on two fronts: authentication and RF signal decoding issues far beyond what is encountered in "IMSI catching" in GSM networks.

In the traditional Harris StingRay assault on a GSM network, the device first emulated a bona fide cell tower to trick the target device into connecting. The StingRay then initiated a man-in-the-middle (MITM) attack to perform GSM Active Key Extraction and gain access to a device's encryption key.

After cracking into the weak A5/1 or weaker A5/2 encryption used in GSM, the StingRay then set the encryption at A5/0 – no encryption. From that point, the StingRay was the MITM. Both the UE (target device) and eNB (base station) continued to communicate, but with the StingRay, not each other.

Newer mobile technologies present more difficult authentication challenges. 3G and LTE require that the device authenticate with the network. Essentially, the tower computes a new MAC (mobile authentication code) that a mobile device must use to confirm that it is connecting with an actual operator base station tower. Mutual authentication changes the cypher keys and is continuous. For example, mutual authentication "kicks in" after an instance of forced 2G circuit-switched fallback.

The other challenge to LTE "IMSI catching" resides in the nature of LTE RF signals, which by definition are dynamic and complex.

LTE RF signals use orthogonal frequency-division multi-access (OFDMA), allowing multiple transmissions on the same band by many customers. They may also use differing versions of the 3GPP's LTE standard, such as LTE-FDD (frequency division duplex), which handles uploads and downloads through paired frequencies, or LTE-TDD (time division duplex), which uses a single frequency and dynamically assigns priority of uploads/downloads based on demand for either. With a system in constant flux, the information contained in signals can change quickly depending on payload.

In addition, information key to an intercept may be partially embedded in multiple downlink and uplink channels. For downlinks: the broadcast, control format indicator, control channel, shared channel and HARQ (hybrid automatic repeat request – for error detection) channels. And for uplinks: the random access, uplink control, uplink shared and sounding reference channels.

LTE RF signal monitoring is a lot of ground for an IMSI catcher to cover. The aforementioned factors explain why LTE IMSI catchers were so long at arriving in the marketplace.

Piranha on the Attack

Vendors' love affair with brand names that evoke creatures of the deep continues with the Rayzone "Piranha." Piranha is noteworthy for its ability to support all IMSI catcher functions in 3G and 4G (LTE) in a single piece of hardware. Lest there be any doubt about this, or assertions that it's simply a CSFB trick, the Rayzone Piranha supports GSM, EDGE, WDCMA, HSPA and LTE.

The Piranha is "dual mode," i.e., it can operate in either HSPA & GSM or in LTE. In LTE mode, Piranha can locate a targeted device from among multiple signals on the same channel and decode the desired signal.

Piranha handles all the usual functions associated with IMSI catchers: recording of all IMSI, IMEI and TMSI numbers within reach of the device; passive detection when a targeted phone enters the region; jamming to hold or release connected phones; and in selected instances the unique MSISDN associated with a SIM card and user account.

One eye-catching feature: the ability to drain batteries on "blacklisted" phones. Another: the power to gain "remote control" and to see results not only from smartphones but also tablets and laptops. This remote control feature points to a key capability hinted at for the Harris Hailstorm, but out in the open at Rayzone – Trojans that take over the device in the same way that offensive cyber firms do.

Rayzone offers a complete set of offensive cyber tools for mobile devices, laptops, tablets and apps. Rayzone's Trojans provide access to any device's emails, cell and VoIP calls, passwords, contact list, photos, files, calendar, web browsing history, SMS, systems information, hosting services used, and GPS data for location. By itself, Piranha lets the agent identify, track and intercept communications of the target from 2G through to LTE. Add Rayzone Trojans, and the user can take full control of a device to provide a complete picture of the target's activity, interests and colleagues.

Arrowcell – For Tuning Out Unwanted Listeners

The downside of IMSI catchers' success is that everybody wants or has one, including entities that wish to turn the tables and track law enforcement, government or diplomatic agencies. Rayzone, like its competitors cellXion and CyberSeal, offers a solution to this problem: the ArrowCell family of hardware that locates and disables IMSI catchers.

ArrowCell devices are designed to pick up on unusual network activity, e.g., signal strength above the norm for a region, or instances of jamming. Special algorithms can discern real from fake radio towers, leverage vulnerabilities to take them out of action, and even determine their precise location down to the exact room in a building.

Rayzone offers three versions. ArrowCell D is a portable hand-held device for "on the go" assessment of IMSI catcher presence. If the unit picks up a signal from a station not recognized as belonging to a network, it alerts the user, who can then opt on further action. A little more versatile, but more cumbersome in a bulky briefcase, is the ArrowCell DP, which singles out a fake base station, alerts the user, grades the level of threat, and then can jam the alien IMSI catcher without compromising the network. ArrowCell DPL does all the above plus precise geolocation of the IMSI catcher.

Rayzone InterApp for Wi-Fi Interception

When the Rayzone InterApp Wi-Fi "cloud and app" interception product debuted at a Paris Milipol event, it was hailed as a game-changing innovation in tactical mobile interception. InterApp can pull any data off a smartphone linked to a public Wi-Fi network, including a target's email, texts, images, files, IMEI and MAC address – and track & identify hundreds of smart devices that walk, drive or sit nearby. At issue: Haven't companies like Wintego already "been there/done that?"

Wi-Fi is a natural evolutionary development for Rayzone, a company recognized for its contributions to the field of mobile location and interception, most noticeably via the Rayzone Piranha LTE IMSI Catcher. And indeed, certain capabilities of the InterApp are reminiscent of the venerable Harris StingRay. That said, it must be recognized that Wintego and other equally aggressive competitors have been sharpening their teeth on the Wi-Fi market well ahead of Rayzone. For that matter, even run-of-the-mill hackers consider targets on Wi-Fi easy prey.

Public Wi-Fi Weakness & Defenses for Beginners

Wi-Fi, short for "Wireless Fidelity," is a subset of Wireless LAN (WLAN) products that provide high frequency mobile connectivity over a short distance. Both Wi-Fi and WLAN adhere to the IEEE 802.11 standards and thus meet the engineering group's strict measures for ensuring interoperability.

Wi-Fi for years was considered the boogeyman of law enforcement and government agencies – and in some quarters, still is. One hurdle is how Wi-Fi changes both the public and local IP address of the user when he or she logs on from a Wi-Fi network versus from home, office or a regular cellular network. For clients, other issues arise from the IEEE's and certain smartphone manufacturers' dogged work at improving encryption – a necessity since the history of Wi-Fi security is a saga of weaknesses exploited, revealed and patched until the next vulnerability or workaround surfaces and the process is repeated.

First came the 2001 exposure of weaknesses in Wi-Fi's original security algorithm, 802.11 Wired Equivalent Privacy (WEP), followed by Wi-Fi Protected Access (WPA and WPA2), and the adoption of the Advanced Encryption Standard (AES). But no sooner had such strong authentication measures taken root in Wi-Fi than attackers simply breached systems from a different angle: common Wi-Fi set-up mechanisms that allowed them to quickly obtain users' passwords.

For dedicated surveillance vendors and hackers, media access control (MAC) addresses that identify specific hardware used by a target have also been a chosen path to device intrusion.

Gamma International (FinFisher), together with its then marketing partner and occasional co-developer Elaman, mastered MAC capture as early as 2007. Partly as a result, smart device manufacturers such as Apple moved to mobile operating systems that continuously scramble MAC addresses.

Another viable approach for the security conscious is to forget public Wi-Fi altogether and move to a more discrete service such as "Private Wi-Fi," which offers the advantages of VPN service, automatically encrypting all network traffic from a mobile device and routing it over the company's servers, without the cost and hassle of setting up one's own dedicated VPN server. In addition to Private Wi-Fi, AT&T and Comcast offer this option.

However, the general populace naturally loves "free," a factor driving preference by more than two-thirds of U.S. consumers for public Wi-Fi versus commercial cellular service itself. Forgotten in the rush to connect anywhere/anytime without racking up paid minutes: hacking users on a public Wi-Fi network can take less than 2 seconds, according to Infonetics. Such hacking occurs with a fair degree of frequency.

Common Public Wi-Fi Hacking Techniques

Every year Google and others issue claims that higher and higher percentages of Internet traffic are secured by HTPPS encryption – 89 percent is the current claim. Others note that only some 50 percent of mobile data is similarly secured. Since the vast amount of traffic is now generated and accepted via mobile, much of it on Wi-Fi, Google's claims might be suspect. While hackers and agents can't have a field day intercepting targets on mobile there are most certainly vulnerabilities. A pair of long-established hacking methods remain relevant to Wi-Fi interception: Rogue Access Man-in-the-Middle (MITM) attacks; and a variant on this method called "Evil Twin."

The Rogue Access MITM attack simply offers the unwitting target what appears to be a more attractive alternative. For example, if the target is sitting in an airport departure lounge and checks Wi-Fi access, he or she might see choices such as "United" and "United Premium" Wi-Fi, the latter with a stronger signal registering on his smart device. Since the two public Wi-Fi services are free, the target is most likely to gravitate toward the premium version, unaware that he's being lured into a hack or intercept. In this case, all the attacking entity has done is set up his reputed Wi-Fi network via a Soft Access Point (Soft AP) that allows the user to easily enter the rogue network.

The Rogue Access attack might initially seem to resemble the way an IMSI catcher works. Both rely on a stronger signal to attract the target and work via MITM attack.

Both connect to the network – the rogue attack to the Internet and the IMSI catcher at least tangentially to the target's mobile network.

But there are key differences. With an IMSI catcher, the network connection is only partial, patching through calls and other connections made by the target, but not handling inbound traffic. Rogue Access is a two-way MITM attack. Another major difference lies in the handling of encrypted traffic from the target. Because in a mobile network the base station selects the type and level of encryption, and the IMSI catcher is mimicking a tower, the IMSI catcher can be set to downgrade encryption to WEP or to none at all. The Rogue Access attack's Soft AP lacks that power, thus the ability to decipher encrypted traffic hinges on the skills and tool sets at hand to do the job.

Evil Twin takes Rogue Access one step further. Here the offered Wi-Fi is identical in name to one known to be commonly used by the target. To carry our example further, the Evil Twin might simply be called "United." How does the attacker know which Wi-Fi access point the target favors? Simple. With commonly available software suites such as Aircrack-ng, the attacker can quickly turn his wireless adapter into an AP. Once the target logs on, Aircrack-ng shows which APs he gravitates to. The attacker can then use the suite to create an exact duplicate of that Wi-Fi AP, emit a stronger signal, "catch" the target and begin the MITM attack. Just two caveats. If the target uses WPA2 with AES, or a VPN, his data is encrypted end-to-end, and the attacker or agent is on his own again.

Rayzone's "Social Engineering" Skills

Rayzone InterApp doesn't suffer the same shortcomings. Rayzone claims its Wi-Fi monitoring product can capture the "user email address and password, contact list, Dropbox, operating system of the phone, photos, Internet history browsing," as well as locations recently visited, IMEI number and MAC address. InterApp works with any Wi-Fi network and delivers captured data in deciphered format. In hacker vernacular, InterApp can "pwn" – own and dominate – any mobile device. Unlike with conventional Wi-Fi hacking methods, the target's cooperation via web surfing is irrelevant. All that is required is that the target's mobile device Wi-Fi be switched on.

The incursion may occur one of two ways: by Zero Days that exploit technical vulnerabilities; or by social engineering, the use of non-technical human interaction as an entree to an attack, be it via MITM attack or malware. Common social engineering techniques include phishing, spear phishing tailored to a group or individual, pretexting, or even a quid pro quo exchange of personal data in return for a reward.

Which does Rayzone InterApp use – Zero Days or Social Engineering? According to one source, Rayzone offers a large collection of Zero Day exploits that leverage the bugs in mobile operating systems and provide an open door for malware that infects the device and begins stealing all its data. That certainly fits the profile of Rayzone, which offers a separate line of malware products.

Zero Days are one mechanism. The company is also on the record with a May 2015 proposal to FinFisher to provide its social engineering skills for projects in Latin America. FinFisher, by all accounts, was favorably impressed – high praise indeed.

How does Rayzone InterApp compare to Wintego WINT? We see them as roughly comparable in functionality. The difference may lie less in technical capabilities than in corporate focus. Rayzone, founded in 2011 and thus only a year older than Wintego, has a far broader product line, including not only mobile interception products but solutions for lawful intercept, big data analytics, satellite and other distinct suites. Wintego, in contrast, is centered on Wi-Fi, 4G LTE and data decoding.

Although younger than Rayzone, Wintego would appear to appear to have the edge its in Wi-Fi interception. If Rayzone has done any ground breaking with InterApp, it is primarily in the area of marketing, not product development.

Saab Medav: 5G Blind Mobile Location

Long before any mobile operator launched 5G, Germany's Saab Medav had already created an innovative way to surmount the challenges of tracking targets on these super-fast networks: "blind mobile location," a technology achievement on par with the sophistication of 5G itself. To understand why, let us first examine the inherent challenges.

5G networks rely on multiple microcells and optimal use of spectrum in extremely high frequency wavelengths measuring 1 to 10 milliwaves (mmWaves).

mmWaves have their downside and upside. Compared to lower bands, mmWaves are susceptible to attenuation due to atmospheric conditions, and as a result are best suited to terrestrial communications over distances less than 1.0 kilometer. On the upside, mmWaves are better than lower bands at permitting frequency re-use over the same short distances by small antennas. This characteristic is a good fit for older 4G and new 5G needs, which shift focus from sending/receiving data via individual signals per channel to handling multiple signals simultaneously in the same channel, optimizing spectrum use for point-to-multipoint applications.

In 5G, point-to-multipoint gets an assist from two sister technologies: MIMO (multiple-input, multiple output) and multipath propagation. MIMO is the use of multiple receive and transmit antennas via multipath propagation, a technique that sends a message via multiple routes to the same destination to mitigate the effect of signal attenuation and data loss.

The end result is radio localization, the use of many small cell sites outfitted with small antennas to efficiently and economically handle massive bandwidth via multiple paths.

In 5G Mobile Location, DOA = "Dead on Arrival"

Radio localization's reliance on multipath propagation is the natural enemy of radio direction finding, often leading to incorrect results in pinpointing a specific emitter. Conventional direction finding relies on tracking a single Direction of Arrival (DOA) for localization. In a MIMO environment, the investigator must contend with multiple signals arriving from a variety of different paths. The difficulty of tracing the signals to one mobile station using DOA increases by an order of magnitude.

Because MIMO has been around for several years and is increasingly used in 4G networks, the challenge to network planners and law enforcement is not entirely new. So-called "blind mobile location" (BML) solutions are available that help to a degree. However, as MIMO becomes pervasive in 4G networks – and ubiquitous in the new 5G – the limitations of current BML will grow. The problem will be particularly noticeable in dense urban markets where physical propagation compounds the problems of reflection, diffraction and scattering, driving higher signal attenuation.

A further challenge relevant to cases of public safety and national security is the assumption that a target perpetrator will be using a non-subscribed mobile device that does not "cooperate" either with a carrier's base station, or with an independent observer.

To explore these problems, Saab Medav and AWE Communications, together with the Fraunhofer Institute for Communication, Information Processing and Ergonomics (FKIE), and the University of Technology in Ilmenau, formed a collaborative group: the Emitter Localization under Multipath Propagation Conditions Project or EiLT. One area they focused on is ray tracing, which measures the dispersion of radio waves through the atmosphere and the impact on ray trajectories.

Together with fusion algorithms that synthesize data from multiple sensors, ray tracing can measure one set of multipaths and produce a second set of predicted multipaths, plus the relative time of arrival for each at an observation post. Normalizing the distances between the two sets enables the assignment of "best matches" between measurements and predictions. The likelihood of one path being the best match for a single path location is found by an estimation procedure. The hypothesis with the highest likelihood is determined as the target location. Standard BML is more or less accurate. Still, the EiLT team saw room for improvement.

Refinements with Monte Carlo Simulation Engines

The team from Fraunhofer Institute followed the path of improving on the BML process by introducing Sequential Monte Carlo (SMC) Intensity Filters, a form of Monte Carlo Simulation Engine.

"Monte Carlo" derives from a specialized niche of probability mathematics known as stochastics, which deals with random possibilities: the uncertain array of values, principles and inputs that may be used to determine multiple diverse signal paths. Note that the emphasis is on "paths" or potential scenarios as opposed to definitive results. Thus a Monte Carlo engine is software that examines and turns out a series of potential routes to an outcome.

Intensity filters (iFilters) are used to track targets that present more than one measurement per radio scan. In combination with Sequential Monte Carlo projections, a series of readings by iFilters can narrow the field of paths pointing to the most probable position of a targeted device. To ensure accuracy, the hypothesis is measured against a ray-tracing scenario.

Fraunhofer's approach delivered comparable results to conventional BML on average, and in specific instances, "even more accurate results." The difference maker: the first ever attempt to apply SMC Intensity Filters to mobile location in a dense urban environment.

AWE Communications' contribution was to create the simulation platform for trial scenarios accounting for different terrain scenarios – buildings, walls, vegetation, etc. – used in measuring the strength, angle of arrival and delay of multipaths between the emitter and observers. To boost performance, AWE developed a real-time ray tracing tool to ensure the most current data input. AWE's work was folded into its WinProp product used for mobile localization. EiLT's results are reflected in Saab MEDAV product development, as well.

Channel Sounding

In MIMO, assessment of channel state information (CSI) is critical to determining channel properties including propagation, scattering and fading across signals, and plays a role in BML. The process of measuring CSI is channel sounding or estimation, which builds a detailed picture of the channel characteristics.

Saab Medav has for some time offered its respected line of RUSK MIMO Channel Sounders used for both military and commercial purposes. Added to the portfolio and reflecting the work of the EiLT is the company's Ultra Wideband MIMO Real-time Channel Sounder equipped "with the capability to connect multiple transmit and receive antenna elements for directional resolution." Read: SMC Intensity Filters and a real-time platform reflecting the work of EiLT.

Promoted for its use in designing UWB communications systems and testing, the Ultra Wideband MIMO Real-time Channel Sounder has one other key application: mobile localization. In Saab Medav's own words: "High-resolution 3-D active localization can be investigated for the localization of UWB terminals with respect to the infrastructure or relative to other terminals." It's only a short hop from there to finding targets' mobile devices, whether they are regular paying members of mobile subscriber services, or users of non-subscribed devices.

While Saab Medav is a pioneer in the field of enhanced BML, they are by no means alone. Contenders in 5G BML include Keysight, a spinoff from SAIC, and also Fraunhofer, which has partnered with Rohde & Schwarz to develop millimeter wave channel sounding products. If Saab Medav has an advantage, it may stem not only from its vision and engineering prowess, but from how it "channels" these assets to strengthen a respected legacy in serving the military and law enforcement communities. Channel Sounder is a practical, forward-looking complement to the company's comprehensive family of radio monitoring, surveillance and "Intelligence Fusion" products.

Septier Mobile Location

Israeli-based Septier is a diversified ISS vendor whose products cover the span of law enforcement and government agency requirements: mediation devices for CALEA or ETSI-based law enforcement; data retention of CDRs and IPDRs for metadata; big data analytics; and Septier mobile location and interception at both the strategic and tactical levels. Each product can operate as a standalone module for its specific function, or be integrated with the other Septier products. But make no mistake: mobile location is the company's core expertise. Septier solutions empower clients with near "bird of prey" accuracy in locating, intercepting and tracking any target operating on a mobile device.

Evidence of the company's prowess: year-after-year growth driven by an impressive array of products: Septier Gateway Mobile Location Center (GMLC), Septier "Passive Massive," Septier Accurate Location, and Septier IMSI Catchers.

Septier GMLC pulls targeted communications via specified queries to a mobile network to track a target's location. Raw data on location is then funneled to a mobile location center, where it is quickly refined to reveal the mobile device's sector location and coordinates. Targets can be pegged by a range of identifiers including IMSI (international mobile subscriber identity), TMSI (the temporary version) and others, regardless of whether the phone number is disguised or the user is roaming on another country's network.

Septier Passive Massive unobtrusively gathers mobile location data on multiple targets in the network with the ability to narrow the field to a single suspect. Using passive probes, the solution monitors links between mobile users and the network to feed a database of precise LDRs (location data records). This high quality record of location data can prove useful to investigators when mapping the relation of a target's location to a suspicious event. Users can query the database, building a case based on the history of the individual's whereabouts. Information is pulled and stored on all subscribers in the network, even before they become a target. Passive Massive scales to handle any quantity of subscribers. Because all data is obtained passively by the probes, there is no added load or impact on the network.

For precise location, Septier's Accurate Location offers three options: GPS, signal cross-referencing, and radio frequency (RF) "fingerprinting." Septier Assisted Global Positioning System (AGPS) uses GPS to locate the target's device to within a few meters. While the solution takes network data into account, its accuracy comes from the data used being handset vs. network-based.

Septier Network Measure Report Processing offers a second accurate location option. Capturing and cross-referencing signals from other devices in the area, software matches each signal to factors such as topographic and geodesic data to follow in the target's footsteps.

RF Fingerprinting produces a "most likely" signal map by assuming the target's location and comparing the result to the known signal of the target's device. The solution then begins a process of correcting the assumptions to quickly sort out the target's true position.

Septier's two IMSI Catchers, Cellular Locater and Extractor, perform similarly to devices by Harris and Rayzone to determine the identity of the sending and receiving parties, their location and any content exchanged.

Septier Cellular Extractor works off-the-air to intercept metadata and content from an entire mobile base station. Septier Cellular Locater, the pocket-sized version, can also detect other devices in the area and intercept, jam or even modify the transmission. That last point is a key add-on. Like competing products by Rayzone and Harris, Septier IMSI Catchers can deliver intrusive capabilities via man-in-the-middle attack.

Wintego WINT for Wi-Fi Interception

For law enforcement and intelligence agents trying to track targets on the move, Wi-Fi networks was once portrayed as an insurmountable challenge. The common theme: Going off-network and onto Wi-Fi in a coffee shop or airport terminal was not just a great way to avoid maxing out carrier data caps, but for criminals and spies, a clever means to evade detection.

In fact, Wi-Fi networks have long been porous to surveillance technology. Large vendors such as FinFisher, Gamma Group, RCS Lab, NICE and ClearTrail offer proven Wi-Fi surveillance products adept at capturing terrorists and criminals when they least expect it – for example, while savoring a cup of expresso at their local Starbucks. All of these vendors and products have strong market presence and brand recognition.

To newcomers in the field, that means most of the real estate is already occupied. Venturing into this crowded landscape in 2016 came a daring upstart from Israel, Wintego Systems, offering its own brand of options for extracting strategic intelligence from targets' devices.

Wintego's core product is WINT for Wi-Fi. Also offered: DEX-Inline for enterprise and ISP networks.

Inside WINT and DEX-Inline

The WINT product is designed as a portable unit primarily for Wi-Fi interception. DEX-Inline intercepts traffic of one or multiple targets operating on corporate, enemy agency or ISP networks.

The core platform for both products is Wintego Data Extractor (DEX), a software engine that uses a variety of methods to isolate a device, overcome network security and encryption, and show the agent all the target's traffic: emails, chats, SMS, social media profiles and communications, files, calendars, list of contacts, web browsing, photos and images, locations and call content.

If these capabilities sound familiar, they are. WINT and DEX-Inline are classic offensive cyber programs, providing the ability to compromise, penetrate, copy and control the target's device on a one-off or multiple target basis – just like FinFisher FinSpy, Memento/Hacking Team's RCS, ClearTrail QuickTrail and comparable platforms made by NICE and RCS Lab. Wintego competes with, and also appears alongside the same companies at industry events, where they all hold forth in sessions on IT intrusion.

Wintego was founded in 2012 at Yokne'am Illit, Israel, in the heart of "Silicon Wadi," a major technology center southeast of Haifa. The company is obviously newer to the game than its more established competitors, which perhaps explains its lower profile. As an offset, Wintego is quick to point to its "30 years of experience" in the technical and operational aspects of intelligence gathering. Since 2013, Wintego has put in a showing at ISS World and Milipol conferences where it aggressively promotes WINT and DEX-Inline.

How WINT Works

All smartphones come equipped with Wi-Fi and Bluetooth capability, transmitting at lower power, for a shorter range – and via different chip sets – than the principal radio signal used for communications.

Wi-Fi operates in the 2.4 GHz UHF band and 5 GHz SHF (super high frequency) ISM (industrial, scientific and medical) band, and Bluetooth in the 2.4 to 2.485 GHz ISM band. Early-on, Wi-Fi was criticized for being more susceptible to security breaches than wireline devices. Newer protocols improved Wi-Fi security, and the claim is often made that accessing transport layer security (TLS) protected web pages over Wi-Fi is now fully secure. Although some tests have shown the ability to transmit (and thus intercept) Wi-Fi and Bluetooth at a distance, for the most part agents or law enforcement officers need to move in close to the target if they hope to capture signals that fade after a few yards.

A common feature of Wi-Fi and Bluetooth is that the signals of both include a device's MAC (media access control) address, a number that is unique to every device. The MAC address of a mobile device is completely unsecured – anybody can pick it up. Moreover, the MAC signal is transmitted regardless of whether the device is in use, idle or turned off. When an agent knows the target's MAC address, it's the same as seeing a "hello, I'm here" signal from the suspected criminal or terrorist.

By definition, being confined to Wi-Fi interception, Wintego's WINT operates at close quarters, perhaps by a user in a car parked outside the building of a free Wi-Fi shop, or in a room next to the target's location. Knowing that the target is present via MAC address signal is one way "in" for WINT. Others include knowledge of the target's IP address, activity status or machine name. Recognition of the target and his presence on the adjacent Wi-Fi network is Step 1 for WINT.

From there, WINT can install malware on the target's device by any of the typical methods: phishing, man-in-the-middle (MITM) attacks that fake credentials of an accepted certificate, leveraging Zero Day vulnerabilities, performing point of sale breaches, and so on.

Wintego's DEX engine can manhandle encryption in several ways, but such capabilities are hardly anything new. Hackers cracked TLS and secure socket layer (SSL) years ago. And there are many other ways around encryption. For example, a flaw in how older iOS operating systems handle SSL certificates while the target is on a Wi-Fi or other networks enables the agent to intercept the device or crash it. Finally, as stands to reason, the content on any device already infected by malware can be intercepted and read "in the clear" regardless of how the target communicates, including on Tor. Intelligence agencies such as the NSA has proven that capability many times over.

The same principles and functionality apply to DEX-Inline, but on a larger scale to accommodate a greater number of targets.

CHAPTER 6

Open Source Intelligence

Overview

OSINT has evolved to become the new SIGINT. By pointing directly to the individuals behind acts of crime, hostile nation state activity and terrorism, social media intelligence (SOCMINT), the Deep Web and other forms of OSINT have earned the respect of law enforcement, government and military intelligence personnel.

Social media monitoring is now an accepted "beat" for law officers who can quickly see the most likely times and places for crimes that are about to occur or recur, and for military users tracking opponents via:

- Troop movements. A combatant posting on Twitter from an occupied village one day, then from a city hundreds of miles away just later days later may point to troop movements.
- Armament Ramp-ups. Images of mobile-missiles and tank brigades posted to social media can provide a clear indication of opposing forces' strengths and readiness for combat.
- Geolocation. Tools by companies such as Snaptrend can precisely position every Facebook or Twitter post to a specific site. Combined with satellite image analysis, other tools can plot posts from any social media site on a high-resolution map.

Police dispatchers now can enter an individual's name into software provided by Tyler Technologies that tells whether that person has previously contacted police and why. PredPol (short for "predictive policing") sells software that scans three data points – past type, place and time of crime – then applies a priority algorithm to help police predict future crimes. Other companies provide mobile apps that scan millions of records to create a unique profile of any individual designated by the analyst.

SOCMINT has attracted not only new players but also established vendors such as FinFisher and Elbit Systems CYBERBIT, which have added the capability to their repertoire of services – and made the market intensely competitive. Competition has also driven some early entrants from the marketplace. Among the best known, BrightPlanet in 2016 discontinued its Blue Jay Twitter monitoring service for LEAs, citing "higher costs and low customer volume." The company remains active in Deep Web monitoring. For all its popularity, critics make a fair point in contending that OSINT is not fool-proof. For example, it is a given that nation states and operatives leverage social media for propaganda and counter-intelligence. At its zenith, ISIS commonly posted images of supposed victories from past or unrelated conflicts, including some that pre-dated the Islamic State.

Other forms of surveillance that fall under the category of OSINT include voice biometrics and facial recognition. While they focus exclusively on target identification, not on interception of communications or location, the intelligence they provide can serve as the doorstep to deploying ISS solutions.

Social Media Monitoring and GEOINT

With a user count numbering billions of people, social media might seem the fastest growing communications vehicle on record, but hand-in-hand with this explosive growth has come another important "follower" nearly as big if far quieter: Social Media Intelligence.

SOCMINT has evolved as a major component of OSINT relied on by military, police and government intelligence, fed 24X7 by streams of social media posts and other public information sources where ordinary mortals, some of them suspects, "tell it all" to the world.

If SOCMINT did nothing more than monitor Facebook, Twitter and Instagram posts it would hardly be the intelligence gold mine many claim it to be. SOCMINT shines brightest when coupled with the full array of ISS technologies. Paired with other tools such as AI and GEOINT, SOCMINT could one day redefine ISS. But at present the field, with but a few exceptions, has not yet experienced the "high highs" of the market giants it tracks.

There is no Mark Zuckerberg of SOCMINT. Nor is there a single company or group that cares to claim they pioneered the art of snooping on tweets. SOCMINT vendors themselves are understandably silent on the topic, and perhaps it is irrelevant at this point. The important takeaway is that most companies involved in this space support three niches: corporate marketing, cybersecurity and ISS.

At first blush, SOCMINT solutions for the three might appear closely similar. Respective SOCMINT solutions all cover identification of social media users and their accounts, location, analytics for trend spotting, real-time opportunity (or threat) assessment, networks of friends, etc. But on closer inspection SOCMINT products vary significantly in some respects, depending on the customer niche and its mission.

Like its commercial counterparts, ISS-focused SOCMINT has the ability to apply advanced analytics and machine learning, identify off-network affiliates of targets and provide brilliant visualization of the data. Unlike the rest, SOCMINT can also serve as the entree to offensive cyber where agents take advantage of made-up hashtags to attract targets with kindred sentiments, or post photos of attractive women, with embedded malware that takes over the victim's device.

ISS applications of Social Media Monitoring can be highly accurate at determining location when deployed with SATCOM GEOINT versus simple geofencing based on geo-tagged posts. Some are casting Social Media Monitoring as central to integrated ISS capabilities that provide a holistic view of threats.

Vendor Snaptrends goes a step further by depicting SOCMINT not merely as an adjunct but an improvement to GEOINT.

GEOINT + SOCMINT: Lengthy Engagement

Geospatial Intelligence or GEOINT is broadly defined as the use of satellite or aerial photography combined with intelligence gleaned from statistical analysis, GIS (geographic information systems) and other tools to provide precise coordinates and maps of human activity. Essentially, GEOINT is the science of plotting a target's precise location based on multiple factors that stand as hard visual evidence of location.

GEOINT has evolved to include a variety of technologies, each of which contributes its part to determining where a subject, movement or other phenomenon is situated or is changing at any given moment. The arcana of GEOINT science can span readings from spectral, spatial, temporal, radiometric and other data for both stationary and moving targets. Some tools can detect changes in ultraviolet light to determine position. Others measure electro-optical data such as infrared and reflected infrared.

GEOINT thus represents a staggering advance over the typical "geofencing" of commercial SOCMINT which, as the name implies, simply renders a geographic zone in which an individual or group is noticeably active on social media. In that vein, geofencing is the SS7 of social media: It gets you within a block or two of the target. And that's only assuming he or she is not posting images or other data that "geo-tag" him to a dated or false location to throw off his trackers – a trick ISIS demonstrated fondness for at the height of that radical Islamic group's career. GEOINT can't be fooled. It takes the user to the precise street corner where the target stands waiting for the stoplight to change colors.

What GEOINT can't do is perceive what the target is thinking or planning while he or she traverses the earth's surface. Advanced SOCMINT can do just that: light up the map via "a little help from its friends" – text, semantic and Natural Language Processing that reveal what the target says, means and intends to do in affiliation with co-conspirators.

Ratcheting Up SOCMINT

In a perfect union, GEOINT would waltz with its partner SOCMINT, showing co-conspirators' precise location via SATCOM imagery data. Around the agents' operation center, the SOCMINT family of capabilities would accelerate the rhythm, tapping out multi-agency intelligence, real-time local video footage, criminal or other archived records, even current weather conditions. Machine learning would distinguish real threats from harmless language while real-time predictive analytics pinned down and prioritized threats. Here we shall examine this union of surveillance technologies among some of the growing leaders in the SOCMINT sector.

GEOCOP

HMS Technologies of Martinsburg, West Virginia USA took an early lead in defining what might be called the Unified Theory of social media monitoring, combining SOCMINT with geospatial intelligence. The end result is GEOCOP, a "ubiquitous, synchronized, and real-time situational awareness regarding any recordable event, activity, or thing that has happened (or is happening now) somewhere around the world."

GEOCOP is sold primarily to U.S. law enforcement agencies, moreover, as a service, not software or a system that the user installs at his own premises. GEOCOP processes data from an array of sources, but events are typically monitored by its own agents versus being fully software-automated or driven by AI. Human error follows. In one instance, a user noted that the intelligence gained from GEOCOP was delivered an hour or more after an incident occurred, which, while helpful in capturing the culprit, was far from being out in front of the crime, thus nowhere near real-time. Finally, GEOCOP comes up short on SOCMINT/GEOINT unification. It does location by geofencing per social media tags, not GEOINT in the true sense of the term.

Digital Stakeout and LexisNexis Accurint

Digital Stakeout, a competitor, has taken SOCMINT a good deal farther. The company addresses the cybersecurity and ISS niches with an SaaS (software as a service) approach that is equal parts big data and AI. It's a winning combination for Digital Stakeout, whose solution is used by hundreds of organizations and has helped prevent loss of life and onset of violence. In a word, Digital Stakeout has a distinctively preemptive focus.

By big data, Digital Stakeout means not just social media but other OSINT sources such as the World Wide Web, Deep Web and the Dark Web. Culling finite intelligence that gives the user a proactive edge, and from such a massive array of sources, involves a distributed computing model with Hadoop and Spark interfaces, or their like. The statistical and scoring components of real-time predictive analytics are the power that lets an LEA, government or military organization see the most likely "futures" and select the next best action.

Digital Stakeout doesn't stop at conventional analytics. The company leverages learning automation and deep learning algorithms to transfer the load of threat analysis from agents' shoulders to machines. The client needn't fret over potentially missing some important detail in the monitored stream of communications. Via AI, any anomaly will pop up for further examination.

With Digital Stakeout, location is determined by a combination of geo, IP, and inference tagging, matched to visualizations of social graphs. This successful package attracted LexisNexis, which for several years has used Digital Stakeout as the core of the LexisNexis Accurint Social Media Monitor product.

Geofeedia SOCMINT

Geofeedia, launched in 2011, is one several of SOCMINT companies that has the distinction of being backed by In-Q-Tel, the CIA's unique venture capital fund, which in the past has helped kick start market winners such as Palantir and MemSQL.

Geofeedia's specialty is tracking multiple social media sites in real-time within a predefined zone. The company claims it is the first to do so, a point that HMS Technologies and Digital Stakeout dispute. In essence, Geofeedia's solution is geofencing by social media tagging in a predefined zone. Geofeedia software won't point to activities outside a zone chosen by the user. That's not necessarily a problem if the client wishes to monitor a zone confined to London, Washington DC or Ramallah, but for those needing to monitor a larger area it might be.

Geofeedia also uses analytics to measure sentiment that might point to a threat. But although the company offers its services to police departments and government agencies, its primary customers are Fortune 500 companies such as Dell and McDonalds.

Walking the Walk: SnapTends

Of the many companies in the social media monitoring niche, Snap-Trends has the best understanding of the need for integrating SOCMINT with GEOINT. The company states that GEOINT professionals must be "trained to analyze the physical, cultural and human aspects of a region while synthesizing this with information gathered from numerous sensors" including satellites, mobile phones and social media. They add that ideally

GEOINT should "take geospatial information gathered from a social media monitoring tool and determine whether or not that social media data is pointing to a potentially threatening situation."

Snaptrends has a product that does just that, or is at least moves a step in the right direction: their Social GeoService, which integrates the company's Social Media Data product with Esri ArcGis, software for creating GIS maps. A second solution, Snaptrends' Social Media, analyzes social media for sentiment, mood and tone in real-time, and also examines the target's web surfing history for valuable heuristic data. Social GeoService automates the analysis on a standard ArcGis map.

GIS mapping is not quite full-blown GEOINT, but it's closer, and enough to help SnapTrends hone in on government customers with deep pockets, such as the U.S. Department of Homeland Security.

Marching to Military's Drum

Military buyers were among the first to show interest in a SOCMINT/GEOINT solution. In July 2012, U.S. Special Operations Command (SOCOM) held a nine-day test of "non-traditional tools" to support the SOCOM mission. Called Project QUANTUM LEAP, the technology trial focused on the integration of GEOINT location with SOCMINT and Deep Web products and services. Specifically, SOCOM lined up various SOCMINT/OSINT products with its AVATAR (Automated Visual Application For Tailored Analytical Reporting) intelligence platform, which includes elements of GEOINT.

Among the more promising tools trialed was Social Bubble, a software plug-in by Creative Radicals. Social Bubble was tasked with tracking social media posts to locate individuals involved in a money laundering operation. The solution met the challenge and performed well. However, the DoD de-funded QUANTUM LEAP six months later, and the SOCOM staff working on the project were assigned elsewhere. Subsequently, the DoD cited its short window for taking action, and sensitivity about the legality of using SOCMINT, as factors in their decision to pull the plug.

Fast forward three years to Washington, DC-based defense contractor Blue Canopy, which in June 2015 selected GEOINT, the U.S. Geospatial Intelligence Forum (USGIF) annual conference, as the venue to announce its Blue Canopy Social Media Analytics solutions. The product is an intelligence platform that balances Blue Canopy's weight in the military market with the strength of key partner ClaraBridge, a specialist in analytics for "customer experience management" and Natural Language Processing.

Although Blue Canopy Social Media Analytics debuted at what its authors called the "preeminent intelligence event of the year" – and one with a strict GEOINT theme, to boot – the company uttered nary a word about geospatial intelligence.

Despite the half-baked efforts of some contractors and its own internal back & forth, U.S. military remained keen on developing an integrated platform. On May 30, 2014, 18 months after QUANTUM LEAP folded, SOCOM issued an RFI for "Computer Systems Design Services" for its AVATAR program. An RFI or "request for information" is Stage One of the DoD's complex process for selecting vendors.

This particular RFI delineated interest in a system that could covertly conduct high-volume data searches, mine and analyze text and other unstructured data, plus provide alerts and visualization – all integrated with AVATAR. The RFI did not specifically reference social media, but "unstructured data" would certainly cover text, images and video, both common elements of SOCMINT and OSINT.

What set that stab at upgrading AVATAR apart: unification of data across multiple sources into a single platform mapped to SATCOM GEOINT imagery and brilliant Palantir-like visualization, all of it made available on-demand to a single SOCOM combatant. Imagine targeted SOCMINT illuminated by real-time satellite imagery.

Further north, Canada followed a similar path. In January 2016 that nation's Department of National Defence announced that it would invest in a platform to collect and analyze social media worldwide and round-the-clock, as well as historic social media and other OSINT including blogs, news sources and comment sections. Details are sketchy, but just as Canada tends to work closely with the U.S. on CALEA lawful intercept and intelligence matters, it is feasible that they received guidance on SOCMINT from their neighbor to the south.

Since these early days of SOCMINT, vendors have accelerated at light speed thanks to machine learning and deep learning algorithms, and have begun to win more U.S. government business either on their own or through partnerships with the array of traditional DoD and Intelligence Community contractors that service the U.S. government. Creative Radicals is one such. Skyris is another. However, one must carefully assess their respective solutions' features to determine what is really there, and what might be missing.

SOCMINT vendors now uniformly tout their ability to glean SOCMINT from big data via from machine and deep learning, and to translate multiple languages. Where many fall down: textual semantics – understanding the meaning of words in different contexts. In far too many cases, SOCMINT companies come up short on the job of understanding a language in terms of its intent. Moreover, they are often oblivious of where to turn for help, for example, to companies such as Italy's Expert System, a renowned textual semantics vendor in business for many years, with offices in many nations including the United States.

Israel's Cyber and SOCMINT HQ

In June 2016 came word that Israel Defence Forces (IDF) had completed a new underground "cyber headquarters," the control center for the IDF's full-fledged cyber intelligence unit under command of a Major General. The IDF's C4I Directorate and Military Intelligence Directorate, responsible for defensive and offensive cyber initiatives, respectively, report there. The mission: Maintain Israel's technology lead in the world's longest running yet most singularly silent conflict: the cyber war.

Israel is well into this game. More often than not, the opponent is not some splendidly outfitted phalanx bristling with tanks and missiles, but the average 15-to 25-year-old male Arab with an affinity for radical Islam, a hatred of Israel and the desire to die as a martyr, which he or she conveys publicly in social media. Many of his kind subsequently launch attacks, either in groups or as lone wolves. Israeli intelligence officers typically know their foe's plans well in advance through an elaborate centralized cyber platform that includes SOCMINT, with semantics that understand Arabic in the target's dialect, and the meaning and intent of every word.

AGNITIO: Pioneer of Voice Biometrics

Founded in 2004 in Madrid, AGNITiO was among the market leaders in voice biometrics until acquired by Nuance in 2017, providing solutions that mapped biometric voice prints (BVPs) – identifiers that are every bit as unique to the individual as the fingerprint. Though totally absorbed by its new owner now, the company's contributions to the field mark it as a major innovator whose work continues to define the field.

At its peak, the company's voice biometrics solutions were used by over 50 police departments in 35 countries. In addition, the company in 2012 won a research and development contract with the U.S. Technical Support Working Group (TSWG), one of three programs overseen by the Combating Terrorism Technical Support Office reporting to the Assistant Secretary of Defense (ASD) for Special Operations and Low-Intensity Conflict (SO/LIC). TSWG works with multiple agencies to develop technological solutions to combat terrorism.

AGNITiO's surveillance suite for government entities comprised three components: BATVOX for law enforcement, BS^3 Strategic for military and intelligence agencies, and SIFT, a portable unit deployed on a laptop.

BATVOX was an advanced voice biometrics tool designed for police technical units, homeland security and intelligence agencies. Based on the company's BS^3 core engine, the diverse functionality of BATVOX provided tracking, positive identification, graphical presentation, organization for case management, and reporting. The pioneering version operated by leveraging a "Bayesian network," a statistical model that draws the links between random variables and conditional dependencies in a graph that determines the highest probability of a proven connection. When deployed, it could identify a voice 99.02 percent of the time in a cross-channel environment based on 60 seconds of audio, regardless of whether the target spoke in different languages or tried to disguise his voice. The final version, BATVOX 4.0, based on the company's iVector algorithm, required just seven seconds of speech to identify the target, and 30 seconds to create a profile of the speaker.

BATVOX also identified voice samples from unknown individuals potentially related to voiceprints of known suspects, empowering users to spread the investigation net to other potential suspects. All captured audios and speaker models were carefully organized for further analytics and forensics. When investigators compiled the evidence they needed, reports could be readily transferred to HTML files for use in court. Graphic representations provided clear evidence that the "likelihood ratios" used in pinpointing the individual by voiceprint were spot-on accurate.

BS³ Strategic for government agencies and the military accurately analyzed millions of audio recordings for mass voice interception and mining. The solution could process 2000 hours of audio per day across 1000 targets, delivering 1.4 million matches per second on a single CPU. The system separated multiple speakers, determined the gender of the target speaker and generated his or her biometric voiceprint, all in real time. Any issues associated with distance, multiple hops or poor connections were eliminated: BS³ Strategic automatically removed artifacts, non-voice events and noises.

Highly scalable, BS³ Strategic was easily deployed in a multi-CPU environment. All voiceprints were stored in the BS³ Master Repository fully interoperable with BS³ Strategic to ensure ready search and access functions. Because the BVP footprint required for voice biometrics is small, systems operators could significantly reduce the number of calls required for speaker identification and thus for storage in the BS³ Repository. BS³ Tactical, a lighter version of the same product, provided similar functionality for users pursuing a smaller, targeted number of calls and callers.

AGNITiO's Speaker Identification Flexible Toolkit (SIFT) provided an affordable standalone speaker ID option for those on the move, and on a budget. Though lower price, SIFT used the same state-of-the-art iVector algorithm found in all late version AGNITiO products. Like BS³ Tactical, SIFT was designed for laptops so that it could be put to work in the field. Being smaller, it was limited to monitoring 1000 speakers per event, but was still extremely effective at identifying speakers even in the briefest of dialogues.

SIFT users could work with stored voice prints, or intercept live conversations of targets at speed – up to 1000 call identifications in 19 seconds, not as fast as BS³ Strategic, but more than adequate. Otherwise, the features of both solutions were quite similar: ability to identify the target from a single voice biometric print; noise blocking; multi-speaker identification and separation; and high accuracy.

BrightPlanet Tor Cracking

Since discontinuing its Twitter monitoring product BlueJay for law enforcement in 2016, BrightPlanet has remodeled itself as a software-as-a-service (SaaS) provider of Deep Web content and global newsfeeds for a broad marketplace including commercial enterprises. The company now focuses on tools designed to help financial institutions and other businesses understand SOCMINT unique to their respective industries. But rest assured that BrightPlanet has not strayed far from its roots in intelligence gathering for law enforcement, government agencies and the military. Three signature aspects of its work stand out:

- The revised suite is based on BrightPlanet's long-standing expertise in Deep Web monitoring via the company's Deep Web Harvester platform.
- While BrightPlanet may have bid farewell to Twitter monitoring, overall it has expanded its OSINT capabilities via access to the spectrum of social media, as well as other open source intelligence.
- Perhaps most significantly, BrightPlanet now provides Dark Web monitoring through proprietary technologies that crack Tor.

This second iteration of BrightPlanet is every bit as interesting, if not more so, to the law enforcement, government agency and military communities seeking topnotch OSINT and cyber capabilities. By continuing to provide services that fit squarely with the interests of these customers, BrightPlanet has remained true to its roots in more ways than one.

BlueJay was far more than just your run-of-the-mill Twitter feed monitor. The product had add-on features with close affinity to offensive cyber capabilities such as the ability to track a target, take control of his mobile device, activate the camera and microphone for real-time monitoring – and link to fixed video cams at street level to follow the target en route to his planned criminal or terrorist activity. What BlueJay lacked was sales. In the end, the decision to drop BlueJay was no big loss to BrightPlanet. With its SaaS approach to data, aptly dubbed Data-as-a-Service, its broad OSINT offering, and more esoteric skills honed to penetrating the Dark Web, BrightPlanet is of far greater value to the Intelligence Community than before.

Into the Deep Web

To fully appreciate BrightPlanet's importance, it is essential to understand the company's expertise in Deep Web querying: the ability to find data on virtually any topic that doesn't show up from perusing the "surface web" with Google,

Yahoo and other commercial search engines. By some estimates, the Deep Web is at least 500 times the size of the surface web, and continuously growing. Compared to the Deep Web, the surface Web is literally the tip of the iceberg of all content available on the Internet if one knows how and where to look.

Typically, surface-focused search engines crawl and index the bare minimum amount of Web content required to catalog a page. Their job is to take the user to the location indicated by a query, not to store every bit of content there. Finding all the content ever posted is the job of a more sophisticated platform: the harvester.

Deep Web harvesting is the main course on BrightPlanet's menu. The company's Deep Web Harvester "extracts every single word every time it accesses a web page" and stores each page as complete raw data in a custom database for the client. But note: Per the company, Deep Web Harvester is not a "client-facing" platform. It is an SaaS solution hosted on Amazon Web Services (AWS) – the Western world's largest cloud service.

Under constant pressure to "do more with less," and because much legacy government IT is outdated and costs upwards of US$100 billion per year just to maintain, the great majority of U.S. government agencies have migrated to this model: outsourcing the high-volume data monitoring demands of Deep Web Harvesting to the comparatively low-cost cloud, the exact model used by BrightPlanet. The user need never concern himself about the cost and complexity of establishing the requisite platforms in-house. BrightPlanet's SaaS Deep Web Harvester handles it all as an end-to-end solution spanning harvest servers, document processing servers, databases and front services. On the tail end of intelligence gathering, BrightPlanet's REST API can easily integrate with an agency's resident intelligence gathering tools, and analytics engines or visualization systems such as Palantir.

The categories of data that Deep Web Harvester processes are limited only by the user's imagination. Each "harvest" of Deep Web data can scale to collect thousands of Web pages and every iota of data they contain that is relevant to a query. For general website harvests, BrightPlanet leverages link-crawling based on the urls of web pages. For Deep Web access, the solution automates queries into Web forms to harvest data. When websites are protected from the outside world by password access, the Harvester uses scripted language to create user names and passwords to get inside. The Harvester can also gather text from emails associated with Deep Web sites to see who is accessing them and the communications generated. Affiliated Twitter, Facebook, RSS feeds and blogs are all fair game.

One of the challenges of Deep Web Harvesting is that so much of the content gathered is harmless and irrelevant, e.g., arcane academic papers of little interest to anyone other than a finite audience. BrightPlanet's curation process quickly determines which content might be high value by using a rules-based engine that uses entity extraction to single out key terms such as target name, criminal or terrorist group affiliation, location, references to weapons, bomb-making, cyberattacks on critical infrastructure, etc.

BrightPlanet's Data-as-a-Service capability then comes into play, providing the intelligence officer with a search dashboard to examine all pertinent Deep Web content collected and stored in a specific section of the AWS cloud, walled-off from all but the agency's own designated personnel. Authorized users can then download, query and apply analytics and visualization to all relevant Deep Web data compiled by the Deep Web Harvester for their specific search interest. If the user happens to be offline when important new data surfaces from the Deep Web, BrightPlanet's DaaS service will send him or her an immediate alert.

Into the Dark Web

As noted in prior chapters, the Deep Web is not to be confused with the Dark Web. The Deep Web is comprised of information that may be extracted from previously published content, but only if the user knows which sites to examine. Dark Web content is deliberately hidden from view and requires specialized tools to access. The bulk of data on the Dark Web lives on the anonymous Internet, most notably Tor, I2P, FreeNet, Disconnect, Whonix and Yandex, which disguise users by passing their communications through a near-endless of array of volunteer routers, masking identities at each step until such secretive intel reaches its final destination.

Previously, BrightPlanet harvesting tools were confined to Deep Web applications, and the company distanced itself from work in the Dark Web. But no longer. Although BrightPlanet does not publicly acknowledge its work in this area, the company actively sells a solution fully capable of harvesting data concealed on Tor.

The process is surprisingly simple. BrightPlanet's own web harvesting engine is designed to work through a standard proxy server, i.e., the exact type of computer used on the Tor network to disguise the identity of users. In other words, to Tor, the Deep Web Harvester looks like any other proxy server.

To BrightPlanet clients interested in hacking Tor, the company simply orders its harvest engine to operate through a proxy and begins crawling links and web pages at scale on the Dark Web. BrightPlanet can also automate searches on the Dark Web, the same way it does on the Deep Web.

Expert System: Deep Semantic Analysis Leader

Often forgotten in today's rush to embrace OSINT and SOCMINT as supplements to or replacements for traditional COMINT and SIGINT is that targets are recognizable not only for what they communicate but how they do so, a field known as deep semantic analysis. Among the leaders in semantic analysis is Expert System of Rovereto, Italy, which provides its Cogito Intelligence Platform semantic analysis products commercially and to government customers in the defense and intelligence sectors for surveillance purposes.

Semantic analysis is the automated process of relating syntactic structures – words, phrases, clauses, sentences, paragraphs, pages, etc., and the author's level of composition – to meanings that are language independent. It involves the textured refinement of cultural and linguistic contexts such as idioms to a concise meaning with no variation.

The field of semantic analysis began with Ludwig Wittgenstein, a mid-20th century philosopher who studied the interrelation of logic, mathematics and language.

Wittgenstein's opus on the philosophy of language is distinguished by two periods – early and late. "Early" Wittgenstein propounded that language consists of propositions and the world of facts, that the two connect, and that language has a single underlying logic that can be understood via the analysis of language, the world, and their relation. But over time, Wittgenstein changed his mind.

Later, in the 1940s, the eminent philosopher – often deemed the greatest since Immanuel Kant – refuted his logic theory and concluded instead that language is a game based upon mutually agreed-upon rules that evolve through collective use. He then proposed that because language is in a constant state of flux, it is neither objective nor fixed. Thus, Wittgenstein concluded, it is impossible to understand the meaning of language through propositional logic alone. His key to learning the games people play with words: statistical analysis. Since that day, semantics analysis has progressed significantly.

Today's semantics analysis, as used in surveillance, enables the extraction of entities such as a target's identity, location, the meaning and intent of his communications and other concepts including sentiment from vast amounts of "unstructured" data. The latter is an important distinction because something on the order of 90 percent of the data generated now is unstructured, i.e., lacks a pre-defined data model.

Data production has reached zettabyte levels. Even so, 90 percent of all textual data breaks down to commonly used words. That doesn't make textual analysis simple – just the contrary. Because a single word can have multiple meanings, understanding its use at a moment in time can be extremely complex. Lacking the ability to fathom conceptual and contextual relationships, commonly used approaches to text analysis based on keywords and statistical or "shallow" semantics can produce wildly inaccurate and misleading results. In that regard, the development of "deep" semantic analysis is a breakthrough. It eliminates all doubt as to meaning. Whether for homeland security, defense, intelligence or commercial purposes, few do the job better than Expert System's Cogito.

The Cogito Intelligence Platform is driven by a semantic network or "disambiguation system" called Sensigrafo, essentially a big data lake like Apache Hadoop providing a conceptual representation of the target's language.

Working in English, German, Italian, Arabic and other Middle Eastern languages, Sensigrafo spans more than 1 million concepts per language, attributes for each concept, and some 4 million relationships between the concepts. In addition to common words, the network also covers detailed dictionaries for verticals including defense, homeland security and telecommunications.

Unlike a dictionary, which simply shows words in alphabetic order, Sensigrafo presents a conceptual representation of the language: a group of strictly interconnected networks forming a complex graph in which each concept is linked to the others by semantic relationships in distinct hierarchies.

Cogito itself is a software platform that pulls data from Sensigrafo to process natural language and perform deep linguistic analysis of text. Cogito uses morphology, syntax, vocabulary and semantic concepts to assign an exact meaning to each word, based on the context. The platform is equally adept at managing and analyzing text derived from documents, multimedia, video, audio, email, web pages, mobile content and social media.

Deep Semantics

To appreciate the above capabilities, consider the difference between Cogito's "deep semantics" and what traditional text analytics does. Deep analytics understands text the same way we learn to read: subject, verb, object, and what they mean. Keyword-based text analysis and "shallow semantics" done by statistical analytics typically stop at a word's common meaning.

With classic keyword-based technologies such as Google, a search for a single query might produce millions of unrelated and ambiguous results – useless to intelligence gathering. Shallow semantics as offered by Enceda, Vivisimo, Autonomy, Microsoft *Fast* and exalead bring it up a notch by applying heuristic (historic) statistical analytics and linguistic morphology.

These approaches identify words and their basic forms, but not their meaning. Compared to Cogito's deep semantics, competing products score uniformly low on key criteria such as categorization, natural language analytics, discovery and entity extraction. And when it comes to disambiguation and sentiment, the grade is zero.

Expert System's Cogito understands that the same word can have different meanings, that different words can have the same meaning, and it does so in real-time across millions of unstructured text files. For the user, the end result is speed, accuracy, and the ability to take action. The "text print" produced by Cogito also serves as a strong indicator of location, and the platform provides a graphical interface to show the current and recent whereabouts of the target on a geographical map. Cogito also integrates with surveillance solutions that log the target's affiliations, past actions, and other data relevant to the investigation.

Other features of note:
- Automatic classification of multilingual text content.
- Built-in alerts and warnings.
- Interactive data analysis and correlation.
- Normalization and storage of unstructured data in structured database format for easy access.
- Customization of rules for data mining.
- Case method analysis via deductive algorithms and neural networks.

One quibble with Expert System. The company claims that Cogito is software that uses artificial intelligence (AI) algorithms to perform cognitive computing in a way that emulates human thinking. Mention of "software" is a dead giveaway that Cogito has nothing to do with true AI. Software is rules-based. AI in its most currently advanced form – neural networks – is not rules-based, just as human thinking is not: AI tools such as machine learning and neural networks learn, adapt and (unless directed) make decisions on their own. When and if technology arrives at true emulation of Man's thought process it will work as the mind does, in flash mode.

Group 2000 Facial Recognition

Biometric identification by facial recognition has evolved steadily since the pioneering 2D efforts of innovators Woody Bledsoe, Helen Chan Wolf and Charles Bisson in the mid-1960s, with rapid advancement in the 21st century by giant tech companies including Apple and Microsoft. Understandably, the technology has generated considerable controversy, as well, with privacy advocates vigorously objecting to its widespread deployment by commercial and government entities, and technology experts faulting the inaccuracy of systems for their inability to identify people of color, and other flaws. Market leaders such as Microsoft have moved back and forth on their promotion of facial recognition, at one point touting its societal benefits, then crying for federal regulation of the technology – and even deleting its own databases holding millions of facial images due to an outcry of objections from privacy advocates.

Regardless, facial recognition is here to stay. In 2020, the best-known and most odious deployment is that of China, which uses the technology to monitor and oversee non-Han citizens such as the Uigher and Tibetan peoples. In the Western world, the use of facial recognition is equally popular with government agencies. The U.S. Department of Homeland Security maintains a database of over 100 million face prints, and American law enforcement makes energetic use of the technology.

The most advanced facial recognition solutions have moved to 3D recognition to improve accuracy, and embraced further enhancements such as use of artificial intelligence, advanced algorithmic analysis of older 2D and new 3D big data video banks, as well as the application of LIDAR (illumination and measurement of facial bone structure via laser light) that provides superb, real-time rendering of facial biometrics. One of the very finest in this space is a European vendor that goes to great lengths to avoid public disclosure of its pace-setting role, or involvement in providing facial recognition solutions to law enforcement and intelligence agencies: Group 2000 of The Netherlands.

Since late 2015 the company has offered a 3D facial recognition system both standalone and in combination with products designed for conventional intelligence work. Group 2000's LIMA Biometric Identity Surveillance (BIS) provides a significant advantage by quickly spotting targets as they move through various locales and is an important add-on to other Group 2000 tools that focus specifically on target communications and location.

It is well-known that targets frequently alter their appearance to elude capture. But LIMA BIS is never fooled, regardless of whether the suspect shaves, grows a beard, wears a hat, sunglasses or hoodie, gains or loses weight, keeps his or face averted when passing by a camera, is visible only in low lighting conditions, or even undergoes plastic surgery. LIMA BIS hones in on the target's facial bone structure, a factor that never changes regardless of the disguise adopted by an individual, and is accurate at ID'ing a face at up to 30 meters' distance from a camera. Moreover, the system captures images in 3D, which is always more accurate than legacy 2D approaches, and stores the images in data retention facilities.

Group 2000 LIMA BIS also holds value for ensuring secured access to secret government facilities. Recall that 2015's hack of the U.S. Office of Personnel Management disclosed the identities and personal data of 21.5 million government employees, including many with top security clearances. One of the worst revelations to emerge came out late in the year: government and contractor employees' biometric data such as iris scans and voice biometric voice prints had been stolen, too, rendering these measures useless for limiting access to sensitive compartmented information facilities (SCIFs) only by those with security or top security clearances.

In the aftermath, the one biometric criterion that has remained unassailable is 3D facial recognition. Hackers, or those they sell stolen data to, cannot fake a 3D facial image when trying to gain access to sensitive government information. Any out-of-sync grain of data in facial bone structure is an alert that those seeking entry are not who they claim to be.

Inside Group 2000

Based in Almelo, The Netherlands, Group 2000 has been a formidable player in the ISS and IT arenas since its founding as a private company in 1979. The company started in its home country, then expanded to Switzerland, Norway and the U.S. Even with this expansion, Group 2000 has preferred to remain a tightly-knit group of professionals.

In a similar vein, while the company sells its wares to diverse users in at least 30 nations, the homeland market of The Netherlands remains central to the company's success. Group 2000 likes to grow its clientele "organically," i.e., by building relationships with communications service providers (CSPs), law enforcement and intelligence agencies who might start with one product and subsequently expand to others. To date, this business model has proved a tried and true formula.

Initially, Group 2000 focused purely on hardware development before evolving to add software solutions and systems integration services. In the ISS arena, the company offers both hardware and software solutions, and its portfolio is quite extensive: active mediation devices; passive probes for IP and VoIP; probes specific to email collection; warrant management; mobile location devices; social media monitoring; data retention; 3D facial recognition; and, "blended" lawful intercept systems that might include any or all of the aforementioned products. On the defensive cyber side of the business, Group 2000 offers an Active Defense network solution, as well as one specifically designed to protect SS7 from intrusion.

Of note, Group 2000 steers clear of the controversy surrounding IMSI catchers. The company does, however, offer the option of solutions that detect IMSI catchers or signal jammers that might be on the prowl in a government, enterprise or commercial network.

By following a modular product development approach, the company assures that the client can purchase products in the knowledge that each can integrate with and support others.

What Makes Group 2000 BIS Special – A Key Partner

Let us begin at the top with 3D BIS facial recognition augmented by AI. Group 2000 is hardly the first ISS vendor to enter this market. NEC NeoFace, for example, relies on a combination of artificial intelligence (AI) and big data analytics to perform detailed facial recognition and is used by agencies in 40 nations. That said, one advantage of acting as a follow-on to a market giant such as NEC is that an ISS vendor can leverage the very latest advances, which Group 2000 does to perfection.

The key to Group 2000 facial recognition's remarkable degree of accuracy is LIDAR and video technology provided by Digital Signal Corporation (DSC), based in Northern Virginia outside Washington, D.C. That fact alone sets Group 2000 apart, as DSC is known for partnering with only a select few commercial enterprises. In the case of Group 2000 BIS, the end result is a winning combination.

First developed in the 1960s, LIDAR (a coined term combining "light" and "radar") has evolved to be widely used as a technology for creating high-resolution aerial maps. LIDAR is highly desirable for this and similar purposes owing to its ability to accurately measure distance and altitude via reflection of a laser image and radar soundings from the earth's surface. The laser provides an exact 3D image, and radar measures distances or depth across that surface. LIDAR is also known as a valued application for 3D or laser scanning.

In a facial recognition application, LIDAR sends a series of laser bursts across a crowd of people in quick succession. The technology then creates a "map" of individual faces, with highly accurate image resolution, gathering more than 40,000 data points per facial image. LIDAR renders a precise facial image, enabling users to accurately correlate this data with images of known targets.

If the client has already invested heavily in older 2D video facial recognition, Group 2000 can improve the accuracy of an extant 2D image database by providing additional detail.

Group 2000 OSINT and SOCMINT

Group 2000 augments its facial recognition capabilities with LIMA Social Media Insights, a robust monitoring system that leverages big data analytics in real-time to glean OSINT from the vast world of unstructured data on Twitter, Facebook, Instagram as well as newer and rapidly emerging social media vehicles. Group 2000 claims the product is "unique" for being an end-to-end solution "in a box" that works from the point of discovery (identification) through data collection and analytics, then provides graphic visualization of the data. Uniqueness might seem a difficult claim to rectify with the presence of other competitors in the marketplace. Is Group 2000 justified for so doing? Well-known peers:

- Taipei's Decision Group offers its E-Detective, which uses deep packet inspection to collect a target's identity, profile and location, but only on Twitter and Facebook.
- Italy's iPS provides its G-SNAKE product to intrude into social media networks.
- The LexisNexis Accurent Social Media line, which includes Accurent LE and Accurent LE Plus for Facebook, Twitter and "other" social media, locates targets and takes control of their social media.

But overall, compared to its competition, Group 2000 would indeed appear to be a step ahead by offering discovery, collection, big data analytics and visualization. The solution IDs a target, tracks his or her activity on various social media, then applies analytics to reveal the target's networks of operatives – including hidden "influencers," ghosts and mediators whose involvement in illegal activities might not be immediately evident. It shows relationships and patterns that point to pending attacks, and geolocation of targets and their connections. All data is presented in brilliant visualizations. Finally, Group 2000 LIMA Social Media provides support in multiple languages.

Knowlesys KIS: Chinese OSINT

When the Peoples Republic of China needs to monitor its citizens it turns to Knowlesys KIS (Knowlesys Intelligence System), a solution that can glean insights from any website, social media vehicle or news engine. The Peoples Liberation Army as well as local government bureaus in China also take advantage of the powerful Knowlesys KIS platform. Building on its internal success, Knowlesys is on the move to add new government clients outside the borders of China.

Knowlesys has a proven solution and an impressive client base to attest to the success of that solution. While U.S. government intelligence agencies would never give Knowlesys a second thought, other nations are less concerned. Knowlesys sells aggressively in Europe and the Middle East, and the commercial versions of its solutions have found ready homes in Western enterprises such as Virgin Media, Ingersoll Rand and Behringer, which do business in China and use Knowlesys products to measure customer sentiment.

At the most basic level, KIS monitors public opinion on the Web to help companies or government agencies immediately spot issues that could turn negative. Moving up the hierarchy, KIS can help its users intervene with customers having a bad experience with products and take measures to improve satisfaction. On the cyber front, KIS can spot hostile nation state, terrorist and criminal activity revealed in the Dark Web's cyber realm so that agents can take immediate action. Or, it can be a vehicle for singling out websites or other online information sources for censorship.

KIS is a real-time analytics (RTA) solution for the biggest of big data fields: the fast moving, ever-changing realm of SOCMINT and other OSINT. The functionality of RTA systems is fairly universal. Knowlesys breaks it down into three parts: Data Extraction Engine, Analysis Sub-System, and Presentation.

Intelligent Data Extraction

The KIS Data Extraction Engine uses a web crawler, html parser and topic detection to scan one or 10s of thousands of websites and social media platforms. Users configure the engine with key words and the system gets to work, finding specific results that adhere to the search criteria. Literally any and every site is up for grabs: Twitter, Facebook, Instagram, as well as bulletin boards, email, chats, news sites, blogs, replies to posts and records of access to search engines. Agents can target specific "authors" and the topics they focus on. All captured material includes its dateline to ensure the data is timely. KIS also can take screen shots of important pages.

KIS is compatible with AJAX, supports popular protocols including HTTP, HTTPS and FTP and any format. The system performs various functions automatically: collection of full content without limit to pages; next page collection; download of images and attachments; and "snapshots" of original images.

KIS stores all data collected in a centrally managed relational database, and is set to automatically avoid collecting data that has already been gathered, a de-duplication feature that saves on storage costs and time spent analyzing data. The engine also deletes irrelevant data such as advertising, copyright notices and column formatting.

Agents need not watch the entire performance. Data Extraction analytics runs 24 X 7, and requires no attendance. All data is classified by category so that agents can select relevant content for analysis.

Rapid-Fire Analytics – Language Agnostic

Post-extraction, stored data undergoes analysis for classification and clustering based on parameters set by the user, beginning with keywords. Parallel processing delivers real-time performance. KIS analytics system can hunt for up to 10,000 key words in a 30,000 word article – or set of articles – in under seven milli-seconds.

The user can weigh in any time, adding new topics, keywords, trends or targets of interest based on visual analysis of a specific website. If a search turns out to be a dead end, deletion of that search is simple.

The user can direct KIS to drill down and look for more. For example, the agent might ask KIS to find the target's full contact info, location, phone number and e-mail address, as well as the account names for specific services such as Facebook, Twitter and Skype. KIS can then create a summary. Alternately, if the new "hook-ups" are irrelevant or harmless, they, too, may be deleted.

Language is no barrier. To name just a few of the languages covered by analysis: Arabic, English, German, French, Japanese, Korean, Chinese and, not surprisingly given China's active repression in the Xinjiang Autonomous Region, the Uighur tongue. Uighurs who follow conservative Islamic doctrine are followed closely by the governments of Turkey and Kazakhstan, as well.

KIS Presentation

As in all RTA systems, new intelligence findings are added to the mix continuously, and key nuggets rise to the top. Agents can program the system to deliver SMS messages for red alerts. For day-to-day operations, KIS automatically prompts the user on topics of interest to national security or crime-fighting, with all assessments of the data based on refined statistical analysis. Any negative information in social media posts is displayed in the exact bulletin boards, titles, texts, along with posting time, view counts, number of comments, and the IP address of the individuals who post and comment.

Because a client might use KIS across multiple departments, authorized agents throughout a law enforcement or intelligence agency may access the system for the latest updates on their specific fields of interest, independent of one another. KIS also facilitates collaboration and alerts the correct authorized users when the data indicates that individual searches overlap, revealing scenarios of inter-related crime or terrorism.

If the agent wants to see more he can do a search in the database by any parameter such as topic, type of social media or other OSINT and keywords. He can then classify the document and add notes for subsequent analysis. In exigent circumstances, the agent can add fresh data from other sources.

Knowlesys KIS provides access to the list of all foreign websites blocked by China. The solution can also provide Chinese law enforcement with the IP addresses of all those who try to access the forbidden websites.

Kofax Dark Web Monitoring

When Lexmark completed its acquisition of Kofax in May 2015, no public mention was made of Kofax Deep Web Harvesting, an invaluable OSINT tool used by military, government and law enforcement agencies worldwide. But Lexmark's acquired OSINT asset is alive, well and going strong.

Kofax Deep Web Harvesting remains an important asset to its hundreds of clients, including the U.S. Department of Defense and Intelligence Community. Adding to the appeal of Kofax is one particularly attractive feature: partnership with Palantir, enabling users to see brilliant visualizations of findings from Deep Web Harvesting.

In the 20 years preceding its acquisition by Lexmark, Kofax and its solutions experienced several iterations. The core technology began life in the late 1990s as a data integration platform for Kapow.net, a Danish e-commerce company. With over 5,000 companies selling merchandise on Kapow.net, data integration proved extremely beneficial for extracting, cleaning and normalizing data. But the company was not to stand still for long. In 2001, Kapow.net was acquired by a bank. Then in 2005, venture financiers funded a third generation company, Kapow Technologies, based on the same platform, but this time selling data integration as a capability unto itself in support of another purpose: extracting data from the Deep Web.

The product was comprised of a software engine (Kapow Mashup Server) and integration platform (RoboSuite), both with a new thrust: harvesting public and private web intelligence, the latter a reference to unindexed content, a domain that in time would become known as the Deep Web. By 2007, following a fresh round of funding and a move of corporate headquarters to Palo Alto, CA, Kapow was used by more than 200 clients including law enforcement and government intelligence agencies.

Rebranded three years later as the Kapow Extraction Browser and Kapow Katalyst Platform, the Mashup Server and RoboSuite together were promoted as "the first and only web browser specifically designed to extract and load data from any layer in the application stack – presentation layer, application logic layer or database layer - without requiring application programming interfaces (APIs)." To do so, the platform relied on a "Synthetic API" that eliminated the need to create a unique API for each data source searched, a significant time and cost savings for users. Pleased with that outcome, customers lined up, more than doubling Kapow's base to 500 clients.

A further rebranding as Kapow Software occurred just prior to the company's acquisition by Kofax. Kofax continues to market its talents at important events such as ISS World Europe, where "Kofax from Lexmark" reps hold forth on harvesting the Deep Web and Dark Web.

Not Just Web Scraping

"Kofax from Lexmark" makes a point of distinguishing itself from competitors who only do "web scraping," that is, browsing the Web, finding sought-after unstructured Web data, and transforming it into structured data that can be analyzed. In industry parlance, this part of Deep Web browsing is known as a "broad crawl." Kofax does that and much more.

At the broad crawl stage, Kofax vacuums big data in bigger doses from Web pages, Word and PDF documents in Word, non-HTTP files such as FTP, as well as video, audio and images. Then its Extraction Browser takes over, using a full Javascript engine that powers its "Katalyst" platform to view and sort all dynamic changes happening in a web page as they occur, picking the desired data while jettisoning the rest. The solution relies on multiple servers to ensure scalability, whatever the load.

The very format of Web pages poses a significant challenge due to widespread use of Asynchronous Javascript and XML (AJAX), wherein various Javascript programs operate a Web page as it loads. Some Javascript programs might load files, while others alter the page, facilitate analytics or load visitor-specific advertising. Typically, each program stems from a different server. The variety of changes in a Web page happening in real time and from different sources can make it difficult for a browser or scraper to absorb all the content or access the exact data specified in a query. But not for Kofax Katalyst, which manages changes in real time.

What about the immense count of data hiding behind apps without API? In a broad search, a scraper can quickly clog and slow down at this point. Result: Important data often goes missing. The deeper the search posed by a query, the faster the problem mounts.

Kofax has a clever solution. Where the engine finds APIs it deals with them in real time. Where it doesn't find them, the "Synthetic API" kicks in, creating one-off applets, or as Kofax calls them "Kapplets," light-weight apps that bore through and obtain the desired data via a simple point & click interface.

After the broad search, the Kofax solution turns "micro." As during a broad crawl, the system must be capable of pulling data from any Web source, whether indexed or non-indexed. The capture of context during the crawl takes on a crucial role in order to determine all possible relationships relevant to a query. Perhaps a target was using Twitter in the same time frame as an event. Contextual data fed into analytics tools might reveal the existence of a previously unknown affiliate who was in the general vicinity of the event and posting relevant content. From there, Kofax might discover an entire network of perpetrators and activities represented in heat maps, timelines and graphs illustrating all primary or secondary level relationships.

The visualization function is conducted with help from Palantir. When the "actionable intelligence" of analytics functions is presented textually, analysts might miss key relationships that are essential to understanding the scope of a problem. But when the same findings are laid out pictorially, they become immediately clear.

Another plus is that the solution can "anonymize" an agent's search to avoid tipping off the target. Among the techniques used: changing HTTP headers between browsers, varying the schedule and frequency of crawls, and working without APIs to avoid registering access. The solution integrates with other tools that ensure anonymity.

It also works in any language including bidirectional Arabic and Hebrew and the multi-bit codings of Chinese. Kofax technology is the basis of platforms developed in collaboration with other tech companies for use in China, India and Afghanistan.

Data captured may be stored and retained in any volume for future analysis. As a backup – and to avoid returning to the data for another crawl – the company offers Snapshot, an exact copy of a site and all its records, including affiliated sites and assets.

Once a crawl is completed and stored offline, the process is repeatable by any analyst approved to do so within the client organization, whatever the location, in the office or on the road.

One of the first companies to make its solution mobile friendly, Kofax Deep Web Harvesting works on any smartphone or tablet. As an offshoot of the mobile revolution, the BYOD movement rightly raises security concerns. However, because the Kofax solution operates in stateless mode, it retains no data for hackers to steal after the fact.

Kofax Versus BrightPlanet

Competitor BrightPlanet maintains that it alone merits the mantle of Deep Web Harvesting, which it defines as the ability to capture "whole text" of content "not just HTML" from unindexed sources known to elude search engine spiders. BrightPlanet argues that firms such as Kofax are merely extraction companies that "use the structure of the web page" to deliver intelligence the agent seeks.

In its critique, the company compares its own versus Kofax's version of a web page data capture. Brightplanet's is whole-page text. Kofax's is in a spreadsheet categorizing key findings. Is one better than the other? No, they simply present the findings differently – BrightPlanet in toto and Kofax broken into its elements.

Is something lost in the Kofax approach? Not at all. The data has simply been put through a first stage of analytics. Kofax retains a complete view of website and affiliate pages that can be converted to whole text.

BrightPlanet's argument may be cutting it fine. In the lexicon of Deep Web search, "harvesting" and "extraction" tend to be used interchangeably. In the past, Kofax has routinely stated that it "harvests" data by "extracting" it, thus the quibble over definitions of Deep Web Harvesting is moot. One feature that helps set Kofax apart from competitors: Palantir visualization.

NEC NeoFace Facial Recognition

The uptick in lone wolf attacks has been a wake-up call for nations that once frowned on electronic surveillance. France, which dismantled Bull Amesys and forced companies such as Qosmos to go quiet on the surveillance apps of DPI, did an about face following the 2015 Charlie Hebdo attack, embracing vigorous new national security laws that favor surveillance.

Other countries are reviving the idea of national identification programs that require every citizen to obtain biometric IDs that include fingerprints, photos and other personal data stored in a universal ID (UID) system. India and South Africa are among the countries that now have adopted UIDs.

One problem: The lone wolf's alignment with a menacing group doesn't mean he shares their affinity for social media and the Web. Too often the subtle trail he leaves for SIGINT, COMINT or ELINT is discovered after the fact. Another: The lone wolf isn't pausing for a fingerprint or iris scan en route to his victims.

An alternative showing promise is the field of biometric facial recognition technology applied in video surveillance systems. Used in conjunction with capabilities that quickly search vast databases for a match, facial recognition holds the potential to single out targets that might not otherwise be recognized before it is too late.

A standout performer in biometric facial recognition technology is NEC, whose NeoFace Watch system is among the fastest, most accurate solutions available.

Where NEC Fits in Facial Recognition

NEC is the Nippon Electric Company, in operation since the late 19th century, and with an estimated US $24 billion in annual revenue, one of the world's largest diversified electronics companies. NEC has been in many businesses over the years, from early telecom equipment and the first digital signal processor to its current emphasis on network solutions and electronic devices including mobile handsets, where it takes credit for advances in color displays, 3G support and cameras.

NEC was not the first to venture into facial recognition technology. That honor goes to American mathematician and computer scientist Woodrow Bledsoe and his team of investigators, who in the mid-1960s pioneered the use of computers to automate identification of images of the human face. Bledsoe and fellow researchers developed algorithms that could use points of the face as coordinates that mapped the face from one feature to another, done at the speed of 40 images per hour. Early systems counted on the subject to sit still and look straight into the camera. Results could be skewed by factors such as the angle of his head, facial expression, or lighting – challenges that hampered progress in facial recognition for years to come.

Gradual improvements emerged through research conducted at Stanford Research Institute, MIT, the University of Southern California, and at government and military intelligence agencies. By 1997 a system developed in Germany had the ability to recognize faces despite facial hair, eyeglasses or changes in hairstyle, and became widely used by banks and airports. Ten years later Minnesota-based Identix introduced FaceIt, a program that could identify a single face from a vast global database.

In January 2015 computer scientists and Yahoo Labs in California announced the ability to sort through millions of images and precisely identify the target from any angle, even when partly blocked – or upside down. Yahoo claimed this as a "first" based on the use of a "deep convolutional neural network," a type of artificial intelligence. However, detractors were quick to point out the existence of other systems that already achieved the same feats, also via AI technology.

Like many another system that leverages machine learning, facial recognition has a long history marked by controversy. Through it all, from the late 1980s to the present day, NEC has distanced itself from debates, quietly concentrating on R & D. The culmination of that effort is NeoFace Watch.

NeoFace Watch by the Numbers

Since its introduction, NEC NeoFace Watch has grown to become one of the most popular biometric facial recognition technologies in the world. Customers include law enforcement and government agencies as well as commercial customers in over 40 countries.

Performance is the key to NeoFace's success. In the first exhaustive tests conducted by the U.S. National Institute of Standards and Technology (NIST) in 2014, NEC's product excelled all others. Testing involved an elite group of 15 commercial enterprises and universities from the U.S., China, Japan, France, Germany and Lithuania. In addition to NEC, the systems tested included algorithms presented by 3M/Cogent, Cognitec, Safran Morpho, Toshiba, HP/Virage and Ayonix.

In NIST's most recent test, whose results were published conducted in November 2018, the agency reported vast improvements in facial recognition with the entry of major new participants such as Microsoft.

However, the results showed that NEC NeoFace continued to lead the market in technical innovation, and the solution was once again judged the fastest and most accurate across all benchmarks and conditions.

Multiple Algorithms Plus Metadata Search

To fathom NeoFace Watch's capabilities it is important to understand what the solution is – and is not. NeoFace is not simply a video analytic content (VAC) system, though separately NEC offers a VAC solution in its Behavior Analyzer software, which provides all the usual functions associated with such products: motion detection, suspicious objects and persons, intrusion into an off-limits area, vehicle monitoring on highways, crowd counts and detection of a fallen or injured person in the crowd.

Neither is NeoFace Watch a camera array, though it certainly integrates with advanced video surveillance systems. NeoFace Watch is something more, a unified AI/big data approach to driving optimal value from surveillance video. The solution comes in two parts: (1) a software platform with sophisticated algorithms precisely geared to facial recognition; and (2) a metadata search system that plows through video data in real-time to match a target's face with its owner from vast troves of video surveillance footage.

In the field, the software can tie-in to one or multiple camera feeds, and follow rules that automate the extraction and processing of thousands of faces, learning and evolving as it goes. If it finds a suspicious match, NeoFace immediately issues an alert.

NeoFace is governed by three overarching sets of algorithms that move facial recognition from the general to the specific in real-time:

- **Generalized Matching Face Detection.** GMFD relies on a neural network algorithm called Generalized Learning Vector Quantization (GLVQ). This algorithm sets up the recognition process first by generating a match with the eyes of the target, then moving on to select the most likely faces in the database that provide a match. GLVQ is immune to disguises such as new facial hair or the sudden absence thereof, or hats, eyeglasses and sunglasses.
- **Perturbation Space Method.** To overcome the problem of identifying a face presented at different angles, NEC developed the PSM algorithm. PSM converts the two-dimensional image in a still shot into three dimensions, then – like CGI – "morphs" that image into various poses, turning the face left-to-right, right-to-left, and tilted up and down at different angles, or even partially occluded. Finally, PSM tests diverse

lighting scenarios on the face, giving a range of looks that create a diverse but very specific "face print."
- **Adaptive Regional Blend Matching.** What if the target grins, frowns, smirks or simply blinks? In older and less sophisticated facial recognition technologies, changes of expression can throw off the system. NeoFace Watch's ARBM algorithm adapts in order to recognize such variations as "local" to the overall data presented by the face map, and still do a match between the new image and its partners registered in the database.

Once a target's image is captured – say, at a military gate or airport terminal – NEC's metadata search system goes to work, performing automated real-time search across the user's video database of captured images, or allowing input from other image databanks. The system can also factor in input from the user including clothing, sex, age or hand-typed descriptions.

Phonexia Voice Biometrics

When ISS World-Europe convenes in Prague every June, one company usually presenting is the Czech Republic's own Phonexia, a provider of voice biometrics and speech analytics. Phonexia is a spinoff from academia, the Brno University of Technology (BUT). In the late 1990s, the college's electrical engineering and computer science faculty took an active interest in speech technology. That idea grew into a program managed by the school's Faculty of Information Technology or FIT. Dubbed Speech@FIT, the group quickly developed a following – and significant funding – for its expertise in speaker and language identification, speech recognition, and keyword spotting.

Among Speech@FIT's backers: the Czech Defense and Interior Ministry and the U.S. Defense Advanced Research Projects Agency (DARPA). The group still maintains close contacts with these organizations and with institutions of higher learning focused on speech technology, regardless of locale or politics.

In 2006, Speech@FIT made its first commercial spinoff, Phonexia, which today provides speech analytics to military and law enforcement agencies, as well as enterprise customers. Not surprisingly, given their pedigree, all four members of the Phonexia management team hold advanced degrees from BUT. The three co-founders also maintain an active role in advancing research not only at BUT, but through engagement with other academicians and scientists. As an example, Phonexia has worked with Johns Hopkins University, the National Science Foundation, Google and others from BUT on the development of Kaldi, an open source tool for speech recognition research.

The close ties at Phonexia between commerce and science are an absolute benefit to the customer of voice biometrics solutions. Rather than cadge another's work, Phonexia does the research itself and with colleagues.

The basis of Phonexia Speaker Identification is the Guassian Mixture Model (GMM) combined with algorithmic vectors to generate a small but precise voiceprint of the target. Accuracy is ensured by applying NIST-tested channel compensation techniques to minimize distortion or degradation. Phonexia's Speaker ID product is compatible with a wide variety of audio sources, making it suitable for law enforcement and government agencies.

Once a Phonexia voiceprint is created it can be used to quickly identify the target against other voices. Phonexia Voice Identification can single out one voice print from a pool of 40,000 in 0.313 seconds from start to finish. The product works on Windows or Linux platforms and can be integrated with software that presents visualization of the target's voice print.

Phonexia can distinguish the target's gender, recognize him or her regardless of device used, or language spoken, and in multiple dialects. The company offers tiered versions of its products tailored to meet tactical or large-scale strategic needs.

Phonexia features for law enforcement and government agencies include high volume analysis, automatic routing of high priority speech records, speaker spotting, pre-selection of records for analysis, dialogue analysis of conversations for semantics revealing a target's intent, and link analysis that points the way to affiliates. Phonexia also offers the ability to archive and mine vast pools of stored records, including not just voice but video, and to provide transcriptions on demand.

Phonexia's products, both enterprise and government focused, have earned broad acceptance in the global marketplace, winning customers in the U.S., Canada, Mexico, Brazil, Argentina, South Africa, most of Western and Eastern Europe, the CIS countries, Saudi Arabia, Turkey, India, Pakistan and Indonesia, to name a few. Phonexia also partners with companies that resell its products, including Netcope Technologies and Flowmon Networks, the offspring of INVEA-TECH, yet another ISS company inspired by innovative research at Brno University of Technology. All Phonexia products work for English, Czech, Russian and Chinese, with other languages available upon request.

In the Shadow of Russia

When Phonexia came into being, the USSR was 10 years dead, Eastern Europe was liberated, and hopes for democracy ran high throughout the region. After nearly a half century under the rule of the Soviet Union, the land called Czechoslovakia was free to go its own way, which it did, splitting into two countries. Western nations were eager to build bridges, not walls, hence the invitations to join NATO and the European Union, and investments by U.S. Defense agencies such as DARPA into tech companies in formerly Communist countries.

Yet, though the Czech Republic joined NATO, many see it as a half-hearted ally. The country has never allowed NATO military forces inside its border. When polled, half the populace harken back to the days when the USSR housed more than 100,000 troops there.

A former head of Czech military intelligence judges that the Kremlin's active intelligence presence in his country is now more formidable than during the Cold War.

Recorded Future: Real-Time GEOINT

Recorded Future is a prominent vendor in the OSINT sector, promoting its machine learning-driven "Threat Intelligence Engine" as the means to continuously scout every corner of the Web for offensive cyber and other threats in the making. Less well-recognized but equally important, the company's many capabilities can be turned to the needs of ensuring national security.

In addition to offering a range of cybersecurity products, Recorded Future also specializes in surveillance, showcasing an "Intelligence Services team" comprised of veterans with experience in SIGINT, network communications, malware and social media analysis, data science and geopolitical risk. The group's formidable experience is the perfect background for running what may be the one of the world's most sophisticated open source intelligence (OSINT) operations. In recognition of the company's prowess, venture capital firm Insight Partners acquired Recorded Future in May 2019 for US $780 million.

The Evolution of Real-Time Threat Intelligence

Recorded Future launched in 2009 with help from the CIA's VC firm In-Q-Tel. In record time it grew to serve "four out of five" of the world's largest commercial enterprises. In the process the company expanded from its headquarters in Somerville, Massachusetts, and opened offices in the Washington, D.C. metropolitan area and also in Sweden.

Recorded Future's core strength is the ability to quickly deliver pertinent warnings on imminent cyberattacks and other threats. It does so by delivering real-time intelligence, analytics and visualization in finite sets of data winnowed from big data.

Recorded Future stays abreast by collecting intelligence from more than 750,000 open, Deep and Dark Web sources in multiple languages. Data refined by machine learning is pumped into its Threat Intelligence Engine to produce visualizations that reveal patterns and relationships, similar to BrightPlanet's approach to Deep Web searches. However, Recorded Future goes one a step further by assessing trends that can point to likely future actions by dangerous elements. It is much more than simply Deep Web analysis.

"Deep Web" analysis connotes the ability to find sites and pages that are not indexed by search engines. Often there is a simple reason why such content is not found by Google, Bing or similar services. Much of the current content on the Web is video formatted in html and unreadable. Other content might be restricted. As often as not, what looks "deep" and hidden has simply fallen prey to corrupted code.

Deep Web searches can be helpful to investigators, for example, when they turn up records that might include court or criminal documents, or foreign language information on terrorist organizations leaking hints of imminent attacks via social media postings. Just as often, however, the so-called "gold" of a Deep Web search might be turn out to be little more than an obscure academic paper or thesis. Given the dead end results that often occur, one might legitimately argue that the many raves over Deep Web searching are exaggerated.

Focus on OSINT

Recorded Future's Threat Intelligence Engine steps it up by delivering extended analytics of OSINT. It separates content from referenced entities, enabling the user or agent to view analysis of events, timing, location, target identification, their URLs, and plans of action. The engine is comprehensive because it reviews all manner of OSINT as well as heuristic information of relevance. It then stores what it finds in a master database. Here the human intelligence aspect of Record Future intervenes via formulation of the right queries to produce the desired intelligence. In so doing, Record Future's approach hits on an important element too often overlooked in many commercial applications of real-time predictive analytics. A system might be very elegant, but if the user does not know how to cast the query, the resulting outputs may be incomplete or even useless. Recorded Future has understood the need for properly structured queries from the beginning.

Recorded Future's work is recognized by other industry leaders including intelligence visualization partner, Palantir, and the two companies often work together. In June 2011 when the Turkish government experienced a DDS (distributed denial of service) attack impacting commercial, government and military agencies, they turned to the U.S. Intelligence Community for assistance. U.S. officials called on Palantir and Recorded Future. Together they learned that the attack stemmed from the Syrian Electronic Army (SEA), and then narrowed the field to focus on key SEA hackers inside the country. The information was delivered to the Turkish government, which quickly arrested the hackers.

Over time, Recorded Future has built a track record of such results. They have also built on the original service portfolio.

Moving into Geospatial Intelligence

2016 saw Recorded Future add an important new dimension for clients through a strategic partnership with Vencor for real-time, multilingual, wholly unclassified collection, analysis, and dissemination capability that currently does not exist on the market." In essence, Recorded Future wed its OSINT capabilities to Vencore's expertise in geospatial intelligence collection to create insights at a level ordinarily accessible only by those with top secret credentials.

As Recorded Future observes, many of the most difficult intelligence challenges involve geospatial data that requires resources available only at the highest classified levels. Applying OSINT tools to the same tough cases enables clients to benefit from real-time insights at a very sophisticated level, without need of classified sources or a security clearance.

Specifically, Recorded Future's OSINT sources can be tuned to GEOINT requirements, producing analysis in 3D. The system will provide real-time alerts of threats pinned to source and impact locations, delivered in seven "high priority" languages quickly at relatively low-cost.

SYSTRAN: OSINT Machine Translation

While monitoring systems can intercept data at unprecedented speeds, merely capturing content is, of course, not equivalent to understanding it. We live in a world of 6,500 languages. Of these, 10 are spoken by half the world's population. Some 2,000 languages are spoken by fewer than 1,000 persons each. In between are another 3,000, any one of which might pose barriers to understanding by non-speakers.

The presumptive process of nearly every field of ISS – lawful intercept, mobile location, advanced analytics, malware – is translation of or to the tongue of the target. The more data transmitted and the faster the speed, the greater the risk that the ability to recognize threats, identify suspects or craft credible methods of penetrating attack vectors will diminish. No fault accrues to interception technology. The greater issue is the simple matter of language, whether we fully understand what is transmitted, or miss key details that hang on the proper understanding of a word.

The ideal in linguistic perfection harkens back to the traditional model used in developing trade craft: ISS systems, like agents, must be able to think in the language of their targets and when they go on offense, speak or write like them, indistinguishably from a native. But when a modern agent enters the big data realm of OSINT, understanding nuance composed in the dialect of social media – mountains of it in real-time – would seem nearly insurmountable.

Stepping up to the task are companies that work in the specialized field of machine translation – the use of rules-based engines, statistics or both plus neural science to deliver a spot-on accurate translation. The number of machine translation vendors has grown significantly in the years since its emergence in the 1960s. Many well-known tech brands are involved: Google, IBM and Xerox, to name a few, and a score of lesser-known players.

Among the most versatile and robust is SYSTRAN, based in Seoul and Paris.

Inside SYSTRAN

Founded in 1968, SYSTRAN is recognized as one of the pioneers of machine translation. In the intervening years, the company has stayed on the leading edge of the technology, with solutions that combine the best features of the most dominant systems: a hybrid solution that leverages statistical and neural network approaches.

Like many an ISS vendor, SYSTRAN arose from technology developed by academia, in their case, at Georgetown University in Washington, D.C. The founder was Dr. Peter Toma, a linguist skilled in the Russian language, whose life mission was to resolve near-deadly misunderstandings between the United States and Russia during the politically-charged era of the Cold War. Precise comprehension of one another's languages, he reasoned, might be among the most effective tools for pulling the two adversaries back from the brink. Dr. Toma's weapon of choice in promoting peace: statistical modeling.

U.S. Department of Defense officials held a less pacifist view. Once aware of the new machine translation capabilities created by Dr. Toma, the DoD immediately latched onto their potential to provide competitive advantage against the Russians. Soon after, SYSTRAN – short for "Systems Translation" – opened for business in La Jolla, California. The company was invited to test its wares for military officials at Wright Patterson Air Force Base. The test was successful, winning SYSTRAN its first client, the U.S. Air Force Technology Division. The assignment: translating Russian scientific and technical documents that might hold secrets valuable to military intelligence and national security.

While the quality of the translations was at the time deemed just adequate, it was sufficient to provide the insights sought by the U.S. Air Force. The ability to automate a process that delivered on the promise of machine translation – and with a vision of strong incremental improvement – proved sufficient. The Air Force would find its trust in Dr. Toma's work well-rewarded. For SYSTRAN, this win was a victory in more ways than one.

SYSTRAN's rise at this time was significant for another reason: negative overhang on machine translation resulting from a 1966 report that dismissed machine translation as premature and ineffective. The report was issued by ALPAC (Automatic Language Processing Advisory Committee), a group established by the U.S. government in 1964 to size up progress in computational linguistics, which at the time meant the use of big iron and specialized software to automate translation.

For enterprises engaged in computational linguistics, the early 1960s was a forerunner of the late 1990s dot.com boom. Euphoric hype dominated the field, attracting private investment as well as major dollars from the government. But questions arose whether the application of computing to translation challenges had really produced significant results.

ALPAC packed significant cerebral power – notable academic figures from the realm of linguistics plus the nascent fields of artificial intelligence (AI) and the quarters of research into machine translation. A group of seven ALPAC scientists went hunting for answers and came back disappointed. Their conclusion following two years of research: For the moment, machine translation was, irony intended, mostly "just talk" and not that far along. The U.S. government reacted by downsizing investment, and the forerunners of the dot.com era went belly up and into retreat. To investors, "machine translation" became akin to what pets.com and drkoop.com would be when the dot.coms went bust in 2000.

The adversity surrounding machine translation makes SYSTRAN's rise just two years after the 1966 ALPAC report all the more significant. Dr. Toma had to overcome headwinds of negative public opinion and government skepticism in order to make an impression and win back their interest, which he did. To the DoD, Toma was onto something big. Following the U.S. Air Force contract win, other members of the DoD, and then the Intelligence Community, climbed on board, and SYSTRAN was on its way as a prime contractor.

Onward and Upward

In 1986, France's GACHOT S.A. purchased SYSTRAN for an undisclosed sum, and moved the machine translation innovator's headquarters to Paris. SYSTRAN went on to add another 88 languages to its repertoire, continued to advance its platform, and to create what the company touted as the "world's most extensive language databases and dictionaries." Along the way it won another key client, the European Union.

Nothing attracts competitors like success. Other new vendors such as Korea's CSLi piled into the market. Liking what it saw in its Parisian adversary, CSLI snapped up SYSTRAN in 2014 for US $53 million, merging the two enterprises under the brand of SYSTRAN International. The deal won acclaim in Korea as the biggest tech sector event in that nation. To CSLi's customer base, SYSTRAN brought its cast of decades-loyal clients.

From its headquarters in Seoul and branch offices in Paris and San Diego, SYSTRAN International today sells in both the commercial and government sectors. It's "quals" are such that the company's government-specific solutions continue to be widely deployed by members of the U.S. DoD and Intelligence Community, the EU and other global markets.

History is all fine and well, but today's buyers are more interested in why they should single out one vendor in making a purchase decision. The short answer: In a crowded marketplace, SYSTRAN is special. Understanding why requires a side trip into machine translation.

Fundamentals of Machine Translation

Machine translation is a branch of computational linguistics, a data science for translating natural (human) language. It is implemented to provide an accurate contextual versus word-for-word translation, including the nuance of dialect, regional differences, slang, idioms, or the cultural inflections that characterize social media.

The most prominent approaches are the traditional methods of rules-based and statistical translation, hybrid versions of the two, and a fourth system that incorporates neural networks.

With rules-based machine translation, words are translated linguistically, replacing a foreign word with its most appropriate counterpart in natural spoken or written language. In approaching a translation, this method is guided by a broad set of rules and lexicons that provide an accurate understanding of how words are formed in a language, and their relationship to other words in a vocabulary. Specific rules take into account how sentences are formed, and their exact meaning depending on all these factors.

The key shortcoming of rules-based machine translation is orthography – differences in spelling, hyphenation, capitalization, word breaks, emphasis, and punctuation that can throw off accurate translation.

The other traditional school, statistical machine translation, is based on mathematical analysis of structured texts such as proceedings of governmental bodies in bilingual states like Canada, or the record of the European Parliament or the United Nations. The statistical method detects patterns in speech from millions of "like documents" previously translated by humans.

Like all such scoring methodologies based on pattern detection, the statistical method at times defaults to intelligent guesswork. The quality of translation can rely heavily on the volume of human translation of structured texts that precede it. Unfortunately, structured texts are not available for all languages, a seriously limiting factor for translating language outside the realm of government officialdom.

That scenario can quickly lead to issues for pure statistical analysis: structured texts are simply not available for all languages. Where parallel texts in multiple languages do not exist, or where verbal morphology introduces variations in meaning, statistical analysis alone can come up short.

When it first set out to perform machine translation, Google relied on SYSTRAN's statistical translation, and improved its outcomes by incorporating over 200 billion words obtained from United Nations transcripts of proceedings. But in 2007, Google rejected statistical machine translation alone and moved to a new approach, neural machine translation. This method uses neural networks, an advanced form of AI. Neural networks deploy artificial neurons, basically the mathematical equivalent of a human nerve cell, which gather information, react to it and learn as they go.

Neural networks operate unsupervised and build intelligence understanding based on prior lessons. The key advantage: Neural translation requires less stored data than a strict rules-based approach and is faster than pure statistical translation.

The statistical system that Google dropped in 2007, in favor of neural networks, was SYSTRAN. Shortly afterward, the Korean/French company went back to the drawing board.

SYSTRAN for OSINT Translation

In 2010, SYSTRAN issued a major release that, in the view of many, continues to dominate the field: a world class hybrid solution that borrows from the other three schools: rules-based, statistical and neural.

How the SYSTRAN hybrid solution with neural network capabilities operates:

- It conducts translations via a rules-based software engine, then applies statistical analysis to adjust output, as needed.
- At the same time, rules tell statistical algorithms what to do, pre-processing data to improve output by the statistical engine.
- After the fact, the rules-based engine processes the translation once more, this time to normalize the result.
- Finally, it leverages a neural network to ensure that the system learns.
- End result: a hybrid system that requires fewer data inputs, and "thinks" in a manner akin to flash mode, delivering results in real-time.

One way of looking at the 2010 SYSTRAN platform: as a prototype big data analytics solution with AI bringing up the rear. While SYSTRAN continues to evolve its platform, the 2010 version was remarkable for pre-dating market euphoria over big data and AI by at least two years.

The company's Google experience played but a minor role in driving SYSTRAN's big leap forward. With the advent of social media, smartphones and near-universal mobile penetration, the world of translation itself was undergoing a radical change. Social media would quickly spur the growth and importance of OSINT as an intelligence source and new form of COMINT.

The challenge: As the most popular form of OSINT, social media is itself a new language, or rather, array of new languages, each with its own morphology, linguistic characteristics, idioms and abbreviations, as a sub-set within a mother tongue. The amounts of data generated by OSINT are literally off the chart.

As a big data solution, SYSTRAN for government clients wields the power to handle accurate, real-time machine translation that learns across the full spectrum of OSINT, from social media to websites, audio transcripts, emails, SMS, public journals and videos' audio – recorded or live – and any captured text or discussions on finance. Some applications are familiar: DOCEX (Document Exploitation) used by military intelligence to interpret captured enemy documents; "Gisting" – ascertaining the gist or main point of targets' verbal and textual communications; counter-intelligence; border protection; terrorist financing; and money laundering activities.

Add to this list the realms of defensive and offensive cyber. SYSTRAN may be deployed as the digital "canary in the coal mine," warning of imminent cyberattacks in the planning stages. Alternately, authors of offensive cyber tools can leverage SYSTRAN to craft malware attacks that perfectly replicate the target's native tongue, whether natural language or social media-inflected, for improved success rates of vector attacks.

SYSTRAN may also be customized per client, e.g, for domain specific systems or target groups. Customization enhances the output of machine translation to include custom language pairs, and to limit allowable word or meaning substitution for a more refined result.

Taming OSINT for ready understanding in the natural language of the agent is perhaps the greatest accomplishment in a set of notable achievements. As a result, SYSTRAN remains a front-runner in providing machine translation for law enforcement, military and Intel customers.

SciEngines Custom Hardware Attacks

Off in the dark realm of hostile nation states, skilled cryptologists train initiates in ways to conceal communications via encryption without triggering detection. All the while, a company named SciEngines is helping intelligence agencies "look over the shoulders" of foes who think they're safe from discovery, and step in to thwart their plans. It's a silent battle between cryptography on the one hand and cryptanalysis on the other, a war without definitive conclusion. The answer to "Who Wins?" is another question, "On which day?" The lead changes depending on who strikes the latest punch.

What exactly is SciEngines? Based in the German port city of Kiel, they make hardware for cryptanalysis. But on occasion they mince words on the purview of this equipment. A disclaimer at the tail end of SciEngines sales literature in one breath states that the company's "products, hardware and software" may be used only for lawful purposes," yet at the same time do not include "decryption algorithms and software." Elsewhere they state that if a specific algorithm isn't ported to a SciEngines device, the company can implement and optimize it.

Strictly speaking, such verbal finesse may be cutting it fine, for if true it might constitute a technology *non sequitur*. SciEngines is and always has been a maker of hardware that enables a custom hardware attack. They don't sell empty boxes. While the algorithms and software inside may not carry the SciEngines brand, the hardware certainly accommodates every element essential to the job of cryptanalysis, including the appropriate algorithms, and it is a job that SciEngines carries out with perfection. The disclaimer about SciEngines hardware not coming with built-in "decryption algorithms and software" means: Users have the flexibility to choose their own, depending on specific requirements.

That is good news for military and government intelligence customers that rely on SciEngines products to support the missions of public safety and national security – particularly as the operative words describing SciEngines products are "powerful," "affordable" and "programmable" (flexible).

Confusion persists over the definition of "cryptography," which is the broad field of writing and solving codes, and "cryptanalysis," which is strictly confined to the latter. Make no mistake: Cryptanalysis is the science of breaking into encrypted communications. It may be performed by mathematical analysis of the algorithms used in encryption, or by "side channel" methods such as man-in-the-middle-attacks (MITMs) that find and exploit vulnerabilities in the manner of an algorithm's use.

The term "cryptanalysis" was coined in the 1920s, derived by combining the Greek for "untying" and "hidden." In the wrong hands, any such uncontrolled prying into the secrets of others is hacking, pure and simple. But in the right, meaning "lawful" hands, cryptanalysis is a vital tool in the battle against crime and terrorism. SciEngines specializes in the "custom hardware attack" variety of cryptanalysis.

In cryptanalysis, the custom hardware attack is performed by brute force. It relies on massive parallel processing: numbers of computers lined up to test one key after another in millions of simultaneous calculations via analysis of cryptographic algorithms. The longer the key, the greater amount of processing and time required to crack it. As cipher keys grow more sophisticated, so does the demand for faster processing.

One method for accelerating performance is to leverage parallel processing via special boxes outfitted with application-specific integrated circuits (ASICs), made of millions of logic circuits (gates) built into printed circuit boards. Each gate might be put to the test against the same algorithm, or alternately to analysis of multiple ones. Cryptanalysis based on ICs keyed to specific applications is highly proficient but also expensive. The cost of circuit design, plus outlays for energy consumption, cooling and, of course, space for the computers themselves, adds up quickly.

Increasingly, cryptanalysts are resorting to an option that is less costly, more flexible and energy efficient, cheaper and very high speed: field programmable gate arrays (FPGAs) that can be quickly tailored "on the fly" to address fresh encryption challenges. SciEngines was one of the first to use FPGAs in hardware designed for cryptanalysis, in its COPACOBANA (Cost-Optimized Parallel Code Breaker) S-3000 released in 2007. Now retired from service, COPACOBANA has been replaced by successors that are even more efficient and high performance. All support cryptanalysis via parallel processing and with comparatively inexpensive FPGAs which, because designed to be multi-purpose for rapid-fire change, deliver fast ROI.

Inside SciEngines

SciEngines itself has an intriguing background. Like voice biometrics vendor Phonexia, which got its start at the University of Brno in Czechoslovakia, and PowerSpy mobile location, which began at Stanford with an assist from a prominent Israeli defense contractor, SciEngines is an offshoot of academia: joint work by IT security specialists, engineers and mathematicians at the University of Bochum and University of Kiel in Germany. A select group of experts including Dr. Gerd Pfieffer, who still serves as CEO, set out to create a special purpose hardware system that delivered high performance at low cost by moving away from ASICs in favor of a then newly-developed alternative: "programmable" ICs, today known as FPGAs.

At the time, ASICS, though without doubt a superb solution for cryptanalytics work, came with a significant downside: high non-recurring costs that placed serious cryptanalytic endeavors out of reach for commercial and research institutions, let alone intelligence agencies on a budget. Dr. Pfeiffer and team sought to bring costs down to earth. End result: the COPACOBANA project, which debuted in 2006 in workshops held by SHARCS (Sensor Hosting Autonomous Remote Craft) and CHES (Cryptographic Hardware and Embedded System) conferences.

COPACOBANA began modestly. It was designed initially for simple cryptanalytic brute force attacks on algorithms such as the Data Encryption Standard (DES) created in the 1970s, a standard that was updated over time, eventually abandoned as "insecure," then replaced by the Advanced Encryption standard in key lengths up to 256 bits.

COPACOBANA would attack encrypted traffic with what was deemed "high computational complexity" for its time, and with minimal requirements for memory and communication. With parallel processing in play it was assumed that communications between computers focused on a cryptanalytic problem would be nominal, and thus that the computers would be processing full time without interference from output and input. Control of the system was to be centralized at one host, which would use a simple UBS or Ethernet interface with FPGAs directing the computation.

With memory requirements for "simple" being finite, given the scope of test attacks, the system could rely on the resident memory within the 120 FPGAs deployed per COPACOBANA UNIT, and as a bonus, parallel processing could be connected to multiple COPACABANAs. For its FPGA, the academics selected the Xilinx Spartan 3-1000, each equipped with 1,000,000 "gates" or logic circuits.

One drawback: Latency. The COPACABANA was slower than ASICS-based systems in circumstances where a new application was to be scheduled. In that event, the control host had to ping all FPGAs connected to parallel processing to assess any changes or new data. The more complex the encryption challenge, or greater diversity of same, the more frequent the need to survey each FPGA for change, and the slower the response time.

With the spin-off of SciEngines GmbH in 2007 from academia to a privately held company, what began as a project morphed to a commercial off-the-shelf product, the COPACOBANA S3-1000. SciEngines went into high gear. Just months after opening its doors, SciEngines debuted the S3-1000's replacement, the COPACOBANA 5000, followed by the RIVYERA S3-5000 in 2008, the first to use a 100 percent API framework for communications. Immediately afterward came RIVYERA V4-SX35, packed with 128 FPGAs in a high performance computing (HPC) cluster.

Time raced on. New and improved versions of the RIVYERA were brought to market in 2009 and 2010. Then in 2011, SciEngines launched what they claimed was the first cryptanalytic vendor offering a Xilinx Virtex-7 2000T based computer with an 80 Gigabyte distributed memory, plus the option of shared memory at 128 Gigabytes, which together represented a significant step up in storage of data for analysis. These were impressive advances in a condensed time frame, but SciEngines never lost sight of its natal memory. Amid one fresh advance after another, the company stayed true its commitment to accelerate power and flexibility without sacrificing affordability.

Even in the workshop stage in 2006, the original COPACOBANA was producing head-turning results. In a series of tests, the company's original workhorse successively mastered the cracking of DES, Norton Diskreet, ePassport used in civil aviation security, the A5/1 Streamcipher then protecting GSM mobile communications, Integer Factorization developed by RSA, and others.

Was SciEngines truly "first" in melding FPGA technology with parallel processing for decryption purposes? Some might debate that point. Hewlett Packard's Helion Enterprise division claims that it was "the first company in the world to offer commercial AES solutions in hardware" for both encryption and decryption. Did SciEngines copy Helion's model?

There are definite similarities. For example, Helion's devices leverage FPGAs by Xilinx for working with up to 256 encryption keys, just like SciEngines does. By now, the issue of who came first is a moot point thanks to the muddle HP created for its Helion division. At issue: HP's decision to diversify Helion into the cloud business – once again demonstrating the Fortune 500's common fondness for never leaving well enough alone. This lame effort at diversification blew up in late 2015 when HP vacated the cloud market, thoroughly trounced by Amazon Web Services and Microsoft Azure.

As of 2016, Helion cloud was dead and gone. By comparison, SciEngines, which has stayed small and tightly focused throughout its shorter tenure, looks the more solid of the two. SciEngines continues to innovate in its chosen niche, and remains true to the principle of "doing more with less."

SciEngines Cryptanalysis Hardware – Current Generation

COPACOBANA today enjoys an honored retirement. SciEngine's current core products are the RIVYERA S6-LX150, and the company's newest ware, the RIVYERA S6-LX150 DDS (for "Desktop Delivery System"). These are special, and deserving of attention both for what they do and for the company's decision to end the debate between ASICS and FPGA by designating these boxes as "application-specific" computing devices outfitted with field programmable gate arrays. In a word, they have combined ASICS and FPGAs.

The RIVYERA S6-LX150 is a state-of-the-art driver of parallel processing for cryptanalysis. As with all prior generations of SciEngines products, it touts extreme flexibility and budget consciousness. The S6-LX150 can be fitted with as few as 16 or as many as 128 Xilinx Spartan–6 LX150 FPGAs, though the latter is the standard model recommended. What it does: high-density supercomputing in rapid fire for "massively parallel set-ups." FPGA is all about acceleration, and with Xilinx inside, the S6-LX150 certainly delivers on that count.

Algorithms vary significantly when it comes to degree of difficulty in cracking, but the S6-LX150 is proven to scale to the performance level of 1,500 to 15,000 CPU cores, i.e., the multiple central processing units that perform computations in parallel processing. This is supercomputer power in a briefcase-sized compartment. Just ten RIVYERA S6-LX150s can provide parallel processing power on par with the largest supercomputers.

The RIVYERA S6-LX150 delivers more memory. Each FPGA offers 512 Megabits of dynamic random access memory (DRAM) in addition to Xilinx's resident block RAM, a dedicated two-port memory containing 2 – 3 kilobits of RAM. The Xilinx FPGA comes with one or multiple block RAMS that can be configured into a larger unit of distributed RAM for logic functions or look-up tables. In addition, the system can accept a Secure Digital High Capacity (SDHC) memory card, one 64 Gigabyte micro Secure Digital (SD) card, and two one-terabyte hard disks in the package.

All these capabilities are built on a standard open source LINUX operating system so that the computer works with ease and reliability.

The last and most powerful member of the suite is the RIVYERA S6-LX150 DDS, which brings supercomputing capabilities to the desktop. This desktop unit can be equipped with four FPGA cards, each holding 32 Xilinx Spartan-6 LX150 FPGAs for a grand total of 132. Alternately, the user can purchase the "starter" version of DDS, set out with a single FPGA card, and upgrade as circumstances require. There is no need to buy a new unit when upgrading. The user simply installs additional cards.

Regarding memory, DDS offers the same DRAM – 512 Gigabytes. Being portable, it is lighter on hard disc back-up than the RIVYERA S6-LX150, providing one disc, not two. The DDS replicates the performance of CPUs by the hundreds, not in the 10s of thousands.

Otherwise DDS is just as flexible, efficient, affordable, easy to set up and kind on energy use as the rack-mounted RIVYERA S6-LX150. The appeal of the DDS: It works "off the shelf" and the user can take it anywhere – to the office, to combat central, or adjacent to the field of action itself.

SciEngines products are military-grade and ideal not only for the armed forces, but also for government intelligence and law enforcement agencies. Military customers from many nations including the largest and most powerful line up to buy them exclusively as evidenced by an award from the U.S. Army Contracting Command for a sole-source purchase order of SciEngines.

S2T: Finding Foes via Cyber-HUMINT

As discussed in Chapter 4, ISS vendors and government agencies leverage numerous tools to crack The Onion Router (Tor) and other Dark Web networks. Among the most sophisticated is a methodology that creates virtual agents that can pose as human members of a cabal and thus gain access to their secrets. Among the leaders in this space is S2T.

Founded in 2002 in Singapore, S2T is actually two companies: Simulation Software & Technology Pte Ltd, which is the enterprise-focused business; and S2T Analytics, which serves the defense and intelligence communities. Both offer the same technology platforms, but with different applications based on the needs of each audience.

As the open-source Web intelligence side of the house, Simulation Software & Technology concentrates on Web crawling, text analysis and mobile data for marketing and customer satisfaction purposes, as well as benign forms of government monitoring. These various data sources are fed into a set of big data analytics and visualization apps that show trending, links and connections, with real-time and predictive capabilities. Clients are heavily weighted toward Singapore government agencies such as Parks & Services, Inland Revenue Authority, the Ministry of the Environment & Water and diverse others, though the company also counts Microsoft, Comverse and Bank of Israel as customers.

S2T Analtyics caters to the defense and intelligence markets with products that include crowd sourcing and security, data discovery, real-time analytics, detailed WEBINT reports and Cyber-HUMINT.

It should be noted that most S2T products, whether emanating from the Simulation Software or Analytics side, bear the brand name of "GoldenSpear" and in some cases are the same product, with modifications. GoldenSpear Foresight from S2T Analtyics, for example, is much the same as GoldenSpear Foresight from Simulation & Software – but only the S2T Analytics version includes Deep Web and Dark Web collection capabilities. In other instances, the products do not overlap. S2T Analytics' Cyber-HUMINT is strictly for military and government agencies.

Virtual Spying in the Dark Web

S2T is quite open about the relevance of classic human intelligence or HUMINT. Its alternative route into the Dark Web is to create "cyber sock puppets" or false identities, combined with social engineering, to penetrate sites on the Dark Web. The product they provide is S2T GoldenSpear Cyber-HUMINT.

S2T is sold to military and government agencies by subscription. With S2T the user is buying more than a list of opportunities. The complete package includes a tailored subscription supported by a range of analysts who, depending on the customer's needs, are fluent in Chinese, Russian, Spanish, Portuguese, Arabic, Farsi, Bahasa Indonesia, Urdu, Hindi, and other languages, and thus able to continuously scout Dark Web intelligence sources in the target's native tongue.

For the most part, S2T's GoldenSpear Cyber-HUMINT is geared toward ferreting out imminent threats from hackers – not surprising, given the rise of cyberattacks and the thriving business of hacking-as-a-service available on the Dark Web. Cyber-HUMINT, in tandem with other Dark Web data hunting tools, can go farther, penetrating terrorist cells and activities by hostile nation states. GoldenSpear Cyber-HUMINT features that enable this kind of activity include:

- **Virtual Spies.** Within specific Dark Web platforms and password-protected forums S2T operates virtual spies that can extract data otherwise impossible to obtain.
- **Active Operations.** The bot agents take on a life of their own, creating trusted interactions with targeted sites, e.g., participating in discussions, purchasing information – or illegal goods and services.
- **Intelligent Data Extraction.** Governed by rules programmed in, the virtual spies know what data to look for, where to find it, how to automate extraction and monitor it to ensure the validity of the results so they know they're not being spoofed by a fake site set up for counterintelligence purposes.
- **Discovery.** S2T monitors many Dark Web illegal sites to keep abreast of what is for sale.

S2T is by no means alone in the marketplace. Other companies such as Israeli-based SensCy offer similar products that use virtual "avatars" to glean intelligence on Dark Web threats and activities. For virtual spies or avatars, success hinges on the ability to get around the equally versatile security methods used by Dark Web residents and visitors.

Dark Web Security

At the most basic level, security on the Dark Web is governed by trust and relationship-building. Entering a hacking site, for example, is typically by invitation only. Those who wish to join must be recommended by someone who is already a member of the community. The invitee is then typically required to begin by establishing credibility through the purchase of low-end products and services that gradually escalate in size and price.

Based on the new member's behavior, he or she builds a ranking. By the same token, the more business a seller does, the greater his score.

The ranking approach is similar to eBay, but for criminals and terrorists. Buyers rate sellers, and vice versa. On the surface the Dark Web might initially seem an easy target for virtual spies.

However, Dark Web users like to leave nothing to chance. Sites such as DeepDotWeb provide extensive information on technical requirements for protecting user identity. One of the more popular threads published by DeepDotWeb.com is "Jolly Roger's Security Guide for Beginners," with 47 chapters on how to safeguard personal identity and thwart law enforcement monitoring. Sample chapters in this "how to" guide:

- Introduction to Tor, HTTPS and SSL.
- General Security Precautions When Posting Online.
- Combining Tor with a VPN.
- Fraudulent Private Messages.
- Learning from Others' Mistakes: Sabu Became FBI Informant and Betrayed Jeremy Hammond.
- Securing Your Account from FBI Monitoring.
- How to Connect to Tor over Top of Tor.
- How to Verify Your Downloaded Files Are Authentic.
- Obtaining, Sending and Receiving Bitcoins Anonymously.
- They Are Watching You – Viruses, Malware, Vulnerabilities!
- Another Scam E-mail – Beware.
- Hiding Tor from Your ISP – Part 1 – Bridges and Pluggable Transports.
- Why You Should Always Back Up Your Drives – Especially Encrypted Drives.

In addition to following these and other pointers, Dark Web sites often use more than one url, or change urls multiple times per day, a common capability of VPNs. These readily available Dark Web security measures pose a formidable challenge to law enforcement and government intelligence.

S2T WEBINT Reports: A Peak Inside Islamic State

Remember that with S2T the user is not just buying software or a menu, he is renting access to top-flight, multilingual technical analysts, plus product. Once retained, S2T analysts use a pair of applications – GoldenSpear Data Discovery and GoldenSpear Foresight System – to comb through social media, Deep Web and the Dark Web.

End result: actionable insights based on detailed WEBINT reports that can identify potential problem areas, and reveal the identity of perpetrators in order to counter a criminal or terrorist threat.

GoldenSpear Cyber-HUMINT virtual spies play an important role in the process, too, gaining access to password-protected sites and building trusted relationships to gain data on targets in the Dark Web. In that regard, S2T logged a major "first": the tracking of commercial activities of a Jihadist site sponsored by ISIS.

S2T cracked a Dark Web site that was used to raise bitcoin donations from U.S.-based ISIS supporters. It was the first documented case of ISIS and affiliated groups using bitcoin – and a major coup for S2T. With GoldenSpear pointed at its throat, suddenly ISIS didn't look so invulnerable on the Dark Web, where it was long rumored to be reaping not only new followers but huge bankrolls.

To its credit, the company modestly acknowledged that it had long been common knowledge that ISIS and other Jihadist groups used Tor. The fact that S2T analysts and systems were able to penetrate one Dark Web site is a certain indicator that many similar sites still remain in operation. At this point, it is no simple matter even to confirm their existence. But S2T's GoldenSpear Cyber-HUMINT, Data Discovery and Foresight are moving in the right direction for military, law enforcement and homeland security clients.

CHAPTER 7

Lawful Intercept Multi-Play Vendors

Overview

Lawful intercept (LI) is the process of using surveillance technologies to monitor targets' voice, data, texting and – depending on the legal jurisdiction – social media on commercial communication service provider (CSP) networks. Modern LI is an outgrowth of decades-old "wire-tapping," wherein law enforcement agents literally placed clips on telephone company copper wires that directly served the business or residence of a suspected criminal, then listened to and recorded voice communications.

In the United States, lawful intercept means just that. To proceed with a wiretap, a law enforcement agency must demonstrate "probable cause" to a judge and obtain a court order authorizing lawful intercept. The only exceptions to the rule are exigent circumstances where lives are at stake and the law enforcement agency (LEA) absolutely must proceed with the lawful intercept for public safety reasons. That said, once the exigent circumstances wiretap confirms the threat, and in some instances the location of the offender, and the law enforcement agency intervenes to capture the dangerous element and stop a crime in progress, the LEA is still required to go to the relevant judge and obtain a court order for the wiretap, after the fact.

Laws outlining the scope of lawful intercept vary from one nation to the next. In the European Union, most member nations's laws permit monitoring of all target communications including social media. In authoritarian states, anything goes: Laws ostensibly authorizing lawful intercept are strictly "for show," and government agencies spy on citizens, at will. One curious exception is Russia. Under SORM, Russian police are required to obtain a court order before proceeding with a lawful intercept; however, they are not required to notify the relevant communications service provider that services the target or obtain the CSP's cooperation in implementing the lawful intercept.

The United States has strict rules for lawful intercept that require a court order, require CSPs to maintain state-of-the-art technologies that facilitate lawful intercept, and require the relevant LEA to protect the privacy of the CSP and of the suspected criminal under investigation. Under the 1994 Communications Assistance for Law Enforcement Act (CALEA), U.S. law enforcement agencies – state, local and federal – are forbidden to name the CSP or the suspect. Lawful intercepts are on the clock, and must operate under court-ordered start and stop dates, which may be continued after the end date only if a court gives approval to do so. One further restriction: Because CALEA became law several years before the advent of social media, in the United States it is not permissible to monitor social media sites or the target's social media posts.

Communications service providers (CSPs) and law enforcement agencies (LEAs) in the hunt for lawful intercept solutions face complex purchase decisions. The global marketplace spans hundreds of vendors. Products can vary by feature, functionality, performance and price. Cloud-based and virtual LI systems are growing in popularity thanks to their ease of use and reasonable pricing. Just-in-Time" (JiT) solutions that are overnighted to a CSP to support court order requests remain popular for those on a tight budget, though many experts argue that JiT is technologically inadequate and leads to delays when installation falls to ill-trained CSP personnel. CSP IT engineers change jobs. All too often, when a court order for lawful intercept is issued, the engineer on hand during the CSP's deployment of its required LI solution has moved on, and his replacement is clueless on how to proceed – or even if his employer in fact has the necessary technology to support the wiretap.

Active vs. Passive

There are two basic types of LI solution: "active" and "passive" – and virtual versions of both. Regardless of what a vendor's marketing ware might have you believe about a given product's uniqueness, nearly all LI solutions fall into one category or the other. Each has characteristics that serve well-defined needs.

The very terms "active" and "passive" tend to confuse the marketplace. A newcomer might conclude that an active solution works hard while the other is lethargic. Rest assured, both do the job. But the two types of system are not interchangeable. Each is designed for specific needs and budget requirements.

The core difference:
- With active solutions, lawful intercept is performed via the internal access functions of network devices: e.g., switches, gateways and mobile base stations.
- Passive systems are overlay solutions that work independently of network equipment, using a device called a probe that, once activated, taps into designated network segments.

Both types of system perform the same function: using technology that mirrors data captured during a lawful intercept and sends it to law enforcement agencies for use as evidence in criminal investigations, without the target suspect's knowledge. Both are highly effective in helping law enforcement agencies track the communications of suspected criminals and either build a case or determine innocence. However, there are departures in how the two types of LI system work and what they cost.

Active Solution Characteristics and Operation

Active LI solutions have two components: (1) lawful intercept modules that perform the surveillance function; and (2) lawful intercept gateways, more commonly known as mediation devices, that control the LI modules. Think of LI modules as the worker bees and mediation devices as the shop foremen.

All major network hardware manufacturers make their own proprietary LI modules, software which can be installed in switches when the carrier owns or wants an active LI solution. In addition, mediation devices are manufactured by and available from scores of companies that specialize in these products.

How an active system works: LI modules are deployed and configured on switches and routers throughout the CSP's network. When the CSP or its "trusted third party" provider receives a court order for lawful intercept, they activate intercept capability via the mediation device, which provisions specific LI modules to be on the lookout for communications traffic generated by a target who is under investigation. When an LI module detects an "event," i.e., a call or data transmission by the target, it routes this data to the mediation system. The mediation system then merges intercepted events and data into the standard format required and forwards them to the designated law-enforcement agency or trusted third party.

A mediation system can be located in the same physical space as the LI modules it controls, or it can be situated remotely. Active systems are thus designated as being either "active/local" or "active/remote."

Passive Solution Characteristics and Operation

As mentioned, the passive approach uses a device called a probe, which operates independently from network equipment. The probe focuses on picking up standard protocols such as Session Initiation Protocol (SIP) and Real-time Transport Protocol (RTP – used for VoIP) and others that facilitate various modes of voice and data communications. The probe uses deep packet inspection (DPI) to capture and isolate specified communications traffic to and from the investigation target. The probe's independence from network equipment ensures extensive computational filtering and manipulation of the data, with no impact on the underlying network.

Probes can be configured to work in any network with any technology. In addition, because some equipment manufacturers do not include lawful intercept capabilities on their network hardware – or do so only as an expensive add-on – the probe in many instances fills a gap enabling the CSP to quickly get up to speed on a given nation's compliance requirements at far less expense than relying on an equipment maker's active approach.

Highly flexible, a probe can attach to multiple points in the network. Typically, a probe is provisioned the same way as a mediation system. It formats events and data according to LI standards and transmits intercepted data in LI standard messages to the designated government agency.

Active systems are typically complex, involving multiple devices and a sizable investment in hardware, as well as in software for the mediation systems and LI modules. Passive solutions using probes are smaller and simpler than active systems. However, users should not make the mistake of assuming that one is better than the other. There are sound reasons why each type of system is designed the way it is – and advantages and disadvantages to each approach that must be weighed in reaching a purchase decision.

The major issues to be addressed include scalability, flexibility, integrity, potential for traffic confusion, security, and of course, cost.

Scalability

When carriers and box makers say "scalability," they're referring to the ability of a network, device or system solution to ramp up and meet sudden infusions of traffic, usually at higher velocity. Here active/local solutions offer a definite advantage over active/remote and passive systems.

An active system's mediation device can easily handle a jump in network speeds, without modification or the need to add another mediation device. Furthermore, because the LI modules are deployed across the network, any change in network speed typically has little impact. While the LI modules will need to ramp up in such instances, too, the cost of upgrades is spread across the network and comparatively minor.

The situation differs with active/remote systems. With any increase in the number of new subscribers or network traffic, the volume of traffic targeted for interception may rise also. These factors can increase pressure on active/remote solutions, and require system upgrades or the addition of new mediation devices.

The probe is what it is. As a standalone box, it must handle the speed of a tapped network. When targeted network segments clock at 10Gb, the CSP must invest up front in a 10Gb probe at a minimum.

To date, Tier I and even Tier II CSPs with large, complex networks have typically opted for an active solution, either local or remote.

Passive solutions are newer than their active counterparts. Starting out, they were most popular with smaller CSPs. However, as probes evolve to become more scalable, larger service providers are beginning to show interest for a variety of reasons. As we'll shortly discover, the probe is almost always less expensive than its active/local and active/remote competitors, and offers other distinct advantages, as well.

Flexibility

Active systems can be inflexible. Recall that in the active approach, LI software modules are embedded in network devices and controlled by the mediation system. If there is a need to watch some different part of the network quickly, the system's static relationship to the LI functions may bog down at any moment of change. In contrast, a probe is flexible – it sits off-network and can be easily connected to different network segments, as needed.

An example of the probe's flexibility is found in VoIP networks, where the routing of audio relies on SIP signaling.

With active solutions, audio content is often seen as a data transfer between two user devices, and is not visible to VoIP equipment.

To resolve this issue, active systems sometimes resort to "forced routing," where audio content is jammed through the VoIP equipment during a lawful intercept. Forced routing makes it easier for the suspect to know he's been tapped. Passive/probe systems typically don't have this issue with VoIP audio or require forced routing that tips off the target.

Integrity and Other Issues

High integrity systems do not lose data. Active/remote systems tend to have problems with packet loss.

In active/remote systems, the LI functions in network devices often route intercept output to the mediation device via the User Datagram Protocol. UDP is not known for its reliability. It can't detect and recover a lost packet.

Another pitfall of active/remote systems: pen registers, one of the most common types of intercepts. Pen registers capture the signaling information of a communication, which includes the day/time that the user device connected to the network and shows the IP addresses and ports his device communicated with. Pen registers, as a rule, don't require much bandwidth.

Where active/remote systems fall down on pen registers: the LI functions lack sufficient intelligence. They make a black & white decision on whether to route a copy of a packet to the mediation system. If a single intercept consumes 50Mbps of download bandwidth when all that's needed is a pen-register intercept, the LI function transmits the full 50Mbps to the remote mediation system. From there, this massive file is forwarded to the law enforcement agency, where its very size renders it unusable.

The same issue undermines the effectiveness of content intercepts via active/remote solutions. In a communications realm now flooded with video, content intercepts consume massive amounts of bandwidth. The active/remote system submits every byte.

Although the active/local approach is less ridden with packet loss than active/remote systems, it is far from perfect in this regard. Problems arise with routers that sit between network equipment and the mediation system. Routers with LI functions for data intercepts have filters that decide whether to replicate a packet and send it to the mediation system. Such packets are stamped with a new IP header and routed as "normal" packets to the mediation device. These packets begin to compete with other "normal" packets. What happens when the router is overwhelmed with "normal" traffic and receives a sudden request for a high-bandwidth intercept? It drops packets. In contrast, packet loss in the independent passive/probe system is virtually nil.

Potential for Traffic Confusion

There are rare instances when a passive/probe solution falls down on the job. Case in point: traffic confusion when handling intercepts on some VoIP networks.

In active systems using session border controllers, the SBC easily identifies a VoIP call. That is not always the case when a passive/probe solution is involved. At times, the probe can't distinguish information about a VoIP call from the same information in SIP traffic.

There are times when the network segment watched by the probe duplicates signaling for the same VoIP call. This duplicate signaling may not be recognizable to the probe. Also, the SBC may pick up on key data that is not evident in the SIP signaling watched by the probe.

Security

In general, probes are more secure from attack than either form of active solution. The interface is configured to respond only to a connection from one specific external IP address, or a from a limited number of such IP addresses.

Active solutions, in comparison, present multiple openings to external attack. The foremost back door: LI functions in the network equipment. LI options in active systems are typically controlled either by Simple Network Management Protocol (SNMP), or by commands sent via UDP messages to the LI module's IP address and designated port. Because tech documentation on these devices is public, these interfaces are fairly easy to penetrate.

Of the two types of active solution, the active/remote approach is notably more vulnerable to hackers because interfaces to LI functions are by definition accessed *remotely*, in other words, through the Internet.

Cost

10 GB passive probes range in price from US $20,000 to US $50,000. Newer cloud-based versions of passive/probe systems drive the cost down. In a typical cloud CALEA arrangement, multiple CSPs have access to the system on an on-demand basis and pay usage.

Active/remote solutions are the next best deal because a single mediation system is capable of covering multiple networks and services, and can even be stretched to cover the needs of more than one CSP.

Prices of active/local systems begin in the low six figures and can rise quickly. Remember that active systems rely on two core components: LI modules throughout the core network, and mediation systems. Building and deploying LI modules involves major costs in development and support. Another element jacking up the expense: Lack of standardization, which inevitably leads to higher R & D outlays by the manufacturer. Mediation devices require significant software investment, as well as upgrades when new versions are released. As a result, the sticker price on a single mediation system might begin at US $150,000.

For service providers whose main interests are flexibility, integrity, security, and low cost, the passive/probe is considered an economical option. The lawful intercept arena is intensely competitive, thus a buyer's market. The wide array of choices available works in the customer's favor. It has also led vendors to diversify into aligned fields. Many ISS companies that began in lawful intercept have expanded their product suites to include solutions that do mobile location, OSINT, GEOINT, analytics, visualization, and both offensive and defensive cyber. Welcome to the complex and competitive world of the multi-play lawful intercept provider.

Aqsacom: Lawful Intercept Used by Verizon

As an active player in the surveillance industry since 1994, France's Aqsacom is among the oldest, largest and most respected lawful intercept vendors. The company has grown from a startup offering a single lawful intercept solution into a recognized industry giant delivering multiple standards-based products that integrate with and expand upon the capabilities of its core product. The model has worked. Headquartered in Paris, the company has regional offices in the Middle East and North Africa, Latin America, India, Australia and the United States, winning recognition as an industry standard bearer.

The proof is in the customer base. Aqsacom has won a sterling roster of clients on multiple continents. Among the most impressive is its agreement with Verizon, the largest telecom and mobile operator in the United States. The ability to boast clients at this rank speaks volumes about the trust that communications service providers have in a vendor, and by inference points to the highest levels of satisfaction by law enforcement agencies.

Product innovation, reliability and performance are key, beginning with the core offering, ALIS. Success also underscores sound and timely product development based on advances that deliver strong value to CSPs and LEAs alike. From its foundation in lawful intercept, Aqsacom has expanded its portfolio into complementary areas that include mobile location, data collection & analytics and data retention.

Yet the company's strength, and where it bases its reputation, remains tied to the lawful intercept business. For years, the company's leading product was ALIS, which Aqsacom claims was the first lawful intercept solution developed – a point its German competitors ATIS Uher and Utimaco would likely dispute.

Such assertions are familiar to those who track Aqsacom, which also credits itself with developing the first ETSI-based lawful intercept solution for GSM, as well as for wireline voice and data. A quarter century after the fact, few would dispute Aqsacom's take on history, and at this stage it really doesn't matter. The fact that Aqsacom helped pioneer the ISS industry is indisputable. In anticipation of its client base evolving to Network Functions Virtualization (NFV), Aqsacom has stepped up to provide its own virtual LI solution, VALIS (virtual ALIS).

Aqsacom Lawful Intercept System (ALIS)

Conventional ALIS is a mediation platform comprised of three parts:

- **ALIS Core**, the basic software for the system, integrating with multiple Network Connectors regardless of network or technology type – PSTN, wireless from GSM through LTE, plus VoIP, cable, Wi-Fi, IMS, SS7, FTTH, and satellite.
- **ALIS-M**, which provisions the intercept for multiple networks.
- **ALIS-D**, managing the data flow between the ALIS platform and the intercept function, whether at switch/router lawful intercept (LI) modules or alternately at the network's edge via passive probes. ALIS-D then formats intercepted data in the proper protocols for routing to designated law enforcement agencies.

Like all active LI solutions, ALIS is installed in a communications service provider's network. To make that easy, ALIS is – again, like all mediation devices – vendor agnostic, able to integrate with any type of network hardware (Acme Packet/Oracle, Cisco, Ericsson, Huawei, Nokia, Siemens and a dozen others) and with RADIUS routers that are used to authenticate and authorize IP connections and facilitate mobile roaming.

ALIS-M and ALIS-D may reside on the same or different computing platforms and perform equally well even when geographically dispersed. When the devices are so configured, data of specific types may be sent to more than one LEA, for example, in instances involving a joint investigation. All data is secured via VPN or secured FTP to prevent leaks or any compromise of the evidence gathered. Once the data is ready, ALIS provides buffering to ensure no loss of intercepted data during delivery. Throughout the process, Aqsacom is attentive to system security. Only authorized parties may use the system, and the company certifies code to guarantee that there are no "back doors" in ALIS.

One feature of ALIS attractive to CSPs and LEAs alike is "hot swapping." To ensure redundancy and flawless operation, ALIS may be configured with back-up power supplies, disk drives, CPU and network cards that can be switched out either to change applications, or as backup in the rare event of a failure.

VALIS, a new virtual (software) offering, is decoupled from the physical assets of the network. It is designed strictly for virtual networks.

Mobile Location – AGES

Law enforcement can purchase specific enhancements to ALIS that provide important capabilities. One is the Aqsacom Geolocation Enhanced Solution (AGES), which provides what the company calls both "active" and "passive" geolocation. The active version performs classic IMSI-catching by "direct interaction with specific network entities and the subscriber handset." AGES in active mode mimics a cell tower, captures the target's IMSI number, then measures the angles from which its signals emanate, and their strength, to determine direction and distance. AGES can then gather ID information on all nearby phones to locate the target.

AGES can also work in passive mode, gathering location from the target's network without extracting data from his mobile device. Aqsacom provides an AGES Mediation Server that consolidates data from diverse operators' Gateway Mobile Location Centers (GMLCs), and acts as a single point of information for all location data available on their networks. AGES essentially pulls geolocation data off the GMLCs. By obtaining the target's IP address via interception at the RADIUS router level, AGES can follow his trail to terminating points. Signaling System 7 (SS7) positions the target in real-time based on distance from a mobile base station.

IP Traffic Filtering

A product used for capturing IP traffic distinctly resembles deep packet inspection (DPI), though Aqsacom prefers the term, "IP Traffic Filtering/Classification," by which the user can see full content intercepts of the target's web browsing habits, CHAT, VoIP, SMS, email texting or other data. The trick, as Aqsacom puts it, is in making a "secure replication of all incoming/outgoing target IP traffic," i.e., silently mirroring the content. In other words, it is deep packet inspection.

This shying away from any hint of controversy betrays lingering political sensitivities in France over government surveillance. During the Arab Spring, many companies, most notably Bull Amesys, were castigated in the media for selling DPI to regimes in North Africa and the Middle East. As a result, Bull spun off its Amesys unit for a fraction of its value, an experience that made other French ISS companies gun-shy. Beginning with Aqsacom, the French ISS sector either disavowed any use of DPI, or in the case of Qosmos – a world leader in the field – stated publicly that its DPI systems could not be used for surveillance purposes, which is, of course, nonsense. DPI may be configured with whatever rules the user establishes or changes along the way, and for whatever application desired, including those in the ISS realm.

AQUMEN and CEMTRIS

Another noteworthy product is AQUMEN, an "advanced data retention" service. AQUMEN expands on lawful interception by collecting data from diverse sources: open source intelligence (OSINT), social media, public records of criminal offenses, and shared intelligence from multiple law enforcement or government intelligence agencies.

Building on AQUMEN, Aqsacom also offers a central monitoring and information system called CEMTRIS, which monitors, captures and retains data from any public wireline or wireless network in a databank.

Together, AQUMEN and CEMTRIS give lawmen and intelligence agents the ability to connect the dots between network intercepts and other relevant sources, see hidden relationships, and reveal imminent threats.

ATIS UHER: Visualizing Lawful Intercept

Think German science & engineering and a number of feats come to mind: the theory of relativity, the modern binary system, the discovery of radio waves, the first telephone transmitter and for that matter the very coining of the word "telephone," both by Johann Phillip Reis in 1861, well ahead of Alexander Bell's contributions to the field.

ATIS UHER, based in Bad Homburg, Germany, added to this roster of achievements with the deployment of one of the first analog monitoring centers in 1980, followed by a digital version in 1983. Today, ATIS serves law enforcement and intelligence agency customers through its robust Klarios suite of lawful interest and data retention solutions.

Of note, ATIS UHER is that rare company in the ISS space that has actually pared back its portfolio. Through 2016 the company partnered with Rohde Schwarz to provide RF monitoring, collaborated with Thales on the EURACAT Air Traffic Management System, and assisted EADS Germany GMBH with radar display systems. Ultimately, the radar systems went to ATIS UHER's Lausanne, Switzerland subsidiary, VoiceCollect, a specialist in voice recording equipment. ATIS UHER then spun off VoiceCollect to atcnetwork in 2017, but retains an interest in voice monitoring via its brand of voice biometrics solutions.

Not ATIS Standards, and Not Alone

Two points of clarification before we begin our analysis.

First, ATIS UHER is not to be confused with another ATIS, the Alliance for Telecommunications Industry Solutions. Located in Washington, D.C., ATIS is a trade group that produces technical standards, including those used for lawful intercept.

Second, while ATIS UHER does manufacture equipment based on its own proprietary technologies, it is not a complete solo act. The company maintains strategic alliances with partners including:
- Saab Medav GMBH, with which it cooperates in integration programs on data analytics and storage; NetOptics for passive monitoring; and IT Software GMBH for visualization.
- Rohde & Schwarz and EADS Defense and Security for application development programs; and Utimaco hardware for lawful intercept.

In a word, while a respected player in the surveillance industry, ATIS UHER primarily acts as a systems integrator, not an innovator. Because platforms invariably rely on apps customized to specific missions, ATIS

UHER plays an important role and is justified in calling itself a provider of end-to-end solutions that can be tailored to all needs.

Defining End-to-End

Many vendors describe themselves as "end-to-end" LI providers, but the claim falls apart on closer inspection. If by end-to-end one means the full spectrum of systems capabilities for law enforcement, intelligence agencies and communications service providers (CSPs), fine.

Depending on need, such a product line could encampass LI hardware, PSTN/IP/mobile "front ends" connecting to Intercept Access Points (IAPs), mediation devices, trunkline monitoring, monitoring center, analytics, GIS, storage, visualization, direct access by the monitoring center to a carrier's metadata, signal and content processing. All of these capabilities are to be found in the Klarios suite from ATIS UHER.

Klarios AIMS For Lawful Intercept Management

ATIS UHER LI products are roughly divided between those designed for CSPs on one side and LEAs/government agencies on the other – with considerable overlap.

For CSPs, the core lawful intercept solution is Klarios AIMS (ATIS Interception Management System), comprised of an Interception Controller for implementing and configuring the intercept in either AIMS LI mediation hardware or mediation devices provided by third party vendors. AIMS connects with LI software modules in circuit or packet-switched networks via front ends that receive and categorize data, then forwards the data in appropriate protocols for delivery to an LEA or government agency. All AIMS products are compliant with CALEA, ETSI, SORM and G10 standards.

ATIS UHER's deep packet inspection decodes and demodulates data traffic whether transmitted by modem, ISDN, xDSL, cable or satellite. The solution receives and de-encrypts data from the network monitoring platform, either for storage or immediate analysis. As with all DPI solutions, decoded data is presented in the format of the original source – email, browser, etc. – so that analysts immediately know the context.

Klarios AIMS Data Retention

Data storage and access continue to be a contentious issue with CSPs, which worry about cost and privacy concerns, and with LEAs, which want metadata and content right away. ATIS UHER provides options that help smooth the process and ease concerns.

Real-time data access is conducted through a network Intercept Access Point to obtain LI-ordered content and metadata. Separately, the system sends a query to ping the CSP's retained data. Data from each source is pre-processed to ensure a correlated "results set" for the user.

This final package – real-time intercept information with metadata context – is then ready for analysis, which may be done by Klarios or by an LEA's own system.

Klarios AIMS Data Retention can handle and export any data medium, and adjust bandwidth and memory depth. One further benefit: Low power consumption helps CSPs keep data storage costs low, while strong cyber protection safeguards the data and alleviates privacy concerns.

With AIMS, when the time comes for analytics, all stored data is indexed by content, metadata and linguistic criteria for easy access by a purpose-built search engine tool that finds the right data for analytics. The next challenge is simplification.

Visualizing Data Findings

Data analytics is by definition intimidating to most, the more so when it requires mastery of SQL syntax. Many an analytics scheme has gone awry thanks to an incorrectly coded SQL query. At the same time, multiple input screens often raise more questions than posed.

With simplicity and results in mind, ATIS eliminates both hurdles to deliver what matters most to the end user: detailed visualization of the data. The Klarios analytics engine is user friendly, and access is not confined to those with knowledge of database management or SQL. A simple GUI delivers the next best thing to plug-and-play ease of use where analysts input queries in clear, then see the results graphically presented on a screen.

As with Palantir and other visualization tools, the user can expand the horizons of an investigation by returning to the data and posting new queries that paint additional findings and context around the target of interest – not just identity but behavior, real-time and recent transactions, affiliates, leadership and, via social media analysis, pending activities and threats.

Sophistication Made Simple

Since the launch of its integrated lawful intercept system, ATIS UHER has built rapidly on a customer base that was initially 50 percent homegrown. They now win new clients throughout Europe, Asia and the Middle East.

Operating from a country once roiled by the 2013 revelation of in-country NSA activities, ATIS UHER emerged unscathed from controversy surrounding domestic spying. And, although its products were definitely deployed in a country subsequently turned upside down by the Arab Spring, ATIS UHER was never mentioned, escaping the censure meted out to others such as Qosmos, Utimaco and the former Bull Amesys.

BAE Systems Applied Intelligence

The United Kingdom has some 20-plus significant vendors of electronic surveillance products and services, surprising for a country Britain's size. Then again, this is the land that developed the world's first modern foreign intelligence agency (MI6) over a century ago, cracked the German ENIGMA machine during WWII – and on a lighter note gave the world James Bond. These public images aside, the array and strength of UK companies devoted to electronic surveillance, and their leadership in innovation, simply astound. The days of empire may be long gone, but the sun never sets on British intelligence.

Among the frontrunners is BAE Applied Intelligence (AI) headquartered in Guildford, UK, and a division of BAE Systems, plc – descendant of the company that made the legendary Supermarine Spitfire that beat overwhelming odds to win the Battle of Britain. *For the purposes of this analysis, and to avoid confusion, please note that use of the acronym "AI" here applies strictly to BAE Applied Intelligence, not to artificial intelligence.*

Today BAE Systems is the world's sixth largest military contractor with some US $22 billion in annual revenue and a global footprint. Its AI unit in charge of surveillance has an impressive record and credentials, as well.

AI's key characteristics: multi-continent presence; a very sizable base of engineering, IT, sales and support staff numbering 85,000; strong R & D driving innovation; keen understanding of the needs of clients ranging from communications service providers to law officers to intel agents; a robust line of products that ranges from data collection at the network level to advanced analytics at the tail end; and, when applicable, the ability to readily complement products produced by its military parent company.

BAE Systems AI, like the parent company, publishes and lives by a Code of Conduct. They actively promote the company's respect for the "right to privacy" in client nations, even when the countries themselves have no such sentiment or supporting laws. There are known instances – e.g., refusal in 2014 to support the government of Tunisia's bid for citizen-wide surveillance being one example – where BAE Systems AI has backed off a deal when it deemed the client's use of their technology as being too intrusive.

In an age when media vilify any form of intelligence gathering – even when it supports the very government that ensures freedom of the press – BAE Systems' approach to this issue is smart business. It is a preemptive PR move typical of the company's finely developed sense for being in tune with the times and the marketplace. One upshot is the complete candor with which the company addresses its communications services provider (CSP) niche, to which it sells CALEA and ETSI compliant solutions for lawful intercept, data retention and analysis.

Acknowledging up front that compliance with law intercept (LI) laws is a requirement, not a profit center, BAE Systems AI appeals to telecoms executives' need to follow the rules, but always in the most cost-effective way. Few sales pitches are better honed on CSPs' exact feelings toward LI requirements, which they view with the same fondness as "death and taxes" – an inescapable fact of life they'd just as soon forego but can't escape. BAE Systems AI sells to that mental attitude when servicing CSPs, providing LI solutions that help the telco or ISP efficiently meet the letter of the law, and be done with it.

Forty-Plus Years in the Making

BAE Systems AI's rise to leadership in what is best termed "upright surveillance" is best understood in terms of its gradual evolution over the past four decades, with a sudden spike in development the past several years. Deciding character builders en route: a series of strategic acquisitions beginning in 2010.

BAE Systems AI began with a tiny consulting firm called Smith Associates opening shop in London in the early 1970s to meet IT requirements of selected UK intelligence agencies. Building on its success, the firm moved to Surrey Research Park in the early 1980s. Then came the dot.com boom of the 1990s, when Smith branched out into the decade's IT showboat, CRM, launching commercial solutions for enterprise customers.

Some may remember from experience how miserably first generation CRM turned out, rife with errors that had the opposite impact on customer experience than the one advertised. Fortunately for Smith, the company had retained its expertise and talent base in security and intelligence. As the dust cleared from the CRM debacle, Smith re-branded as Detica, with its core focus on homeland security and what the world would eventually would come to call big data, in this case, the ability to manage and perform analytics on massive amounts of data for national security interests.

Success resumed. In 2003, Detica opened its first office in the U.S. The timing was right on both sides of the Atlantic. Government contracts soared. Detica introduced new security and surveillance solutions such as NetReveal and became part of the Trusted Borders consortium of companies led by Raytheon Systems, long recognized for expertise in advanced radars and perimeter security.

Potential buyers took note. Before 2008 ended, BAE Systems made an offer and bought Detica for $US 1.0 billion. Integrated with the acquiring company's existing in-house talent, the new company took on the name "BAE Systems Detica."

Then the real shopping spree began. In short order, BAE made a series of strategic purchases aimed at looping in highly desirable technologies that complemented and enhanced the company's resident capabilities and solutions suite. Among the top outfits BAE Systems Detica courted and won:

- 2010: Australia's Stratsec (cybersecurity).
- 2010: Denmark's ETI (full suite of lawful intercept capabilities).
- 2010: The USA's SpecTal and other divisions of L-1 Intelligence Services Group (intelligence and counterintelligence, homeland security, law enforcement and support to military operations).
- And in January 2011: Ireland's Norkom Technologies (real-time predictive analytics)

Add it up and they had assembled a formidable arsenal weighted to the surveillance side of the business. Among the capabilities inherited from this growing clan, perhaps most notably from ETI, which had been in the surveillance business for a quarter century and called itself "the world's leading supplier of state-of- the-art monitoring solutions":

- Active mediation and lawful intercept modules for network hardware.
- Passive probes as an alternative or enhancement to active mediation.
- Deep packet inspection determining types and content of targeted IP traffic.
- Normalization of intercepted data into designated lawful intercept protocols.

All would prove applicable to any network or type of service:

- PSTN, IP, Mobile, SATCOM.
- Switched voice, VoIP, mobile, SMS, social media, video.

Intercepts on any of these networks could be rendered into actionable intelligence in the present and stored for subsequent analysis via:

- Mobile location tracking.
- Metadata analytics for trending, identity plus network of target relations.
- Real-time predictive analytics to model, view and preempt target actions.
- Data retention and automatic purging per country-specific requirements.

By 2014 the company was big, with 3000 employees and worldwide presence. Befitting its image, they picked a new name, becoming today's BAE Systems Applied Intelligence.

NetReveal Suite + SIBA

BAE Systems Applied Intelligence is at once a brand and a statement of the powers they can put to work for law enforcement and intelligence agencies, most notably analytics, BAE Systems AI's sweet spot.

The company's NetReveal suite works at multiple layers allowing law enforcement or intelligence agents to sift through data, find connections, build profiles, target individuals, assess risks before they turn into threats, and take the appropriate action. Core functions include:

- **Identifier:** Applies search criteria to the target entities and their connected network data as a whole, collating against the target entities to ensure an accurate match.
- **Visualizer:** Lets investigators view complex data and associations that might not be explicit in the underlying data, and highlighting patterns or other behavior of interest.
- **Analyzer:** Provides a rich graphical user interface and query builder to identify targets, associates, terms, times and locations. Highlights the connections between the search results.
- **Real-Time Scoring and Scenario Building:** A robust engine that builds target and network profiles in real-time and models their behavior patterns, then executes potential scenarios against events.
- **Predictive:** Creates robust profiles that span "drift detection," showing changes that may impact the accuracy of a prediction, as reflected in real-time data evolving by the minute or second.

An important capability – dubbed SIBA – promotes real-time sharing of data between users to improve collaboration and performance on investigations that involve multiple agents or agencies.

SIBA enables rapid, secure transit of intelligence information to shared repositories that can be easily accessed via Microsoft Office and SharePoint. Technology agnostic, SIBA does not require additional investment in new inter-agency clouds. Customers can use existing legacy platforms to share information.

With NetReveal and SIBA, agents can fuse data from multiple sources to gain a complete intelligence picture, automatically track and score thousands of criminal networks, detect previously unidentified links and relationships between targets and events, be immediately alerted to changes in risk levels, and quickly share data across regions to collaborate and close the loop on potential threats to public safety.

BAE Systems Applied Intelligence solutions have been used to coordinate the sharing of intelligence between US, UK and other European police and intelligence agencies to track down malware attacking financial institutions. The ability to leverage cybersecurity skills, malware investigation and criminal monitoring shows the value of offering complementary products and services.

ARGUS-IS: Many Eyes in the Sky

Crossovers with BAE Systems, plc, the larger military contracting division, are equally formidable. Of particular relevance: ARGUS-IS, short for Autonomous Real-Time Ground Ubiquitous Surveillance Imaging System.

Commissioned by the U.S. Defense Advanced Research Projects Agency (DARPA), ARGUS-IS is an advanced aerial reconnaissance system that integrates hundreds of tiny cameras to create a single mosaic view of a ground area covering up to 36 square miles.

ARGUS-IS allows users to monitor targets and events and create a long term chronological video record that can be stored and analyzed in detail, providing a "life pattern" of the target.

Intended for military use, ARGUS-IS delivers "eyes-on" persistent wide area surveillance to monitor enemy troop movements or observe in tactical battle scenarios.

Multiple cameras let the user see close-ups on up to 100 detail areas while still viewing the broader view. For example, detail screens might show enemy troops setting out on an attack, while the broad view reveals one's own air strike on a separate screen.

That ability to zoom in to granular detail while maintaining the big picture seems a fitting image for BAE Systems Applied Intelligence itself. It suits their approach to the needs of customers, their investment in technology, and an overarching sense of integrity that guides common sense, and plain spoken decisions on some of the most sensitive issues of the day.

Cisco Routers' Wiretap Role

For all the public criticism Cisco levels at government surveillance, a key product – the Cisco 6500 Router – remains a bulwark of lawful intercept in the U.S. As the dominant provider of Internet service provider routers Cisco has little choice in the matter. As long as the Communications Assistance for Law Enforcement Act (CALEA) remains the law of the land, all networks that intersect the public switched telephone network (PSTN), including Internet broadband and VoIP, must be technically compliant.

By extension, the same technical mandate applies to the network equipment that carriers purchase and operate. For Cisco, as for any network equipment vendor, making CALEA-compliant network hardware is an inescapable fact of business.

In keeping with that requirement, Cisco in 2007 began publishing its "Lawful Intercept Overview" outlining how wiretaps are executed via the 6500 Router. That document, together with a description of Cisco's lawful intercept architecture, is unique in the network hardware industry for being open to peer review and scrutiny.

As products reach the end of their lifecycles, Cisco is also mindful to issue "end of sale" plans. On Dec. 22, 2015 the company issued one such plan for the 6500's line cards, power supplies and accessories, with an order cut-off date that came and went on March 31, 2016. Does that mean the 6500 itself is now history? Yes and no.

Cisco's gradual wind-down of the the 6500 is tied to the company's strategy to migrate current users of that unit to the Cisco 9500. Yet the embedded base for the Cisco 6500 of some 50,000 owners and 1,000,000 units would dictate that the device remains in service for some years to come. While Cisco stopped serving and supporting the 6500 on January 31, 2018, it is highly doubtful that all 50,000 users immediately switched to the 9500.

Cisco Routers for Active and Passive Lawful Intercept

Cisco routers such as the 6500 perform classic lawful intercept, acting as the Internet Access Point (IAP). The device may be configured either as an Identification IAP or a a Content IAP. In the prior configuration, the router is programmed to collect specific intercept-related information or IRI as specified in a court order, e.g., target name, IP address, or VoIP call agent. IRI also shows which routers are carrying the target's traffic.

As a content IAP, the router can capture actual communications to and from the target and make an exact copy which is forwarded to the service provider's installed mediation device or probe.

A third party mediation device or probe is used to provide most of the management functions of lawful intercept: provisioning the intercept on the Cisco 6500; ensuring that other relevant devices in the network are similarly configured, and converting the intercepted traffic from the Cisco IAPs into the appropriate protocols for forwarding to the law enforcement agency.

As with conventional wiretaps conducted on the PSTN, lawful intercepts using the Cisco 6500 on Internet traffic can be done in both the input and output directions, and are completely unnoticeable to the target, as well as other called parties. Because taps on the 6500 have zero impact on network performance they are completely invisible to the network administrator. For that matter, one or more LEAs can perform lawful intercepts on the same IAPs, even without knowledge of one another. New features added to the 9500 model include Wi-Fi 6 capability for reduced latency, multilingual support of application hosting for Internet-of-Things (IoT) protocols, and traffic offloading at 10 Gbps to reduce or eliminate bottlenecks.

Security is provided through two secure interfaces, and Cisco claims the ability to restrict access to intercepted metadata and content only to authorized individuals. However, that said, Cisco and other routers have in the past come under fire for including "back doors" that might allow anyone with sufficient skill to access non-targets' confidential call, data and content information.

Backdoor Issues

In her 2010 book, *Surveillance or Security*, cyber expert and activist Susan Landau raised concerns that the mechanisms of lawful intercept by definition open compliant network devices to higher vulnerability risks. The argument made was that lawful intercept solutions designed to provide restricted access to network data, as well as underlying network hardware, can just as easily be exploited by unauthorized parties such as hackers and political activists.

As evidence, Landau pointed to a pair of well-known, and by now well-worn case studies: (1) the 2004-2005 "Athens Affair" where hackers leveraged backdoors in Vodafone's Ericsson-equipped network to spy on the Prime Minister of Greece and other political figures; and (2) a similar hack occurring in the same time frame in Italy.

Closer to home for Cisco was the 2008 discovery by the FBI of fake Cisco equipment "sold to the U.S. Naval Academy, the U.S. Naval Warfare Center, the U.S. Undersea Warfare Center," as well as U.S. bases overseas and a prime defense contractor, Raytheon. Once discovered, the problems caused by cloned Cisco routers and switches were determined to be minor and confined to bugs that led to outages. However, as Landau correctly observed, the fake equipment could have been used to copy and transmit confidential data. Though authorities have likely identified the source of the equipment by now, that information has never been publicly released. Not long after, more damning revelations about vulnerabilities in bona fide Cisco equipment took center stage.

In a presentation at the Black Hat security conference in 2010, IBM Internet Security Systems researcher Tom Cross revealed that the lawful intercept function in Cisco's IOS operating system made it easy to pluck data from service providers' network routers. Multiple failed password attempts didn't trigger an automatic block or notify administrators, meaning that hackers could keep plugging away until they broke into this back door.

Cross first discovered the problem in 2008 and quickly notified Cisco, which released a patch for the failed password glitch. But many administrators ignored this warning. As a result, routers operated by large service providers remained vulnerable to breaches that leveraged the lawful intercept component of the router to hack customers' confidential data.

Another problem: Because lawful intercepts were invisible to all but authorized service provider employees – both to ensure privacy and avoid alerting targets – the process was impossible to audit. Any credentialed employee who knew the password could use it, launch a tap and see customer data.

To its credit, Cisco has always made its lawful intercept architecture public and open to peer review as recommended by the Internet Engineering Task Force. Cisco is one of the few network hardware manufacturers that follow IETF's process, thanks to which, Mr. Cross was able to discover Cisco's router flaws in the first place. Other vendors by and large ignore peer review. Thus it is more appropriate to think of Cisco's one exposure for vulnerabilities, and their rapid response, as a badge of honor versus a black mark on the company's record.

Vendors that fail to publish their architecture, and as a result cannot be publicly audited, very likely have security issues that have never been recognized. While Cisco may decry law enforcement's call for back doors that it considers infringements of privacy, the company has met requirements to ensure that the lawful intercept component of its network hardware meets the highest standards of security.

ClearTrail: Lawful Intercept and Offensive Cyber

Western surveillance leaders that follow the call of "Go East, Young Man" to India and South Asia have likely encountered a surprise welcoming committee: India-based competitor ClearTrail, well entrenched on the subcontinent and expanding outward in all directions.

ClearTrail is among the foremost participants in India's growing surveillance marketplace, and in a short time has built mass through expansion into adjacent regions including the Middle East and Southeast Asia.

In part, the company's success owes a debt of gratitude to the Mumbai terrorist attacks of 2008, which awoke India to the need for a new centralized approach to intelligence gathering. Caught off-guard once, India promptly enacted rules establishing a National Intelligence Grid (NATGRID) providing integrated access to intelligence data by eleven investigative and Intel agencies, to track and preempt future attacks.

It was an environment where home-based companies such as ClearTrail and Paladion would thrive. Western surveillance companies saw the opportunity, too, and soon well-known global vendors including FinFisher and Utimaco would open offices in India.

Stair-Stepped and Integrated

From its early days of offering an assortment of lawful intercept products, ClearTrail has evolved to a strategic, integrated approach. The end result is the ComTrail Interception Suite, whose elements include mass and targeted interception together with important adjuncts in traffic filtering and terabyte storage – and the payoff of advanced analytics capabilities either real-time or after the fact. Special capabilities include "inline" interception for censorship purposes, and offensive cyber that infects and takes control of a target's device.

The mass and targeted modules are familiar ground.

Using passive probes, ComTrail provides interception and monitoring of all voice and data services. At the mass intercept level, ComTrail may be used by nation states to intercept high-speed links in any network environment. Both metadata and content can be intercepted in the world's second most populous nation or anywhere else, and ComTrail provides storage facilities to hold every byte.

For offshore sales, targeted intercept is designed for law enforcement agencies operating under CALEA, ETSI or similar regulatory mandates wherein the LEA acts under court order to intercept traffic and content of a specified target. Here a passive probe is installed in a service provider's network, and intercepted data and content, plus the target's location, are forwarded to a monitoring center managed by the LEA.

Depending on the nation state, ComTrail can deploy probes in multiple CSPs on a distributed architecture basis to provide centralized monitoring countrywide for any number of LEAs. As with cloud communications, respective clients pay for what they use versus invest in hardware, thus benefit from systems upgrades at no additional cost. Each investigation by a separate LEA is isolated from the rest, ensuring the privacy and security of all parties.

In a distributed architecture arrangement, LEAs may also use different features that apply to their needs. For example, one LEA might wish to use filters based on keywords such as target name, while another might be more interested in IMSI numbers. A client can also select filters for social media, webmail, chat, VoIP, User Info or file transfer protocols.

Filtering "Junk" from Potential Evidence

One of law enforcement's biggest headaches in monitoring networks is having to deal with content that takes up tremendous amounts of bandwidth but is worthless to an investigation: video, software upgrades, content and games.

The ComTrail Traffic Filtering feature provides a helpful solution: pre-filtering of high-speed networks to isolate relevant data. The solution accomplishes this by analyzing links to control the types and volume of data permitted to enter a passive probe. An LEA in the monitoring center can see exactly what is transiting the network and decide what is to be accepted for review or dropped.

Filtering provides a significant capital advantage to the client by improving control over the bandwidth requirements of the probes used. This is a highly desirable alternative to the "throughput inflation" seen with most hardware. Rather than push the client into using a piece of interception hardware – merely to deal with an influx of network trash – ComTrail Filtering Solution attacks the problem at the root and dispenses with the irrelevant.

Such "outside the box" thinking carries over into ClearTrail's family of more intrusive surveillance solutions. ComTrail offers two varieties of intrusive surveillance product: its "inline interception" platform, and a malware product that goes head-to-head with FinFisher's FinSpy.

ComTrail Inline is a deep packet inspection tool that intercepts, monitors and blocks SSL traffic in real-time. Designed for carrier grade networks, ComTrail Inline mirrors, captures and decrypts SSL traffic, sending a mirror image to the monitoring center with no effect on the network. The platform can zero in on a target to execute an intercept on https.

With ComTrail Inline, the user gains complete control of SSL-encrypted traffic in a network, intercepting and even modifying communications and content, or blocking selected content to prevent criminal or terrorist attacks on data essential to national security.

ComTrail's Astra is classic offensive cyberware, providing remote control of a target's device. Astra may be uploaded via conventional methods such as physical interface or social media engineering, as well as by bot.

ComTrail calls its Astra bot the "SEED." Astra SEED has been successfully tested in breaking multiple commercial antivirus tools, after which it can do all do the usual tasks associated with a malware attack: ID the target's location, log keystrokes, do a full history of web searches, take control of the device's microphone and camera, listen to conversations in a room, control or modify content within the device, or even send fake messages from the target's device to his cohorts.

SEED is remotely activated, requires no backdoor intervention and, to date, has proven to be completely undetectable to targets. ClearTrail also has a solution for Wi-Fi: QuickTrail, their interception-in-a-suitcase solution that can be set up in a flash to break into all Wi-Fi channels, as well as wireline and IP networks.

As analytics systems go, ClearTrail's entry, ClearInsight, is top notch. Completely flexible, the system integrates with the company's mass, targeted or inline interception solutions – or all three plus OSINT – and sorts through it all to find insights from terabytes of data.

Metadata and content are stored in logical grids that can be readily accessed through keyword search. By assessing OSINT and other records that relate to the investigation, ClearInsight can turn up leads that might escape investigation from intercepted data alone, such as associates, leaders or correlations not previously inspected. Explorative link analysis and visualization make the evidence crystal clear.

Other helpful features include analysis of voice biometrics and text to pinpoint the target's language, gender, and often, his or her identity.

Elbit Systems CYBERBIT Multi-Play

When Israel's Elbit Systems acquired CYBERBIT, the Cyber and Intelligence Division of NICE Systems in 2015, their mission was to elevate the new subsidiary into a global powerhouse providing cybersecurity for the enterprise, intelligence collection for government agencies, lawful intercept for police and C4IS for the military.

Not surprisingly, given Elbit's core market focus, military thinking and needs have had a profound influence. Effective January 2018, the government-focused piece of CYBERBIT reports to Elbit's defense and C4I unit.

On the law enforcement side, the end result is a product suite lightyears beyond the capabilities of the typical lawful intercept solution found in the United States or Canada. In those nations, lawful intercept is confined to court-ordered network monitoring of commercial communications networks, and only for targeted suspects. Clients in Europe, the Middle East, Africa, LATAM and AsiaPac generally have far broader latitude in surveillance. CYBERBIT can easily accommodate their needs. The company's layered suite lets the law enforcement or government agency client choose solutions that span real-time and predictive analytics, OSINT including social media, GEOINT with precise location, speech analytics and offensive cyber. In sum, CYBERBIT is multi-play ISS at its best.

Elbit's military products can come into play as a vital adjunct when needed. For example, CYBERBIT's lawful interception solutions can easily accommodate intelligence gleaned from aerial reconnaissance, SATCOM surveillance, border patrol sensors, IP videocam from any location, or solutions specifically designed for and used by troops to gather intelligence in the field. Elbit makes products and solutions in each of those areas. This is lawful intercept that can, if the client so wishes, pack a military-grade wallop.

CYBERBIT's approach to the marketplace is based on the premise that traditional boundaries between national security, defense and local policing are rapidly disappearing. The driver: recognition that the three formerly discrete areas are increasingly interleaved and should work together. Making that happen relies on a strategic, full-fledged and fully integrated intelligence system.

"Wising Up" to Broad-Based Lawful Interception

CYBERBIT markets its product on multiple levels – bulk metadata collection and analytics, malware, OSINT, SATCOM, and lastly a unified platform for all these capabilities. The best way to understand CYBERBIT's unique offering is to view it as one big system with multiple capabilities that can be used in toto, or broken out and purchased as separate modules.

The core product is CYBERBIT's "Wise Intelligence Technology" (WIT), a high-speed bulk metadata collection and analysis solution for HUMINT, SIGINT, OSINT, GEOINT and IMINT (imagery intelligence from satellite and aerial photography and video). Basically, WIT is a data collection and analytics suite that processes, analyzes and stores data on a massive scale from any and all sources, each facilitated by specific CYBERBIT, Elbit or legacy NICE Systems products.

WIT follows the classic step-procedure of any big data analytics system. Data is collected via a "Dynamic Gateway" that can be programmed to amass data according to pre-set parameters, for example, all the social media, wireless comms or imagery of a particular population niche and their affiliates, known and unknown, within a large pre-specified geofenced zone. The data is ingested, converted to a single consistent format and stored. The data collected may be both structured and unstructured.

In Step 2, WIT extracts and identifies entities from the bulk metadata, provides updates per entity/group based on real-time update feeds, and supports translation from multiple languages.

Next up is analysis and evaluation of the data, presenting identification of target entity and affiliate identification, historic activity, trending activity and current threats, and providing further real-time updates including precise geo-location plus imagery of the targets. Visualization reveals hidden patterns and imminent threats.

Finally, WIT presents a set of "next best actions" for preventing events with the highest likelihood of occurring.

Taking Cybersecurity the Other Way

CYBERBIT likes to tout its "elite group of PhDs, cybersecurity researchers and analysts, and former intelligence tacticians" who spend their days and nights scouring the Web for malicious patterns, tactics and Zero Days that can bring down a corporate network. This same group works heavily in R & D on producing the very same kind of threats, albeit for a different purpose: hacking hostile nation state intelligence and military targets. While helping enterprises safeguard their data with "battle hardened" cybersecurity, CYBERBIT also goes in the other direction – with offensive cyber for law enforcement clients.

Called CYBERBIT PC Surveillance System or "PSS," this solution provides Zero Day exploits and malware for "monitoring and extracting information from PCs." PSS is similar to ethical malware products made by companies such as Trovicor and FinFisher. The user can bypass encryption and firewalls to discretely enter a target's devices without being noticed, and intercept any communication, download any content on the device and/or take control. PSS lets an LEA listen to the target's VoIP conversations, make audio recordings, view emails in real-time, see historic communications, check urls visited, open address books and files, log keystrokes, take screen shots, extract passwords and activate still or video cameras, and determine location based on proximity to Wi-Fi network access points.

Unlike traditional lawful intercept, CYBERBIT PSS does not require the foreknowledge, cooperation or any involvement whatsoever with the target's communications service provider. Variations in nations' surveillance laws pose no barrier – PSS works anywhere in the world. As long as the target connects to a network, he or she can be easily "captured" online by PSS. The solution is highly flexible. Extraction can be scheduled, performed on-demand or even triggered by target actions or communications set up by the agent in advance to initiate surveillance.

How does the law enforcement agent determine who to monitor in the first place? That step is made easy by ready integration of PSS with other CYBERBIT solutions, such as WIT, that filter bulk metadata to determine eligible targets for PSS.

Note, as mentioned earlier, that PSS is exclusively the work of CYBERBIT malware experts. PSS is a turnkey in-house solution that has been developed and tested in CYBERBIT's own laboratories. This is a big operation with more than enough bandwidth for R & D, and they use it to the hilt. The chief benefit of being in charge of its own R & D: the company is not beholden to or reliant on a third party provider. As a result, CYBERBIT can offer truly elegant features such as the unique "agent factory" of PSS: Clients can create their own virtual agents that meet the needs of a given assignment in intelligent, automated fashion.

In the case of PSS, CYBERBIT goes far beyond the prior capabilities of NICE Systems, which served as a reseller of The Hacking Team's RCS suite.

With PSS, CYBERBIT clients are free of headaches such as worrying about whether their malware provider's license is kaput and upgrades impossible. And there's no deception along the lines of the former Hacking Team's practice of marketing and exhibiting in nations where it is not allowed to actually sell the product. CYBERBIT is 100 percent above board.

Low-Earth Orbit SATCOM for GEOINT and Images

Through partnership with Israeli satellite company ImageSat International, CYBERBIT uses the latter's low-earth orbit EROS satellites for high resolution imagery (IMINT) and geospatial intelligence (GEOINT).

EROS satellite facilities are specifically designed with intelligence and national security in mind. CYBERBIT can provide iSi services that include real-time imagery of designated areas downloaded to a ground receiving station for high priority tasks. Clients can choose Autonomous Mission Planning capability with Exclusive Pass on Demand service to select satellite passes (fly overs) on a weekly basis, with immediate transmission of imagery intelligence detecting any changes. Or the client can purchase full-time observation.

With CYBERBIT's iSi arrangement, government and military agencies can achieve persistent wide area surveillance from space providing notification of any perceived abnormalities in ground activity, delivered in real-time. The IMINT intelligence is integrated with key findings from CYBERBIT's other capabilities in COMINT, HUMINT, SIGINT and OSINT including social media – plus findings from the PSS malware.

Down-to-Earth Lawful Intercept

If the client wants a full-fledged lawful intercept solution along the lines of those used for CALEA, ETSI and SORM – but minus the "eye in the sky," CYBERBIT can bring things back to earth with its Target 360° solution. Modules begin with classic lawful intercept and branch out with functions not typically associated with an old-fashioned wiretap.

For starters, Target 360° powers active (mediation-based) and passive (probe-based) lawful intercepts conducted on the Internet, wireline and wireless networks. Beyond these basics, the solution can also be configured to capture social media or be integrated with special Web crawling engines to ferret out intelligence from the bigger cosmos of OSINT.

The Target 360° "Telephony" module understates what the system can do. To be sure, CYBERBIT Telephony performs conventional intercept and monitoring of voice communications, but once again the company adds an unusual twist. The solution also comes with biometric voice capabilities that identify the target, determine gender, age and social group regardless of language, and also trigger alerts on threat level via key word analysis or measurement of voice stress.

CYBERBIT also provides Target 360° Location, an end-to-end mobile location solution with 3D Accurate Positioning (X,Y and Z) axes in real-time. Target 360° is scalable and uses CYBERBIT WIT as the master platform for data ingestion, analytics and visualization of intelligence.

iPS: From Lawful Intercept to Dark Web

Founded in the year 2000, iPS is one of two companies operated by the RESI Group, a holding company based in Aprilia, Italy. The second is RESI, a network management systems and cybersecurity business dating to 1987. The fraternal twins share a single campus in Aprilia, clients such as Telecom Italia and Deutsche Telekom, and a sizable in-house R & D team that develops leading edge products.

With this close working relationship it is not uncommon for iPS to turn up at cyber events where RESI's capabilities in cybersecurity and network penetration testing are showcased. iPS also borrows on its collegial company's expertise when expanding into new areas such as Deep Web and Dark monitoring. When you operate in a marketplace where cyber expertise is increasingly intertwined with conventional lawful intercept solutions, it is most convenient to have an elder brother like RESI.

Regardless, iPS is a big league player in its own right, offering a comprehensive suite of products built around a centralized platform that can manage, integrate and analyze data gathered from every conceivable form of appliance, all made by the same company, and at the same time integrate smoothly with solutions by others.

Italy's iPS is favored by CSPs, LEAs, intelligence agencies and military clients worldwide – and they have the proof points, a list of named customers that includes Italy's Ministry of Justice, the Ministry of Defense and the Presidency of the Council of Ministers supporting the Prime Minister. Outside Europe, the company has a broad footprint in the Americas, Africa, the Middle East and AsiaPac.

Part of iPS's appeal is the company's versatility at offering literally any kind of ISS solution a communications service provider, LEA, government intelligence agency or military client could desire – including offensive cyber.

For law enforcement, the best starting point is the collection device, the GENESI Probe.

GENESI Probes for Lawful Intercept

The focus of iPS lawful intercept is the probe. iPS groups its probes under the banner, "GENESI Network Interception Platform." All iPS probes are designed to be compliant with CALEA, ETSI and similar laws and technical standards used by other nations for lawful intercept.

iPS probes can be installed without any changes to the network, and immediately begin to intercept traffic without the knowledge of the target. If the user so wishes, iPS probes may also be deployed as in-line active devices that connect directly to network hardware such as switches and routers. This is an important distinction, as active LI solutions are highly scalable to serve the needs of Tier 1 operators, whereas passive probes are more typically used by small to mid-sized carriers. The flexibility of iPS probes means that the company can serve CSPs of any size.

There are two broad types of iPS probe: the GENESI NIP and the GENESI NIP Packet Switched Probe for IP traffic. The GENESI NIP is used to intercept traditional voice calls. An added feature, GENESI Multiconference Gateway (MCG), delivers the capability to intercept group calls, whether the targets are using fixed wireline or mobile devices. The iPS Probe for LTE can be configured to capture specific IP traffic, and to segment or factor out data not relevant to an investigation, for example, if an LEA wishes to see only LTE.

The GENESI NIP Packet Switched Probe uses DPI to inspect and isolate targeted IP traffic. It can intercept traffic by user name, MAC address, static IP address or an address range. Using parameter-based interception, the devices can identify traffic by specific text strings within a protocol header, or the application content.

All probes are provisioned for the intercept and fully administered by a Web server app (called the Provisioning and Administrative Center or "PAC") that hosts management of the intercept. A second function called the Configuration and Administration Center (CAS) displays all devices in the system and their status, and triggers alarms to the user in the event of any malfunction of a device that might impede an intercept.

iPS probes are available in 10 Gbps and 40 Gbps models, intercept in real-time, can store data for later analysis, and are offered in rack-mounting or portable models.

GENESI Monitoring Center

Next in line, for LEAs that need a central management system, is the GENESI Monitoring Center. This platform comes with multiple interfaces to receive, record and store information of diverse types – voice calls, metadata, video, text or other content – both on and off-network. Users may access the data either real-time or off-line, edit the data, and apply analytics to glean intelligence. The data can also be exported to portable media for remote use.

If the client is not using iPS probes, GENESI Monitoring Center may be easily configured to operate with other interception hardware such as mediation devices and routers.

In operation, GENESI's unified platform supports key capabilities that include voice and data lawful intercept, voice biometrics, geolocation, cell positioning, geofencing, private social media for agents that incorporates real-time data and video of targets, relation building, semantic text analysis and what iPS calls "unconventional" surveillance.

Predictive Analytics, Social Media and Dark Web Insights

Intelligence gathering relies on the ability to perceive hidden links in network traffic that lead to connections and relations. To support that function, iPS offers G-WISE, an advanced analytics tool that reveals relationships, and singles out trends that enable pattern analysis. Plowing through immense volumes of social media, email and voice calls, G-WISE quickly identifies the target and his relationship to other parties in the stream, revealing those that may have been previously unknown. As importantly, the solution applies predictive analytics tools to provide the most likely future scenarios of activity by the target.

In conjunction with G-WISE, a separate app called GENESI DATI provides a portal to the target's Facebook account. The app builds a relationship chart of the individuals who have connected to the target's Facebook account, showing their profile photo, Facebook name, user ID, and the number and content of messages. All intercepted data can be exported for further analysis.

Also of note in the company's portfolio is the iPS HUMINT Module, a specialized search engine that can fathom surface Web social media as well as the Deep Web and Dark Web. Virtual HUMINT does it all, end-to-end, harvesting data from multiple sources across the Deep and Dark Web for real-time updates.

Social Media App Plugs Into GPS and Video

Like all popular movements, public social networking has engendered an opposite reaction, a phenomenon that might be termed the "antisocial" social network – essentially a private social medium intended for a selected few. iPS's version, for law enforcement only, is G-SMART.

G-SMART is commonly used by lawmen and government agents when they are in the process of stopping a crime or covert act in progress in close proximity and require real-time coordinated communications with their cohorts. As a secure, private social network, G-SMART combines social media capabilities with GPS mobile location so that investigators can track a target's smartphone or tablet and immediately share information on the next steps. The solution can interconnect with video cameras in a local environment and capture images and audio of the suspect moving by, and also tag his exact whereabouts.

For LEAs and government agencies that need to keep tabs on a target, iPS offers its G-TRACK system – beepers that can be hidden in the suspect's vehicle or in multiple targets.

G-TRACK provides real-time tracking as well as historical analysis of the target's movements. Two key features that add power to the solution: (1) a 3-axis accelerometer that accurately determines the magnitude and direction of changes in movement; and (2) an automated "freezing system" that puts the beeper into stealth mode so that a target's own anti-scanning detection equipment can't reveal its presence.

If the target leaves his vehicle, iPS geofencing and cell positioning take over. Solutions using GPS and radio frequency identification (RFID) can quickly track the precise whereabouts of the target's mobile device and begin intercepting communications traffic.

In conjunction with or separate from beepers, at the user's discretion, the Monitoring Center can also be programmed to intercept audio and video from an array of hidden microphones, cameras or motion detection devices.

Voice Biometrics and Semantics

Voice biometrics is a proven method to identify the target. iPS has its own approach to voice biometrics based on mathematical modeling of phonemes. Called G-SPEECH, the solution collects conversations and creates a unique biometric voiceprint of the target. Like other popular systems such as the former Agnitio (now part go Nuance), G-SPEECH is language independent and can quickly distinguish the speaker's gender and identity.

Currently G-SPEECH works with over 60 languages, and its modular framework makes it easy to add new languages to its biometric vocabulary.

When the target's intercepted content says "bag him," what does that mean exactly – capture, kill or simply forget about it? Interpreting the significance of content can depend on a conceptual and contextual understanding of what is meant by a specific word or phrase, measured against the backdrop of every instance in which a target or his accomplices have used it. As big data analytics challenges go, this one is right near the top.

Using semantic analysis, iPS's G-SEARCH scans the intercepted data pool in real-time, searches files in multiple sources and identifies the precise meaning of a target's turn of phrase. Any changes, including those that arise in social media, are immediately updated and presented to the agent via simplified graphic interface.

"Unconventional" Solutions: MITM & Malware

With the rapid spread of encryption and HTTPS-based protocols for secure communications, many fear that criminals and terrorists are gaining the upper hand – able to communicate free of lawful intercept or any intrusion into their plans.

LEAs call this trend "going dark." iPS's solution is to provide law enforcement, government agencies and military intelligence with what they term "unconventional" tools that crack the dark vault of covert signals and let in the light.

The essence of these products are technologies that power man-in-the-middle (MITM) attacks that enable the user to surreptitiously enter an HTTPS channel to mirror then pass on subversive communications. Among the top sellers:

- **G-SNAKE.** For intrusion into social networks. Designed for intelligence agencies. Provides direct access to the target's IP content, independent of a service provider.
- **G-SEC.** MITM (man in the middle) traffic management. Intercepts and decodes encrypted traffic.
- **ITACA.** Malware that takes remote control of a target's device.

How often do LEAs, the intelligence and military communities need the full range of capabilities offered by iPS? More often than many might suspect.

IP Probes are the "meat and potatoes" of lawful intercept, and location-based tracking of mobile devices is the most commonly cited use of lawful intercept by U.S. law enforcement agencies.

Predictive analytics, social media tracking, Dark Web data and semantics analysis can be essential to determining the plans, location, and accomplices involved in potentially illegal and dangerous activities. Voice biometrics provides positive ID.

When all else fails, doing the unconventional may be the only course that remains to prevent horrific events that lead to loss of life, destruction of property and destabilization. Hence the need for an intrusive solution such as ITACA malware.

Iskratel: SORM & ETSI Hardware

One of the constants of ISS World Europe is Iskratel, a quiet giant that has maintained a presence at this exclusive spyware event going back to 2014. Iskratel's presence at an ISS World conference might come as a surprise to some. The Slovakian company is best known as a producer of high quality hardware for telecom, wireless, broadband and FTTH networks, as well TETRA networks for first responders, cloud and video monitoring services.

Tucked away in the company's versatile portfolio one finds the "regulatory solutions" niche where the company's products for lawful intercept reside. Let us say up front that lawful intercept is far from being Iskratel's main source of revenue – and that ISS World Europe is the company's only known venue for surveillance solutions. Iskratel puts far more emphasis on mainstream conferences such as Mobile World Congress where it showcases its virtual IP Multimedia System (IMS), or FTTH events where it champions GPON vs. DOCSIS, or Smart Rail Europe where the company's specialized mobile system for railways is showcased.

But commercial solutions aside, Iskrakel also sells a very serviceable family of mediation devices to communications service providers, as well as video monitoring systems to government agencies and public transportation clients.

Of note, while the company's mediation devices meet ETSI standards and Iskratel sells in major Western European nations including France, Germany, Spain and the UK, they call their lawful intercept hardware "SORM" mediation devices, and for good reason. Iskratel has a big market presence in Russia and the Commonwealth of Independent States (CIS) – former boundary nations of the USSR – from which it generates 80 percent of the company's annual revenue. Russia and Kazakhstan contribute the lion's share. The Slovakian tech enterprise also works worldwide through partners that sell its solutions in Europe, the Middle East, China, India, Nigeria, South Africa, New Zealand and Latin America.

SI3000 Mediation Device & Concentrator

Iskratel's premier lawful intercept tool is the SI3000, a multi-purpose mediation device and concentrator, equally adept at capturing content and metadata whatever the format – conventional voice or VoLTE, LTE, SMS, email, video, social media OSINT and video. Iskratel provides the necessary LI "network elements" relevant to virtually any network device for interconnection with the SI3000. Each mediation device can be equipped with deep packet inspection (DPI) for gathering targeted data. Alternately, the SI3000 can be used in modular mode, and operate like a passive probe, or stepped up for larger networks to work actively via switch interconnection like a true mediation device.

Such versatility is a desirable feature in the realm of lawful intercept hardware, where the buyer is typically confronted with an array of boxes, each dedicated to a specific function: mediation devices for active LI, LTE devices for 4G networks, and probes for data. With the SI3000, the user has the flexibility to build on and modify the equipment to serve his or her purpose for interception of any communications type.

Given Iskratel's primary geographic focus, such versatility is no surprise. Russian law sets forth requirements for all types of communications under SORM-1, SORM-2 and SORM-3, and these standards are generally the rule in CIS nations, as well. Moreover, because European and other nations that adhere to ETSI typically have a broader mandate than the U.S. (under the Communications Assistance for Law Enforcement Act) on the types of communications eligible for lawful intercept, the SI3000 is an attractive option far beyond the pale of Russia. Technology agnostic, the SI3000 can be used nearly anywhere and interface with virtually any network hardware.

Law enforcement appreciates such versatility. During the lawful intercept process, once a mediation device collects metadata and content, the information must be formatted in the proper protocols for online delivery to an LEA monitoring center. The SI3000 line covers off on that requirement, amenable to the specific protocol requirements of any nation's police or government intelligence agency. Buffering ensures that no data is lost during the intercept or transport process, thus the end product is complete in every sense for subsequent analysis. For back-up, the SI3000 provides its own storage capability.

Not withstanding this rich feature set, Iskratel nonetheless clearly had cost-efficiency in mind while designing the SI3000. Multiple LEAs can access the device to simultaneously conduct separate investigations, and in diverse ways. For example, one LEA might be focused on mobile call content and metadata, while another uses the resident DPI capabilities to measure trends and perform risk analysis of targets using IP communications and social media, while a third is interested in reviewing live video at an airport or train station if suspicious SMS traffic points to an imminent threat.

As with all mediation devices, use of end-to-end encrypted VPN connectivity safeguards the transmission of intercepted data from loss or interference. The Iskratel SI3000 is an impressive device – CALEA compliant, thus technically capable of supporting lawful intercept on U.S. networks.

NETI: Hacking and Mobile Location

NETI IT Consulting Ltd. is a versatile surveillance vendor run by the government of Hungary. While the company is ostensibly an exclusive holding of a non-profit entity – the Theodore Puskás Foundation (PTA) – the two are one and the same entity, share identical corporate officers and one website, and report directly to the e-Government Center in the Office of the Prime Minister. In that regard, NETI, although incorporated, is not a corporation, at all, but simply a branch of government, much like NSA or GCHQ. The "Ltd." is for show.

Founded in 1993 and based in Budapest, NETI provides an array of IT services including those of interest to national security organizations and law enforcement agencies: a mix of strategic monitoring suites and tactical surveillance products. The umbrella suite is BONGO, which is offered in both a master and a tactical version covering deep packet inspection (DPI), bulk metadata collection and data retention. Also offered: offensive cyber capabilities through a resale agreement with Gamma Group for FinSpy products, and third party IMSI catchers.

NETI's BONGO Monitoring System (BMS) is the embodiment of bulk metadata collection, similar to NSA's former PRISM and GCHQ's work with SS8. Advertised as a comprehensive system for monitoring and culling data from what it calls "the massive and ever-growing jungle of networks and data masses," BMS deploys classic DPI to intercept IP traffic including data, video and VoIP via servers. Data is mirrored and decrypted through man-in-the-middle attack by the DPI system prior to forwarding to a monitoring center. BONGO also accommodates conventional wiretaps of fixed wireline and public land mobile services, as well as "EDGE" (Wi-Fi) traffic. Once data is received, the Monitoring Center can then offload to NETI data retention for storage and subsequent analytics – the latter provided by a different vendor, not NETI.

BONGO may be configured for a single workstation with resident DPI for lawful intercept purposes and can also scale to accommodate bulk metadata collection for national security.

The system can also be integrated with third-party devices, peripherals and interfaces such as IMSI catchers and offensive cyber, the latter using Zero Days to exploit weaknesses in target systems. As a result, BONGO outputs include not only intercept-related content and call or IP metadata, but also IMEI, IMSI, radius identification, IP address, e-mail address, MAC address (identifying a specific device used) and other data such as mobile location inherent in commonly-used third party tools for tracking and taking control of targets' communications instruments. All data is exportable to external drives – DVD, CD, flash memory sticks – for access and analysis by any approved user. For tactical applications, NETI provides BONGO COMPACT, which performs the identical functions as big BONGO, but in a portable unit.

NETIMON Mediation Device

NETI also offers NETIMON, its own mediation device for lawful intercept. NETIMON is a standard-fare mediation device, designed to sort data gathered from network hardware and categorize the data in the appropriate standards for the monitoring center located on premises at a law enforcement agency.

Mediation, as opposed to probes placed at a network's edge, is an "active" approach to monitoring wherein the device is directly connected to special intercept modules embedded in network hardware, which it configures and controls for the duration of a lawful intercept, or in the case of bulk metadata collection, to an array of servers that perform the same function more broadly for national security.

Like all mediation devices, NETIMON monitors public-switched and IP voice and data, as well as mobile. The lawful intercept is completely invisible to the target, and the device can be partitioned to serve the LI needs of multiple LEAs with secure access available to each only – or to one LEA if permission is granted by the others. Captured data is encrypted, typically on VPNs, before it transits to the waiting LEA monitoring center. "Unique" features claimed by NETI such as assigning single identifiers to target data and providing access by multiple LEAs, are in fact fairly commonplace in the mediation device arena.

Secure Communications

Outside the narrow scope of intelligence support systems, but still of potential interest to law enforcement and government agencies, NETI also sells its own encrypted NETIPHONE, fine for another era, but far below what an agent, law officer or common citizen finds available in any "strong encryption" iPhone or black phone. Another product, ONEWAYER, vies to safeguard centrally stored intelligence by separating "low level" (sender) from "high level" (recipient) and confining all communication strictly to the former. ONEWAYER is literally one-way only.

NetQuest: Policing 100GB DWDM Fiber

Monitoring DWDM (dense wavelength division multiplexing) presents a challenge familiar to law enforcement and government agencies vying to intercept target communications on 10GB/40GB and 100GB fiber optics: the high cost and complexity of using conventional methods to conduct surveillance. NetQuest Corporation is among the companies resolving these issues. To understand the importance of their efforts, it is first essential to know the basics on DWDM: how it works, why it is favored by service providers, and the factors that make DWDM at once a salvation for fiber-based network operators and a headache for law enforcement and government agencies.

DWDM Fundamentals

DWDM is a common solution for fiber-based network providers overwhelmed by bandwidth demand. DWDM enables optimization of network loads and speed in the existing fiber optic infrastructure and is economical versus the cost of deploying more fiber to keep pace with bandwidth expectations.

DWDM places data signals from multiple sources onto a single fiber strand in C-Band optical networks, each channel with a separate wavelength transiting fiber infrastructure at GB speeds.

A DWDM system can provide up to 96 wavelengths, also known as "lambda" circuits or channels.

For law enforcement agencies, monitoring all 96 wavelengths to determine the one or ones that carry targeted traffic can be prohibitively expensive, and also elevate the risk of disrupting data flows in a manner that alerts the monitored target. Cost and risk problems escalate in lock step with the rise in bandwidth demand and service providers' quest to meet that demand cost-effectively with higher speed and better optimized networks. As a direct result of DWDM's performance in supporting these objectives, lawmen and agents find themselves battling not just targets and foes, but the fundamentals of network economics in a service provider market driven by pressure to cut costs and improve margins.

DWDM Conundrum: Simplicity vs. Cost?

In the early days of fiber optics, conventional switching of optical signals was handled by converting optical inputs into electronic signals at each network node, then converting the signal back to optical model for transmission to the destination point. This process, known as OEO (optical-electronic-optical) was cumbersome, costly and confined by bit rates.

A better approach emerged in the 2010-2011 time frame with ROADMs (reconfigurable optical add-drop multiplexors), optical devices able to switch DWDM at the individual wavelength (lambda) level without the back and forth of electronic signal conversion. Among ROADM's benefits: the ability to switch wavelengths selectively and to automatically balance power requirements so that switching was unnoticeable to the network, a pair of features also beneficial to inline remote monitoring of DWDM traffic.

Initial or "classic" ROADMs represented an important evolution in wavelength switching, eliminating the need for OEO conversion, and providing switching capabilities through a smaller and less costly device than that required for OEO.

However, while initially viewed as an improvement, ROADMs introduced their own level of complexity as increased network loads led to new issues such as channel spacing, occasional "contention" resulting when the same frequency was assigned to wavelengths arriving from different ports, and the need for manual intervention to fix these problems.

Augmenting ROADM were a new set of architectures including: CDG (colorless contentionless gridless); CDC (colorless directionless contentionless); and CDCG (colorless directionless contentionless gridless). Here "colorless" applies to the ability to add/drop wavelengths assigned the same frequency aka "color"; "gridless," to flexible grid channel spacing; and "contentionless," to the assignment of unique frequencies regardless of the source from which a DWDM signal emanates.

How did these evolved architectures help?

Classic ROADM spanned multiple components including: AWGs (arrayed waveguide gateways – for separating signals); optical switching and amplifiers; mux/demuxer; mux transponder cards; and transponder. But the setup was far from perfect. For example, the AWG's (gateway's) job was to assign a wavelength to a specific port. The hang-up: Any transponder reallocation to a different wavelength required manual intervention to handle the unplug from one mux/demux port to the plug-in of a new one. Enter ROADM CDC, which automated wavelength-to-port assignment, thereby eliminating the need for manual intervention. But ROADM CDC came with its own pitfalls – it added complexity and cost at the node level.

Similarly, each new ROADM architecture introduced not only improvements, but also unresolved challenges that needed to be fixed on a one-off basis.

Wavelength contention, which occurs when lambdas of the same frequency land at a node and thus look indistinguishable, remained an issue with ROADM CDC just as it had been with classic ROADM. The solution came with ROADM CDCG (with the second "C" standing for "contentionless.") ROADM CDCG addressed the issue with modifications that made same-frequency wavelengths assignable to distinct ports on the "drop" side of the switch. But again, fixing a problem led to larger, more costly nodes.

Today, ROADM architecture selection always presents a choice between make-do accuracy and complexity/cost. Classic ROADM is the most problematic, but least costly. Succeeding architectures resolved issues with earlier ROADM generations, but invariably at great complexity and cost. Monitoring solutions hit the exact same obstacles and choices in endeavors to intercept DWDM.

The Role of NetQuest OptiCop Interceptor

NetQuest Corporation works both sides of the marketplace, providing optimization systems for fiber optic-based communications service providers (CSPs), as well as interception systems that leverage the same technology for law enforcement and government intelligence agencies. One of the company's key contributions is technology that obviates the need for ROADMs when conducting surveillance on DWDM-facilitated networks.

The concept of clean, accurate and economical interception on fiber nets begins with NetQuest's SONET products and reaches fruition with Wavelength solutions for 10G/40GB and 100GB networks.

Netquest's flagship product family is the OptiCop Interceptor suite. OptiCop displaces multiple devices that sit between a mediation device or probe and T1 – T2 SONET networks where a target might reside: Taps, add-on deep packet inspection systems and Ethernet switches. As a result, OptiCop Interceptor serves as an intelligent "middle man" that integrates the full functionality of traditional SONET interception devices in a single unit, providing better performance at lower cost.

NetQuest describes the heart of its Interceptor line as the company's Auto-Discovery Process, a rules-based engine. Fully automated, as the name implies, the end-to-end Auto-Discovery Process adjusts and scales to variations in data streams, discovers and intercepts targeted traffic (ATM, POS, PPP, Frame Relay, HDLC), pre-filters and finally translates these diverse streams into GB Ethernet packets for routing to the relevant agency.

All monitoring is passive to preclude any notice by the target. Results can be forwarded directly to the law enforcement or carrier interception device, or to separate tools such as deep packet inspection for metadata and content intelligence. Depending on input requirements of the interception system, OptiCop Interceptor comes in versions that can segment traffic, for example sending 1G and 10G traffic loads to respective probes/mediation devices capable of accommodating either load. For tight integration between the OptiCop Interceptor and the service provider's or LEA's monitoring system, NetQuest also offers a purpose-built interface that delivers machine-to-machine unified integration for optimal performance.

OptiCop Wavelength Interceptor

NetQuest took a major step forward with the launch of its DWDM product, OptiCop Wavelength Interceptor I-9196, touted as having the ability to provide LEAs and government agencies "real-time, targeted access to customer data."

The I-9196 integrates the functionality of multiple conventional modules used in fiber interception end-to-end, thus reducing capex for hardware with varying lifespans and higher replacement costs. NetQuest's machine-to-machine interface ensures seamless integration with LEA and CSP interception systems.

The I-9196 can monitor DWDM across all wavelengths and traffic types at speeds ranging from 2.5 GB through 100 GB. Attaching direct to a C-Band fiber network, the I-9196 uses NetQuest Auto-Discovery to discretely examine related lambdas and instantly notify the investigator of any changes in bandwidth allocation that might look suspicious.

Another key achievement: the I-9196 can use Auto-Discovery's automated intelligence to intercept DWDM without reliance on costly ROADMs. The combination of these features makes the OptiCop Wavelength Interceptor a worthy competitor in the DWDM monitoring arena. As optical amplifiers move fiber from C-Band to L-Band, expect NetQuest to play a leading role.

NICE: All-in-One ISS Vendor

Launched in 1986 by a group of former Israeli soldiers, NICE (short for "Neptune Intelligence Computer Engineering") began operations offering products for defense and intelligence before branching out into the broader commercial market realm, applying the same technologies to its newer pair of core businesses: Customer Interactions, i.e., customer experience management (CEM), and Financial Crime and Compliance. The company's intelligence and surveillance work is managed by the Security business unit, which also supports a range of other niches such as public safety and airport security.

Now, as in the beginning, NICE Systems' focus is helping its customers do business smarter, whether that involves presenting the perfect treatment to ensure a superior service experience that boosts satisfaction, loyalty and sales, or to provide real-time insights that pin down a target and show exactly what he plans to do next, and where.

The NICE Systems platform for commercial accounts: real-time predictive analytics (RTPA) to capture, analyze, and apply insights from both structured and unstructured data on phone calls, mobile apps, emails, chat, social media, video, records and transactions.

The surveillance solutions package includes lawful intercept mediation, voice biometrics and speaker identification, text analysis, mobile location, video systems, satellite-based communications surveillance and image reconnaissance, and end-2-end metadata collection, trending and analysis driving actionable insights based on the same RTPA platform.

Evidence of NICE Systems' success abounds. The company's commercial solutions are used by over 25,000 organizations in 150 countries, including 80 Fortune 100 companies. Off the books, NICE Systems' intelligence division covers the same global footprint serving LEAs, intelligence and the military from offices in the Americas, Europe, the Middle East, Africa and throughout the Pacific.

Though NICE Systems draws the lion's share of annual revenue from its two newest units, the company's roots in aiding law enforcement and intelligence agencies run deep, feeding a powerful set of state-of-the-art surveillance products. LEAs, intelligence and military customers can use NICE to create a granular view of the target in real-time.

The NICE Systems surveillance suite scales from tools that serve simple law enforcement needs for location data, up to big data analytics systems that collect and crunch data of all types to provide actionable insights for nation states. The salient point about NICE Systems: predominance of the phrase, "intent based." On all three sides of the business, NICE serves the user's intent by determining the target's thinking, plans and next action. This is the heart and soul of all modern RTPA-based systems designed for commercial, law enforcement and intelligence needs. NICE Systems utilizes the phrase at every turn so there is no mistaking the intent of their solutions.

NSA-Level Capability

The centerpiece of NICE Systems' high-end surveillance solution is a pair of products called NiceTrack Mass Detection Center and Pattern Analyzer. The joint capabilities of these systems bring to mind the power of NSA's PRISM or Palantir: omniscient gathering of all relevant data and connections filtered and pruned via multivariate analysis to deliver a detailed snapshot of the target at that moment.

NiceTrack Mass Detection Center (MDC) can be programmed as a wide area collection system for all manner of data – voice, mobile, social media, video, email, SMS, webmail and images. NiceTrack MDC also integrates with other systems such as databanks storing criminal records.

Streaming data from live communications services, including mobile location tracking, language identification, speaker recognition and automatic text translation is standard with NiceTrack MDC, and can be complemented by street level and aerial video from another product, NiceVision, which puts the user virtually at the target's side that second. Like other real-time predictive analytics programs, NiceTrack Mass Detection Center delivers real-time actionable insights driven by a 5-level architecture: data ingestion; data normalization and univariate analysis; heuristic and predictive statistical modeling and analytics; integration with RESTful and Service Oriented Architecture; and the end result – data visualization of insights via dashboards, coupled with the power to pose queries or run KPIs to correct or adjust models.

NiceTrack Pattern Analyzer is a sister system that finds trends and builds profiles. Pattern Analyzer's forte is legacy data – call detail records (CDRs), IP detail records (IPDRs), historic mobile location data, text content and financial records. Pre-existing models are mapped to already-analyzed data and hone in on suspicious activities, or new potential targets, and then alert the agent.

NICE Systems' a la Carte Menu

At NICE, the range of solutions maps to the spectrum of customer needs. A large organization might opt for the full metadata collection/analysis capabilities of NiceTrack Mass Detection Center and Pattern Analyzer. Or they might buy that, then purchase specific add-ons, or opt for standalone solutions on a one-off tactical basis. Among the options:

- **Lawful intercept.** Collecting, filtering and mediation on any network: voice, mobile, IP, social media, intercepted at the switch or router, then passed to a collection center for storage and analysis by NiceTrack MDC and/or Pattern Analysis, or both.
- **Voice Biometrics.** NICE Systems' creates biometric voiceprints, quickly determining the language, dialect, gender of the target, and the intent signaled by the spoken content.
- **Mobile Location.** NICE Systems provides hybrid location tracking – a mix of SS7, Geofencing, RFID and other technologies – to find the precise location of any device on any mobile network. The hybrid approach first uses SS7 and Geofencing to determine the target's general location, then applies RF analysis to nail the exact position. NICE calls the end result, "3D Accuracy."
- **Video Surveillance.** NiceVision Net comes complete with smart video cameras, a control center and analytics. NiceVision SDK software lets users integrate NiceVision intelligence with legacy video systems to improve performance; users simply "drag and drop" the NiceVision Player into their current system.
- **Satellite Surveillance.** Full interception capabilities for Inmarsat and Thuraya systems and devices, and for very small aperture terminals (VSATs) deployed either as ground or maritime antennas. Picks up any communications transiting a satellite network – switched and IP voice, mobile communications, data, email and social media – both on and off-the air.
- **Text Analysis.** NICE Systems text analysis automates the process of searching for keywords and patterns that can identify a known or potential target, and trigger alerts on possible events that require preemptive action.
- **Open Source Intelligence.** The NiceTrack Monitoring Center provides continuous access to other intelligence sources including human (HUMINT) and visual (VISINT) for integration with insights gleaned from metadata analytics by NiceTrack MCD and Pattern Analysis. In addition, 24 X 7 web crawling scans the Internet for all publicly available data relevant to a target to further enhance intelligence.
- **Strong Decryption.** Although the precise technology for doing so is closely guarded, NICE Systems experts have developed decryption techniques that can work around and penetrate virtually any current

form of decryption, rendering content and networks protected by Hypertext Transfer Protocol Secure (HTTPS) over Secure Socket Layer (SSL) an open book for the intelligence practitioner.

Paladion: From Malware to Cybersecurity

Founded in Bangaluru, India in 2000, Paladion began its corporate life as a provider of lawful intercept and offensive cyber solutions, helped by an early infusion of US $500,000 from Nadathur Holdings, and fueled again by a second cash injection from the same investor. From the start, Paladion proved to be a VC's favorite kind of investment: cash positive through a strong pipeline of business, in their case with Indian law enforcement and intelligence agencies. Paladion committed Nadathur's cash to international expansion. The strategy worked. Before long, Paladion was selling outside the country and at the highest reaches, to customers including the U.S. Department of Justice.

By the early years of the decade Paladion could tout a full suite of surveillance products: lawful intercept for law enforcement agencies and communications service providers, deep packet inspection for Internet monitoring and censorship, SSL hijacking and decryption, portable Wi-Fi monitoring, device forensics, analytics, visualization and "remote desktop/laptop monitoring," the latter akin to what FinFisher and ClearTrail offer. The company was a regular participant in venues such as ISS World, and also worked with partners such as MBIS to provide on-site training in Paladion tools for customers in Algeria.

The outcome of this concentration in lawful intercept and spyware: a stepping stone to bigger things. By 2010 Paladion was off the surveillance trade show circuit and focusing – with an with excellent sense of anticipation, as it turned out – on products that "detect malware and phishing." Paladion had turned the page. With offices in Europe, North America, the Middle East and AsiaPac, Paladion is now a market leader in information security and defensive cyber.

However, as witnessed by job postings for individuals who can "write exploits, malware, Trojans," Paladion's surveillance business remains very much alive. Cybersecurity may occupy the pilot's seat, but there's little doubt what skills help inform and navigate Paladion's course.

Paladion's Black Bag

Paladion goes head-to-head with the best in its field – not only other home market players such as ClearTrail and Vehere – but with established malware specialists like FinFisher, visualization pioneers like Palantir, and broad-based competitors including BAE Systems, SS8, and others that offer end-to-end surveillance suites for diverse market segments.

We're not going to devote time here to the lawful intercept line, which is all standard fare – "universal probes" for voice/data content and metadata, a centralized monitoring center, etc. Of greater interest are three specific sets of tools of use to law enforcement and government agencies: DPI, SSL interception/decryption and offensive cyber.

Paladion DPI

Under the banner of "Internet Monitoring Systems," Paladion offers a sliding scale of hardware/software DPI products for tactical, distributed and national operations. The standalone version is ideal for tactical needs when tracking a target or targets on a single operator network. Version 2 for distributed requirements gives a single user site "command and control" of deep packet operations aimed at targets who may be on multiple networks. Paladion's "national platform" provides DPI for multiple agencies across diverse networks and regions. Intercepts from any of the three are routed to data retention facilities for analysis.

Paladion DPI systems may be programmed to look for a specific target, keywords or other rules-based factors, and counted on to operate at carrier grade speeds. Like all top DPI products, Paladion's attack the payload at Layer 7 of the OSI stack to identify applications and reveal full content – voice, email, SMS, video files and so on. Output includes behavior analysis and link affiliates of the target.

LEAs can also turn Paladion DPI to another purpose: granular webpage blocking that lets the user prevent access to selected pages without impacting the rest of a website. While DPI in this capacity has attracted unfavorable attention for its use in censorship by China and certain Middle Eastern states, elsewhere there is arguably merit in a tool that precludes criminal or terrorist access to data on how to purchase weapons, make explosives or hack IDs to set up illicit banking activities.

A key selling point of Paladion DPI is this very flexibility, allowing the user to pick the appropriate application and scale of product required, and ramp-up when needed. The same client-focused thinking applies to pricing. Paladion DPI is not strictly a low cost or "working man's solution" by any means – it is full-fledged and bows to no one on capabilities. But given the robust market and many alternatives in DPI, Paladion makes a point of keeping its prices competitive.

Man-in-the-Middle Attack by SSL Proxy or Diversion

Paladion bluntly acknowledges that LEAs often hit a wall when target communications are encrypted via the Secure Socket Layer (SSL) protocol.

The company was among the first to move quickly in capitalizing on new approaches that help defeat SSL, and offers a pair of classic man-in-the-middle attacks. Either one can filter and capture the target's IP communications, financial or other transactions by designated IP address or IP address ranges, emails, keywords or other signposts, all without raising red flags for the target.

Paladion's front runner, SSL Interception, works inline to stealthily intercept targeted traffic via proxy. It presents an accepted Certificate Authority that facilitates access, then proceeds to decrypt content. The intercept occurs unobtrusively, copying and decrypting signals and content before routing the originals to the intended router.

In the second option, the system is deployed parallel to an edge or gateway router to divert designated traffic to an LEA or government agency for decryption and analysis. This approach is just as fast as the Proxy method. Similarly, the sender's original transmission arrives at its intended destination without lag time so that the interception is unnoticeable.

Offensive Cyber

It is no mystery why Paladion is adept at helping government agencies and mission-critical enterprises safeguard sensitive data troves. This expertise is years in the making, and aided by work in the offensive cyber arena.

Paladion's Remote Monitoring product provides comparable functionality to that of FinFisher. Controlled from a central server, it provides the ability to: take charge of the target device, access all communications in real-time, view stored documents and files, retrieve deleted files, examine browsing history, monitor keystrokes, alter communications to and from the target, activate the device's still or video camera, and see all encrypted content "in clear."

Paladion ethical malware may be installed directly by thumb drive, or remotely through the very techniques that the company defends so well against on the other side of its business: phishing, malware, Zero Day exploits, and exploitation of behaviors. Once installed, the software can monitor one or multiple targets simultaneously.

The aforementioned Paladion job postings are clear evidence that the company is interested not only in creating malware and Trojans, but in analyzing, root causing and mediating a security attack. Is it wrong for a company to excel in both realms? Not at all, in fact any such assumption reveals a profound naïveté about the realities of the marketplace and the world we live in. The disciplines work hand-in-hand, and Paladion – and its cybersecurity and intel clients – are the better for it.

Pen-Link: Moving Beyond Pen Registers

Mention "Wizard of Omaha," and everyone knows you mean Warren Buffett, the genius who redefined investing, became one of the world's wealthiest men, yet remains noted for his unassuming manner. Just a few miles away in Lincoln, one finds his corollary in a quiet but extremely successful pioneer in the surveillance industry: Michael Murman of Pen-Link.

What the "Wizard of Lincoln" has accomplished with Pen-Link, off radar to most, is in its field just as impressive as Berkshire Hathaway. Pen-Link products are often deemed the gold standard of pen register trap & trace, lawful intercept of metadata and content, data analytics and data woven in from OSINT sources. The suite is popular with police departments, as well as the FBI, the U.S. Secret Service and other high level government agencies throughout the U.S. Since 2005, Pen-Link has been a global player, as well, drawing the greater part of its revenue from offshore sales.

Over the last two decades, the market for such products has grown immensely competitive. As Pen-Link has proved repeatedly, however, being "first out of the gate" is a significant market advantage.

The upside of having the brand-name "Pen-Link" is that potential customers in law enforcement immediately understand the basis of at least one major facet of the company and its product line. The downside: The brand has become interchangeable with the product area itself, much as "StingRay" is now applied to all IMSI catchers.

However, Pen-Link Ltd. is about much more than just pen registers. The company's evolution to an end-to-end provider of intercept and analytics tools sets the bar for what customers demand and expect, attracting attention not only of the marketplace but other vendors that benefit from partnerships with Pen-Link. Pen-Link software is not only sold direct to clients, but also – in a form of "coopetition" – embedded in the products of other surveillance vendors as a proven tool for reliably automating the collection, storage and detailed analysis of intelligence. By "proven," we mean established over a period that spans more than three decades – the kind of longevity that stands out in any industry.

Murman, a native "Cornhusker" with a business/marketing degree from the University of Nebraska, launched Pen-Link in 1987. At the time, law enforcement agencies needed support for pen registers issued under authority of the Electronic Communications Privacy Act (ECPA). Pen-Link's arrival in the marketplace was timely for other reasons that were to follow: advent of the World Wide Web in 1992, passage of the Communications Assistance for Law Enforcement Act (CALEA) in 1994, and, further down the road,

Section 215 of the Patriot Act, which updated ECPA to ensure that LEAs could capture data on a target's communications regardless of the network or type of service they use, and on a cross-jurisdictional basis. Pen-Link's trick was to stay ahead of policy change through this period.

The company's first product was the eponymous "Pen-Link," software that automated the process of wiretap metadata collection. Pen-Link the product is still around, but the company's full line of products and services illustrate its successful diversification.

From Metadata to Content and Analytics

Pen-Link's top-of-the line product is the LINCOLN Collections System, a CALEA and ETSI-compliant solution that meets the full lawful intercept requirements of the LEA. The system has two fundamental components: LINCOLN LAN Server and PEN-Link software.

The LINCOLN LAN Server is designed as a CALEA-compliant passive device that terminates targeted network connections. Just like a probe, the server is configured to collect call information in real-time directly from the network, as specified by court order. In a significant departure from the pen register days, the server collects both metadata and call data, which can include content. The LINCOLN Server copies intercepted metadata, call content, images and more, forwards data to the pertinent LEA monitoring center and also provides data storage.

The process of managing data collection is the job of Pen-Link 8 software, which is installed in workstations in the LEA monitoring center, and provides sophisticated analytics capabilities that narrow the field of possibilities across the vast amounts of data that might be accumulated in an investigation. The end product is evidence that may be used to apprehend criminals and terrorists in the act of planning an attack. Pen-Link 8 and LINCOLN are valued by LEA and intelligence agents for their versatility and reporting/analytics functionality. The integrated tools comprise a complete end-to-end solution for any type of network: wireless, satellite, or wireline.

Next, Pen-Link's supplemental features come into play:
- **Xnet**, specifically designed to provide analytics of a target's IP communications: emails, messaging and Web browsing.
- **PLX**, a database platform that shows the findings of analytics in formats that include spreadsheets, charts, GIS maps or other graphic visualizations.
- **Pen-Proxy**, which provides tailored feeds from OSINT sources directly into Pen-Link PLX.

Prospective customers may, if they wish, purchase individual products as modules to be integrated with legacy lawful intercept systems, or opt for the full Pen-Link suite.

Training

For newcomers, the company offers CATS (Call Analysis Training School) for a primer on how Pen-Link products work. One step up from CATS is PLX 101: Basic Analysis class, described as the "foundational course for all Pen-Link users," with focus on basic functions and hands-on practice with simulated real-life scenarios.

Experienced Pen-Link users can further develop their skills at PAC (Pen-Link Advanced Class) where they learn to leverage every feature of Pen-Link products for enhanced results in case management, database management, and sophisticated analytics techniques.

Further advanced training in pen register and content intercepts is showcased in the Pen-Link LINCOLN Administrative Training course. Pen-Link also offers in-house training, goes on the road to provide special classes on products such as Xnet, and hosts training at events held specifically for law enforcement and intelligence specialists charged with surveillance.

RCS Lab MITO³: Unified LI Management

RCS Lab, a maker of diverse lawful intercept solutions and related spyware, is a respected member of the ISS community. Based in Milan's Caldera Business Park, with branches in France and Spain, RCS Lab is an active participant at ISS World conferences and claims that its systems conduct and manage more than 10,000 intercepts per year for customers.

It is pure coincidence that the company operates but a short drive from The Hacking Team (now Memento), maker of the RCS (Remote Control System) offensive cyber suite. There is absolutely no relationship between the two enterprises.

RCS Lab opened for business in 1993 offering a mediation device for wireline voice, and in rapid succession began to produce complementary units for monitoring mobile voice, UMTS/GPRS data, Skype and intercept support for ISP architectures of Cisco, Lucent and Cisco, as well as probes used for RADIUS authentication detection.

Today their mediation device suite is a robust portfolio, fully compliant with ETSI, CALEA and other national standards, but if that were all RCS Lab did it might be little different from the other three score mediation device manufacturers. RCS Lab goes far beyond, providing a master solution that integrates the key branches of surveillance – lawful intercept, mobile location, social media monitoring, offensive cyber, analytics and visualization – in one master management suite, MITO³.

MITO³ Capabilities

In the lawful intercept arena, storage and analysis of evidence is typically the responsibility of the law enforcement agency. A perennial challenge is maintaining a data center to ensure it is current and able to upgrade systems without generating significant capex outlays.

Based on a flexible, modular platform architecture, MITO³ provides consolidated, centralized management. The device retrieves, decodes, processes and stores both intercepted metadata and content from any type of network: wireline, mobile, SMS, multimedia messaging service (MMS), and fax. It operates out-of-the box with other RCS Lab equipment and tools, and also integrates readily with those of other manufacturers. As innovations emerge in the surveillance field, the future-proof MITO³ is ready to adapt.

As is true of many advanced monitoring centers, the MITO³ decodes most content quickly. For those using the RCS Lab dedicated Internet mediation device, the system also accommodates any protocol or app including VoIP, email, webmail, VoIP, chats, peer-to-peer file sharing and social media, without the need for third party software. MITO^3 also can integrate with associated software-based Internet data probes, accumulating additional evidence keyed to the rules set in the DPI engine.

A separate analytics module available with MITO^3 builds a comprehensive profile of the target and her or his relationships, revealing links to other entities. The user can perform searches to add more data, then perform statistical analysis of group or individual characteristics. The solution also shows the target against demonstrated patterns of behavior and connections, and reviews actions against geographical maps by date and time.

MITO^3 tags the perpetrator's whereabouts through a combination of functions: mobile location shown via analysis of metadata, plus SS7 coordinates in relation to cell tower proximity. The use of these data sets can help connect the target to an incident. The data is historic and does not necessarily lead to the target's doorstep; however, when the target's location records map to the scenes of criminal acts, the association can help build a case.

Special Tactical Tools

When time is of the essence RCS Lab helps the user step up, with a variety of tools that perform: GPS localization, Wi-Fi monitoring, an active mobile location and device penetration by RCS IMSI Catcher, and an RF monitoring tool for determining location by strength of a radio signal as measured from more than one point. Of interest to those that want visual evidence: RCS also offers hardware that provides video surveillance for input and analysis.

It is not often that one sees a company touting its "Trojans," but RCS Lab does just that with a line of malware that can penetrate the target's device, monitor keystrokes, check messages sent and received, listen in on voice conversations, capture downloads and uploads, review data stored in device memory, and take over the camera – the same as FinFisher FinSpy.

Inside the Caldera

Lingering confusion over who owns the RCS brand is understandable. There are many companies called "RCS," including a Russian specialist in click-2-call software and an Italian sports media firm.

Rather than clear the air, RCS Lab would seem to prefer having this cloud of mystery hover over its mission and operations. But those who attend ISS World in Dubai or Johannesburg, INTERPOL in Cartegena, Monaco or Prague, MILIPOL in Paris, or a Home Office Security and Policing event in the UK, will see the company's footprint.

Roke Manor Research: LI with a Military Twist

Say the words "Roke Manor Research" and many industry observers immediately think in terms of advanced radars for the military, smarter air traffic control and state-of-the-art manpacks that outfit today's soldier for electronic warfare, but not necessarily the sophisticated array of surveillance technologies expressly designed for law enforcement and intelligence gathering made by the same company.

Roke's success in the military and commercial IT arenas tends to obscure the company's contribution to the surveillance field. But lawmen and intelligence agents in the hunt for surveillance solutions hold Roke in high regard for technological expertise that matches its capabilities in the military sphere.

Roke's suite of surveillance products covers the gamut from probes, mediation devices and deep packet inspection (DPI) capabilities for law enforcement agencies, to sophisticated mobile tracking and SIGINT that help ensure on-the-spot positioning of civilian and military targets – all with an eye to meeting the needs of customers mindful of capex.

Nestled in a 19th century estate near the village of Romsey, Hampshire, UK, the company looms in the shadow of the still-standing manor house. The residence and 22 acres were purchased in 1956 by Plessey Company, an electronics, defense and communications firm dating to the First World War. Plessey launched Roke Manor Research shortly after buying the estate, focusing the new company on military communications systems.

Time passed, the company grew, was acquired by Siemens and Britain's GEC (the General Electric Company, renamed Marconi plc, whose defense unit was subsequently sold off and merged with British Aerospace to form today's BAE Systems).

In 2010, Chemring Group plc acquired Marconi and revived its original name, Roke Manor Research. But well before becoming an operating company of Chemring, Roke Manor Research had already established its presence as a provider of complete electronic surveillance solutions for the digital era.

Vanguard – Big Performance from Compact Devices

The work horse of Roke Manor Research's lawful interception suite is Vanguard, a scalable modular system that delivers high performance at an economical price.

Used by law enforcement and intelligence agencies worldwide, Vanguard comprises turnkey solutions that include: warrant system management (interception management system or "IMS," a soup-to-nuts approach from data collection to LEA protocol interpretation); a mediation device that actively integrates with any network switch, router or call server; and, integration with passive probes.

Roke Manor Research offers its own line of probes that, utilizing the companies' skills in miniaturization, have been reduced to the size of a 1990s era flip-phone, for plug and play interception capability. Called the "Pico Probe," the device scales to 100 Gbps, and is ideal for tactical interception scenarios such as Wi-Fi hotspots in public places. Pico's tiny size makes it perfect for covert operations, as well.

All Vanguard interception devices, whether active mediation or passive probe, cover the field of digital communications: broadband, mobile data, Wi-Fi, WiMax, VoIP, email, SMS and social media.

The sweet spot of Roke's surveillance package is deep packet inspection. Roke's patented approach to DPI dates to the 1990s, and has been regularly updated in new patents since the company's initial involvement with the technology.

Roke's Packet Capture and Analysis software is compatible with any platform, from low cost servers to carrier grade systems, and is scalable to wire speed on local, national or global networks. Ready integration with all commercial equipment also speeds ROI, eliminating expenses such as application-specific integrated circuits.

Ease of use and reliability are built-in. Users can configure DPI from Roke's Web-based front end and quickly proceed to capturing targeted data by user name, IP address, portal, email address, text field or application. Collected data may then be sent via mediation device or the DPI system itself to a database for subsequent analysis. As with all DPI, the intercept of content and signals is completely invisible to the target.

Mobile Surveillance and Location Monitoring

Roke's own IMSI catcher lets law officers and intelligence agents collect full cell site data. Completely automated, the system intercepts data samples from the site, matches them to GPS coordinates, and shoots the findings to a database for analysis.

Roke solutions also keep agents abreast of the latest mobile apps used by targets so there is advance warning of apps being used to plan and execute a terrorist or criminal event – and at the same time, insight on opportunities that the same apps may present as evidence that puts suspects in lock-up.

With jamming and spoofing incidents on the rise, national security agencies need a reliable backup to GPS not only for navigation purposes, but for monitoring and surveillance of target positions, as well. Enter Roke's MILOR eLORAN receiver, a miniature antenna able to provide reliable location and tracking of targets, and time-stamping each instance.

MILOR's ultra-high signals can penetrate any obstacle – buildings, foliage, garages or poor weather conditions – to find and maintain contact with the target's position with perfect clarity even when GPS signals are blocked or go awry.

MILOR's antennas operate at 100 kHz at performance equivalent to standard eLORAN, and with extreme miniaturization: up to 100 times smaller than standard versions of the technology. All MILOR units come with GPS, too. When GPS goes down, the eLORAN function takes charge seamlessly.

On the Battlefield: Radio Interception with RESOLVE

No discussion of Roke Manor Research is complete without mention of systems designed expressly for military SIGINT. Here Roke provides RESOLVE, a modular, scalable and compact system ready for combat scenarios where instant interception of enemy RF signals is mission critical.

RESOLVE provides interception and exact geolocation of tactical signals across high frequency and ultra-high frequency bands to pinpoint the source and ensure accurate response in combat. RESOLVE is equally effective at signal interference to interrupt enemy communications in a tactical environment.

LOCATE, a lightweight, ruggedized version of the company's original Roke Electronic Warfare Manpack, works from a knapsack, providing infantry with a clear vision of what lies ahead. The system delivers accurate location of hostile targets via cross-reference of intercepted signals and has racked up sufficient success to spawn an array of imitators.

Russia's SwitchRay Masquerades as U.S. Company

Of the various Russian lawful interception vendors, few are as mysterious as MFI-Soft. Like a phantom or a "Leshy," the creature of Slavic mythology that can change shape at will, MFI-Soft has time and again shown the ability to vanish then reappear in a new form. In its present shape, the company is among the more prominent players – some would argue the leader – in Russia's burgeoning market for lawful intercept solutions.

Based in Nizhny Novgorod 260 miles due east of Moscow, MFI-Soft has been in business since 1989 and today sells network hardware/software and lawful intercept solutions to some 400 Russian communications service providers including Rostelecom, MTS, Megafon and Beeline, as well as leading regional service providers.

Customers for the full range of MFI-Soft products also include former satellite countries now part of the Commonwealth of Independent States (CIS), including Belarus, Kazakhstan, Kyrgyzstan, Ukraine and Uzbekistan. MFI-Soft lawful intercept products are deployed in at least two: Kazakhstan and Uzbekistan.

And yet to view MFI-Soft solely in one guise or place is to completely misunderstand its nature and reach. Through merger, spin-off, "partnership" with a Canadian shell company that posed as sales agent then was subsequently revealed as MFI-Soft itself, followed by divestiture of certain assets to a California startup named SwitchRay that weeks later launched with a sizable employee and customer base – under a Board majority-ruled by the self-same Russian company – MFI-Soft is a fluid corporate entity.

Today, for all purposes, SwitchRay is a bona fide U.S. tech company with a legitimate presence throughout Canada, Latin America and the U.S. But the presence of Board members with ties to Russian surveillance does raise questions.

The pair of Russian Board members have been affiliated with MFI-Soft or its partners since the company's earliest days. They comprise two-thirds of the Board of Directors at SwitchRay, which is based in Mission Villejo, CA, and also has a branch in Nizhny Novgorod for research and development – the same location as MFI-Soft's headquarters. One of SwitchRay's Board members is the registrant for MFI-Soft's website.

Yet publicly, SwitchRay has no official or even informal ties to the Russian lawful intercept and surveillance giant.

Still, if the lawful intercept business has a chameleon in its midst, that company is MFI-Soft. They've certainly done a good job confusing Privacy International and other groups of that ilk, which variously refer to MFI-Soft as "now known as Mera Systems" or "currently doing business as ALOE Systems." But truly, it's just MFI-Soft. For a closer look at the tangled tale of this shape-shifter, see the timeline that closes this analysis.

MFI-Soft Lawful Intercept Products

In its role as lawful intercept vendor, MFI-Soft produces a suite of products that Russian communications companies may purchase to be in compliance with SORM, that country's body of laws governing domestic surveillance. The company also sells general purpose network products which are used by service providers in more than 80 other countries.

Core products are shown below. Note: MFI-Soft products such as SORM 1, SORM 2 and SORM 3 are not intended to correlate with identically named provisions of Russia's lawful intercept technology requirements.

- **Sormovich E1T Probe.** "Son of SORM," as the literal translation goes, is a passive probe that like all such devices sits on the edge of the network – not part of inline network hardware – and is activated by a law enforcement agency or communications service provider upon receipt of a court order. Sormovich E1T is the "entry level" model of lawful intercept product offered by MFI-Soft. It accommodates common protocols including SS7 and PRI on any network and intercepts IP traffic including VoIP.
- **SORM 2 Mediation.** Moving upstream, MFI-Soft offers its own mediation device, the SORM 2, built to reside in close conjunction with switching and routing equipment within the network. SORM-2 may be configured to intercept the full range of targeted traffic – wireline, IP and mobile. One SORM 2 can handle traffic processing at rates of 4 Gbps to 10 Gbps. Used in cluster, the SORM 2 can scale to 320 Gbps.
- **SORM Converters.** A great many Russian CSPs rely on network hardware manufactured in Western nations. Typically such hardware is made to be compliant with U.S. and European lawful intercept requirements outlined in the Communications Assistance for Law Enforcement Act (CALEA) or the European Telecommunications Standards Institute (ETSI) – but not for Russia's SORM. MFI-Soft provides converters designed to ensure the interoperability of its SORM products with switches and soft switches made by vendors located outside of Russia.
- **Perimeter Deep Packet Inspection.** For small to mid-sized CSPs, MFI-Soft offers a low-cost, DPI product that enables Russian law

enforcement to monitor targeted IP traffic for content deemed counter to government interests. In Russia, DPI is also used for censorship.
- **SORM 3 Metadata Retention.** All Russian law, communications companies including Internet and social media vendors must store customer records on in-country databases and data centers for a period of three years, and make data available to law enforcement on demand and in real-time. MFI-Soft's SORM 3 serves the purpose. The combined hardware/software solution provides data acquisition and processing, storage module, and a control unit for handling searches. The unit can be designed for smaller local needs and also scale to support metadata generated by Tier 1 service providers.

Now to the curious history of this latter day Proteus.

The Mera Systems, MFI-Soft, SwitchRay Timeline:
- 1989: Dmitry Ponomarev co-founds and becomes president of Mera Systems, an IT company located in Nizhny Novgorod and specializing in solutions for telecom operators.
- 1989: MFI-Soft, an IT company also based in Nizhny Novgorod, launches.
- 1995: Konstantin Nikashov joins Mera Systems as VP – Business Development.
- 2001: Sergey Drozhilkin named VP at Mera Systems.
- 2007: Sergey Drozhilkin named CEO of MFI-Soft, by then a maker of VoIP solutions as well as lawful intercept and other surveillance products.
- 2008: Mera Systems merges its "Russian branch" with MFI-Soft.
- 2010: Mera Systems is renamed ALOE Systems, with new HQ in Canada and a "branch" office in Nizhny Novgorod. Konstantin Nikashov Named CEO of ALOE Systems.
- 2011: ALOE Systems reports 35 percent annual increase in sales, with 1000+ deployments in 79 countries.
- 2013: MFI-Soft website registered in the name of Konstantin Nikashov.
- 2013: SwitchRay Systems founded in Mission Viejo, California with 10 employees.
- 2014: SwitchRay Systems acquires intellectual property rights of ALOE Systems products with existing software license agreements and associated maintenance and support services agreements. ALOE Systems said to continue as "authorized sales agent" for SwitchRay Systems.

- ALOE Systems disappears. Searches for ALOE Systems default to SwitchRay Systems.
- Dmitry Ponomarev and Konstantin Nikashov named to Board of Directors at SwitchRay Systems.
- SwitchRay Systems jumps from 10 to 150 employees and 500 deployments in 32 countries.
- SwitchRay Systems' R & D is located in Nizhny Novgorod.
- 2016 - Edmund, Oklahoma-based 46 Labs acquires SwitchRay Systems.
- The labs in Nizhny Novgorod, and many key personnel convey to 46 Labs.

As of 2020, MFI-Soft is still doing a boom business from its Russian sales office in Moscow, its R & D center in Nizhny Novgorod, and in the U.S.

Sinovatio: China's Global Surveillance Player

Sinovatio, based in Shenzhen, China, is an established maker of surveillance products: lawful intercept devices, deep packet inspection, decryption, mobile location, forensics, social media, OSINT harvesting and big data analytics. But is a comprehensive portfolio sufficient to overcome the poisonous environment surrounding China's record of hacking everything – including via Sinovatio solutions purchased by naïve or ambivalent customers in other nations?

Concerned customers face difficult decisions when considering Sinovatio products. Sinovatio is a holding of ZTE, one of China's largest makers of telecom network equipment. Since 2012-2013, the U.S. has banned the use of ZTE, Huawei and Lenovo hardware by government and military agencies on the suspicion that it contains back doors that enable the theft of sensitive secrets. On the same grounds, the U.S. actively discourages American businesses from buying Chinese networking equipment.

While high profile Zero Day attacks such the infamous hack of the U.S. Office of Personnel Management tend to dominate headlines when such events occur, official concern remains high that the most pervasive risk comes from vulnerabilities deliberately built-in to Chinese tech hardware. At least one recognized intelligence authority believes that Huawei, ZTE and Lenovo act as government agents that conspire with Beijing's top brass to steal secrets, using network hardware to leverage back door entry to government and commercial systems in over 150 nations.

In all fairness, no one has yet found concrete evidence linking any of the three companies to involvement in cyber espionage. Huawei's CEO has refuted the allegations and countered that if the U.S. feels it can't trust Huawei or other companies based in the PRC, the American government might as well sanction all tech equipment made in China. Unfortunately, there is an impeccable logic to that argument that might work against Chinese tech enterprises hoping to end the U.S. embargo.

Thus far, the U.S. ban has had little impact on Chinese tech companies' financials. Huawei hummed along in 2019, earning US $122 billion in annual revenue – a 15 percent gain over the prior year – and surpassed Apple to become the world's second largest maker of smartphones, after market leader, Samsung. ZTE ranks seventh among the top 10 smartphone makers. Yet at the same time, like the flip side of a coin, the Chinese hacking community is arguably among the largest, best-funded and most sophisticated in the world, with proven connections to China's military.

Considering that all three – private industry, the military and black hat adepts – are closely linked in a totalitarian regime, it is hard to refrain from drawing a connection.

Inside Sinovatio

Sinovatio emerged over a period of 10 years through a series of shell, subsidiary and holding companies. In its first iteration, in 2003, the company was called Shenzhen ZTE Special Equipment Company Limited or simply ZTEsec, a holding of ZTE.

Four years later in 2007, ZTEsec created a subsidiary, Nanjing ZTE Special Software Company Limited, or Nanjing ZTEsec, specializing in software and communications products including surveillance solutions.

Then in September 2012, ZTE transferred 68 percent of equity to shareholders including Shenzhen Capital Group. A name change followed. ZTEsec emerged as Sinovatio, and Nanjing ZTEsec became Nanjing Sinovatio.

Appearances aside, the two companies are really one – Sinovatio – in different locations: Sinovatio is situated in Shenzhen's huge tech hub on the coast beside Hong Kong, and Nanjing Sinovatio lies northwest of Shanghai. For simplicity's sake we'll refer to both as Sinovatio. Whether using the name Sinovatio or Nanjing Sinovatio, they address the surveillance market, and speak to the same core interests when appearing at industry events.

Sinovatio is a primary supplier of surveillance solutions, with locations throughout China and deployments in over 70 countries, with particular depth in Asia, the Middle East and Africa. The company maintains its market position through a heavy emphasis on original research and development.

DeepInsight

Sinovatio offers a sizable product portfolio, DeepInsight Solution, for national security spanning the strategic and tactical needs of government agencies and law enforcement. DeepInsight is an umbrella that covers multiple product modules that may be used in modular or integrated fashion. Its principal focus is Internet, telecom and mobile networks, where it provides products that intercept both metadata and content.

Some of the products have multiple purposes; for example, Sinovatio's deep packet inspection may be used both for interception and content blocking. Others are single purpose, such as special modules for OSINT and SOCMINT (social media intelligence), or password detection to aid the process of decryption. Tactical products are offered for active mobile location and monitoring (IMSI catchers) and passive mobile location and monitoring.

Sinovatio also provides complete video monitoring systems together with video analytics. All the intelligence gleaned from such tactical tools moves upstream to Sinovatio's big data analytics and visualization system.

Lawful Intercept

Sinovatio supports the lawful intercept needs of police and the compliance requirements of communications service providers (CSPs) with active and passive solutions.

The active solution is Lawful Interception Gateway (LIG), a mediation device that interconnects directly with a CSP's network switches and routers. The device has been tested and proven interoperable with hardware such as that made by Nokia Networks (including the acquired assets of Alcatel-Lucent), Ericsson, Huawei and many others, of course including lawful intercept hardware from Sinovatio's parent, ZTE.

The LIG system is made in-house by Sinovatio and is compatible with all common standards including CALEA, ETSI and 3GPP. Like any high-end active solution, LIG works equally well with wireline and wireless networks, the latter including 4G LTE, and can integrate with passive probes and other devices made by third party vendors. Sinovatio counts CSPs, law enforcement agencies and government agencies among its clients for LIG.

As a complement or alternative to LIG, Sinovatio also offers its Internet Monitoring Center (IMC) system based on passive probes outfitted with deep packet inspection. IMC can be deployed at any point in virtually any type of network – Internet gateway, backbone network, MAN, WAN, Wi-Fi, PON, ADSL – and may also be used to intercept social media including Twitter, Facebook, LinkedIn, Skype, and others.

IMC's DPI capabilities can open packet payloads at OSI Layers 2 – 7 to reveal targeted content. The DPI engine may also be configured for purposes other than surveillance, such as content filtering and blocking for censorship.

Sinovatio markets the identical probe-based DPI product to mobile operators as Mobile Internet Monitoring Center (MIMC). MIMC does everything IMC does, but targeted at mobile data, social media and other IP-based apps. MIMC offers one additional feature of interest to law enforcement and government agencies: a rudimentary form of mobile location based on SS7, tracking proximity of a targeted device to a mobile base station.

Tactical Mobile: IMSI Catcher and Passive Intercept

Similar to its approach to lawful intercept, Sinovatio also offers both active and passive tactical mobile interception and location.

An early player in the IMSI catcher business, Sinovatio has offered its own "Mobile Catcher" product since the early years of the century. Mobile Catcher can quickly identify the target, capture the IMSI or IMEI number of a mobile device, determine its location and also intercept content. Like the Harris Hailstorm, Sinovatio's Mobile Catcher can perform man-in-the- middle attacks to intercept content, review data on the device, control downloads/uploads and activate the device's audio, camera and videocam.

Sinovatio's Mobile Catcher is designed with more than one type of mobility in mind: not just the target's device but also the user's ability to move in close. Mobile Catcher comes in vehicle-mounted and inconspicuous manpack versions. The latter permits the analyst or agent to locate and intercept the target's mobile device on the street or inside a building. The one snag with IMSI catchers is that, by interrupting the true base station tower signal, they can be easily exposed by inexpensive detection apps.

Sinovatio's passive solution – Mobile Signaling Interception System (MSIS) – provides an undetectable alternative. Deployed via optical splitter on the mobile operator's network, MSIS operates without any impact on the network, and is completely discreet and unnoticeable to the target. Agents can quietly intercept signals and full content, and determine location.

MSIS provides analytics capabilities that assess targets' activities, residency, links, behaviors, emerging trends, networks of affiliates, and immediate location. It can also send alerts on a target's actions if they become dangerous, or on any traffic analysis that points to abnormal behaviors. All of these features work in real-time.

The analytics apply to scenarios with known or suspected targets, and thus to finite data gathered for assessment. MSIS analytics can also be correlated with the broader findings of Sinovatio's big data analytics suite, which includes all data gathered by tactical products, as well as SOCMINT and OSINT.

Big Data Analytics

Sinovatio's interception and mobile specialty products are all standard fare, not strongly differentiated from the dozens offered by other companies in the same areas. While the Sinovatio Mobile Catcher, for example, performs all the basic IMSI capture functions, it really can't compete on the same level as Harris, which provides a dozen variations on the product depending on tactical requirements. With Sinovatio, the problem may be a matter of trying to cover too many areas, and as a result not excelling at any one in particular.

Recognizing this issue, Sinovatio devotes the greater part of its current marketing effort to presenting an integrated view of interception and big data analytics – with emphasis on the latter. When Sinovatio appears in public at events such as ISS World, the company touts its DeepInsight suite as an end-to-end approach that incorporates intelligence at the network level via probes, through to analysis of big data.

Sinovatio DeepInsight applies analytics tools to deliver insights from its surveillance products as well as SOCMINT and OSINT. As an integrated solution, DeepInsight draws on intelligence from Sinovatio's lawful intercept products as well as the tactical data gained from its active and passive mobile interception and location, and video monitoring solutions. DeepInsight also integrates with third-parties whose products include geofencing and data retention.

Is that big data analytics? Not when intercepts are confined to one or a few targets. In the latter scenario the data gleaned is small versus on a scale requiring massive parallel processing. But the picture changes when Sinovatio applies its DPI-enabled Internet and mobile Internet suite to capture and decode large volumes of structured and unstructured data in real-time at a national level. With these capabilities, the user is pointed to in-depth findings on timelines, links, behaviors, target identity and mobile location of others in a network of targets.

An important feed in this data stream is provided by yet another Sinovatio solution, DeepInsight Web Opinion monitoring, which scans social media to pick up similar intelligence – hidden identities of targets, trending topics, risk potential, timing, location and networks of affiliates, including those who might not even be on social media.

Lastly, DeepInsight can wrap in data from its own Digital Forensics Solution, which uses passcode-breaking technology to crack smartphones, tablets, PCs and other devices.

China's bid to build intelligence databases for future mining continues apace. The sale and placement of Sinovatio products beyond its borders might well be another driver of such intelligence gathering initiatives. The ability to leverage back doors in lawful intercept and other surveillance solutions, wherever they are produced, is well documented.

If you are a government agency monitoring the Far East or a business competing there, and your IT infrastructure, network or end user devices include or touch Sinovatio and ZTE products – then whatever data you communicate or store, Chinese authorities can and will access in real-time.

SS8: Evolution from LI to Commercial Cyber

SS8 is a multi-play lawful intercept vendor that also offers offensive cyber capabilities, and in recent years has diversified by entering the cybersecurity business for commercial accounts. From its headquarters in Milpitas, California USA, SS8 operates branches in London, Dubai, Tokyo and Merida, Mexico. The dual-market focus is evident from the company's multiple taglines: "Safeguarding Societies," "Comprehensive Communications and Visibility" and "Actionable Insight and Intelligence for the Enterprise."

SS8 did not start out as a surveillance company. They were, from the mid-to late 1990s, primarily providers of a signal and service platform called ServiceSwitch, which acted as a bridge between Signaling System Seven (SS7) used by the public switched telephone network (PSTN) and the then relatively new Internet. Services made possible by this platform, for marketing purposes dubbed Signaling System Eight (SS8), included unified communications, one number follow-me/find me, Internet roaming and virtual private networks (VPNs). Hence the origins of the company name, "SS8."

Events transpired quickly to turn the signaling system company into a surveillance player. In June 2001, SS8 hired a new CEO, Grant Wakelin, who came from the Enhanced Services Division of ADC Telecommunications. The division Wakelin led at ADC provided SS7 and softswitch applications. ADC Enhanced Services also produced software used to meet the monitoring requirements of CALEA.

Three months after Wakelin's arrival at SS8, jet airliners crashed into the World Trade Center buildings and the Pentagon. Wakelin immediately opened talks with his former employer, ADC, and on on November 1, 2001 SS8 announced the US $45 million acquisition, funded by Warburg Pincus, of ADC's Enhanced Service Division.

In all, Warburg Pincus raised US $62 million for SS8. As Wakelin noted at the time, the US $45 million piece was done for the express purpose of buying ADC Enhanced Services. In retrospect, the total sum raised by Warburg Pincus is impressive given the economic downturn that followed the year 2000 dot.com bust. For months after, Wall Street's sour mood on tech companies made funding for any company affiliated with the Internet extremely difficult to obtain.

Although Internet soft switches were the public focus of SS8's acquisition, something vastly different and more far-reaching drove Wakelin's thinking. He was set on re-framing the company's foundation, staking its future on the about-to-blossom markets for lawful intercept and government surveillance. After paying US $45M for the ADC division, SS8 had US $17 million in reserve to see it through the tech downturn, retain a research center in Canada and move Wakelin's company in a new direction.

SS8's Rapid Evolution

From those early days, SS8 quickly evolved to become a formidable presence in LI. The assets from ADC were just the start. SS8 went on to develop a line of products that spans the needs of law enforcement in LI, social media monitoring, analytics and offensive cyber. SS8 hired many of the best and brightest in the industry, including CTO Faizel Lakhani, who on his own and with partners held patents in data mining, security policy and Online Analytical Processing (OLAP), often a synonym for DPI.

In the ensuing years, SS8 picked up additional funding from an array of financiers that included Warburg Pincus, Kleiner Perkins Caufield and Byers, Goldman Sachs, Intel Capital, Novak Biddle, ONSET Ventures, Protostar Partners, W Capital Partners and Woodside Fund.

SS8 began to experiment. In 2009, the company's move into one of the more controversial aspects of surveillance – offensive cyber tools – gained unwanted exposure for involvement in hacking Blackberry phones offered by the United Arab Emirate mobile operator, Etisalat. At the 2014 ISS World - US, SS8 shared exhibit space with BT, the UK's largest communications service provider. The companies had paired to provide offensive cyber solutions for GCHQ. However, while such ethical malware always generates interest, the company's primary focus remained lawful intercept (LI).

Inside SS8's Xcipio Mediation and AXS Probe Devices

The iron horse of the SS8 LI platform is the Xcipio mediation device. Like competing mediation systems, the Xcipio is compatible with surveillance software modules installed in network switches and routers made by leading hardware manufacturers: Nokia Ericsson, Huawei, Cisco and many others.

Most impressive about the Xcipio, perhaps, is the amount of thought put into its design for a mobile market that leapfrogged to 4G LTE, and now 5G monitoring. SS8 showed a profound understanding of the challenges – and openings – posed by the current mobile standard. As correctly observed by SS8, the 4G standard was designed from the ground up as the sole channel of all voice, data, web surfing and SMS on mobile networks. But 4G LTE is not all roses for law enforcement and intelligence agencies trying to find and monitor targets on 4G networks.

The move to 4G led to an unprecedented leap in data speed and throughput, often leaving legacy LI systems in the dust. Services formerly confined to PSTN, DSL and cable carriage migrated en masse to 4G mobile, and inspired countless new apps and devices used on mobile networks.

SS8's Xcipio masters the 4G challenge, able to handle metadata and call content at any speed across four key points: (1) the Mobile Management Entity (MME), which serves as the primary control node on LTE access networks; (2) the Serving Gateway (SGW), which routes traffic between LTE and 3GPP elements and can be used to replicate traffic; (3) the PDN Gateway (PGW) which is the interface between a mobile device and the network, and provides a point of attack for packet filtering and DPI; and (4) the Home Subscriber Server (HSS), which establishes sessions on IP networks and is thus integral to the interception of metadata.

The Xcipio mediation device integrates with all leading vendors of MME, SGW, PGW and HSS, and also acts as a single platform integrating monitored calls from legacy 2G and 3G networks, as well as with IP multimedia subsystem (IMS) sessions on 4G LTE. SS8's Xcipio offers full IPv6 support for 5G lawful intercept, able to intercept a totally new type of mobile that is extremely low latency, and beam-centric versus cell-centric.

SS8's second piece of hardware is the AXS Probe, which like all such devices operates from the network's edge to passively monitor traffic designated by a lawful intercept court order. The AXS accommodates acquisition of CDR and IPDR metadata, full content of voice, as well as full content of data, the latter via DPI. The AXS may be used solo in a passive system or be integrated with the Xcipio in a hybrid approach.

SS8 Intelligo Visualization

Perhaps the most intriguing product in the SS8 line is the Intelligo platform which provides visualization of the data provided via the collection systems. Like Palantir, it works with assembled data pulled on known targets, in this case the intercepted metadata and content collected by Xcipio. Visual reconstruction lets the user see exact images of the target's Internet activities – Web browsing, email, SMS and social media. Intelligo's Social Network Analysis (SNA) feature maps connections in a link chart that looks much like a Palantir or similar visualization. Users can change formats to look at leaders, communications gateways or simply participants, and also review link charts of events and time of day. They can also use the system's Query Engine for searches of indexed metadata, full content and time-based behavior of targets.

With Intelligo, no target using social media is an island unto himself. Analytics builds patterns between the target and his associates, leaders and related organizations, creating unique Internet profiles of targets that may be correlated to real life identities. Designed with law enforcement in mind, Intelligo supports CALEA and ETSI as well as other international standards of lawful intercept.

SS8 Insights Platform for Enterprises

SS8's product line for lawful intercept is comprehensive, providing metadata and full content interception via mediation device, probe and/or a hybrid approach for CSPs, coupled with sophisticated analytics and visualization for law enforcement. However, they face tough competition from many other surveillance vendors offering similar products.

SS8's solution to this dilemma has been to diversify beyond the lawful intercept arena into enterprise cybersecurity, a huge and very alluring field for a broad market with urgent needs to safeguard company and customer data. SS8 made the leap to this lucrative niche with its SS8 Enterprise Solutions suite, essentially its LI products, re-branded for the commercial sector.

The "Xcipio Controller," and "Xcipio Sensor," are re-named, slightly modified versions of the company's mediation device and probe. SS8 also provides analytics for the enterprise, and in this instance the company's marketing department was candid enough to retain the name of the original LI product, Intelligo.

TelcoBridges Lawful Intercept for VoIP

TelcoBridges of Canada makes hardware and software systems for communications service providers, with deployments in over 100 nations. Established in 2002, the company specializes in session border controllers (SBCs) and VoIP media control gateways (MCGs). Privately held TelcoBridges competes with renowned brands such as Sonus, Juniper, ASTRAN, Cisco, Oracle, Edgewater Networks and Metaswitch Networks, to name a few. While the competition is in many ways outsized, one characteristic sets TelcoBridges apart from many in its class: lawful intercept hardware that comes equipped from the factory to do the job.

Having the foresight to include a capability that many SBC and MCG vendors omit or defer to partners, yet all providers of VoIP are required to provide by law in the majority of nations, is a key differentiator. Carriers that buy TelcoBridges hardware have the assurance that they will be in compliance with lawful intercept requirements – without the additional integration, testing, certification and costs that often arise from reliance on dedicated monitoring equipment. Compatible with ETSI and other globally recognized standards, the TelcoBridges solution is ready for action right out of the box.

With the growing importance of VoIP, the value of VoIP hardware with built-in lawful intercept capabilities cannot be overstated. Owing to its prominence, VoIP is now a principal target for law enforcement agents.

Operators are attracted to VoIP for the same reasons that lure customers: economies of scale and better use of bandwidth. VoIP makes full use of the "empty space" in conversations, using compression and eliminating redundancy in speech patterns. Carriers can fill that space with other profitable services.

In a word, VoIP is here to stay. With growth comes responsibility. Service providers can no longer look the other way on compliance with national surveillance laws. For many VoIP operators, such legal compliance can be a rude awakening.

In some ways, lawful intercept on VoIP networks is identical to the process used on legacy PSTN networks. Under the Communications Assistance for Law Enforcement Act (CALEA) and similar laws in other nations, the intercept must be confined to the target, be applied only to the time period specified in a court order, and must never impact other users of the network or jeopardize their privacy. Depending on the type of surveillance requested, the solution must be capable of intercepting signaling data (metadata), intercept related information (content), or both. The intercept must be invisible to the target, either through change of path via rerouting or other methods that use existing network elements.

But there are specific and unique challenges to doing lawful intercept on VoIP networks.

Compared to the PSTN, there are relatively few points in a VoIP network where both metadata and content can be monitored reliably and consistently.

VoIP service providers let customers choose their area code, which makes it easy for targets to conceal their true location. Some VoIP carriers provide the underlying network infrastructure for voice communications but not the service itself. In such situations, the identification of the caller may range from difficult to impossible.

Traditional lawful intercept hardware such as the mediation device formats metadata and content in the proper protocols required by law enforcement agencies before transmission to a monitoring center. Most VoIP network hardware is not designed to accommodate this essential feature, nor can it buffer interceptions to prevent loss of data.

VoIP implementations vary across the span of service providers, increasing the difficulty of meeting the full requirements of CALEA consistently. Unlike the easy days of the PSTN, there are not just half a dozen monopoly carriers with fairly uniform networks, but hundreds of service providers and networks.

Mobile operators that offer VoIP have special challenges – and risks. With the explosive growth of LTE and resultant demand for broadband, the trend is toward deploying femtocell base stations to reduce costs and ensure ample bandwidth on demand. For large operators, deploying monitoring equipment at each node is feasible, but not financially attractive. For smaller operators, the cost is prohibitive.

The final twist: It is all too easy for a target to adopt multiple identities on a VoIP service.

TelcoBridges *T*media to the Rescue

Can a VoIP provider solve these problems and be in "safe harbor" with CALEA compliance requirements simply by purchasing either active (mediation) or passive (probe) monitoring equipment? Certainly, and there is a wide range of choices. With dozens of LI equipment vendors to choose from, and a growing number of trusted third parties (TTPs) that provide end-to-end service, competition has driven prices down. In addition, new cloud-based and virtual lawful intercept solutions provide fresh choices that bring down the price. But these emerging cloud/virtual solutions still require additional outlays, either for shared hardware or software licensing. In sum, there is no free ride for any carrier. CALEA compliance solutions nearly always involve new capex, monthly charges and maintenance. This is where solutions such as those of TelcoBridges come into play. The ideal points for VoIP lawful intercept are session border controllers and media control gateways, where metadata and content converge for management, routing and security reasons. For lawful intercept purposes, the TelcoBridges *T*media family has the VoIP provider fully covered.

All too often, hardware companies produce a panoply of solo function devices that not only come at a stiff price, but require specialists for deployment, systems integration, testing and certification. When problems arise, vendors are quick to blame the customer for failure to provide sufficient data on network type and requirements, or to point fingers at the client's internal staff for not being available or knowledgeable. Personnel changes can lead to issues, too, for example if the techie on duty during deployment leaves the company, and his or her replacement is clueless about lawful intercept.

With TelcoBridges, everything the VoIP services provider needs, both to run the network and be compliant with lawful intercept requirements, is provided in one box from the *T*media family, and at no additional cost.

*T*media is that rare breed of network hardware that offers the benefits of a "two-fer" product. These gateway products come with ETSI 201 671 v2-1-1 standard compatibility built-in so that the user automatically meets the requirements of a lawful intercept court order for metadata and content, whether land or mobile VoIP.

Highly flexible, a *T*media gateway can also work in a non-ETSI environment. Of course, the main day-to-day function of the *T*media is to serve as a gateway for trunking, media, signaling and VoIP service connectivity and routing – where the functionality of most VoIP media gateways begins and ends, and where third party LI hardware is typically interconnected. But in this instance, no mediation device is required because the *T*media gateway includes that function. That adds up to simplicity and savings.

Where Lawful Intercept is Headed

At this point it is natural to ask, "Why don't all SBC and VoIP media gateways operate on this model?" and "How did the bifurcation of network and lawful intercept hardware begin in the first place?"

In the case of VoIP, the answer is that the two sectors grew up independently. Mediation devices arose in the PSTN era when there was no Internet, let alone VoIP. Probes came along later as a lower cost alternative built for the Web but also able to accommodate PSTN interception.

In the United States, with enactment of CALEA in 1994, hardware manufacturers began to produce switches and routers with lawful intercept software modules added on, for interconnection to lawful intercept hardware. Then, as VoIP gained momentum, a new generation of hardware manufacturer arose to provide SBCs and media gateways. Exactly like switch/routers makers, the vast majority of vendors that set out to make VoIP hardware never bothered with providing lawful intercept capabilities. To this day, while some SBC and media gateway companies partner with LI hardware companies, most operate on the principal that legal compliance is the network operator's problem.

TelcoBridges is the rare exception. For that reason, the company occupies a unique position in its industry, and is very well-positioned. The TelcoBridges model might be viewed as the conceptual forerunner of Network Functions Virtualization (NFV): a single all-purpose solution. In the case of NFV, any modifications or additions to a system will be software changes – easy, inexpensive and reliable. TelcoBridges does the same thing with its hardware. For the present, until NFV becomes a reality, *T*media gateway will continue to be "one box that rules them all" in VoIP.

Telesoft Technologies: Probes for Voice and Data

Serving more than 250 communications companies in 100 countries, Telesoft Technologies is one of the world's most formidable and prestigious providers of systems used by law enforcement and intelligence agencies to intercept voice, data and packet content.

Located in Dorset, UK – and not to be confused with another Telesoft, a provider of fixed and mobile telecom expense management solutions based in Arizona – Telesoft Technologies Ltd. has for more than 30 years specialized in telecom traffic monitoring, analytics and management solutions for communications service providers.

For a company with that range of expertise, surveillance is a natural fit, and Telesoft Technologies does not disappoint, offering an assortment of hardware that meets the needs of CSPs for compliance with court-ordered lawful intercept, and of intelligence and other government agencies for tracking and gathering evidence for the apprehension of criminals.

The Hinton Interceptor – PSTN and Mobile

The Hinton Interceptor is a probe that accommodates PSTN and all forms of mobile traffic, with the ability to provide "targeted or mass capture of tens of thousands of simultaneous conversations" from wireline and mobile networks. Originally designed to monitor 2G and 3G, the Interceptor has kept pace with advances in mobility's evolution to 4G LTE.

Standout features of the Interceptor:
- Ensures accurate interception on up to 100,000 targets simultaneously – the first passive probe to reach that level, and for a long time, the only one to do so.
- 100 percent capture of all voice and data calls monitored.
- Recording of calls, SMS and location data.
- Automatic identification of the target's circuit identification code (CIC) used in SS7 location monitoring – plus mapping.
- Conversion of intercepted data into correct protocols for transmission to a law enforcement monitoring center for analysis.
- Use of standard interfaces to transmit all data collected to the relevant LEA.

The market has spawned many competitors and imitators since Telesoft Technologies entered the field with its first probes: ClearTrail, Datakom, Emulex, Group 2000, INVEA-TECH, IP Fabrics, INNOVA Spa, iPS, Pine, Protei, Qosmos, Roke Manor Research, SS8 and Vision Group – and as many or more makers of active mediation systems that compete on the higher end. But strong competition has not hurt Telesoft Technology's market position. The Hinton Interceptor is still considered first rate.

Other Telesoft products dubbed "probes" go beyond the conventional definition of such devices.

Hinton Abis Probe – Geofencing

With Harris StingRay and other "IMSI catchers" much celebrated in tech media, geofencing – using intercepted cell phone signals to determine when a target is in a general defined area – is not as sexy as it once was, but it still has its value. Without geofencing via GPS or RF monitoring and triangulation, IMSI catching is all but useless: You must know the target's general location before an IMSI catcher can take you in close. So the Hinton Abis Probe's geofencing capability still supports an important niche need.

As a probe, the Hinton Abis connects passively to a mobile network, monitoring the radio network access interfaces. As the target moves, his or her mobile device shakes hands with the nearest cellular base station with the strongest signal, as well as with signals picked up by adjacent towers. Triangulating the signals allows the Abis to provide the target's rough position, generally within 100 – 500 meters. A third party geolocation app then converts radio measurements to a geofenced location that can be mapped.

Deep Packet Inspection Choices

Telesoft Technology takes a strong position on the importance of deep packet inspection to lawful intercept and other surveillance, stating "Network operators and law enforcement agencies need the ability to identify, sort and block selected packets in IP and converged networks." To back it up, the company offers a trio of products that can accommodate differing aspects of DPI:

- **Packet Probe.** Uses auto-discovery and DPI to deliver either metadata or targeted content from Gigabit Ethernet and SDH/SONET networks. The Packet Probe applies hardware acceleration technology to multiple 1 – 100 Gbps streams to filter Layers 2 – 4 (datalink, network, transport) of the Open Systems Interconnection (OSI) model, the basis of IP communications. Advanced DPI applied to Layers 5 – 7 (session, presentation, application) then identifies and captures the targeted content. The Packet Probe extracts both voice (VoIP) and cellular data, and also produces call data records (CDRs)

for later analysis. Very price-friendly, the Packet Probe also integrates with other collection and mediation devices already in place, using standard ETSI interfaces.
- **Packet Extractor.** To accommodate today's broadband glut, Telesoft offers the Packet Extractor, a field programmable gate array (FPGA) that pre-filters packets before sending them to another third party DPI solution. Packet Extractor does all the heavy lifting first – scanning multiple 1 GB – 100 GB streams to sort out data relevant to the target, as directed by an external DPI engine. By winnowing specific traffic from jam-packed networks, Packet Extractor reduces the load on the DPI engine (and the cost of investing in multiple DPI probes). Law enforcement and intelligence agencies appreciate this device, too. Reason: They can bypass the service provider's in-house DPI engine and have the data sent via protected line straight to their own monitoring center for further analysis.
- **MPAC-IP Monitoring Cards** – or **Packet Acceleration Cards**, as they're also known, deal another winning hand, depending on the user's needs. The cards can be installed directly in a law enforcement or other agency's existing servers to provide much the same function as the Packet Extractor: high-speed and intensively accurate packet filtering so that the engine receives only the data needed for analysis versus the entire mountain of broadband transiting a CSP's network. The user has full control of the filters for Layers 2 – 4, then a second stage filter grabs the target's content, and from the start of a message, not just from a match to the target. Fully buffered, the cards guarantee zero data loss.

Of all the aspects that make Telesoft Technologies products attractive, flexibility is tops. With Telesoft the customer is not required to buy one master product that may offer features that are seldom or never used. Telesoft has a product for each surveillance need. Customers can buy the Interceptor for fixed and mobile voice, and the Abis for mobile geofencing. Customers can buy a full-fledged DPI engine, a filter alone, or just the cards to handle filtering on their own DPI engine. In today's budget-conscious environment, Telesoft's "a la carte" approach has solid appeal.

TraceSpan: Where Fiber and LTE Monitoring Meet

Are LTE and fiber network surveillance worlds apart, requiring separate high-end monitoring appliances? Not when the solutions provider is TraceSpan. While some ISS hardware vendors specialize in surveillance of fiber networks and others in 4G LTE, TraceSpan goes where these seemingly disparate worlds cross paths: the network-agnostic realm of telemetry. Its core products are GPON Phantom and LTE Advanced Xpert.

Headquartered in Ra'ana, Israel since its founding in 2002, TraceSpan has grown to become a significant provider of network telemetry solutions for many of the world's Tier 1 communications service providers including BT, AT&T, Deutsche Telekom and T-Mobile, as well as "big box" makers such as Nokia, Cisco and Huawei. Concurrently, TraceSpan also makes specialty lawful intercept equipment for use in ADSL and VDSL networks (though seemingly obsolete, they still exist in 2020), as well as GPON (Gibabit Passive Optical Networks).

The company's line of commercial products for network monitoring used in performance testing purposes demonstrates facility with every network type up to and including 4G LTE. The architectures for testing and lawful intercept applications are fundamentally similar, enabling TraceSpan to compete in both arenas. Moreover, because wireless operators have turned to GPON as the optimal approach to mobile backhaul – for both performance and economy – TraceSpan GPON Phantom is applicable for lawful intercept at that juncture and can play in the LTE realm, as well.

As originally designed, GPON Phantom had two primary applications: as a "remote site solution" for interception of specific targets on GPON fiber-to-the-home (FTTH) deployments; and as a centralized solution for monitoring multiple targets. For remote site intercepts, a law enforcement agency taps directly into the FTTH connection – a GPON link – between the target and a service provider's central office. The service provider is not involved. With centralized monitoring, the GPON Phantom is situated in the service provider's central office, connecting to an optical switch that does the job of tapping however many fiber lines the law enforcement agency designates.

As a passive tactical device for solo work, GPON Phantom may be deployed from any secure location to collect data between the target and the provider's central office, without alerting either. Centralized monitoring offers the advantage of intercepting any number of targets, but with the service provider's knowledge and cooperation, as with a classic CALEA or ETSI lawful intercept. As a passive optical monitoring device, the GPON

Phantom provides simultaneous upstream/downstream interception, captures data in real-time, and time stamps it. The device can store data or forward it in encrypted format to an LEA monitoring center.

A key advantage of optical taps is that they provide highly granular data on the traffic transiting a fiber link – a precise duplicate of the signal. However, care must be observed in deploying a fiber tap. Unlike conventional wiretaps, whose basis is electricity, fiber taps rely on light, which moves in only one direction. When deployed, a fiber tap diverts a tiny fraction of the light signal from the intended receiver to the monitoring equipment. Adequate optical power is essential within the tapped port to ensure that the telemetry equipment produces an accurate copy of intercepted data for analytics. Also, because tapping can cause slight "jitter" in the light signal, particularly when monitoring a high-speed GPON link, care must be taken to minimize jitter that might affect the quality of service and thus inadvertently alert the target.

With the predominance of mobility, rapid growth of broadband and demands on network infrastructure, service providers naturally have a keen interest in monitoring traffic to determine customer needs, forecast trends and head off problems that might arise due to inadequate support. Monitoring traffic at the protocol level provides invaluable data on trends, content types, and bandwidth availability to ensure high levels of performance. Telemetry devices perform this intelligence-gathering function.

In a conventional pre-5G mobile network, cell towers route traffic to master switching centers via "backhaul" facilities. Initially, backhaul was handled by wireline or microwave facilities. In the early 1990s facilities-based operators began introducing fiber optics. Everyone raved about the quality, but bemoaned the high cost. As an alternative, carriers migrated to an equally fast but more cost-effective option: Passive Optical Network or PON, which uses fiber optic splitters enabling a single optical fiber to serve multiple end points. PON and then Gigabit PON (GPON) were immediate favorites with telcos pushing FTTH services for video, and for mobile operators looking for better ways to meet accelerating broadband demand while at the same time reducing backhaul network costs.

GPON today is among the fastest-growing segments of the fiber optic marketplace, both in mobile and other types of networks. To monitor high-volume, high-speed LTE traffic for network performance purposes, mobile operators use commercial GPON telemetry devices that tap the network at the switch level, mirror traffic flows, filter flows as designated and send replicated packets to server-based analyzer tools.

GPON-based lawful intercept tools use the same architecture as their network performance telemetry peers, and work on similar principles. Fiber is fiber, whether deployed between central office and residence, or in the mobile backhaul network, and GPON telemetry and lawful intercept solutions are network-agnostic in that regard.

As mentioned, TraceSpan has established its credentials on the commercial side with another product. LTE Advanced Xpert, intended for network performance telemetry, captures and provides real-time analysis of the downlink and uplink on both LTE and LTE-Advanced networks, providing a complete view into the elements of the radio access network. Xpert works either through direct tap of the network or "over the air" to probe any LTE band. In addition to the usual areas of interest for tuning networks, fathoming trends, and preempting customer problems, Xpert provides one other service favored by law enforcement and government agencies: LTE protocols and "upper layers" – Layer 7 of the OSI stack.

Trovicor: Spyware's Black Panther

Trovicor, known for its black panther logo, is the classic multi-play spyware vendor, offering lawful intercept solutions, mobile location, semantic technologies, offensive cyber, analytics and OSINT. Solutions leverage the company's own strengths, and also incorporate best-in-class tools of other market leaders to evolve quickly with the needs of the marketplace.

Trovicor traces its roots to German multinational conglomerate Siemens, and to a subsequent joint venture with Nokia Siemens Networks (NSN), the latter now better known as Nokia Networks.

Beginning in the early 1990s the company that would become Trovicor was part of the Siemens Intelligence Solutions group's seminal Voice and Data unit. With the rise of the Internet and creation of diverse new communications mediums, the surveillance end of the business accelerated. While many in the business began churning out boxes or software, Siemens adopted a more holistic approach with its Intelligence Platform. Built on a monitoring center developed by NSN, the Intelligence Platform was among the first big data analytics plays designed to incorporate feeds from multiple sources.

Classic lawful intercept of content and metadata, together with mobile location, were included. But Siemens opted to go further, adding intelligence modules that could incorporate and apply analytics to data feeds from a wide variety of sources: financial transactions, air flights, car rental, voice recognition, fingerprints, DNA, police and criminal records, to name a few.

With this platform, intelligence agents in many lands could for the first time develop a layered view of the target. Heuristic analytics created target profiles and patterns. Link analysis revealed connections to other potential members of a criminal or terrorist cell. Predictive analytics pointed to near-term threats. Real-time analytics brought threats into the present, providing "next best actions" for tactical preemption.

For agents who had long been forced to rely on multiple tools that at best delivered a fragmented picture, the idea of gaining a complete detailed view in a single solution proved very attractive. Siemens' Intelligence Platform quickly won customers in Europe, the Commonwealth of Independent States (CIS), the Middle East, Africa and Asia Pacific.

For reasons that remain unexplained but subsequently proved a blessing for the company, Siemens management steered clear of the lucrative North American marketplace and – with the exception of the UK – instead focused on Tier II and Tier III nation markets. By 2007, Siemens Intelligence Solutions and its Intelligence Platform had won approval from numerous customers.

Government agencies that had struggled to manage multiple challenges – not just terrorism and organized crime, but also smuggling, illegal immigration and threats to critical nuclear and energy infrastructure – now had access to one system that could handle them all.

Siemens and NSN Take Heat

In 2005, German investigators began to explore rumors that Siemens was involved in the bribery of government officials to obtain contracts. Because the company was traded on the New York Stock Exchange, the investigation in 2006 spread to the U.S., involving the Securities & Exchange Commission and Department of Justice.

Reviewing millions of documents, criminal investigators discovered that since at least 2002, Siemens had operated a bribery unit that annually doled out $US hundreds of millions in payoffs. The purse was staggering. Between 2001 and 2007 Siemens reputedly made $US 1.4 billion in illegal payments, with the principal recipients being China, Russia, Argentina, Israel, Venezuela, Greece, Norway and Nigeria.

Neither Nokia Siemens Networks nor the Siemens Intelligence Solutions unit were specifically referenced in the suit. Bribes were aimed principally at winning deals in the aligned field of telecommunications equipment. Bribes funneled through offshore bank accounts typically cost Siemens five percent of the contract value, though in some instances the payoff figure went much higher.

One German investigator characterized the company's offshore operations as follows: "Bribery was Siemens' business model." He may have been playing to the media. While the evidence against Siemens was strong, it's just as likely that the company was simply dealing in the realities of many of the markets it entered, where the practice of paying "baksheesh" has been part and parcel of business dealings for centuries.

Regardless, when the suit came to settlement in late 2008, Siemens ended up paying $US 1.6 billion in fines – at the time, the largest ever penalty of its kind – and $1.0 billion as reimbursement for government investigations in the U.S. and Germany. The cost of restructuring and cleaning house at Siemens was estimated at another $US 1.0 billion. Siemens paid up, cleared the decks and moved on.

Then in mid-2009 came the first hint of troubles in the Arab world. Reports surfaced that NSN technology was deployed in Iran to monitor protesters. The company abruptly halted all work related to its monitoring center in that country.

In 2011, North Africa exploded with the Arab Spring. Appended to that story: the revelation that NSN tech was used by regimes throughout North Africa and the Middle East.

NSN did not wait for an investigation. They quickly spun off their intelligence unit to an investment group based in the Channel Islands. That entity is now Trovicor, providing the same technologies and services as before, now under the sign of the black panther.

In a world where AI has become a buzzword for everything imaginable, it is almost refreshing to find a vendor that doesn't utter one syllable about machine learning or deep learning. Trovicor sticks to solutions that have proved their worth in helping law enforcement and government agencies track and capture criminals and terrorists.

While Trovicor's product suite may look diffuse at first glance, each module has a role that makes an important contribution to the big picture law enforcement needs to see to make informed decisions and avoid false positives. Data from one source is integrated with the whole to reveal new information that might not have been previously recognized. All modules work in tandem to build, complement and enhance intelligence gathering.

The solutions suite, in full:

- **IQ Engine** for Deep Packet Inspection (DPI). The Trovicor IQ Engine, loosely patterned after the world-leading Qosmos ixEngine, collects and makes exact mirror copies of targeted – or all – data transiting an IP network. Trovicor's full DPI product suite offers models that can scale up or down to any law enforcement agency's specific requirements. The company's top-of-the-line DPI product, the IQ100 Engine, works in real-time, able to decipher encrypted data and provide both content and metadata – in the clear – and can handle over 1,000 applications with ease. E-mail and SMS are a specialty. By fully examining the metadata in messages, Trovicor's IQ Engine quickly discerns the addresses of each party, determines their personal identities and MSISDN numbers, then can jump to content and attachments to see a target's plans in the moment. Detailed findings are then routed to the Trovicor Monitoring Center and Intelligence Platform.
- **Geo-Location.** Trovicor relies on the power of not one, but multiple approaches to targets: for general location, network-based data from call detail records plus SS7 tracking for target distance from the nearest mobile base station; and for positioning, radio measurement analysis coupled with the use of passive tactical direction finders. Together, these techniques let the agent see targets as they move. The user can

choose solutions that track target movement both in-country and internationally. Key augmentations let the user jam or spoof the target's communications.

- **Cell-Based Monitoring.** To keep an eye on specific zones, whether borders, airports, harbors, train stations, high risk areas, military complexes, buildings, sporting events or any public gathering, Trovicor's Cell-Based Monitoring lets agents zero in on cells and micro-cells in mobile networks. This solution collects all communications content and metadata in a designated area, and enables the user to conduct analytics based on specific speech patterns or parties involved. Criminals or terrorists lurking on the fringe of or mingling with a crowd are revealed as being connected when they speak or text, and singled out for rapid response action.
- **Passive Voice Interception.** This capability can capture every word that transits a fixed wireline or mobile network, both domestically and internationally. "Passive" means that the solution operates independent of the CSP to avoid detection. The system comes with an embedded analytics suite, and can also integrate with the Trovicor Monitoring Center and Intelligence Platform for detailed analysis. Alternately, the data may be routed to Trovicor's Data Retention system for cataloguing, filing and subsequent analysis.
- **Social Network Crawler** – for public or restricted social sites. Bad actors don't always cooperate by operating on publicly-known social media sites, and instead use Dark Web options. The Trovicor Social Network Crawler can find them wherever they might lurk, be it on a public or secret social site. Based on the chosen search mechanism – social ID, keyword, trend of specific social media site – the Crawler proceeds via automated processes to find the target and his/her affiliates all the way to the Dark Web. As often as not, the solution will find previously unrecognized players for monitoring, and learn their backgrounds, relationships and plans with others. Crawler can also geo-locate the targets using a solution from Gamma Group/ FinFisher, which also supplies Trovicor on another important front: Offensive Cyber.
- **Cyber Investigation Solutions.** Trovicor offensive cyber tools include Zero Day attacks, malware and non-malware endpoint attack capabilities provided by Gamma Group/FinFisher. Anything FinFisher can do, Trovicor can also, from social engineering and phishing to gaining access to a target device, capturing all content and stored files, monitoring all voice and data communications in real-time, and

taking control of the device. As with data captured by other Trovicor modules, intelligence collected by the company's Cyber Investigation Solutions integrate with the Monitoring Center, adding fresh insights.
- **Interception Monitoring Center for Mobile.** With wireline telecom rapidly fading from view, the vast majority of lawful intercepts target mobile networks. Trovicor's dedicated solution for this big niche is the Interception Monitoring Center for Mobile, designed to keep a close eye on any and all mobile communications. Just in case, the company throws in wireline network interception from both public switched telephone (PSTN) and cable networks. As an "active" solution designed to interconnect with lawful intercept software in networks, the Interception Monitoring Center is fully tested and operational with hundreds of physical switches as well as the virtual variety – Session Border Controllers and Media Gateways.
- **Monitoring Center.** The Trovicor Monitoring Center serves as the meeting point for intelligence gathered by other modules. The Center provides a single platform and central hub for collecting, collating and connecting evidence from each source including Trovicor's Gamma Group/FinFisher offensive cyber suite. Centralization of data means users have one convenient place to upload and hear captured voice calls, or view metadata, files, images and video. The Monitoring Center organizes evidence, and is the prime data source for the final step of Trovicor intelligence process.
- **Intelligence Platform.** Formerly known as the Fusion Center, the Intelligence Platform provides the best of heuristic, real-time predictive analytics (RTPA) and visualization. Pulling huge volumes of data from the Monitoring Center, the Intelligence Platform filters, applies statistical analysis to and scores data to instantly determine the significance of intelligence and accurately intuit the immediate future of a potential high-risk scenario. Where classic RTPA ends, however, Trovicor's Intelligence Platform goes into rocket mode displaying RTPA's findings with high quality, stunning visualization. The platform's graphics make the next step clear – whether to act at once or go on hold and wait. The patterns in Trovicor's visualization tell all.

Utimaco: Flexible, Top Quality Interception

In business as an IT security company since 1983 and a surveillance vendor beginning in 1994, Aachen, Germany-based Utimaco shares honors with Aqsacom as a pioneer and ongoing innovator of surveillance technologies deployed by law enforcement and government agencies in all corners of the world.

One notable aspect of the company: Utimaco is very open about the business it's in and why lawful intercept is vital to society. At the same time, Utimaco does not – unlike many vendors – assume that its audience understands the business or their responsibility to be compliant with surveillance laws. They speak in the language of the service provider customer, explaining how the growing complexity/diversity of communications has not only benefited the consumer of these advanced services, but also provided new tools with which dangerous elements can thwart and avoid the law.

When Utimaco speaks, it resonates with carriers. No sales spiel, just straight talk in telco lingo, as exemplified here:

The interception of a single e-mail message can pose a major challenge to an Internet Service Provider because of the high volume of IP traffic handled by a typical large Internet Exchange, such as the Internet Exchange DE-CIX. This organization calculates the average throughput of the 175 Internet Service Providers it carries at 41.3 gigabits/second, and spikes in traffic range to nearly 70 Gbps. Clearly, state-of-the-art technology is required to handle lawful monitoring activities that involve this level of data throughput.

Utimaco's mission is to ease the job of lawful compliance for carriers and ensure the collection of incriminating data across the spectrum of advanced voice and data services used not only by the average citizen, but also by criminals and terrorists.

The workhorse of Utimaco's line of lawful intercept products is the Lawful Intercept Management System (LIMS) which provides active, passive or hybrid solutions used by service providers in accordance with national laws, benefiting lawmen, agents and analysts on the receiving end of the evidence stream.

Inside LIMS

LIMS is a flexible hybrid solution comprising both "active" and "passive" components for fixed wireline, mobile, Internet and next generation network service providers. Whatever the need or type of network, Utimaco LIMS covers it – one reason LIMS has been deployed in some 40 nations to date.

The basic operation of LIMS is a simple concept. In active versions, LIMS serves as a mediation system between lawful intercept modules resident in network switches and routers, and LEA monitoring centers that administer collection and analysis of intercepted call data and content. Utimaco refers to these bodies of collected data, respectively, as IRI (Intercepted-Related Information) and CC (Content of Communication).

The LIMS Management Server is the centerpiece of the system, programming and administering surveillance software modules resident in network switches to single out calls and voice content from the target as designated by court order. The LIMS Management Server maintains a complete database of all intercepts and permitted LEAs, and also maintains security.

In the active version, a LIMS Mediation Device is deployed to handle functions relative to converting intercepted IRI and CC into the correct protocols, format and Handoff Interface (HI) for the law enforcement agency involved in the investigation. Utimaco covers the field on communications hardware, offering mediation devices that work easily with any of some 250 different network elements by well-recognized manufacturers. Furthermore, because Utimaco devices are compliant with major standards including CALEA, ETSI, ATIS and 3GPP, they can be used by LEAs in virtually any nation.

In the simplest active setup, the LIMS Management Server programs surveillance software modules resident in each network switch. Once activated, the Mediation Device collects and formats the IRI and CC intercepts in the proper format and HI for the LEA.

The day-to-day reality for many service providers is that they offer multiple services – mobile, VoIP, Internet, SMS, e-mail, and to some extent traditional wireline voice. The basic process of lawful intercept does not change, regardless of the type of communications being monitored. However, the system could involve multiple mediation devices dedicated in sufficient quantity to monitor each service. In the case of lawful intercept for Internet service, an active solution would also involve a LIMS AP Radius Router to perform full content intercepts using deep packet inspection.

Another option with Utimaco is to go the passive route. In that event, Utimaco provides LIMS access points in the form of probes that work independently of other network hardware to filter, decode and forward intercepts directly to the LEA monitoring center. Utimaco offers a variety of probes, including models outfitted with DPI.

Depending on the complexity of the network, Utimaco might recommend a hybrid approach that uses both probes and mediation devices. In a hybrid network, probes report to a mediation device, which handles formatting and delivery of intercepted data to the LEA.

Feature rich, all LIMS systems scale and load balance, automatically reroute traffic in the event of bottlenecks, provide remote provisioning, decoding of data intercepts, and can be integrated with third-party monitoring centers. Utimaco's IT security heritage comes into play via hardware-based encryption that protects LIMS systems from intrusion, and authentication that works either standalone or integrated with an existing internal system.

Performance-wise, Utimaco LIMS can scale to support service providers with any size subscriber base, from a few thousand to many millions. The LIMS active solution can handle up to 400,000 intercept targets, and the LIMS 10 Gb probe up to 25,000 per LIMS access point. LIMS is offered at a fair price. Over time good design means low maintenance costs.

During the Arab Spring meltdown, Utimaco was among those named and subsequently condemned by Privacy International and the media. In the aftermath, British IT company Sophos, which had acquired Utimaco just three years earlier, issued a public apology and promptly spun off the German company to private investors. Now that the tempest has subsided, Utimaco goes about its business and shows every sign of prospering.

Vehere Takes on The West

India-based Vehere is a provider of diverse lawful intercept products, as well as spectrum monitoring and cryptanalysis specific to military and national security needs. Founded in Kolkata in 2006, Vehere offers a lawful intercept suite that includes an enriched monitoring center, DPI probes, mobile location solutions, data retention, and capabilities in voice biometrics, analytics and visualization – sold through sales offices that extend to Singapore, Dubai and Kansas City and supported by a global network of partners.

Credit for this success goes to Vehere's versatility, technology and aggressive sales and marketing. But a signature event also played a role: the 2008 terrorist attack on Mumbai, often called India's version of 9/11. Following Mumbai, India enacted laws that created a framework for ISS sector growth: pro-national security policy that quickly propelled India's private surveillance sector from nascence to maturity. Vehere was one of many local companies quick to capitalize on the opportunity, committing research and development to products specifically designed to meet the needs outlined in the new laws.

Vehere still makes a significant investment in R & D, reflected not only in its product line but in original research that is internationally recognized by IEEE, NASA and CERN, and by an academic-based research community including Cornell and Harvard – something of a rarity in the private surveillance sector.

It's the work of a company with big ambitions. Geared not only to India's needs but those of the broader world market, Vehere's lawful intercept products are compliant with international standards including CALEA, ETSI, SORM for Russia and CIS countries, and can be adapted on a one-off basis for other country-specific technical requirements.

India Responds to Mumbai

Prior to the Mumbai attacks, lawful intercept and intelligence activities in India were carried out under authority of the 1885 Indian Telegraph Act and the 1950 Telegraph Rules, with two government security bodies placed in charge: the Research and Analysis Wing (R&AW) and the Intelligence Bureau (IB). In the hours leading up to the Mumbai attack both the laws and their administrators showed themselves seriously out-of-sync with the times.

A RAND report issued in 2009 concluded that although advance notice of a possible attack on Mumbai was provided by the United States, and Indian intelligence was itself aware of the potential threat, there was little coordination and local police in Mumbai were not notified. Coordinated attacks by four terrorist teams led to the deaths of 172 persons.

RAND cited failures from the highest levels of intelligence down to the street level response, where Mumbai police were characterized as "poorly trained and lacking ability to collect, store and analyze forensic evidence in accordance with international standards."

After the fact, India moved to correct the problem, passing four laws:

- Information Technology Procedure and Safeguards for Intercepting, Monitoring and Encryption of Information Rules (2009).
- Information Technology Procedure and Safeguards for Monitoring and Collecting Traffic Data or Information Rules (2009).
- Information Technology Intermediaries Guidelines (2011).
- Information Technology Guidelines for Cybercafes (2011).

Among the highlights:

- All wireless customers must register with the government.
- All telecom, wireless, Internet and providers are required to deploy and maintain technically compliant equipment that will facilitate an intercept.
- Indian authorities have the right to obtain any evidence available on a communications network.
- CSPs must store all communications metadata for a minimum of one year.
- Cybercafes must be capable of providing name, address, photo ID and any relevant video surveillance data on customers to government agencies on demand.

The laws quickly came under attack from the privacy community, claiming that the government had made the Indian intelligence community master of all communications metadata and content within India. Despite legal challenges, however, the laws stand.

High Throughput DPI Probe

For Vehere, lawful intercept begins with a sophisticated passive device, the EI Probe, capable of 100 Gbps throughput. The probe can handle monitoring of any wireline, wireless, or IP network.

Outfitted with deep packet inspection, the Vehere probe filters traffic based on rules established by the client, decodes payload, then performs full packet content inspection at the Layer 7 application level to identify protocols and applications.

Designed for today's next generation networks, Vehere's probe intercepts both IP data records (IPDRs) and IP content, and can be programmed to select data by parameters that include keywords – single or multiple, message content or attachments, in any volume – plus by special characters or case sensitivity.

When used by a communications service provider operating under CALEA, ETSI, SORM or other nation state standards, the probe captures data in the appropriate format per LEA requirement. LEAs that purchase the Vehere probe for direct use simply set up the formatting in advance.

As with any DPI product, taking packets apart is one matter, reassembling them for the LEA in the exact format used and seen by the target, quite another. Here Vehere's packet reconstruction tool "Replay" steps in, reassembling packets in the original protocols so that agents see what the target sees.

Centralized Monitoring

The centerpiece of Vehere's product line is the CommuLIM Monitoring Center, a turnkey solution for law enforcement and government agencies. CommuLIM performs centralized provisioning, monitoring, processing and distribution of voice and data communications, either from a switch-based mediation device or a passive probe positioned at the edge of the network. As an end-to-end solution, CommuLIM performs all the standard functions expected of any LIMS system, intercepting, organizing and formatting metadata and content for analytics, and more.

Where CommuLIM takes off is in its built-in analytics capabilities. With the core analytics tool the user can submit queries that reveal trends, patterns, social links, networks of cohorts previously not recognized, principal leaders and – via GIS – their locations. The data is culled and viewable through a user-friendly GUI and dashboard that provides detailed visualization of the data from any facet of interest to the agent or LEA officer. CommuLIM analytics work in real-time, and also deliver alerts in the event that monitored content points to an imminent threat. As with the Vehere Probe, CommuLIM is CALEA, ETSI and SORM compliant and can be adapted to any protocol for lawful interface handover.

CommuLIM integrates with a sister tool for CDR analytics tool called vCRIMES for metadata analysis of call detail records, subscriber data, passport information and other records. Based on Vehere's Distributed Mass Data Processing Engine, vCRIMES is fast – clocked at processing 40 billion records in three seconds. Like CommuLIM analytics, vCRIMES reveals patterns, networks and sleeper cells hidden in network and social media connections. One added feature: dynamic probability analysis of threats, i.e., predictive analytics. vCRIMES also puts metadata to work for target positioning, using GIS data captured within CommuLIM to map the record of target location.

Voice Biometrics and IMSI Catcher

Because mobile location based on GIS alone can be limited to heuristic findings or general position of the target relative to a mobile base station, Vehere offers two further products that help make location pinpoint accurate.

The first is Vehere's voice biometrics product, VSIS. Like Nuance's AGNITIO, VSIS is a language independent system that creates a unique biometric voiceprint, splits conversations by multi-speaker separation, and singles out the target and his/her gender, regardless of device, network, location – or whether the target uses multiple languages even during the same intercept. VSIS works in real-time across multiple channels, and can be used with CommuLIM or other existing systems including electronic warfare (EW).

To eliminate all doubt on location, Vehere also sells its own IMSI/IMEI catcher. Similar to the Harris StingRay, it emulates a network cell tower, captures one or an entire group of mobile devices' IMSI/EMEI numbers, and can find the desired target to within one meter of geographical location, again, in real-time. As with other Vehere products, the client has choices: one model that alerts the user to mobile activity in sensitive areas of interest such as airports, railway stations, border perimeters or military bases, and another version with specific tracing/locating capabilities.

Making Big Data Manageable

Recall that Indian law now requires CSPs and ISPs to maintain all CDRs and IPDRs for a minimum of one year, a data retention challenge of immense proportions given India's population and the fact that carriers stimulate usage by providing LTE calling free of charge. For CSPs, Vehere offers its own branded data retention system, the vCRIMES DRS Intelligent Data Retention suite.

vCRIMES DRS is a high performance solution that brings data volume to heel, storing it in accordance with the A3 standard (Authentication, Access and Audit Trail) so that evidence-rich data may be readily accessed when needed, and in a format responsive to queries by law enforcement and national security agents. With trillions of records to contend with, this is no small feat. Like the vCRIMES metadata analysis solution, Vehere data retention relies on the company's Distributed Mass Data Processing (DMDP), which in this instance stores and provides rapid access to IPDRs, CDRs, full content intercepts, and location data.

Taking on The West

When ISS World convened its first conference in Dubai, surveillance gunslingers of "The West" jumped on the opportunity, arrived en masse, and comprised 100 percent of show sponsors and exhibitors.

It is a very different picture now, with companies that include Dubai's Stratign, South Africa's VASTech and Seartech, and a full cadre of players from India – Vehere, ClearTrail and Paladion, to name a few – occupying prominent places on the roster. Today, companies adjacent to a regional conference, but with decided global ambitions, are apt to constitute 30 to 40 percent of the program agenda. With R & D-based technical sophistication driving a versatile product line, Vehere is among the frontrunners giving North American and European-headquartered competitors healthy competition in lawful intercept solutions.

One plausible advantage held by India's home-grown surveillance leaders is simple geography: operating in the shadow of major terrorist havens to the North and West.

Verint's Big Iron STAR-GATE

Now in the third decade of production, the Verint STAR-GATE is remarkable for its longevity as an active lawful intercept solution still popular with large Tier 1 operators. Though destined for discontinuation as virtual solutions take over, the STAR-GATE is worth examining here for detailed understanding of how mediation devices and their support systems work, end-to-end.

STAR-GATE Legal Interception Gateway, as it's known by its full name, is a comprehensive access and administration system for intercepting content and data of calls and data transfer sessions. STAR-GATE is sold to communications service providers, ISPs and broadband companies worldwide. Verint also offers a STAR-GATE Lite version designed expressly for rural and other smaller rural broadband operators.

STAR-GATE serves as a turnkey solution for CSPs of all stripes. It handles virtually all forms of network traffic: legacy circuit-switched PSTN, VoIP and IP and broadband in every way, shape and form. Compliance with CALEA in the U.S., ETSI in Europe, and other standards makes it easy to install anywhere.

STAR-GATE's open architecture operates independent of switch type, model or software version, is easily customizable and future-proof, i.e., adaptable to new technologies. The system scales to meet the capacity requirement of Tier 1 CSPs.

There are three components:
- **Access Device (AD)** provides taps and filters network traffic, and forwards intercepted data to the Mediation Device.
- **Mediation Device (MD)** delivers intercepted communications to predefined LEA sites.
- **Surveillance Administration Subsystem (SAS)** provides administrative functions to both circuit-switched and packet data networks.

The dual heart of the STAR-GATE system is its mediation device and SAS.

The mediation device handles all tasks needed to deliver intercepted communications to the appropriate law enforcement agency accurately, securely and in compliance with the law. The SAS and AD parts of the system define target assignments and the destination parameters of law enforcement agencies, control system security and monitoring system activity, and ensure that monitoring systems are properly maintained.

Mediation devices can be added as new switches join the network, and a central SAS can be expanded to support more users, targets and mediation devices as the network grows.

That idea underscores a key advantage of STAR-GATE – its diversity. The system works in multiples:

- **Networks:** STAR-GATE supports all existing circuit-switched networks and fixed voice and data networks.
- **Protocols:** STAR-GATE provides full support for all call content and data protocols, and readily translates the data to the requisite lawful intercept standard interface for delivery to an LEA.
- **Switches:** Each STAR-GATE Mediation Device (MD) can be connected to several switches concurrently, even if the switches are of different types or versions. Multi-switch support reduces initial setup costs and simplifies deployment by offering capacity management that is unavailable with switch-only solutions. The enhanced capacity can be deployed efficiently in a regional framework that adheres to the published configuration guidelines.
- **Multicasting:** The MD has the power to deliver intercepted call data and content to multiple law enforcement agencies simultaneously and in different delivery protocols.
- **Combined Content Delivery:** When a network switch sends separated content, the mediation device combines and delivers integrated content to the LEA, eliminating the need for additional switch hardware to perform this function.

STAR-GATE Mediation Device

The STAR-GATE Mediation Device (MD) has three functional components:

1. **Switch Interface System:** An integrated adapter to the switch, designed for a specific switch type and software version.
2. **Kernel:** The nucleus where most of the MD functions are performed.
3. **Output Formatter:** An integrated output formatter module, customized specifically for LI standards, to convert the switch proprietary message format to the required lawful intercept format.

The STAR-GATE Mediation Device serves as the center point of lawful intercept between the switch, the law enforcement agency and SAS that provides centralized management and systems maintenance.

The mediation device receives intercepted information from the switch and converts the collected data to the standard format required by law enforcement. It then delivers the data directly to the appropriate law enforcement agency. Data is sent via LAN to the site router and then to the collection module at the LEA's office.

At the same time, the mediation device sends and receives administrative data to and from STAR-GATE's SAS. The means: a WAN connected to the site router that includes information such as target assignments and management instructions.

SAS Part 1: The Global System Administrator (GSA)

The GSA provides centralized control over system administration, target allocation and security functions. The GSA features an intuitive graphical interface based on standard Windows conventions. Among the application's highlights is a search feature that lets the systems administrator quickly retrieve data on targets pulled by system users. The GSA application also enables back-up of all target information with removable media.

One central GSA can administer all network switches. Or alternatively, several GSA units can be situated regionally, and additional GSA units can added to an SAS site for load sharing.

Administrative features of the GSA include:

- **Remote Access:** The GSA can be controlled with remote dial-up connections, enabling authorized administrators and security officers 7 X 24 secure access to the system. Configurable security restrictions limit remote administrators to specific, predefined tasks.
- **Reports:** System activity is easily monitored with the generation of reports. The GSA can issue reports on the operations of the entire system, on specific targets or on particular users. Reports can be printed, saved and exported to other applications.

Target allocation functions of the STAR-GATE GSA enable the accurate, rapid transfer of intercepted data. A Target Administrator defines information about switches, targets and law enforcement agencies, including:

- **Territory Data:** Divides network switches into regions, enabling the GSA to determine which switches should monitor particular targets.
- **Target Data:** Defines and distributes actual targets for surveillance. Target data includes a number, delivery information, and warrant restrictions.

Warrant restrictions may include time and date limitations and can specify surveillance type – whether content and call data or call data only. The GSA distributes the targets to the appropriate mediation device in the CSP's network and switches, then monitors and logs the distribution process.

The mediation device automatically provisions the switches with the target data and warrant restrictions.

The GSA provides security features to ensure data confidentiality and grant access to authorized personnel only. From the top:

- **Controlling Data Access:** The GSA ensures the confidentiality of both target information and intercepted data. Access rights defined by the System Administrator apply to files, network connections and database tables. SAS can also operate with any firewall programs provided as part of the network design.
- **Monitoring System Activity:** Operator actions and system activities are recorded in a log file, called an "audit trail." Information in this file includes the operators' access to resources and information changes.
- **Transfer Encryption:** The GSA provides encryption between the Mediation Device and the System Administration Subsystem (SAS) or between the Mediation Device and the collection facility.

SAS Part 2: Maintenance and Fault Management Server (MFM)

Within the SAS, the Maintenance and Fault Management Server (MFM) oversees system operations and controls system maintenance, including configuration and software upgrades of units throughout the STAR-GATE system, providing a base for monitoring network-wide operations.

The MFM continually receives event information from other units in the system such as the unit source, description of event and a time stamp. In addition, the MFM regularly contacts system units to gain operation information. Based on this data, the MFM constructs a "picture" of the system and its status.

Using an intuitive graphical interface, the MFM creates a visual representation of the entire surveillance system for users, with each user site represented by an icon. The display is color-coded according to current status, giving users an immediate grasp of overall system operations.

Highlights of MFM operations monitoring include:

- **Rapid Navigation:** The user can quickly and easily gain a broad view of the entire system, or a microscopic view of the operations of individual units. By clicking the site icons in the visual representation of the system, the user can move through different subsystem levels. This feature enables the user to see everything from the "big picture" to detailed information about individual cards and channels.
- **Remote Access:** The MFM can be controlled via remote connections, giving an authorized user round-the-clock access to the system.

Configurable security restrictions can limit the user to specific, predefined tasks. In addition, the MFM can be configured to page an engineer or other user when defined events are generated. The paged message includes a code indicating the type of event.

Clearly, a system this large and complex is primarily targeted to Tier 1 carriers. STAR-GATE remains one of the most popular and widely deployed surveillance solutions deployed by CSPs, and for that matter, re-sold by other providers to carriers in need of a scalable solution for high traffic volume needs.

Insights

Like many tech companies that emerged during the late 1990s dot.com boom, Verint has a checkered history. Initially a subsidiary of Comverse Infosys, the company started out in Israel. Even though Verint's official HQ is now Melville, NY, the company to this day still keeps half its employees in Israel – a major sticking point for Arab countries, which as a result typically do not buy from Verint.

Verint and its parent attracted negative notoriety in 2006 over a stock options backdating scandal, and both companies were subsequently de-listed. But Verint sprang back, re-listing with the NASDAQ in 2010, then buying Comverse's share and winning full independence in February 2013.

In its early days, Verint focused on a mainstream commercial business, call recording, then branched out into surveillance in the early 2000s with the launch of the STAR-GATE product line for communications service providers. While Verint also expanded into cybersecurity and customer experience management, and in December 2019 announced plans two split into two separate companies by January 2021 – one focused on cyber – in the law enforcement realm it will remain best known for the STAR-GATE.

CHAPTER 8

Military Intelligence

Overview

Worldwide, military departments play a central role in intelligence. In the U.S., nine out of 17 member agencies of the Intelligence Community report to the Department of Defense: the National Security Agency, the Defense Intelligence Agency, the National Geospatial Intelligence Agency, the National Reconnaissance Office, and the separate intelligence divisions of all five military services.

The The U.S. intelligence budget has two principal components: the National Intelligence Program (NIP) and the Military Intelligence Program (MIP). Both report to the Office of the Director of National Intelligence (ODNI), which is charged with managing strategic military intelligence. Tactical intelligence, however, is the business of the respective military branches. Control and administration of MIP falls to the Under Secretary of Defense for Intelligence.

In the field, military services often have the final say. In the hunt for IEDs in Iraq, DCGS-A was known to pronounce bomb-laden roads as being "clean." For that reason, U.S. Marines and Special Forces replaced DCGS-A with Palantir, which proved its ability to pull together 10s of thousands of data points in real-time and provide life-saving intelligence.

The same structure holds true in other nations. The largest contingent in the Israeli Armed Forces is the Central Collection Unit of the Intelligence

Corps, also known as the Israeli SIGINT National Unit, or simply Unit 8200 – some 3,000 strong. Reporting up to Unit 8200 is Unit Hatzav, responsible for collection of OSINT. Israeli military intelligence and ISS vendors are closely intertwined, both for business and as mutual spawning grounds of technical advances.

Germany provides yet another example of this symbiotic relationship. It is no accident that key German ISS firms – FinFisher, Datakom, Dreamlabs, Elaman, Rohde & Schwarz and others – are centered around Munich. Just down the road is the Bad Aibling Station (BAS), a former military base run by the NSA through 1994 and for years the third largest site of ECHELON, the global SIGINT network of "The Five Eyes" –- the U.S. UK, Canada, Australia and New Zealand – until turned over to German foreign intelligence in the early 2000s. Through 2014, Pullach, a suburb of Munich, was also the central location of the Bundesnachrichtendienstn (BND), the Foreign Intelligence Service of Germany.

Just as with U.S. intelligence agencies, the U.S. Department of Defense and the US Army, Navy, Marine Corps, Air Force, Space Force and Coast Guard each rely heavily on surveillance solutions and training sourced from government contractors under separate "SAP" (special access program) awards. All too often, the end result is a barnyard quilt where respective intelligence gathering capabilities interconnect poorly, if at all. One further offshoot of this lack of cohesion is inadequate cybersecurity for intelligence transmissions between one military branch and another. Vulnerabilities are rife, most notably in the one sphere meant to link them all: Space.

Airborne ISR: SAR, Electro-Optical/Infrared Fusion

SAR (Synthetic Aperture Radar) is an active sensor ISR (Intelligence, Surveillance and Reconnaissance) solution, using its own independent energy source.

First developed during the Cold War, SAR is still going strong in various evolved forms, as is its one-time competitor, Electro-Optical/Infrared (EO/IR). But top government and military clients such as the National Geospatial Intelligence Agency (NGA) are trending toward solutions that combine the best of both: fusion systems that complement and augment the capabilities of each technology.

Nearly as old as SAR technology itself, the broad marketplace of airborne ISR or ISTAR (Intelligence, Surveillance, Target Acquisition and Reconnaissance) capabilities is packed with competitors, large, mid-sized and tiny.

Established names in these fields include government contractors Airbus, Boeing, GE Aviation, Israel Aerospace Industries (IAI), Lockheed Martin, Northrop Grumman, Raytheon, Thales and ViaSat. Most adhere to a strategy offering SAR, electro-optical and infrared capabilities, but in discrete ways that may not integrate.

SAR 101

Synthetic aperture radar was "invented" in 1951 by Goodyear Aircraft mathematician Carl Wiley while at work on a guidance system for the SM-65 Atlas, the first US intercontinental ballistic missile system. SAR is a variant of Side-Looking Radar – radar affixed to a satellite or other aerial vehicle that passes over and views targets on earth from a perpendicular angle, hence the term "side-looking."

With other side-looking or conventional radar, the resolution of the image captured is a function of the aperture size. The larger the antenna, the finer the resolution. SAR gets around this challenge by providing the same high resolution, but from a smaller antenna. How: by taking full advantage of imaging across a target as the satellite passes by, or as the SAR antenna moves to focus on an area. Radio waves are constantly beamed on targets below, and the received data is processed, as with ordinary radar. Because the antenna position relative to the target changes with time, however, captured echoes are combined, essentially delivering a big picture from a small lens.

End result: image resolution as good as or better than that of a conventional large radar antenna, at lower cost. Although as we shall see shortly, subsequent tour-bus size SAR antennas with multi-million dollar prices tags would arise as SAR developed.

Wiley's discovery would over the long term prove as valuable as missile defense: a way to create accurate 2D and 3D images of targets/areas from a satellite or other airborne vehicle as it passes overhead – and regardless of weather conditions. SAR "sees" right through clouds, 'round the clock and with or without sunlight.

Wiley's revelation led to a quiet explosion in further research, with close involvement by the U.S. military and permutations of the original technology by other contractors. Compared to the speed of today's typical technology revolution, getting SAR right involved a long, slow process. The first successful military test was not completed until 1957. Public acknowledgment of SAR's use by the U.S. Department of Defense came in 1960 with an announcement by the U.S. Army.

Notwithstanding Wiley's 1954 patent, government and military contractors then piled on by the dozen, each with its own variation on the original solution. In a word, SAR was hot.

How SAR Works

SAR measures: the "backscatter" or reflection of radio waves to determine the size and shape of objects; the "dielectrics" or transmitted electric force of terrain, water, and specific metallic or mineral objects; and the motion and direction of moving objects. Importantly, SAR shows changes in the targets viewed. Any variation in beams gathered pops up immediately. In addition, data from one sweep may be compared to heuristic views of the same object. Direct comparison of such "phase" information can spot the tiniest change.

SAR is not perfect. Certain features of side-looking radar can influence how a SAR solution captures data and is subsequently processed and analyzed. For example, targets hidden behind objects such as houses aren't "seen" by a SAR beam, thus produce no backscatter for capture and analysis. Objects can be shortened when the time required to scan both the top and bottom is equal to that of the time the satellite passes by.

In general, the longer a target is in view, and the greater the distance to the SAR antenna, the higher the resolution of the captured images.

Electro-Optical/Infrared

Electro-Optical and Infrared systems are "passive" solutions that rely on natural energy radiated by an object. Electro-optical systems measure electromagnetic waves of reflected sunlight. Infrared sensors measure infrared wavelengths emitted by live or natural objects. Optical and Infrared sensors may be offered separately or combined in one solution.

There are two common variants on optical sensors. Multi-spectral optical sensors separate wavelengths into the most common color categories – blue, green and red – plus a fourth wavelength for near-infrared. Hyperspectral sensors extend into smaller color fields. Hyperspectral sensors cover the full spectrum of light plus near-infrared and medium-infrared and offer the highest resolution of the two categories.

Military clients prize satellite-based optical sensors for their proven contribution to situational and tactical awareness. Highly maneuverable and equipped with control moment gyroscopes (CMGs), high-end optical satellites can be moved quickly to collect intelligence across multiple targets on a single pass, in consecutive order. When outfitted with hyperspectral sensors, the satellite can provide outstanding resolution on target images at the sub-meter level.

The key limitation of straight optical sensors is their dependence on sunlight. Bad weather conditions can make use of optical sensors a no-go. To compensate, operators must often rely on more frequent tasking to acquire data to bypass cloud conditions. Alternately they can switch to help from another source, Infrared Radiation (IR), electromagnetic radiation in longer wavelengths than visible light.

Thermal infrared sensing provides an important assist to electro-optical solutions. Thermal infrared sensors measure waves radiated from land surface and other heat-generating objects, including live ones. Cameras that detect radiation in the infrared range of the electromagnetic spectrum (900–14,000 nanometers) produce images of that radiation, making it possible to "see" without visible illumination.

Thermal Infrared works in any light or atmospheric condition, overcoming the principle challenge to electro-optical ISR systems.

Fusion Systems

Combining SAR capabilities with EO/IR, thereby gaining two signatures on an ISR event or target, carries obvious advantages:
- SAR provides a major advantage when cloud cover or other weather systems block access to intelligence by optical solutions.
- SAR and optical are skewed to detect different types of surface characteristics. Combining them enhances the ability to discern targets.

- When persistent monitoring is a must-do, but weather blocks optical, or busy conditions consume SAR's resources, the operator has back-up with Infrared.
- Infrared provides the ability to distinguish live targets including people, vehicles, military weaponry and missiles.

The challenge of fusion ISR systems resides in making disparate data sets compatible, and integrating data for ready access by the user.

Lockheed – ISR via Pigeonhole

Lockheed is a long-time player in both the SAR and EO/IR arenas, with work on the former dating back to the 1980s. Early on, the company's stellar SAR vehicle was the "Lacrosse," a satellite that was used by U.S. military from 1988 through 1995.

The Lacrosse was typical of the bus-sized, multi-million U.S. dollar market entrant. A total of 5 vehicles were launched during Lacrosse's long life span. Three remained in operation in 2005. The replacement for Lacrosse is the NROL-41, the first of which was launched in 2010 and also uses SAR. To this day, Lockheed continues to place heavy emphasis on SAR as a vital tool of national security.

Not to imply that the company has never branched out. Far from it. Lockheed offers the spectrum of ISR technologies for space, aerial, ground force and naval requirements. Technologies are often used side-by-side. But are they integrated?

Case in point. When launched in the late 1990s, Lockheed's RQ A-1 Predator included both SAR and EO/IR capabilities in an unmanned aerial vehicle. Today their RQ-170 includes infrared sensors, SAR, and an electro-optic camera to route streaming video of battle sites to command centers. Each technology solution handles a discrete intelligence function.

Lockheed's Space Based Infrared System (SBIRS) follows a similar path, providing a constellation of satellites for the U.S Air Force that uses scanning sensors for wide area surveillance, and sensors for smaller regions of interest. SBIRS is thus a two-fold solution that patrols the earth for what traditionalists consider Infrared's core use – missile launch detection – while also using Infrared for tactical reconnaissance purposes.

Battlefield Awareness is clearly a sidebar, promoted as an Evolving SBIRS function that complements situational awareness by providing additional intelligence and tactical alerts.

Note that fusion doesn't enter the picture. SAR, EO and IR capabilities are applied frequently, but separately and for distinct purposes.

MDA – Fusion Masters

For years, Colorado, U.S.-based DigitalGlobe was a distinct presence in airborne ISR, some said the foremost electro-optical ISR vendor in the business. Among its many showcase government clients: the National Geospatial Intelligence Agency (NGA).

But on the enterprise side of the business, DigitalGlobe was feeling pressure from mid-sized competitors. These newcomers capitalized on commercial enterprises' willingness to settle for medium resolution satellite imagery if offered at a discount price.

Come February, 2017, in stepped MacDonald, Dettwiler and Associates (MDA) of Canada with an estimated US $3.1 billion dollar buyout offer that DigitalGlobe could not refuse.

DigitalGlobe today retains its current location and brand, and reports as a division to SSL MDA Holdings, MDA's U.S.-based branch. SSL is a well-known maker of satellites. As the company notes, some 25 percent of U.S. government-leased commercial transponders use satellites built by SSL.

With operations in the U.S,. MDA can continue to meet legal requirements for selling to government agencies and military. MDA's strategy in acquiring DigitalGlobe proved equally savvy. DigitalGlobe's prowess in EO ISR is complementary to MDA's expertise in other areas including SAR.

Like Lockheed, MDA offers the full kit of airborne ISR capabilities both separately and combined. The Canadian/American venture also provides one of the market's best fusion solutions: MDA Multi-Sensor Data Collection and Fusion, an integrated multi-sensor solution.

Each sensor is multi-channel, providing electro-optical, infrared and synthetic aperture radar sensors. Working as an integrated whole, MDA's fusion ISR device simultaneously collects and integrates images and data from each source to provide a synthesized whole.

Bad weather? Lack of sunlight? Not an issue. MDA's product is combat-ready, delivering the cloud-thwarting virtues of SAR, high resolution of electro-optical, and the discovery functionality of infrared for any scenario. The solution is also ideal for identifying and tracking targets.

Multi-purpose in this case by no means equates to "watered-down." MDA's fused solution hits the mark across the board, with the added advantage of lower cost as a direct result of fusion. In a marketplace long dominated by heavyweight contractors, but loosening up with the advent of lightweight, cheaper competitors, MDA has struck a sound middle course – focus on all-purpose, affordable top performance in a fused ISR solution.

Signal Jammers and Anti-Jamming Systems

Imagine an ingenuous tool that can defeat any IMSI catcher, threatening to undermine law enforcement's ability to track and intercept any mobile device – and turning their high-end Harris Hailstorms and Rayzone Piranhas into bread boxes. It is for real, inspired by the National Security Agency, and completely impervious to radio signals.

The deadly counter-surveillance system? It's a simple cloth smartphone case lined with copper, a highly conductive metal that blocks electromagnetic fields and radio signals. The copper-lined case, named "Tunnel," takes a cue from the NSA, CIA and other branches of the U.S. Intelligence Community, where office windows are fitted with copper-lined glass to prevent any similar form of intrusion of an agency's sensitive systems.

While the Tunnel might appear to have little to do with the subject of this analysis – GPS and other RF signal jammers and their opposite, anti-jamming tools – it is emblematic of market conditions in these fields. Every new advance on one side triggers an equally clever method of evasion.

GPS or Global Navigation Satellite System (GNSS) jammers and other signal interference devices are widely used for counter-surveillance, communications disruption, and personnel security against explosive devices designed for activation by radio signal. Many respected IT/ISS brands such as Keysight manufacture RF signal jammers. Jamming is also a well-known side feature of IMSI catchers.

On the opposite end of the spectrum, low-cost GPS jammers – which are illegal in the U.S., parts of the European Union and many other nations – are often used by criminals and terrorists to evade capture following a crime, or to temporarily disguise their location and movements.

GPS jammers can also be used by war fighters to interfere with commercial aviation, military avionics systems, advanced combat maneuvering instrumentation (ACMI) and weapons guidance systems.

GPS is by now such a universal technology affecting so many branches of communications, commerce, navigation and the military that the opportunities for thwarting it are limited only by the imagination. GPS is at once a wonder for precise mobile location – and, due to its vulnerability, a sore point for defense and security.

So are jammers a good or bad thing? The answer: It depends on who's using them. In either case, it is wise to understand such tools and why they create concerns.

How Jammers Work

A good starting point is to consider the example of jammers used on a conventional mobile phone network. Conventional mobile communications is provided in cell-based networks such that users transiting geographically-designated cells are handed off to a new tower and base station – like passing a torch, though in this case the torches are radio signals available in various frequencies depending on the type of network.

A jammer disrupts a cell phone network by transmitting on the same radio frequencies that are being used as these wireless devices. The resulting "white noise" breaks communication between the mobile base station and the user's handset.

In a classic denial-of-service attack, a jammer overpowers and cancels out a signal within the radio spectrum used by handsets. Low-end jammers might have a range of 30 to 50 meters. Military grade jammers could cancel out signals in an area hundreds of meters in diameter.

There are commercial ISS jammers for every occasion:
- Compact hand-held and high-powered jammers for network mobile and Wi-Fi.
- High-powered pocket jammers for 3G, 4G LTE and Wi-Fi.
- Video jammers that disable IP cameras operating in the 900 – 1000 MHz bands and 1200 – 1300 MHz bands.
- Wideband RF jammers.
- Audio jammers for recorders and room conversations.
- Walkie-Talkie jammers.
- Adjustable Four or Five Band Mobile Network Jammers.
- 8-Band Jammers for all Mobile Networks + UHF/VHF, Walkie-Talkie and audio bugs.
- Long-Range (up to 500 Meter) jammers for any mobile network.
- Bomb jammers for remotely controlled IEDs.
- Jammers for disabling vehicle Lojack systems.
- Drone jammers for interfering with unmanned aerial vehicles (UAVs).

US/UK Military GPS Jammers That Rule the Sky

Of greater strategic interest are GPS jamming systems currently under testing by the U.S. Department of Defense and British military that can jam GPS and GNSS signals over hundreds of square miles.

The first hint of this capability came in June 2016 when the U.S. Federal Aviation Administration issued a warning to all commercial aircraft that might be flying within a 400-mile range of the U.S. Naval Air Warfare Station at China Lake, California. The base is located approximately 125 miles north/northeast of Los Angeles. The concern: Secret tests of a massive GPS jamming system that would disrupt GPS and GNSS satellite signals region-wide and possibly interfere with flight stability controls.

The testing, spread across five different dates in June of that year, was remarkable not only for the intensely busy air traffic area that it overlapped but also for the fact that it made GPS readings either unreliable or completely unavailable at predesignated zones of 50 feet, 4,000 feet and 10,000 feet, respectively. The UK military reportedly conducted similar wide area, selective altitude GPS tests the following month.

IAI ADA Anti-Jamming

Oher nation's military organizations are, of course, equally interested in anti-jamming technologies. No sooner did the U.S. and Britain offer a peek at wide area GPS jamming than Israel unveiled its latest advance in protecting systems against attacks on GNSS.

The relative ease of jamming weak GNSS signals has given impetus to new "anti-jamming" systems from Israeli Aerospace Industries (IAI), owner of ELTA Systems. As IAI points out, a jamming signal as weak as 1 kilowatt can compromise GNSS in a radius of 40 kilometers.

IAI's ADA GNSS Anti-Jamming System has been successfully tested for its ability to defeat deadly jamming efforts.

ADA GNSS AJ leverages advanced digital processing to provide full immunity against multi-point jamming and deliver assured position, navigation and timing (PNT). The system supports GPS-m and multi-GNSS code, and is available in three airborne models: ADA-MS for helicopters; ADA-M for fixed wing including drones; and ADA Block II (satellite).

A fourth model, ADA-O, is designed for surface applications including nautical vessels, tanks, trucks or fixed position. IAI's ADA GNSS Anti-Jamming is integrated with the avionics platform of the Israeli Air Force.

Military Alternatives for GPS Location

Since deployment of the first constellation of 24 Global Positioning Satellites in 1994, GPS has grown to become the go-to location technology for commercial enterprises, health care, first responders, consumer smart devices, automobiles and the military.

GPS is such a huge success that most people are oblivious to the downside of relying on it as completely and unquestioningly as mankind does: the vulnerability of this amazing technology to interference and disruption via signal jamming.

In a way, GPS is a victim of its own success. The very attributes that contribute to the technology's popularity – open access, global availability, and adaptability to serve a wide area of applications – also play a role in its vulnerability. Yet even as concerns over GPS jamming increase, GPS continues on a path to 100 percent market penetration. The story of GPS is thus a tale of both opportunity and risk.

By some estimates, GPS directly impacts nearly 6.0 million industries and annually produces U.S. $125 billion in economic benefits. Applications of GPS include search and rescue, train and auto location, underwater hazard location for shipping, accurate mapping and personal location in outdoor "extreme" sports. Even in farming, GPS is relied upon for soil sampling and mapping.

On the battlefield, on and under the ocean and in the air, GPS is mission-critical in ensuring the position of combatants and the accurate delivery of supplies. For its precise timing capabilities, GPS is today an essential component of financial trading, communications networks, banking and power grid supply. GPS has reached a state of ubiquity.

That doesn't mean that the technology is perfect. One downside is that GPS satellites are expensive. The U.S. Department of Defense's GPS satellite network cost US $12 billion to deploy and has an annual budget of US $750 million. Signals are weak, by design, thus poor in remote terrains and blocked by other radio transmissions in large metro areas. GPS is also hard to secure.

In tandem with normal disruptions is the graver concern of deliberate GPS jamming by hostile nation states. GPS jammers, ranging from low-cost models available on Amazon to more sophisticated purpose-built units, can easily be used to disrupt military, civilian, power grid and first responder networks.

Multi-purpose devices may include GPS jamming in their repertoire of services. IMSI catchers, for example, can in addition to tracking and intercepting mobile calls also be used to jam radio signals.

Jamming on the Rise

Considering the public fear and government concern over cyberattacks on critical infrastructure, it is ironic how little attention is paid to the equally insidious threat of jamming attacks on the same targets. Jamming requires far less expertise and investment.

What if hostile nation states used the same capabilities to manipulate power grids, undermine safety in shipping lanes, or create an economic meltdown by scattering GPS time-stamped stock market trades to the four winds? None of that has happened yet, but analysts are spot-on correct that it easily could.

It is well-known that North Korea routinely hampers the GPS networks of its southern neighbor via truck-sized jammers for periods lasting from hours to more than two weeks, seriously impacting the navigation of commercial and military airplanes and vessels. Similarly, during the wars in Iraq and Afghanistan, U.S. military forces' reliance on GPS for delivering essential supplies via parachute was regularly undermined by terrorist groups' use of jammers.

The Deliberate Built-in "Weakness" of GPS

In simple terms, conventional GPS networks are comprised of large satellites filled to the brim with atomic clocks that record time with extreme accuracy. Each satellite spans the globe twice per day, sending out coordinate signals and the exact time as it flies by at 13,000 feet over the earth. Down below, users can see their precise position thanks to triangulation between three or more satellites and the user's exact reckoning and distance from each. Location data is shown instantly on the user's GPS device map. The satellites' atomic clocks mark the correct time to within 100 billionths of one second.

Signal strength of GPS and Global Navigation Satellite Signals (GNSS) is weak – just 1.5 decibels (dB) – by intent owing to the complexity of the signals and the need to conserve energy in solar-powered satellites. But one potential end result of this built-in "flaw" is the ease with which GPS signals may be disrupted. Jamming is defined as transmission from external sources of any noise or external signal across a GNSS or GPS frequency, thus interfering with the location and timing "lock" between the satellite and on-the-ground users. At 1.5 dB it does not take much to overpower GPS satellite circuitry.

Making GPS More Jam-Resistant

Solutions to GPS jamming span systems and devices that make satellite positioning more secure and resilient, as well as alternate "GPS free" alternatives. Thus far, the greatest successes in anti-jamming reside in military applications.

Among the more popular solutions are the new generation of compact anti-jamming antennas that make GPS safe, reliable and available in combat scenarios. Two that stand out are the palm-sized Novatel GAJT anti-jam antenna and the Chemring GINCAN.

The Novatel GAJT is credited with nullifying jammers and guaranteeing the availability of GPS signals without interruption to ensure accurate, real-time calculation of position. As a COTS (commercial off the shelf) product, the GAJT provides comparable performance to heavyweight anti-jammers but at significantly lower cost. GAJT is also systems agnostic, offering the flexibility to integrate quickly either with legacy or new military GPS receivers.

The miniature GINCAN anti-jammer from Chemring offers comparable performance in a small, affordable "jam nulling" unit. Reduced size and weight make the GINCAN a potent adversary to today's low-power jamming devices, whether for vehicular, cellular, troop or infrastructure protection requirements. Based on adaptive antenna technology developed for military use, the GINCAN is widely regarded as one of the most effective anti-jamming devices for GPS available on the market today.

GPS-Free Alternatives

Notwithstanding the viability of solutions that protect extant GPS systems, innovators have moved in the direction of abandoning established approaches altogether.

One "GPS free" technology gaining a fresh look is Long Range Navigation (Loran). Loran is a radio navigation system that allows users to fix their position via low-frequency signals from fixed ground antenna. The first generation, Loran-C, was introduced in the late 1950s and by the 1970s virtually dominated the market for maritime geo-positioning. However, Loran-C popularity fell sharply in the 1990s with the arrival of GPS. Most Loran-C systems in the U.S. and other nations were completely decommissioned by the year 2010.

Loran has made a comeback in recent years through new advancements in its successor, eLoran, which uses a stronger signal than GPS and is more resilient against jamming. By adding a data channel to the traditional Loran-C transmission, eLoran guarantees signal integrity for location and precise time and frequency essential to critical applications. eLoran systems are now deployed in South Korea as an alternative to the older GPS networks so easily foiled by North Korea, and the U.K. has also installed eLoran as a back-up to GPS for nautical navigation purposes.

One unique alternative is the "joint precision airdrop system" or JPADS, which can determine location with pinpoint accuracy. Currently JPADS is used to ensure reliable parachuting of supplies to ground troops. Completely GPS-free, JPADS works by direction from a small computer and a camera employed on the cargo. Data generated by the camera is fed into the computer, which matches what it sees to terrain maps, then uses motors to direct the parafoiled cargo to its destination. Unlike higher-cost GPS JPADS such as those used in Afghanistan, new "vision centric" JPADS is immune to jamming and can be used in terrain such as canyons, valleys and large cities that are inhospitable to GPS.

Early tests of GPS-free JPADS in Arizona proved successful, and if trials continue on the same track, the technology may soon find its way into military aircraft, drones, vehicles and then commercial applications, as a new standard in jam-proof positioning.

On an up note for the military, Lockheed Martin put its first next generation GPS satellite into orbit for the U.S. Air Force on December 23, 2018. The GPS III system, once fully deployed in 2023, will feature satellites with a life span of 15 years, and a signal that is up to 100 times stronger than its predecessors, making it much harder to jam.

Airbus Zephyr: Satellite Surveillance Supplement

Military analysts are contemplating the future of what may prove to be the most exciting aerial surveillance innovation since the satellite and U-2: the Airbus Zephyr, a lightweight, solar-powered drone that can soar to 70,000 feet, stay aloft for 45 days, evade detection or bad weather, and carry all manner of surveillance equipment.

Designed for military use and successfully tested in 2018, the Zephyr is seen as a supplement to C4ISTAR [Command Control Communications Computers Intelligence Surveillance and Reconnaissance] with more flexibility than conventional satellites, and at lower cost. Airbus has coined the acronym "HAPS" (high altitude pseudo satellite) to describe the unusual aircraft.

There are two versions, the Zephyr-S and the larger Zephyr-T. The Zephyr-S has a maximum payload of 5 kilograms. The Zephyr-T can handle heavier payloads of equipment with great demands for power. Both can support high definition optical and infrared video, automatic identification of oceangoing vessels, Tetra narrowband, as well as monitoring by RADAR, LIDAR, ELINT and other electronic surveillance niches.

Whether the Zephyr and other similar aircraft ultimately replace or simply assist satellite surveillance, there is little doubt that the new craft will find a welcome home. And like many an innovation that first sees light for military apps, Zephyr will almost certainly draw an eager audience among government agencies and law enforcement.

The Zephyr's flight time is not limited by fuel. Solar power keeps the bird aloft during the day even in cloudy weather, while back-up batteries that recharge in daytime hours assure night time operation. It can be landed quickly for modifications that might change with an assignment, then relaunched with new marching orders.

Alternately, the Zephyr can be flown continuously for a month and a half, either over one region or multiple areas. One particularly appealing aspect is that the drone can be launched by one ground crew over a range of 250 miles, then directed to continue to a destination thousands of miles away. In autonomous mode, Zephyr flies itself without being managed by a ground crew, at all.

At maximum altitude of nearly 13 miles Zephyr can see to the horizon and convey ground images over an area of 400 square miles, providing a remarkable 15cm resolution of objects on the ground.

The Airbus craft is considered ideal for monitoring and providing Special Forces with real time intelligence on enemy troop movements or an enemy action in the making. Interested buyers include customers on both sides of the Atlantic.

Zephyr has been successfully tested by the U.S. Army. In the UK, The Ministry of Defence (MoD) has purchased three of the unique craft for a total of US $16.0 million. British military note that the Zephyr is more like a satellite than a conventional unmanned aerial vehicle. Not surprisingly, the unusual craft can be used as a communications relay satellite, as well.

Elbit Systems Airborne and Ground Surveillance

Elbit Systems' creation of Cyberbit – where it houses the NICE Systems intelligence assets acquired in May 2015 – tends to foster the impression that the parent company's range of ISS solutions is confined to that market only. It is essential to remember that Elbit Systems, a division of Israel Aerospace Industry (IAI), is first and foremost a military contractor. The intelligence gathering capabilities of solutions designed for use in combat, both ground and air, are every bit as important as those designed for conventional ISS and cyber purposes.

Elbit Systems goes both ways: innovating in military technologies and adapting proven ISS solutions for military purposes. The company offers the full palette of intelligence solutions that provide a real-time view of the battle theater, protect and inform the warfighter, whether in the sky or on the ground, and keep command centers in the loop.

Strategic and Tactical Command Centers

The starting point is the Elbit Intelligence, Surveillance, Target Acquisition and Reconnaissance (ISTAR) Management Center, or IMC. Think of it as a strategic "command central" for any client army – the hub that manages all military intelligence solutions in the field, integrates their findings and provides actionable intelligence to commanding officers at the highest level.

The tactical wing is the Elbit System Mobile Arena Dominance ISTAR Center (MADIC). MADIC is mobile and on the ground, able to move with the battle and direct force movements based on real-time inbound intelligence. As a centralized and integrated management center for local forces, MADIC coordinates all external forces, including airborne and ground.

Airborne Intelligence

The bulk of airborne intelligence systems designed by Elbit Systems focus on combat scenarios, with the goal of protecting pilots and aircraft. Such solutions include digital radar warning receivers, missile and laser warning systems to provide a real-time tactical picture of the battlefield. But Elbit Systems goes far beyond warning systems, using airborne intelligence solutions that draw on techniques common to the ISS sector.

Elbit Solutions Airborne Reconnaissance commonly fixed to or embedded in aircraft include:
- **Wide Area Airborne Persistent (Video) Surveillance System**. WAAPS continuously monitors, records and analyzes large area footprints,

intercepts events and maintains multiple regions of interest (ROI) under constant surveillance with high spatial resolution and video.
- **WAAPs Storage and Analytics.** Elbit Systems also provides storage for video collected from the airborne platform. Users can access video archives of aerial missions, picking queries sorted by time, location and event. The system comes with video processing and analytics capabilities including Video Motion Detection (VMD) and Virtual Perimeter, both with alert mechanisms that point out anomalies.

Elbit Systems also delivers on Electronic Warfare Systems for mobile communications, turning aerial craft into intelligence gathering machines. The core tool is the company's CELLACTIVE solution. CELLACTIVE is an airborne cellular network COMINT solution that monitors, records and analyzes mobile phone conversations and text messages, and also provides mobile location of targets to ground commanders.

Once CELLACTIVE locates the targets, another solution – SKYJAM – can take over, blocking their inbound or outbound mobile communications. Alternately, CELLACTIVE may be used to meddle with and take control of calls and data communications occurring between targets – acting like an airborne IMSI catcher.

Client armed forces may opt to use Elbit's electronic warfare for cellular communications as a separate package, or as part of the "WiT" (Wise Intelligence Technology) suite:

- **Locked Wave.** For RF interception of radio networks, Locked Wave is a cyber RF Intelligence system that operates in active or passive mode, providing access, inspection and manipulation of encrypted airwave-transmitted communications such as Wi-Fi. Locked Wave decrypts Wi-Fi standard encryptions including WEP, WPA and WPA2, and cracks encrypted data stored in Wi-Fi communications. The solution operates in real-time to intercept and deliver enemy communications "in the moment," and the data may be stored for further analysis.
- **Miniature Reactive Jammer (MRJ).** Designed to counteract remotely-activated roadside bombs, the MRJ family of protection systems jams signals used to detonate Remote Controlled Improvised Explosive Devices (RCIEDs). Able to to jam RCIEDs in either reactive or active modes, or both simultaneously, the MRJ is popular with ground forces, penetration units, small vehicular convoys, border patrols and homeland security applications.
- **GroundEye Video Surveillance.** For panoramic video surveillance at ground level, Elbit Systems GroundEye is topnotch. The system provides round-the-clock, persistent wide area video monitoring and

recording. Offering total flexibility, the system may be deployed on masts, towers or vehicles in single unit or multi-unit configuration. Intelligence collected by GroundEye provides instant access to threats so that users may take immediate action. The system also provides advanced analytics enabling agents to see and understand more subtle changes occurring over time. End result: outstanding situational awareness.

- **Long View Short Wave (LVSW).** For long range observation of targets, LVSW acts like a zoom lens on a wide angle camera, except that this solution sees thermal images. LVSW combines the best of Forward Looking Infrared (FLIR) and Shortwave Infrared (SWIR), adding GPS and magnetic compass for pinpoint accuracy location, and "eyesafe" laser rangefinder for exact distance. Working in SWIR mode, LVSW provides perfect images even when targets are obscured by weather conditions or smoke. Where LVSW comes into play: ensuring the accuracy of long range fire against enemies and other targets.
- **SupervisIR.** SupervisIR takes infrared imaging to the next level by providing wide area persistent observation, day or night, of enemy movements on a tactical basis. The system is user-activated, as opposed to LVSW, which is reactive, and the user may feed close-up IR intelligence back to command where it is integrated with LVSW.

ELTA Systems Defeats Drone Bombs

With the uptick in the use of drones by nefarious elements, U.S. military officials have taken action: a contract with the North American division of ELTA Systems, among the best-known military equipment and ISS solutions providers based in Israel.

Following the use of bomb-laden drones by ISIS during the siege of Mosul, the U.S. ordered twenty-one ELTA "Man Portable Aerial Defense System kits" for US $15.6 million. Just what are these devices? Speculation has run rampant, with some speculating that ELTA's solution fits the usual definition of a "Man Portable" aerial weapon, meaning a missile. Fired like a rifle, such missiles are used to bring down enemy aircraft.

But the customer name on the purchase order points in a different direction. The buyer was the Air Force Life Cycle Management Center (LCC), in charge of communications and electronics. The LCC doesn't buy missiles, they buy communications tools, including RF monitoring and jamming and spyware – and devices that enable them.

ELTA is in the jamming business, and has a product that can interfere with and take down drones. Their product: Drone Guard, already in use by other ELTA clients to detect, track and jam small drones. ELTA debuted the Drone Guard in late 2016 at an aerospace and defense convention in Asia.

But the solution purchased by the Air Force goes a step beyond the conventional Drone Guard. It not only jams an enemy drone, it takes control of the flying menace by hacking in.

U.S. troops have since that time used the portable device to single out drones, take control away from the "pilot" and either crash or send the flying bombs homeward bound.

Inside the Drone Guard

The Drone Guard works in two fundamental ways. First, the device uses 3D radars and Electro-Optical (EO) sensors to detect and identify predesignated threats such as drones. The challenge lies in identifying a target as small and low-speed as a drone. ELTA resolves that issue with special algorithms that enable 3D radar to track small objects in motion. Next generation EO sensors add certainty to the ID process via precise visual identification. In field operation, Drone Guard gives the trooper the option of monitoring the sky in long (20 kilometer), medium (15 kilometer) or short (10 kilometer) range.

The second function is a jamming system that disrupts radio signals that control the drone. ELTA's advanced jamming systems can be set in automatic mode to hunt for enemy drones 24 X 7 with 3D radar and EO sensors so that specific drone threats can be spotted quickly and brought to an end. The Drone Guard operator can opt to either turn off the enemy drone's engine and send it into a tailspin or fly the bomb back to its point of origin.

One Drone Guard can do the work of many. For example, if an enemy force launches a squadron of drones, Drone Guard can take them on and destroy or re-route them.

The 3D Radar Difference

Classic parabolic radar uses a curved dish with the cross-sectional shape of a parabola to direct or capture radio waves. In operation since World War II, parabolic or 2D radar remains popular for radar applications in aerial surveillance. The downside of 2D radar is its broad elevation beam, which is unable to detect the altitude of a flying object. Early attempts at 3D radar tried modified parabolic reflectors with multiple feeds that in turn generated multiple receive beams in a stacked formation. However, the height calculations remained highly inaccurate.

A breakthrough in precise altitude calculation came with development of the phased array antenna, comprised of hundreds of tiny radiating elements, each capable of sending guided "pencil beams" into space while the antenna itself remained stationary. By applying a phase shifter, a microwave technology that can adjust the phase of each of the radiating elements, the phased array antenna can essentially point and focus on specific targets, determining their position in space in 3D, by precise altitude.

Modern 3D has advanced even further, making it possible to assign specific RF values to different echoes captured by radar. A radar signal processor (RSP) uses special algorithms to classify radar echoes by power and spectrum. As a result, the RSP makes it possible to discriminate between relevant signals and noise that is irrelevant to monitoring. Aiding in this procedure, a Moving Target Detector is used to classify the spectral properties of an echo into distinctly recognizable characteristics that indicate probable targets. These signals are routed to the radar system's computer to filter out false targets and determine the direction, altitude and velocity of the correct ones.

In the case of the ELTA Drone Guard, identifying and tracking inbound drones is instant.

New Co-Pilot: Drone Guard with Hacking

The compelling feature of the ELTA Drone Guard is a radio receiver with the ability to take total control of the winged device while in flight, hijacking it from the ground pilot. As the uninvited pilot, Drone Guard can then govern the drone's velocity, altitude and direction. It performs this feat by hacking the drone's DSMx, the world's most popular remote control system used with garage-grade drones such as those operated by ISIS and other terrorist groups during the tenure of the Caliphate.

Conventional DSMx operates "in the clear" without encryption. Thus, the RF exchange between the controller and drone is easily compromised by intercepting the protocol used and conducting a quick brute force attack. The ELTA Drone Guard synchronizes to the target drone's frequency, impersonates the pilot control's device, and emits a malicious control packet in the target's path. The drone's receiver then accepts the attacker's signal as genuine. From that point on, the enemy operator is locked out and the friendly force attacker owns the drone.

DSMx is, of course, a hobbyist tool. But as drone control systems grow more advanced, ELTA already has the sophistication to master standard military RF protocols, as well.

ELTA Systems LTE and SATCOM Interception

ELTA's s Tac4G system, in combination with data from the company's Black Granite tactical radar data, on-network mobile and IP, and off-the-air mobile and satellite interception, presents a formidable power-pack against enemy troops. Using a device that looks every bit like a smartphone, troops can see videos of opposing unit formations, while command centers parse signal intelligence to determine what the foe has planned next.

On another front, military can also benefit from ELTA's advances in passive mobile interception/location and SATCOM cellular interception. But let's start with a look at ELTA's militarized version of 4G LTE.

Note that as of July 2020, 4G LTE is still state-of-the-art for Israel. Unlike other Middle Eastern nations, Israel is behind the curve on 5G. In fact, the relevant government authorities did not issue tenders for 5G frequencies until July 2019, and have only begun to announce winners. Therefore, we shall confine this discussion to military apps for 4G LTE.

Tac4G

In Israel, over three quarters of the population of seven million owns a smartphone, with the highest concentration among men and women under the age of 25. Military conscription is mandatory. Developing a tactical military communications device that is second nature to millennial soldiers' preferences would seem a sound practice. But there are distinct challenges to overcome.

Commercial standards for 4G LTE collide with military systems' requirements, bumping the latter out of commercial spectrum, and forcing changes to make the square peg of civilian technology fit the round hole of military frequencies. Another downside: spectrum overload and interference that impact the reliability of mission-critical communications.

Then, too, military operations often take place in zones with limited network infrastructure that doesn't support 4G LTE smartphones.

Finally, there is mobility to consider. Commercial networks rely on fixed base stations that link directionally to their peers to ensure integrity of the network. When troops are on the move with 4G devices they need portable base stations with the same robustness as their commercial counterparts.

To address these issues, ELTA looked to the cloud to support a new service that would complement Israel's current military communications system, the Digital Army Program, commonly known as "Zayad." While leading edge in many ways, Zayad's primary benefit is the ability to support multiple technologies including legacy radio technology still in operation. As with U.S. first responder networks, which for years were forced to support decades-old P25 Land Mobile Radio systems as well as emerging technologies, Zayad's move to 4G LTE has been evolutionary, not flash cut.

Bringing that future into the present day, the Tac4G system creates a "cellular cloud" based on a militarized version of 4G LTE. ELTA's system offers many advantages. It leverages commercial infrastructure, provides advanced mobile networking capabilities, uses portable base stations that move with the troops, and is fully secure, creating an encrypted private network for all wideband communications. The system cannot be jammed.

Tac4G's ruggedized hand-held device supports 3G and 4G, is available in Android and as special ops devices go, looks, feels and provides the same features and performance as a smartphone. Those in command can instantly convey relevant content to those in the field for immediate action, and know that the device in hand is thoroughly familiar to the user.

Because the Tac4G builds on commercial-off-the-shelf (COTS) technologies it is future-proof and easy to upgrade as new advances emerge.

ELTA also has an "eye in the sky" – ELK-7099CL – a product that can intercept, monitor, decipher and geo-locate satellite phones operating in the C-Band and L-Band, automatically mapping between the two channels for full duplex monitoring.

ELTA's satellite monitoring system provides full deciphering of signals encrypted in GRM-1 PCS, the common standard used by satellite phones, handles multiple calls simultaneously, and may be used remotely by multiple field personnel to track calls and capture call data and target location.

ELTA's satellite phone monitoring system is passive and completely undetectable by the target. The system integrates with analytics tools that quickly produce actionable insights to be relayed to troops or agents.

Satellite monitoring is, of course, a crowded arena. South Africa's VASTech Strategic Satellite Monitoring offers a small footprint that provides unified intelligence across a broad field including content. Verint RELIANT combines intelligence gleaned from fixed satellite interception with data from IP and mobile networks.

ELTA, too, can combine satellite interception in a broad solution that spans passive mobile interception tools, microwave, Web, social media and standard telecom network ISS solutions for use by the military.

Keysight Signal Analyzers and RF Sensors

Agilent's 2014 spin-off of Keysight unleashed a giant in the military RF monitoring marketplace, with a dozen R & D centers worldwide and 9,500 employees including the largest sales and support staff in its sector.

RF signal monitoring enables the user to spot aberrant signals called "emitters" in the radio spectrum and determine the geo-location of the target's transmitter. RF monitoring can be done short or long range and track even the tiniest signal through walls and floors. With decades of experience, Keysight occupies center stage in this arena.

At its genesis, RF monitoring relied primarily on standalone spectrum analyzers that worked from a central location, with later versions adding digitizers. Spectrum analyzers still perform the basic functions of displaying amplitude and frequency.

Today's new breed of RF monitoring devices goes by the name "signal analyzers," an important distinction from "spectrum" analyzer. Signal analyzers add an element vital to surveillance – vector analysis, which enables a system to capture broadband signals over a large frequency range in real-time.

With the proliferation of broadband and wireless, another option growing in popularity is to build a network using inexpensive RF sensors that monitor signals across a large area.

Whether conventional bench top or RF sensor-based, RF monitoring systems have similar characteristics:
- Ability to cover a broad range of frequencies.
- Channel scanning.
- High-speed frequency resolution.
- Data storage.
- Automated alerts on signals of interest.
- Analytics that recommend actions.

How Signal Analyzers and RF Sensors Work

In a commercial RF monitoring transaction, the system looks for and reveals the type of signal, its duration, frequency and bandwidth, type of modulation, and number of occurrences.

In surveillance, certain refinements are required. First, the system must be configured to monitor and measure signals that occur infrequently, are of short duration, and possibly have low received power. Also important: direction-finding (DF); the ability to monitor both indoors and outdoors using mobile sensors; and finally, signal demodulation to extract intelligence.

DF can be accomplished in several ways including use of highly-directional antennas, triangulation, GPS, Angle-of-Arrival (AOA), Time Difference of Arrival (TDOA – processing a signal arriving at multiple receivers at the same moment), and Received Signal Strength (RSS) – distance from a target measured by the decrease in signal strength.

Given the short duration, infrequency and low power characteristics of radio transmissions by unknown and possibly unfriendly emitters, the speed of RF monitoring frequency scanning is critical to an intercept, as is the ability to monitor a high volume of signals simultaneously.

How the RF signal analyzer handles these challenges: First, RF signals are filtered and converted to "intermediate frequency," enabling the system to scan more signals.

The analyzer tunes the system's local oscillators to pin down a suspicious signal, often with an assist from "Fast Fourier transform" (FFT) digital processors. FFT, an algorithm that converts time to frequency, finds the target emitter instantly. Rapid-fire channel scanning by the local oscillators achieves the desired result: a higher probability of intercept.

Government agencies and military units can go one of two ways in deploying RF signal analysis: traditional bench top or RF sensor. Rack-mounted RF surveillance systems are built around big box signal analyzers installed in a building and connected to one or more antennas.

RF sensors, lighter and lower cost, perform much the same functions as conventional receivers. Slightly larger than an iPad, RF sensors are inconspicuous and can be easily affixed to outdoor poles, walls and vehicles or used portably. Configured in a network across a broad geographic area, RF sensors take signals from antennas and pass them to a central control unit that stores, processes and demodulates signals of interest.

Keysight X-Series Signal Analyzers

Keysight calls its flagship X-Series signal analyzers an "evolutionary approach to signal analysis that spans instrumentation, management and software." The X-Series can meet the user's RF signaling needs today, and as new technologies emerge, the user can upgrade the same box to meet any challenge simply by purchasing hardware and software upgrades.

There are six models in the X-Series line, from simplest to most advanced: the CXA, EXA, MX, PXA, UXA and the state-of-the-art N9041B-RT2. All come with upgradable CPU, standard fast sweep and enhanced phase noise, real-time bandwidth booster and real-time spectrum analysis as options. A shared library of common applications is available for signal monitoring cellular communications, MILCOM or SATCOM.

Note that all X-Series products come with Keysight PowerSuite Third Order Intercept (TOI) capability. PowerSuite TOI provides a simple one-button measurement of a two-tone signal, and highly accurate (zero-span) measurement of the same signal. TOI works by sweeping the center frequency inherited from the signal or spectrum analyzer measurement. It then identifies the two highest peaks for upper and lower signal frequencies and measures the power not only at the signal frequencies, but also at the third-order intermodulation frequencies.

Why TOI is important: It ensures accurate measurement of low-power intermodulation distortion signals and can track signals so weak that they are typically below the intercept threshold.

All members of the X-Series are powered by Keysight's 89600 VSA and WLA software packages. VSA can measure and analyze some 75 different signals types – LTE, LTE-Advanced, CDMA, HSPA+, all versions of 802.11, WiMax, Bluetooth, etc. – instantly verifying the signal. WLA is the media access control (MAC) complement to VSA, decoding and verifying MAC messages and correlating them with the physical layer, the basic transmission hardware.

Top of the Line N9041B-RT2

Keysight's top-of-the-line X-Series model for intelligence and military purposes is the N9041B-RT2, designed for the most elusive and challenging signals: fast-hopping, wideband and transient in applications that include electronic warfare, radar and counter-terrorism. The N9041B-RT2 can capture elusive signals as short as 3.52 milliseconds for real-time spectrum analysis. The new unit is built to intercept and monitor the most sophisticated radio technologies including 5G, 802.11ad for high-speed Wi-Fi, and automotive radar.

The core of Keysight's highly flexible RF sensor system is its N6854A geolocation server software, able to track and map the real-time RF geolocation of an emitter. The server software works in conjunction with Keysight's N6820ES "Surveyor 4D" software, designed exclusively for use with Keysight N6841A RF sensors.

The Surveyor finds the signal while its software partner geolocates the emitter. The Surveyor 4D and N6854A geolocation server can detect and locate even the shortest duration signal.

RF sensor systems start out lower cost than signal analyzers, but the price rises depending on the size of the area to be monitored, from a city, county, state to nation state.

Lumacron: Lighting Up Enemy Forces at Wire Speed

Ask any military intelligence or law enforcement agent what their greatest challenge is, and they might respond, "packet loss on high-speed networks." By the same token, the switch from 1G to 10G, 40G, 100G or faster networks can be a problem of comparable magnitude for communications service providers. With bandwidth demand rising, they have little choice but to move upstream to faster networks.

Until they do so, CSPs also have trouble monitoring network activity accurately. Massive amounts of data dumped at wire speed on processing systems undermines the ability to gain useful metrics on network performance, service usage, problem spots in the network and other issues that can beset carriers making the transition.

The risk of packet loss is of concern to military intelligence and law enforcement agencies, as well. In moments of candor, even the elite will acknowledge that the growing gap between the need to capture data on high-speed networks and the reality of slower connections and processing systems on the receiving end means that no solution is perfect, and some data may always be lost.

One solution to this challenge is to simply go around it, offloading wire speed data streams and sizing them up on separate, FPGA-powered solutions before the load swamps downstream processing tools. Using such hardware accelerators is an efficient solution that handles wire speed data analysis reliably, never losing a packet.

Elsewhere in this book we examine a company that handles the two sides of this job – IP flow monitoring and hardware accelerator solutions – Flowmon Networks. Another ISS vendor that also competes in this space is Scotland's Lumacron, a company that brings an illustrious parentage plus fresh perspective to the challenge.

Situated in Dunfermline, Fife, just across the "Firth of Forth" (inlet) from Edinburgh, Lumacron launched in 2014 via "spin-off" from another Scottish company, Aliathon. Aliathon, founded in 2001, marketed optical monitoring, test and transport solutions to the broad world of optical network service providers. Lumacron does that, too, but the company's main foci are the lawful intercept and military venues.

To that end, Lumacron is a regular presence, both as exhibitor and speaker, at ISS World events in Europe and the U.S., where it holds forth on "100% Signal Visibility" in next generation networks.

Lumacron and Flowmon are roughly the same age, and both draw on a strong academic heritage. While the two companies define excellence in their niche, Lumacron stands out in several ways: guarantee of data capture; a product life cycle that surpasses others; and one of the best-known and most successful players in this sector, Dr. Ian Graham, Chairman of Lumacron.

The Endace Connection

Dr. Graham is renowned for his seminal role in developing the first tools used for intercepting data traffic on the Internet, beginning in 1994. As Dean of Computer Science and Mathematics at Waikato University in Hamilton, New Zealand, Graham led a team of academic data scientists studying the challenges of what was then a new field, data interception. The work was so successful that – much like the Brno University professors who formed Flowmon's former parent, INVEA-TECH – Graham and fellow PhDs left university in 2001 to form a new company.

That venture, Endace, quickly earned fame as a provider of IP traffic monitoring. What made Endace special: incorporating field programmable gate arrays (FPGAs) as hardware accelerators that could leverage IP flow monitoring technology to capture data at wire speed before it entered a processing system. The basic selling point: capture of all packets transiting an IP network efficiently, reliably, and at far less cost than investing in massive parallel processing systems to compensate for beleaguered in-house processing systems.

Government and military intelligence agencies took notice. The first to buy from Endace was Great Britain's GCHQ, which would become a steady customer as the solution evolved. Over the next decade, other nations' intelligence and military agencies followed suit and purchased heavily from Endace: the Israeli Ministry of Defense, the U.S. Naval System Warfare Command; the U.S. Army; Denmark's Defense Intelligence Service; Morocco's civilian surveillance agency; Spain's Ministry of Defense; and the Defense Departments of Canada, Australia and India. Even competing ISS vendors got in line to purchase Endace solutions, including South Africa's VASTech.

On the service provider side of the business, Endace won major contracts with major U.S. telecom operators including AT&T, Verizon and Sprint, and with the largest carriers in Europe – France Telecom, Deutsche Telekom, Belgacom, Swisscom and Telstra.

During this time, another development came along that would significantly influence the future of network infrastructure: soaring bandwidth demand driven by smartphones, social media and streaming video, fueling the need for faster fiber optic backbone and local facility networks. Endace was on top of that movement, too, and in 2011 introduced Medusa, what may well have been the world's first FPGA-powered card capable of capturing data on 100G networks.

And so the FPGA boom met and conquered the bandwidth boom. It was a crowning achievement for Endace, and like many such an event attracted interested buyers. In 2013 Emulex bought Endace, and Dr. Graham officially retired. But not for long. He took the intellectual "jewels" with him and became principal investor and Chairman of a new venture, subsequently named Lumacron. Key Alathion executives with similar backgrounds followed him to Dunfermline, where Lumacron resides today.

Lumacron Solutions for Military, Govt. and Law Enforcement

Lumacron offers a trio of FPGA-powered hardware/software products that, curiously, nowhere make reference to the use of FPGAs. Regardless, the devices work from a platform that is "reconfigurable in the field," and takes advantage of FPGAs in its packet capture cards. The basic premise is exactly the same. As Lumacron puts it, "providing the raw packet data," e.g., via IP Flow Monitoring, "is not enough." Lumacron products "also provide acceleration functions by offloading tasks from the downstream processors that are more efficiently solved as data comes off the wire."

Detailed information on the operation of FPGAs is available in Chapter 2 on packet monitoring. But to briefly re-cap, FPGAs are circuit boards that accelerate the capture of unstructured data including SMS, video, HTML web pages and text, the dominant forms of traffic on today's communications networks. Principal vendors include Xilinx and Altera. Endace used Altera and Xilinx FPGAs in its advanced packet capture cards. Today, so do Alathion and Lumacron. FPGAs are a commodity.

The Lumacron products are, respectively, named LUMA-MON-ARIES, LUMA-MON-CHRONOS and LUMA-MON-DRACO, which here we will shorten to ARIES, CHRONOS and DRACO for simplicity's sake. The fundamental differences derive from client needs, depending on speed of the network and the age/type of the protocols involved. Move up the food chain, the newer and faster the Lumacron product:

- **ARIES:** For legacy protocols. Captures bidirectional traffic at up to 10G, and terminates 2 X 10 or 8 X 2.5 channels.
- **CHRONOS:** For evolved protocol networking. Sees bidirectional links at up to 100G, and terminates 1 X 100G channel or 10 X 10G channels, or both.
- **DRACO:** For next-generation protocol network monitoring. Delivers 100 percent visibility of network links up to and including 100G,

and terminates 2 x 100G channels, or 2 x 10×10 G channels, or 1 x 100G + 10x10G channels.

As Lumacron observes, some networks fall into the legacy category, making ARIES the applicable solution. Others that use evolved and next generation protocols would need to move upstream to CHRONOS or DRACO. Because network upgrades are seldom if ever performed on a flash cut basis, the usual deployment scenario involves a mix of advanced with legacy network protocols. Depending on the network, the appropriately protocol-aligned Lumacron device would apply.

Each product uses a Linux-based CPU, and all can capture a wide range of packet, OTN and SDH/SONET protocols. Speed of interception and processing varies by device. On the high end, DRACO provides up to 28 bidirectional optical interfaces, each able to independently support data rates from 1G to to 100G, CHRONOS up to 14 at the same speeds, while ARIES provides up to 12 bidirectional interfaces and taps out at 12G. Lumacron claims that in each category, its products provide "significantly higher port density than competing solutions."

Worth noting: The platforms also integrate. Thus, for example, DRACO might capture an Ethernet packet stream at 100G and ship it downstream to CHRONOS or ARIES for further analytics. CHRONOS can perform the same function, transmitting to ARIES.

Each platform is designed to recognize the type of packet stream it is working with, and to instantly adapt to different ones. Operation is completely hands-free, eliminating the need to pre-configure for one type of protocol then reconfigure for another. The process is automatic, with dynamic detection applied to identify and treat any incoming data stream by hierarchy, type and rate.

As an FPGA-based system, a Lumacron product can be field-configured for any need deemed non-standard, or to focus on specific types of data, at specified speeds, structure of frame or packet, and protocol.

To a great degree, the ability to customize ARIES, CHRONOS or DRACO helps make them future-proof against network changes. Added to configurability is the benefit of interoperability, where the user can program one Lumacron device to offload traffic to another. Such flexibility means they can meet the challenge as network speeds grow faster and more complex.

Lumacron platforms and processing power will handle the load as fiber and 5G networks take wire speed to the next level.

Parsons OGSystems: Disruptive GEOINT

OGSystems, acquired by Parsons Corporation in January 2019, is a GEOINT technology enabler. Its core expertise lies in developing real-time 3D GEOINT solutions that merge with intelligence from other disciplines – OSINT, COMINT, ELINT, HUMINT and MASINT (measurement and signal intelligence) – to bring all key elements of a mammoth big data picture into a well-defined frame in real-time, and trigger alerts when that frame differs in the slightest respect from its predecessors.

In the simplest terms, OGSystems is about location and all related intelligence that imbues specific locations with the meaning, significance and intent of people, places and events being monitored constantly from above, on the ground and through the air. OGSystems' approach is highly directed. What made it an attractive target well worth the US $3 million Parsons paid: expertise in supporting The National Reconnaissance Office (NRO) and the National Geospatial-Intelligence Agency (NGA), which in turn provide vital GEOINT for U.S. military forces as well as two other agencies: the CIA and NSA.

The NRO and NGA, which together count a century of experience, perform two basic functions: the gathering of satellite imagery the world over (by the NRO), and the conversion of this raw imagery data into usable visual intelligence (by the NGA). Back in the day, as agencies and contractors grew and profited together, one undesirable offshoot of government-run or sponsored R&D was the development of heavyweight, slow-moving bureaucracies. Even on the tech side, small innovators were quickly swallowed by big contractors and made to heel to the business-as-usual motto of *Ergo in occursum ego sum* – Latin for "I meet therefore I am."

It was one such meeting that led to the formation of OGSystems in 2005. A pair of mid-20s engineers, Omar Balkissoon and Garrett Pagon, found themselves in a day-long meeting held by a large contractor in charge of a huge, and from the sound of it, directionless intelligence initiative. The confusion betrayed by that 8-hour session could have only one conclusion: a proposal for another day-long meeting leading nowhere...except to more billable hours for the contractor. That was enough for Mssrs. Balkissoon and Pagon, who struck up a friendship, quit their day jobs and took a chance on a new venture that would be tightly focused and purely results-oriented. The name: "O" (for Omar) and "G" (for Garrett) Systems – OGSystems. OGS, as they abbreviate it, which grew to become an enterprise of some 300 experts based in Chantilly, VA.

OGSystems' success was built on its resident 3D GEOINT talent, a key acquisition, close working relationships with companies whose talents complement its own – and changes in the way GEOINT is now collected from the get-go.

OGSystems set the stage in 2015 with the acquisition of Urban Robotics, a provider of airborne military Intelligence, Surveillance, and Reconnaissance (ISR) applications. Based in Portland, Oregon, Urban Robotics contributed by providing TerraFlash – cluster computing and cognitive computing (AI) capabilities that massage complex 2D and 3D GEOINT imagery to deliver answers that the naked eye might not see.

Urban Robotics was no stranger to the skies, either – or to government clientele. Along with the added data crunching capabilities of TerraFlash, OGSystems acquired two other key Urban Robotics properties: PeARL, an aerial digital imaging sensor and camera system; and WulfPack, a computer cluster for processing both aerial and ground images.

At the time of the acquisition, Urban Robotics was already tight with many of the U.S. government clients served by OGSystems including the DoD, Department of Homeland Security, and other agencies that favor GIS and remote sensing. In acquiring Urban Robotics, OGS played a familiar hand called, "get more hooks into the customer."

Next up, OGSystems built on what it had made and bought, partnering with other innovators to polish its big data platform, BlueGlass. Some helped flesh out analytics capabilities. Others contributed to factoring in intelligence from multiple non-aerial sources, performing real-time predictive analysis or adding the ability to send red flag alerts of risk.

One such partner was noted for leveraging a core change in the collection of GEOINT itself. For years, elaborate high-end satellite systems ruled GEOINT. While these embedded networks still play an important role, their capabilities have been surpassed in key ways by younger, more flexible approaches using multiple low-orbit cube satellites that collect data round-the-clock. The advantage of these newbies: They do not require the last-minute positioning of a satellite to handle a real-time need for intelligence.

OGSystems' partner in this arena was Planet, a company that provides persistent monitoring via satellite of highly-localized, subnational, national or global targets for both government and commercial clients. Always on, the Planet platform guarantees real-time satellite imagery any time, anywhere. As importantly, the company maintains a voluminous database of captured imagery dating back to 2009. That means the user can time-map current images against the record by day, month or year and instantly see changes.

As good as these capabilities may be, a key differentiator that helped set OGSystems apart is keen focus on "humanizing" intelligence from a variety of sources that can be fed into the system. Partners play an important role here, too. Prescience shoots rapid response alerts of on-the-ground risks in any area under aerial surveillance. Narrative Science converts unstructured data – notably visuals, but also text – to reveal insights on meaning and intent. Real-time predictive insights gleaned for actionable intelligence can be the contribution of yet another partner, Transvoyant, one of the best RTPA vendors in the business.

To this day, all continue to convene at the Parsons OGSystems' Viper Labs to produce solutions tailored to serve real-time and long term missions.

Palantir: Visualizing the Future of Warfighting

At center stage in big data's surveillance realm stands a single giant: Palantir, of Palo Alto, California. The company debuted in 2004 with $30 million in seed funding from In-Q-Tel, the CIA's venture capita group, and also from Peter Thiel, co-founder of PayPal. The quid pro quo of In-Q-Tel support is commitment to delivering technology solutions that support the U.S. Intelligence Community and the intelligence breaches of the U.S. Department of Defense, a mission that Palantir has served to great effect.

The value of Palantir's technology quickly spread to other customers. Today, Palantir's customers include not only the IC and DoD, but state and local law enforcement, first responders, financial institutions, and members of the Fortune 500. All benefit from Palantir's skill at integrating data from vast sources and delivering mission-critical intelligence to understand the data via insightful visualization that helps ensure informed decisions.

With Palantir's growth to multi-billion dollar annual revenue, the company has augmented its brand and message, and today opts to call itself a "data integration" specialist. Curiously, in presentations given at 2020 conferences where IC and defense customers collect, Palantir scarcely references the term "visualization," either. Regardless, visualization remains the company's key differentiator.

Palantir's two technology platforms are Gotham, the basis of solutions for government, and Foundry, for the enterprise. For government ISS customers, the solutions of key interest include Palantir Intelligence, Defense, and Law Enforcement.

All three solutions apply relevant information from diverse sources – in many cases, freeing it from silos where it previously saw only limited use – optimizing the value of the data by creating a much more detailed picture of the chosen topic, event or target.

The data can be either structured, which is database management jargon for "organized in a predefined way," in a record or file. Or it can be the opposite – lacking any tag or field that facilitates easy storage and access, hence unstructured.

In big data, the core questions are always: "What happened?" and "What's happening now?" Based on the answers, a system can build models that will help crack the third riddle: "What *will* happen?" Each question is addressed by a respective field of data analytics: descriptive, real time and predictive. Descriptive analytics, also called "heuristic" analytics, pertain to the study of past events logged in data: billing records, purchases, calls, emails – anything historic relating to the subject. From descriptive analytics, Palantir's products single out trends that point to repeatable behaviors and actions. Whatever happens multiple times is a pattern. Real time analytics involves the use of live streaming data to gain detailed insights on a topic or target "in the moment." In most big data systems, real time analytics apps review as many as 50,000 variables instantaneously.

In a counter-terrorism mission, Palantir Intelligence collects data from multiple sources to compile a portrait of the chief plotter and the threat. The data incorporated in this picture might include airline tickets, restaurants visited, types and locations of living accommodations, vehicles used, suspicious purchases, presence at training camps, military background, complete personal history, communications records – linked to similar data for all known accomplices. End result: an intensely detailed and completely logical visualization of all the previously missing pieces of the puzzle.

Palantir Defense merges findings from big data with sources that include HUMINT, SIGINT, GEOINT, IMINT (imagery intelligence) and SOCMINT to provide high-value insights to warfighters at the tactical edge. During the worst of Iraq's IED epidemic, for example, U.S. military analysts used Palantir Defense to assess and plot the safest routes through Baghdad, and to change them continuously to stay one step ahead of enemy combatants. Injury and death rates from roadside bombs, which at the worst point accounted for over 70 percent of casualties, dropped markedly. While there were many factors behind the decline, Palantir certainly played a role.

Among the biggest challenges in law enforcement is the problem of accessing data stored in multiple databases – with no way to access, analyze or view all relevant information. Palantir Law Enforcement provides a single unified platform for case management. LEAs use a simple interface to instantly access all data that is available on the target, from a single location and from one system versus multiple sites. Lawmen can update the data remotely from the field, view geophysical data and its relation to crime scenes, review a target's history, and apply what they learn to stop crimes in progress – or preemptively.

In all, Palantir is an enabler of intelligence, with the ability to make in-house systems work better at finding and using not only big data streams, but their own data. The NSA has used Palantir to improve analysis by the agency's KEYSCORE program, pulling profiles and records of targets of interest and correlating those elements with data gleaned from communications and networks of affiliates. Similarly, the UK's GCHQ has deployed Palantir to enrich social media OSINT.

For the military, Palantir serves as a data fusion system that provides real-time intelligence on enemy movements, force tracking of allies, and integration of ground, air and other support. Troopers under fire at the tactical edge do not have the leisure of perusing laptops for bar charts. They need concise data that is easy to understand and act on. Defeating enemies in combat in the 21st century relies heavily on situational awareness assisted by real-time visualization of battlefield conditions – still Palantir's ace capability.

PLATH Group: Military and Malware

As an acknowledged leader in military RF monitoring and analysis for more than 60 years, Germany's PLATH Group surprised many analysts in 2012 with its sudden dip into markets of primary interest to intelligence agencies and law enforcement: cybersecurity, forensics and malware. Was PLATH changing course? No, that is hardly the case. PLATH simply took advantage of an opportunity to diversify. The company's business is still almost entirely focused on radar and RF signal acquisition and monitoring for military clientele. The only difference between the conventional and new PLATH is that the company, through strategic partnerships and acquired properties, can now step up to meet a broader array of ISS needs.

Circularly Disposed Antenna Arrays

PLATH's roots in military SIGINT run deep, dating to the early days of radar during World War II. Germany's contribution to the field was the "Wullenweber," a vast circular antenna sometimes as large as three or four U.S. football fields in diameter. The Wullenweber consisted of two concentric circles of antennas, one for high band, the other for low band. The Wullenweber would continuously scan the horizon in a 360 degree arc with a direction-finding beam. It was very effective at triangulating radio signals both for direction finding and intelligence.

When the war ended, Allied Forces studied a surviving Wullenweber, which stood in Denmark, then destroyed it. However, the technology was too valuable to be left idle. By the late 1940s a team of German scientists was assembled to build a new Wullenweber, this time in southern Germany, and the technology took on a new name: the Circularly Disposed Antenna Array (CDDA). One of the key engineers involved was a Dr. Maximilian Wächtler. After brief operation, this facility was disassembled and moved to the U.S. for further study. However, Dr. Wächtler retained an interest in CDDA technology in the homeland that would carry over to his next venture, the founding of PLATH, which today serves military clientele in over 60 nations.

Signal Acquisition, Analysis, Evaluation, Visualization

PLATH products are offered through the corporate holding company, PLATH GmbH, and a set of five subsidiaries located in Germany, Switzerland and France. The principal RF divisions and their locations are:

- **PROCITEC** (Pforzheim, Germany)
- **innoSysTec** (Salem-Neufrach, Germany)
- **PLATH EFT** (Norderstedt, Germany)
- **PLATH AG** (Bern, Swizerland)

The fifth subsidiary is where PLATH focuses on surveillance:

- **NEXA** Technologies (Paris, France)

PLATH RF products cover four functions: signal acquisition, analysis, evaluation and visualization offered by the holding company, PLATH GmbH, and its PROCITEC subsidiary, as follows:

- **Signal Acquisition.** PLATH GmbH is home to the company's CDDA products and services for military clients. On the signal acquisition front, PLATH GmbH offers the ASM 4221, a descendant of the classic Wullenweber antenna array with up to 32 inputs and outputs, and non-blocking switching for monitoring in the 1 MHz to 30 MHz range. The product's modular design allows easy integration with modules that cover the other three capabilities.
- **Signal Analysis.** PLATH's signal analysis product is PROCITEC PROCEED, an Integrated Circuits and Signal Processing (ICAS) desktop software tool for automated recognition and decoding of signals. Using an automated production channel, PROCEED ensures against loss of signaling data via buffering, then checks signals to isolate predefined modem types, capturing signals matched against patterns. The system then displays signal characteristics and decoded text. PROCEED also: detects speech signals and their characteristics including voice pitch and modulation; records all signals; and provides complete decoding. Users can switch to manual mode to make precise measurements of signal characteristics. Specific solutions are offered "cafeteria style" so that military and government agency customers can choose from six different models of PROCEED, which offer features/ functions tailored per need, e.g., military or non-military decoders.
- **Signal Evaluation.** Based on data findings, the user may then proceed to evaluation with a third product, Bit Stream Processor (BSP), which extracts metadata, content and technical parameters. Basically, BSP acts like a big data analytics and retention system, scoring and sorting data via algorithms, finding connections, and storing the data for playback in a readily searchable format. The user can work in real-time or offline, accessing various inputs, as well as external data sources, to perform the most detailed evaluation of the data.
- **Signal Visualization.** The final stage is signal visualization, performed by ICAS Traffic Analyst, PLATH'S Palantir-like product, COPIN

(Communications Profiling Intelligence), and complemented by the company's COPS (Communication Profiling System.) COPIN illustrates the BSP-driven data in 3D, showing relevant metadata and content in a geospatial context. The system enables the user to quickly single out data relevant to an incident or investigation regardless of frequency range (HF, VHF or UHF), signal type, time reference or the specific RF monitoring system deployed. COPS can absorb and analyze network metadata, including from off-network signals, and issue alarms when patterns point to anomalous behavior.

Special Needs

Beyond these four fundamentals of PLATH Group RF monitoring, PLATH provides an array of products and systems designed to meet nearly every conceivable RF monitoring need of the military or government intelligence agency. Several stand out.

For rapid deployment:

- **AAU-7480.** PLATH's "Adcock" antennas, named after the British inventor who first patented directional radio signaling systems in 1919, are available in both rapid install and long term configuration versions. The AAU-7480 Adcock series is ideal for military applications, enabling troops to deploy radio direction finding equipment of exceptional sensitivity in under 30 minutes. It's as simple as pushing half a dozen antennas into the ground at set intervals. Invaluable for picking up both ground and sky signals in the 1 – 30 Mhz range, the AAU-7480 performs on par with fixed DF CDDA "Wullenweber" systems.

For direction finding:

- **DFA 2405 with DFP 2400.** Integrating PLATH's DFA 2405, which acquires signals in the 20 MHz to 3 GHz range, with the company's DFP 2400, which captures both stationary and frequency-agile emissions across seven channels.
- **DFP 5400.** A digital HF radio direction finder for narrowband, the 5400 is designed for automatic search in the 1 – 30 Mhz range. Of note, this RD finder handles both standard and LPI (low probability of intercept) signals. The DFP 5400 leverages spectral data analysis as an added advantage over conventional direction finding and locating systems.
- **PDV 5400.** Designed to work in conjunction with the DFP 5400 HF direction finder, the PDV 5400 provides automated detection and classification of emissions in real-time.

For military RF monitoring mounted on vehicles:
- **CMA 2400.** PLATH's CMA 2400 Active VHF/ UHF DF antenna supports mobile or semi-mobile signal acquisition in the 20 – 3000 Mhz frequency range. The PLATH CMA 2400 is mounted on combat vehicles for battlefield communications and reconnaissance.

NEXA Adds Russian Malware

The rapid devolution of France's Bull Amesys following the Arab Spring is a matter of public record. More interesting, and less well-known, is how the company was saved from extinction by the intervention of PLATH GmBH, which snapped up the remains of Bull Amesys for pennies on the dollar. Amesys, which won unwanted notoriety for selling deep packet inspection (DPI) systems to the governments of Libya and other North African/Middle Eastern nations, lives on in PLATH's subsidiary NEXA Technologies and in of all places, Paris, where negative publicity and court actions brought operations of the original Bull Amesys to an end. NEXA also operates an Amesys office in Dubai, United Arab Emirates (UAE).

NEXA Technologies is where PLATH Group houses the cyber, forensics and malware components of its product portfolio, which includes the original Amesys Eagle DPI solution, as well as products from its partners:
- **Nuix:** an Australian company that specializes in systems for the collection and assessment of unstructured data.
- **BERLA:** a provider of Blackthorn forensics software for acquiring, examining and analyzing data from aviation, maritime, portable automotive, and handheld GPS devices.
- **BlackBag Technologies:** complete forensics for iOS, Android and Windows devices.
- **Micro Systemation:** advanced mobile forensics covering over 600 mobile apps.
- **Videntifier Technologies:** automated fingerprinting of videos from as little as one image.
- **CRU Weibetech:** devices for digital asset security and forensics.
- **Oxygen Software:** Moscow's version of FinFisher – the Oxygen Forensic Extractor is malware for penetrating and taking control of mobile devices, remotely or locally, providing access to target files, documents, metadata, web surfing history, and sending fake messages on behalf of the target, or turning device audio, video and still camera on or off.

Providence Group: SOCOM Tracking

Among the leaders in ISS solutions catering to military clientele are Providence Group and its affiliated companies, founded by CEO Michael Davies and his partners Stephen Turner, Director of Providence and CEO of its Counter Threat Solutions division, and Peter Stolwerk, Providence Group COO. The trio share an unusual but highly desirable background and skill set: prior careers in military Special Operations, surveillance, intelligence, counter-terrorism and rescue.

Fresh out of school, Davies entered the British Army Corps of Royal Signals where he became expert in telecommunications networks. Next up: Five years in British Special Forces specializing in electronic surveillance, and with deployments in some of the world's toughest hot spots, from the Balkans to North Africa and the Middle East. Following military service, a brief but highly successful stint in defense contracting led to healthy cuts on deals in the U.S. that would ultimately inspire funding for The Providence Group. Coming to this venture, Davies would prove well-rounded in all key respects: first-hand knowledge of networks, ISS and special ops, which combined to form perfect understanding of what agents and operatives need tactically in order to succeed on missions.

Davies' partners are equally qualified, adding complementary layers of expertise. Stephen Turner's quarter-century career includes 10 years in the British Army Corps of Royal Signals, followed by seven years in 22 SAS Regiment including a tour with the Special Reconnaissance Regiment. Turner's specialties: surveillance, expeditionary intelligence gathering techniques and dynamic targeting. As CEO of the Providence Counter Threat Solutions group, Turner leads the company's work in counter-terrorism training. On a larger scale, Turner also heads Counter Terrorism Equipment Procurement services provided by all Providence Group of companies.

COO Peter Stolwerk is Providence Group's resident expert on counter-terrorism from the law enforcement SWAT perspective. Following a stint with the Royal Dutch Military Police, Stolwerk worked in Dutch law enforcement including its SWAT Special Operations Unit, where he ran counter-terrorism and hostage rescue operations. Stolwerk retired from law enforcement in 2006 and spent the next three years developing relations with agencies throughout the European Union that would ultimately become high value to Providence.

In 2009 the three like-minded, similarly prepared individuals linked up to form Providence Group. Their specialty is intelligence training and counter surveillance training, and the equipment for both: covert ISS solutions, and on-the-ground tools for apprehending and fighting criminals and terrorists.

They produced a winning formula. First, consult on what the client needs in either or both fields. Then choose the appropriate ISS and counter-intel or counter terrorism tools from the host of those available, and modify as needed for the specific job. If the necessary tool doesn't exist, Providence can design and create a first-ever solution that fits the bill. Alternately, in keeping with the company's individualistic instincts, it might abandon convention and experiment with modifications to COTS products.

Most important is the activity that crowns service selection and customization. Before any client hits the ground with a new solution, the company makes it clear that proficiency hinges on training and practice. Providence provides both at their special facilities based in Hereford, UK and Haarlem, Netherlands, as well as mobile training capabilities that teach, coach and grade clients where they do business.

Satisfied clients include law enforcement, border patrol and military units in the 28 nations of the European Union, plus the United Kingdom, Malaysia, Singapore and the United States.

The Providence Group has the ability to customize, integrate and enhance any surveillance or counter-intelligence product/service so that it fits the client like a tailored suit. Thanks to a competitive marketplace and the divine hand of Providence's global procurement operations, the prices are right for the client, too.

Products range from low-end (e.g. audio) to very high-end items such as unmanned aerial vehicles – not simply drones but true pilot-less aircraft. Groupings give a good general idea of the capabilities available:

- Covert Audio and Video.
- Tag, Track and Locate.
- Long Range Video.
- Tactical Intercept Systems.
- Tactical Cyber Solutions.
- Store & Forward Audio/Video Solutions.
- Covert Method of Entry (CMOE) Solutions.
- Alarm Defeat.
- Data Forensics.
- Infrared and Night Vision Cameras.
- RF Jamming.
- Technical Surveillance Counter Measures (TSCM).
- Tactical and Covert Radio Systems.
- Unmanned Aerial Vehicles (UAVs) – pilotless aircraft.

Providence also provides deployment kits for the majority of products used for tactical surveillance.

What about when the agency is ready to move in on the target? Here another Providence subsidiary can step in to help.

The company's Tac-Up, based in the Netherlands, offers a variety of products for use in capturing targets, controlling unruly groups or other special needs specific to monitoring and apprehension. On the list: tools for climbing or quickly breaching hideouts; Tac-Trigger and Tac-Extend tools for safely stretching the reach of stun grenades, or creating a diversion; the light-weight Tac-Focus that delivers powerful door ramming in tight operating conditions; and, the patented Take-Down stick for controlling suspects without use of a taser. Tac-Up even produces its own special Cold Fire Tactical fire extinguisher.

As with everything offered by the Providence Group, all Tac-Up products come with extensive training. And if the need is for a highly specialized solution, Tac-Up can design and build what's needed in its on-site laboratory.

User Focus Based on First-Hand Experience

Providence Group stands out from many companies in the ISS sphere for its understanding of a client's tactical end user needs. We're talking about the law enforcement, border patrol, Special Ops, SWAT and similar battle front personnel at the point where intelligence is not just gathered, but quickly converted into action. The company's leadership knows this space because they've lived there themselves. We've included non-ISS tactical tools by Tac-Up in our analysis to make the point: In understanding and delivering on client needs, Providence thinks it through to the end.

Are Providence Group solutions meant for broad monitoring and analysis on the scale of an NSA or GCHQ? No, and to confuse their purpose with that of broader strategic tools is to misunderstand the core thrust. Providence solutions are about prevention, to be sure, but at the same time they are what Special Forces need for location, apprehension and capture.

Raytheon 3G Forward-Looking Infrared

As a battle-tested technology, thermal imaging is vital for the warfighter, and few solutions do it better than Raytheon Forward-Looking Infrared (FLIR). FLIR's ability to single out targets on the ground, from the air and over the sea, delivering detailed thermal images of hostiles in any light condition without revealing the presence of the user, is a must-have advantage in combat.

The competition is tough and very good: FLIR Systems, BAE Systems, Lockheed Martin, DRS, Israel's Elbit Systems, Tamam (a division of Israel Aircraft Industries) and Sagam (part of France's Safron). Technical advantage is often a back-and-forth game, with companies like L-3 communications constantly nipping at Raytheon's heels. But with its advanced electro-optical/infrared (EO-IR) technologies, Raytheon remains the decided leader with a major segment of the U.S. military, including the Army, Marine Corps and Special Operations Ground Forces.

As in any field of technology, you find "coopetition" alongside competition, the joint venture "ThalesRaytheonSystems" being a prime example. Competitors also buy the work of others, as BAE Systems-Sweden did in 2012, contracting for $US 11 million of thermal imaging and sighting equipment from FLIR Systems for BAE combat vehicles sold to the Norwegian Army.

Or one company may supply critical components to another, as was the case with the Raytheon ATFLIR, which combines a trio of functions critical to aircraft – navigation, weapons targeting and laser designation – with help from components made by BAE Systems.

The market for FLIR products is a vibrant one, and the story of how Raytheon came out on top, well worth the telling.

Ballad of the Ho Chi Minh Trail

The tale begins with a visit to Vietnam. Not the tourist attraction of today, but the land where some 58,000 U.S. troops lost their lives in the Second Indochina War from 1955 – 1975. As U.S. involvement ramped up in the early 1960s, one of the biggest problems was the inability to track enemy troops filtering by the tens of thousands into Vietnam from the north. At night, when most troop movement occurred, aerial reconnaissance was a wash – North Vietnamese regulars and Viet Cong moved, at will. But hope was in the offing.

9,000 miles to the east, the Defense Systems and Electronics Division (DSED) of Dallas-based Texas Instruments (TI) had been experimenting with infrared thermography, aka thermal imaging, using new thermographic cameras that could detect and create images of objects in the infrared range of the spectrum.

While TI DSED was running tests on what would become FLIR, the U.S. Air Force had been working on thermal imagery for aerial reconnaissance over Vietnam, as well. But the military's own solution was cumbersome and the results mixed. The basic problem: shooting images on film which then had to be flown back to base for developing and analysis. By the time photographic prints emerged from the darkroom, any enemy was long gone.

TI had a better idea: replace film with photonic detectors and translate the images to TV in real time right on the airplane. Thus was born the first FLIR. It weighed half a ton and required equally heavy equipment to keep the system from overheating due to massive power requirements. But it worked. By the late 1960s, the U.S. Air Force was routinely using FLIR to accurately capture the movement of troops and trucks down the Ho Chi Minh Trail under any conditions.

In the coming decades, TI continued to lead further development of FLIR and coin other advanced radars and electronic weaponry: inverse synthetic aperture radar (ISAR) which captures high resolution two-dimensional images of the target based on its movement; and in 1991, through a joint venture with Raytheon, the Military Microwave Integrated Circuit (MIMIC) program to improve the efficiency and cost of gallium arsenide radar modules.

As importantly, TI's Defense Systems and Electronic Division had caught Raytheon's eye. In 1997, with TI looking to streamline its business, Raytheon bought the division for just under $US 3.0 billion, followed by the purchase of Hughes Aircraft, which also had an attractive infrared and detector group.

The combined assets of DSED and Hughes, together with Raytheon's assets, became the platform for what would become one of the largest and most successful military FLIR businesses in the world. Along the way, a tool that once weighed as much as a cart of bricks has been reduced to a few pounds – in some cases, ounces – and is used in fighter aircraft, satellites, naval vessels and thermal weapon sights carried by soldiers on the front line.

How FLIR Works

All objects with a temperature above absolute zero emit infrared radiation. The amount of radiation varies by object, with warmer ones emitting more, making it possible to distinguish between animals, humans, trees, rocks and ground, producing a picture every bit as detailed as that made via film or today's digital equivalent.

The difference is that FLIR works with or without light and "sees" through any method of concealment. Darkness is no barrier, nor is camouflage. Individuals hiding under tree canopies or in earth tone fatigues and invisible to the human eye stand out plain as day. Smoke and fog in the way? Not a problem for FLIR.

Another advantage: Because FLIR is a passive system that relies only on radiation emitted by the target, it is completely undetectable. One shortcoming: FLIR systems can't distinguish between foe and friend. To prevent "friendly fire" taking out the wrong people, it is essential to equip one's own troops with distinguishable heat beacons.

FLIR is not to be confused with night vision systems. FLIR operates in the mid- and long-range infrared band of the spectrum, creating 2D images from infrared radiation and their associated heat, whether from a tank muzzle, guard dog, band of troops or solo sniper. Night vision devices work strictly in visible light and near-infrared portions of the electromagnetic spectrum. FLIR systems can detect the target in all conditions up to a distance of several kilometers. Night vision technology? Better hope for a full moon.

Generation Gap

FLIR military applications have evolved in three waves.

First generation FLIR, used with broad success by U.S. troops during the Persian Gulf War, relied on scanning an area via long-wave infrared, which was fine for its time – able, for example, to pick out an Iraqi tank hiding behind a sand bank. The one problem: background radiation from other objects might easily confuse the picture.

Second generation FLIR represented a significant improvement, switching from long-wave to mid-wave infrared, and "staring" versus scanning. The change to mid-wave infrared meant gaining images with significantly higher resolution. 2G FLIR is still the dominant type used by the U.S. Army.

Third generation FLIR takes it up a notch, providing dual-band detectors and an optical system for each. Multi-sensor configurations include both infrared and visible spectrum sensors.

3G FLIR is the current state of the art. Most would argue that fourth generation FLIR is still over the horizon. But is that true?

4G FLIR is broadly defined as including improved sensor fusion, high-definition two-color imagery, and a greater standoff range from the target.

Into the mix steps Raytheon "Next Generation FLIR." For whatever reason, the company doesn't come out and call it 4G FLIR, but their next generation product certainly looks like the real thing. However they choose to name it, Raytheon's current entry in the FLIR market sets the standard for meeting the military's key threats.

Overmatching the Opponent

For military in the U.S. and other advanced nations, the threats are two-fold: conventional and hybrid. On the conventional side, many nations equip their armies with 2G FLIR. In terms of what the U.S. military can do in response, the threat is now at parity.

The hybrid threat – foes that can readily blend with the civilian population – is another problem. Hybrids may have access to military 2G FLIR as well as commercial infrared products. In either case, the risk is that identical technologies create a stalemate.

Raytheon is driving the evolution of FLIR to "overmatch" both threats. Raytheon's 3G FLIR products can integrate with and update any prior generation of sensors. That makes the upgrade a no-brainer: The bulk of the U.S. military already uses Raytheon 2G FLIR, thus the transition to the Raytheon 3G FLIR system is seamless.

The U.S. Army's FLIR status provides a case in point:
- Infantry: Already in place are the Raytheon Improved-TOW Acquisition System (ITAS), Long Range Acquisition Scout Surveillance System (LRAS3), and Fire Support Sensor System (FS3).
- Stryker Combat Vehicles: All use the 2G Raytheon LRAS3, FS3, ITAS, and Minuteman Guidance System (MGS) FLIR.
- For Armored Units: Raytheon's 2G Commander and Gunner systems is on Abrams and Bradley tanks.

Raytheon Visual Analytics Fiasco: DCGS

With just a handful of U.S. troops left in Afghanistan on the tail end of America's longest-running conflict, there remain important lessons to be learned about major flaws in delivering intelligence. Here we will depart from the practice of focusing strictly on market leaders. It is just as important to reveal the ineptitude of laggards.

At the head of our list: the Distributed Common Ground System (DCGS) used as the principal intelligence, surveillance and reconnaissance (ISR) platform by the U.S. military.

Originally produced by Raytheon with help from Lockheed Martin, DCGS (pronounced "D-sigs") has repeatedly failed U.S. troops, yet the Pentagon continues to turn a blind eye to evidence proving top military brass's pet ISR initiative as slow and ineffective.

DCGS is meant to support all branches of the military, condensing data from multiple sources into useable intelligence for troops on the battlefield. But in reality, the system is so complex, user-unfriendly and buggy that it is best known for missing dangerous targets head on, or crashing and erasing all data in the midst of combat scenarios.

Following unsatisfactory experiences with the DCGS in Afghanistan and Iraq, U.S. Special Forces and Marines switched to Palantir, which produced brilliant, easy to understand visualizations of threats in their path. Palantir repeatedly surpassed performance of the Raytheon platform. While the Pentagon has bowed to legal pressure to allow Palantir the opportunity to be considered for top ISR slot – and awarded the leading data integration & visualization giant several contracts – military brass still have an undisguised bias toward DCGS. Why cleave to such an obvious loser? Simple: When senior jobs are on the line, executives generally choose to live with (and hide or deny) failure than own up to it.

Ironically, Raytheon passed on an opportunity to upgrade DCGS with visualization capabilities that might have put its solution's capabilities on par with Palantir. In 2012 the company acquired Visual Analytics Inc., a company whose Data Clarity product created detailed images of complex data. But any hope for beefing up DCGS with Palantir-like visuals evaporated when the Pentagon locked in on the DCGS clunker, as is.

Encouraged by the customer, Raytheon buried Data Clarity in its commercial cyber unit, where the product's potential value to the warfighter merits no mention. Never mind that forward-thinking analysts deem visualization the future of intelligence. The company continued to build, and the military to buy, a DCGS system condemned by users as a faulty anachronism, and by Congressional leaders as a US $10 billion "boondoggle."

The decisions to stick with DCGS and forego improving it are a shame and a waste. Understanding why requires a digression into visualization – what it is, and what it does.

Visualization: Simplifying the Complex

Big data has evolved into such a hot draw that interested parties sometimes overlook the principal challenge associated with the field: mainly, how big it really can be. Mountainous amounts of data are meaningless without analytics to "ask the right questions" and deliver useful, actionable insights from the databank. Analytics itself, typically heavy on algorithms and statistical analysis, can be daunting to those without advanced degrees in math. When looking down the barrel of an M4 rifle, there is no time to scroll through equations and statistics.

The field of visual analytics (VA) was developed to help resolve this issue. VA has been defined as the science of analytical reasoning facilitated by interactive visual interfaces. Simply put, VA provides easy-to-understand visual representations of complex issues that might otherwise be impossible for the human mind to fathom.

The underlying data may have thousands of dimensions, making numbers alone incomprehensible to the user. VA converts it into an intelligible form. By presenting images of analytical findings, VA can immediately show the user key trends and patterns in a simplified format, even as the targets alter their behavior or shift position.

Such flexibility is key to Palantir, which early-on eschewed artificial intelligence as too limiting. The founders observed that criminals, terrorists and other opponents tend to adapt quickly to surveillance and change their tactics to elude capture. They believed that computers alone – even those that "learn" – could not defeat fluid adversaries. Any system, it was felt, must embrace the human touch. Palantir's solution: "intelligence automation," which combined the power of automated analytics with human analysts. It's a bet that has paid off handsomely.

The company's Palantir Defense product uses big data analytics techniques to fuse SIGINT, HUMINT, OSINT, GEOINT and other sources, integrating structured, unstructured, relational, geospatial and temporal data into a single model. A soldier on the front line can use his or her laptop to access the same data available to a commander and customize it to his own location, in real-time. Granular detail on every enemy individual or event at that specific site is funneled to the user, not as numbers but as visualizations. End result: accurate situational awareness based on the data, but modified by a human agent to meet his precise needs and empower him to anticipate a threat and take the right action to beat it.

For example, a Special Forces unit might use Palantir to glean up-to-the minute data on terrorist activity in a Niger village, together with the individuals involved, collaborators and specific events. In the process, Palantir can call on data from the Tactical Ground Reporting System (TIGR), Multimedia Message Manager (M3), Biometrics Automated Toolset (BATs), Cellular Exploitation (CELLEX), Unattended Ground Sensors (UGS), Unmanned Aerial Vehicles (UAVs), as well as SIGINT and OSINT including intercepted enemy messages and social media tracking. Visualizations would display the findings, reveal patterns of activity that SOCOM could expect to encounter, link to other threats not immediately visible, then point to the best course of action.

Importantly, with Palantir the SOCOM warrior will see not just the target but the network of targets and the real dangers: recent improvised explosive device (IED) strikes, IED distribution across a territory, and the names and last locations of insurgent leaders. The data is pulled from thousands, millions or even billions of records and visualized for simplified understanding. The Palantir difference: data is not just from the past, but from the moment.

Welcome to the Tar Pit

Despite the fact that Palantir's capabilities move well beyond what DCGS can do, the Pentagon long remained firmly wedded to its Raytheon system. In part, military leadership's loyalty to the DCGS stems from its significant monetary and technological investment in the old system. But thus far it's a case of unrequited love. Almost from the beginning the program experienced fits and starts that hampered adoption.

The U.S. Army was first on-board, followed by the Air Force. As of April 4, 2018, the U.S. Navy has made DCGS its primary tool for ISR.

The fact that the Navy calls its ISR "DCGS-N" indicates not only that this platform is for their branch, but that the Pentagon continues to operate separate DCGS systems for each military branch. Army's version, by no coincidence, is called DCGS-A. Thus a model intended for the sharing of intelligence across all services of the military continues to be challenged by differences in data formats, amounts of data and collection platforms, plus varying degrees of complexity in exploiting data. Not to mention the balancing act of fixing problems while trying to keep pace with rapid advances in technology. With every step forward, DCGS seems to relinquish two steps, leaving a trail of blunders.

Matters have come to a head not once but many times. One of the more embarrassing moments occurred in August 2012 when officials testing DCGS detailed multiple malfunctions. A memo from Army Test and Evaluation Command (ATEC) Major General Genaro Dellarocco to Army Chief of Staff Raymond Odierno criticized "server failures that resulted in reboots/restarts recorded every 5.5 hours of test" and workstation failures every 11 hours.

General Dellarocco cited extreme user frustration when DCGS required multiple open screens to perform a single task, often resulting in "workstation freeze-ups." Other critics piled on, characterizing DCGS as "exceptionally complicated," requiring special training not ordinarily provided to combat troops.

Raytheon responded with efforts to simplify and de-bug the platform, releasing a supposedly new and improved DCGS-A in December 2012. But problems continued.

In late 2013 a memo from five Army units in Afghanistan called DCGS-A "unstable, slow, not friendly and a major hindrance to operations." Upgrades meant to improve DCGS-A ended up destroying users' data. The five units had no choice but to switch to a commercial product, Palantir.

One year later, software glitches in DCGS-A forced the Army to drop the system during a major training exercise. Even with new enhancements cited by the Army at a hearing before the U.S. Senate in early 2015, those forced to use DCGS knew it was a technological dinosaur in a tar pit.

Raytheon Visual Analytics

Raytheon was certainly aware of the issues surrounding DCGS. In December 2012 – four months after the incendiary Dellaracco memo and concurrent with the release of DCGS-A – Raytheon made what many analysts deemed a right turn into visualization with the aforementioned acquisition of Visual Analytics Inc. At the time, the move was considered fresh and original for Raytheon, which had previously focused IT investment on cyber threats and protection. With the addition of Visual Analytics' Data Clarity product, Raytheon appeared poised to take on Palantir.

Visual Analytics and its core product Data Clarity certainly had – and still have – all the right stuff. Founded in 1998, five years ahead of Palantir, Visual Analytics was an early player in big data analytics and visualization that grew to serve an eager client base with capabilities developed completely in-house.

Like Palantir, Visual Analytics developed its product on Java. Using open standards, Data Clarity proved itself capable of retrieving data from disparate sources including structured and unstructured data, enabling the end user to perform collaborative analysis of difficult problems. Of note, Visual Analytics ensured that it always had the bandwidth to meet client needs by operating its own data centers. Furthermore, by adhering to industry standards, Data Clarity was easy to expand without the cost and hassle of forklift upgrades. In all, Data Clarity was a true end-to-end service, providing access, integration, analysis, reporting and visualization.

Visual Analytics designed Data Clarity for commercial enterprises and financial institutions, but with primary emphasis on law enforcement and government agencies. Quickly proving its value in revealing money-laundering, performing criminal analysis, countering terrorism and narcotics trafficking, preventing fraud and ensuring cybersecurity, Data Clarity won over 300 accounts. Like Palantir, Data Clarity drew on multiple sources to reveal telling patterns, behaviors and networks of affiliates to stop bad actors, all visualized for easy comprehension by the user.

Post-acquisition, Raytheon didn't alter the fabric of Visual Analytics one iota. Today's Raytheon Data Clarity product is faithful to the original and targets the same types of customers. Raytheon shops Data Clarity two ways: via its Raytheon Cyber Products division located in Herndon, Virginia USA, and as part of the company's Intersect family of data analytics products. Data Clarity is strongly marketed to law enforcement agencies (LEAs), offering advantages that include temporal, link, statistical and geospatial analytics *with* visualization.

Curiously, Data Clarity is an exact match of another Raytheon big data product also marketed to LEAs: Raytheon SureView, run by the company's Websense division out of the same Herndon, Virginia address. Not only are Data Clarity and SureView twins, they bear an uncanny resemblance to Palantir Defense: using visualization to show patterns and connections derived from big data analytics. But Raytheon does not promote the use of SureView for war fighters.

By now, the oversight is a moot point. In 2017, Palantir won a lawsuit proving that its exclusion from the DCGS-A contract blatantly violated DoD procurement policy. Then in 2019, Palantir's challenge paid off when the company won an $880 million award to revitalize DCGS-A and make all systems and components interoperate seamlessly. In sum, Raytheon lost the recompete and is off DCGS work for the time being. However, following the April 7, 2020 conclusion of its merger with United Technologies, look for the new Raytheon Technologies to renew its pursuit of DCGS work. As time passes, the company's DCGS boondoggle will become a distant memory.

Sepura COVERT SRC3300

Cambridge, UK-based Sepura, a subsidiary of China's Hytera since 2017, is the maker of the COVERT SRC3300, the world's best-selling TETRA terminal for covert and special forces missions. The Sepura SRC3300 Covert TETRA Radio was designed by proven experts in covert operations for clients on the front line of monitoring and apprehending criminals and military foes. It is designed not only to assist surveillance, but also to avoid counter-surveillance measures by sophisticated opponents. First introduced in 2007, the SRC3300 has evolved to become the best-selling TETRA covert phone in the world.

The SRC3300 comes with impressive features: an end-to-end embedded encryption platform, a "Zeroise" button that instantly erases all cryptographic content/data on the phone, a built-in Direct Mode Operation (DMO) repeater that extends the range of the phone, and Bluetooth to accommodate add-ons. There are also things the phone does *not* come with, by intent. No screen, sound or other features that distract or give away the user, or anything that detracts from the device's simple-by-design man-machine interface.

The SRC3300 is very compact – same size as an iPhone – and completely unnoticeable: the perfect device for discreet communications when working up close and personal with targets.

The rugged SRC3300 works under tough conditions encountered in the most demanding covert operation: underground, in tunnels, air shafts or other places where mobile coverage is generally off the map.

In buying Sepura, Hytera gained control of a powerful force in TETRA with near total dominance of the market in much of western Europe, notably the UK and Germany, and growing business in the Middle East, Eastern Europe, LATAM and AsiaPac.

Overview on TETRA

TETRA is a standard introduced in 1995 by ETSI (the European Telecommunications Standards Institute) for two-way radio transceivers used by first responders. Subsequently, TETRA-based radio was embellished with features that made it a suitable tool for covert operations in the field, enabling agents and military special ops to communicate safely and secretly in close proximity to targets or enemy forces.

TETRA works for direct communications for emergency scenarios where it might be essential to reach one party, or alternately, broadcast to a larger group via trunked communications.

Trunking makes TETRA systems a computer-based radio network, with the usual advantages and efficiencies of packet switching: the ability, in an era of competition for precious spectrum, to share limited public radio frequencies as they become available on vacant channels. Basic idea: conserving frequencies.

TETRA networks automatically let the user default and connect to commercial networks, albeit with the usual security vulnerabilities that can arise therefrom. But while operating strictly via TETRA base stations, the system is ostensibly ironclad, using authentication and end-to-end encryption to help ensure security (though as we will see shortly, there are issues here not only for TETRA but for DMR or any trunked radio service).

TETRA's signature features:
- Designed for high volumes of radio traffic in smaller areas such as dense urban markets.
- Requires fewer discrete radio channels.
- Accommodates multiple users privately.
- High quality voice and data. Modern TETRA has overcome early limitations on data speed & now supports higher bandwidth and real-time capabilities for short data (SMS, packet & circuit data).
- As an open standard, TETRA enables interoperability between different manufacturers' networks and devices. Communications are stable and resilient when the user moves between TETRA networks.
- Devices comply with CALEA, ETSI and other legal/technical standards for lawful intercept. In other words, police can conduct lawful intercept against targets that use TETRA phones.
- Most (but not all) TETRA devices support both one-to-one and one-to-many comms.

All the aforementioned indicate why TETRA has enjoyed popularity in selected markets for so many years. But TETRA has issues.

Downsides of TETRA

Some of the minor issues:
- Reliance on fewer base stations means TETRA needs help from linear amplifiers that power loads delivered between base stations.
- Some TETRA phones cannot be programmed to do direct mode operation (DMO).

But the key concern can be a show-stopper if not addressed: security. TETRA phones' air interface is vulnerable to hacking/eavesdropping and requires AES (Advanced Encryption Standard) level security to prevent snooping and to safeguard invisibility of traffic and anonymity of the user. AES must be implemented in a way that does not interfere with interoperability between TETRA networks.

With AES 256, a TETRA system can defeat attackers and prevent them from accessing or reading content. However, the authentication protocol is still vulnerable, leaving a TETRA network susceptible to breaches that can disrupt communications at any critical juncture – such as during a terrorist attack.

In an academic paper published by the Norwegian University of Science and Technology-Trondheim, researchers demonstrate the ease of breaking into a TETRA network via three separate attacks on the authentication protocol. In the simplest of the three, an attacker flips the interface bits to mimic a base station and take control of the radio link. As a result, "a targeted mobile station may falsely show that it is connected to the network while, in fact, the mobile station is unable to receive network communications." The attack is a form of anonymous jamming capable of interrupting critical communications.

The problem, according to the Trondheim researchers, has to do with standards updating. While GSM evolved to support UMTS then to LTE, the TETRA standard has essentially stood still since 1995. With arrival of a newer ETSI standard – DMR – the marketplace has moved on.

Is DMR more secure than TETRA? Not necessarily, though DMR-based devices were the first to embrace AES 256. TETRA has since caught up on AES. DMR has advanced, too, with Level 3 capabilities that use trunking. In so doing, ATIS standards masters have inadvertently opened the gate for new vulnerabilities. It is well recognized that any radio trunking network is fairly easy to disable by a denial of service attack. With its move into trunking, DMR is now as much a target for DDoS attacks as TETRA is.

Did DMR's Advantages Disadvantage TETRA-Focused Sepura?

Digital Mobile Radio or DMR is an open standard that initially covered non-license radio communications in the European 446 MHz band for consumer apps (Tier I), and conventional licensed professional mobile in the 66 to 960 MHz band used by first responders (Tier II). The latter segment immediately won fans for its spectrum efficiency.

With the launch of the next phase – Tier III in 2012 – manufacturers began using DMR to go after TETRA-based systems. As a trunked service, Tier III was exactly on par with TETRA, feature-wise. And DMR now could lead with two significant differentiators: the ability to handle intense volumes of traffic over a much broader area, with far fewer base stations or repeaters; and the ability to do so at one-third to one-half the cost of a similar TETRA network.

In addition to greater efficiency and savings, DMR offered other advantages:

- **Ease of Migration.** The user can leverage existing network infrastructure and trunking systems in a DMR buildover – a thing not possible with TETRA.
- **Ease of Deployment and Use.** Compared to highly complex TETRA, DMR is basically plug & play.
- **Flexibility.** DMR can be deployed in any band above 136MHz used by analog FM land mobile radio. TETRA is confined to portions of spectrum in 25kHz channels.
- **Analog FM Fallback.** DMR supports analog narrowband FM mode, allowing users to upgrade in stages or talk on analog FM in the same band. TETRA does not provide such fallback capabilities.

Radio equipment manufacturers that eased into the market in the early days of DMR piled on with the enhanced standard of 2012. The most prolific was China's Hytera, which makes nine device models for DMR Level II and another 12 for Level III.

Sepura did not join the DMR party until November 2013 with its SBP Series, and then only for Tier II support. The company's final DMR product, the DMR Level III Controller, came out in 2014, and was short-lived, as the company closed this line of business not long after and returned to its TETRA roots. By acquiring Sepura, Hytera hedged its bets by playing both the DMR and TETRA markets. The merger helped ensure the future of a device many consider the best covert radio in the world: the SRC3300.

Thuraya: Terrorists' Favorite Satellite Network

Another day passes with another terrorist attack, as like as not orchestrated via Thuraya, long favored by terrorists as their preferred mode of secure satellite communications. Thuraya's location in the heart of the Arab world, and the alignment between its network map and known centers of terrorist activity, are known for attracting unsavory clientele of the bomb-and-behead variety. The threat is heightened by the arrival of "Satsleeves" that can make virtually any mobile device network compatible, lowering the cost threshold for personal SATCOM to Islamic militants, who can now go BYOD.

The ability to sign up for Thuraya service through its partner "Morsviazsputnik," the Russian federal agency responsible for that country's segment of INMARSAT, adds another hurdle between investigators and targets. Russia is not noted for cooperating with intelligence efforts outside its own sphere of operation.

The bigger challenge is that Thuraya is a bona fide company whose services are used by many thousands of harmless customers ranging from corporations to first responders, and even backpackers while trekking in remote regions. It is difficult to pin a black hat on a company over the comparative handful of hard cases who use the service without Thuraya's detecting them. But bad apples they are, and all too eager to take advantage of Thuraya for their own nefarious ends.

Inside Thuraya

Based in the United Arab Emirates (UAE), Thuraya Telecommunications Company started out in 1997 with the mission of providing geostationary SATCOM services to businesses, government agencies and individuals. Initially, Thuraya operated one satellite serving the Middle East, parts of Europe and much of Africa, but quickly expanded to cover all of Europe and South Asia. With the addition of a second satellite, the service expanded to support Australia and much of East Asia.

Certain key geographic zones remain untouched. Thuraya does not service the Americas. In addition, commercial SATCOM in general is banned or tightly monitored in authoritarian nations including Russia, China, Burma, Cuba and North Korea. Notwithstanding corporate claims to the contrary, these nations remain out of bounds for Thuraya.

More than two decades after its launch, Thuraya now operates in 116 countries, with roaming agreements that increase its reach to 161 nations and over 350 mobile network operators. Thuraya offers nine handsets, three made by Hughes Corporation and the balance branded by Thuraya, all of them dual-purpose to support either Thuraya SATCOM or terrestrial mobile service, with GPS built-in.

In 2013 the company began offering SatSleeves that can be fitted to quickly convert most Android and iOS devices for use on the Thuraya network. The company claims to have been the first to debut this "bring your own device" or BYOD model in the SATCOM marketplace. Thuraya also now offers SatSleeve Hotspot, a portable satellite Wi-Fi hotspot that lets the user place voice and data calls from indoors – provided they leave the hotspot outdoors or in a window.

All in all, this is a healthy, growing and respectable company with a dedicated clientele in the global marketplace, and a steady revenue stream in the low seven figures, which appears to satisfy the 10 principal shareholders of this private joint stock company: Boeing, hedge fund operator Perry Capital and investment bankers Jefferies and Third Point (all from the U.S.), and another six closer to home – etisalat, Invest AD, ARABSAT, Dubai Investments, Gulf Investment corporation, and Qatar's Ooredoo. Day in and day out, Thuraya meets its mission of providing reliable SATCOM and mobile communications wherever, and however remotely located the client.

Although Thuraya is spread out across the globe, the perception of its traditional client focus being in the Middle East and Asia remains strong. And there, perhaps, lies the problem for national security interests.

Thuraya phones were used in the Nov. 26, 2008 terrorist attack in Mumbai, and these devices as well as those used for Iridium and all other SATCOM services except Inmarsat have been banned in India ever since. In 2015, the Taliban's documented use of Thuraya with all the latest technological enhancements posed a challenge for government forces in Afghanistan.

Dust-proof, shock resistant, lightweight and with a glare-deflecting screen, several Thuraya handsets would appear custom-made for the harshest conditions, climate- or combat-wise.

How Thuraya Works

Thuraya hasn't changed in terms of the radio frequency (RF) bands used since Day 1. It operates in a combination of the L-Band and C-Band RFs.

As an example, Thuraya 2, hovering roughly over the Sudan, communicates with handsets via the L-Band (1.525 to 1.559 GHz and 1.625 to 1.6605 Ghz), which is low frequency compared to C-Band, and in ascending order, also to the Ku-Band and Ka-Band. In SATCOM, L-Band is commonly used for military satellites and low earth orbit satellites, as well as mobile phones.

For connection to ground or marine satellite stations, Thuraya goes one notch up and uses the C-Band, where it receives at roughly 4 Ghz and transmits at 6 Ghz. C-band is good for dedicated, continuous transmission/reception, and for the higher pointing accuracy required of "spot beams" – highly concentrated signals that cover a tight geographic area and allow re-use of a frequency to send different signals to varying spots on the globe. As a network map shows, Thuraya offers very extensive spot beam coverage, with the highest concentration in North Africa, the Middle East and South Asia, but tapering off to much larger, hence less accurate spot beams in Europe and Russia.

One caveat for Thuraya as well as other SATCOM services: security. Although all can be outfitted with ostensibly secure encryption, versions in use such as A5/2, both for satcom and GSM, have been easy to penetrate for several years.

Thuraya Monitoring in Action

Thuraya monitoring systems offered by providers including Germany's Rheinmetall Defence Electronics and PKI Electronic Intelligence, Israel's Ability and the UK's Delma MSS operate on much the same premise.

The Delma MSS model is a good example. For Thuraya tactical monitoring, the system receives in the L-Band from satellites, and in L-Band line-of-sight from a target's handset. Strategic monitoring intercepts targets from both the L-band and C-band of the satellite. Delma can also intercept at line-of-sight to the handset and via direct downlink from the satellite.

The basic set of tools are a laptop and a 2-channel downconverter, with one channel for the target's handset antenna, and another channel for the satellite antenna. The intercepted data includes:

- Selected Thuraya spot beam, ID and GPS location.
- Adjacent spot beams.
- Type of call and type of terminal.
- GPS/Region/Country of Mobile event and the target's phone.
- Telephone number dialed.
- IMSI numbers.
- Ciphering key sequence number.
- RAND (128-bit key generated by Home Location Register) & SRES (32-bit Signed Response generated by the Mobile Station, based on encryption of the RAND).
- Cipher algorithm.
- Voice, data and SMS decoding.

Ability's ATIS intercepts tactical L-Band and strategic C-Band with multiple L-Band links, adding features like real-time decryption.

Rheinmetall Defence Electronics offers notable features, differentiating its Thuraya monitoring system in distinct categories of field and urban use. Included in the mix are the Argos L-Band system for straight L-Band interception accompanied by the appropriate L-Band downconverter, plus storage server and uninterrupted power supply; the Artemis for L/L band interception with decoding and storage, for urban settings but also "climatized" for rough conditions; and the Ares, which intercepts both L-Band and C-Band and whose downconverter includes a sophisticated decryption unit for real-time deciphering and 1.8 terabytes of storage.

Ultra Electronics C5ISR

As a mid-sized multinational defense contractor, the UK's Ultra Electronics has come to the ISS sector along a path marked by strategic acquisitions, each designed to add a specific technical strength and talent pool. The company counts the U.K. Ministry of Defense, U.S. Departments of Defense and Homeland Security and the Canadian government among its many clients on six continents. Add law enforcement agencies that favor Ultra's turnkey lawful intercept solution. Even defense contractors that compete with Ultra in one niche – BAE Systems, Raytheon and Thales, to name a few – line up to buy Ultra's products in areas that complement their portfolios.

Today Ultra is a US $1.0 billion company serving eight defense sectors including underwater warfare, land and maritime systems, aerospace, communications (SATCOM, radio, wireless), nuclear (sensors for safety and radiation leaks), infrastructure (transportation and border security), and C2ISR (UK military jargon for "Command and Control, Intelligence, Surveillance and Reconnaissance" and except for the digits on par with the US military's "C4ISR," or most recently, "C5ISR").

Each of Ultra's acquisitions has contributed to one or more of these areas. Unlike defense behemoths that absorb acquired companies until little of the original entity remains, Ultra grants its new members a measure of independence. This is true to the point of tacking on the acquired company's name to the Ultra brand. It's smart business, a move illustrating how the master company has incorporated the strengths of others while demonstrating respect for the reputations that made them worth acquiring in the first place. For example, years after its merger with Ultra, DNE Systems does business as "Ultra Electronics DNE Technologies." The former ProLogic, since acquired in 2008, is now "Ultra Electronics Prologic."

In the C2ISR arena, this eclectic, semi-autonomous mix provides an ISS solution for every need: mass and targeted IP monitoring including VoIP, cellular specific intercept solutions, SATCOM and radar interception, mobile location, RF signal monitoring and direction finding, voice biometric identification, full-motion video, analytics and visualization.

Thanks to tight integration, these and other solutions typically operate on a "plug and play" basis, one with the other. Like building blocks, the C2ISR modules can be added or omitted depending on the needs and interests of the client. All operate covertly and independently of the network under surveillance. When used in tandem they provide a complete picture of the target and his or her cohorts.

At its century mark, Ultra Electronics resembles many a company that started out small in one field, rode the wave of a major trend, then gaveway to corporate sprawl before regaining its sense of direction.

The company began in London in 1920 under the name of its founder, Edward Rosen, and in a field completely unrelated to defense: the manufacture of headphones. The wave Ultra caught of that era was radio, the first global electronic medium and for its time as popular as Facebook or Web surfing are today. 1925 saw the name change to Ultra Ltd. Then came the company's expansion into radio receivers in the early 1930s, and even an early television set for the BBC in 1939.

Post-WWII, Ultra sidelined into television and other consumer electronics devices until thinking better of being one among so many in that field and splitting off its consumer business in the late 1950s. From that point on, Ultra focused on the defense sector, with occasional excursions into seemingly unrelated fields such medical monitoring where the company's expertise in sensors could be exploited. Otherwise, the emphasis has been on acquiring smaller military contractors for their leadership in emerging technologies.

Ultra Electronics DNE Packet Monitoring

Packet monitoring plays an important role in both defense and lawful intercept solutions, and Ultra made a serious move on selling DPI in both markets beginning in the mid- to late 1990s. The company's commitment to packet monitoring solutions got a major lift from its purchase of DNE Systems in 2004 for US $40 million. At the time, Ultra explained its interest in DNE by pointing to the smaller company's expertise in integrating "battlefield communications networks," and close DoD ties that would help further Ultra's sales ambitions in the U.S. But packet monitoring came along for the ride, too. Today, DPI hardware manufacturing rolls up to Ultra Electronics DNE Technologies.

The lead product in this niche is the PacketAssure iQ1000, described as a "high performance, Layer 2 switching platform for voice, video and data service delivery" with "Cisco-like" functionality.

Designed for military users, or as part of lawful intercept solutions, the PacketAssure iQ1000 sits atop existing network architectures to handle all the usual jobs of IP network management, packet grooming, quality assurance and monitoring. It comes embedded with thousands of service classifiers and policies that enable the user to quickly create APIs for virtually any mission.

As a result, the iQ1000 may be put to work in nearly every market niche served by Ultra – from underwater to aerial and land warfare. Use of the

1Q1000 in Ultra's C2ISR marketplace has led to development of special products for ground troops:

- **Ultra-Vu:** A hand-held device that can receive voice, data and full-motion video to provide combat and special forces units with situational awareness information.
- **ForceWatch:** Android devices for secure voice, data and video, carried by troopers to enhance situational awareness.
- **RockPhone:** Secure "through the earth" (TTE) two-way wireless using low-frequency radio spectrum for audio and data literally through the ground versus over the air. Originally developed for the mining industry to locate miners trapped in an underground accident, TTE relies on a special in-ground antenna and has found its way into military apps with solutions such as Ultra's RockPhone.

Though initially developed for military purposes, DNE's packet monitoring capabilities would soon get a hand from other capabilities that pulled data from commercial networks and open source (OSINT) sources.

Following the purchase of ProLogic in 2008, Ultra in 2011 acquired Zu. Both companies were respected for their work in emerging technologies, and being U.S.-based added strength in selling to American military customers. With ProLogic, Ultra gained new talents in developing tactical data links, cryptographic remote rekeying, advanced battle space management systems with a focus on geospatial data, and the integration of unattended sensors used for infrastructure security.

Zu's Integrated Integrated Surveillance & Intelligence System (ISIS) provided the ability to filter, monitor, mirror, process and store data, including voice, text, e-mail and metadata from commercial networks including mobile, fixed line and broadband, as well as open source intelligence – a good fit for the work already underway at Ultra DNE.

The new twist that came with Zu: voice biometrics. Zu's Voice Print Analysis (VPA), a voice biometric technology providing precise identification of a target's voice and gender, regardless of language, location or switch-off between different communications devices.

Zu's solutions quickly became part of the ProLogic portfolio where they make a valuable addition when integrated with other Ultra solutions. ISIS's perfect fit with DNE's packet monitoring hardware proved a case in point. Soon video monitoring joined the mix under the Ultra Electronics ProLogic banner.

In the aftermath, one nomenclature tweak took place. Remember that Ultra acquired Zu in 2011. When Islamic State came to life in 2014, ProLogic found it needed a better name for one product. "Zu ISIS" became Ultra

Advanced Retrieval Information Exploitation System (ARIES), a popular packet monitoring and intelligence gathering platform at Ultra.

Ultra's Current C2ISR Suite – End-to-End Coverage

ARIES is advanced DPI, available as software or in a DNE-made hardware unit such as the PacketAssure iQ1000. As a DPI solution, ARIES takes its lead from a rules-based engine that can be configured to conduct mass or targeted surveillance based on criteria set by the user: IP and MAC address, keywords, events, known figures, transactions, metadata on the parties targeted, as well as full audio and video recording.

It is at this stage that many makers of interception solutions consider their job done – the capture of targeted information formatted into the correct protocols for hand-off to government or military customers. But ARIES goes far beyond the basics thanks to complementary solutions that increase its utility for LEA and military users alike:

- **End-to-End Communications Analysis System (ECAS):** ECAS is Ultra's take on real-time predictive analytics or RTPA. ECAS uses commercial off-the-shelf (COTS) solutions to refine massive amounts of data gathered by ARIES and OSINT (social media and web surfing), then applies scoring and statistical analysis to isolate the target and forecast his or her likely next actions. ECAS can also fold in data from other members of the Ultra analytics family, as follows.
- **Voice Print Analysis:** VPA provides positive identification of the target via a biometric voice print (BVP) as unique as his fingerprint. Used in real-time, the BVP can point to the target's location in that moment. For back-up, ECAS and VPA can gain an assist from a geolocation app.
- **Signal Analytics and Geospatial Exploitation (SAGE):** SAGE begins with RF signal analysis to determine the short duration, infrequency and low power characteristics of radio transmissions often used by unknown targets. It does so *en masse* and employs rapid channel scanning to avoid missing any stray signals. Direction finding is then accomplished with highly-directional antennas, triangulation, GPS, Angle-of-Arrival (AOA), Time Difference of Arrival (TDOA) – processing a signal arriving at multiple receivers at the same moment), and Received Signal Strength (RSS) – the distance from a target measured by the decrease in signal strength. For added flavor that reveals more about the source and the terrain, SAGE also incorporates GEOINT, satellite imagery (from its own SATCOM network), plus communications network and OSINT data gleaned by ARIES.

- **Miderva:** One of the biggest challenges and costs associated with big data analytics solutions is storage. Ultra's Miderva provides a unified recording and storage platform for all types of data captured – audio, video and screens from IP, mobile and wireline networks, and can be easily accessed in any standard media format. Web-based replay brings the cost down.

No discussion of Ultra Electronics is complete without mention of intelligence products designed exclusively for the military and for infrastructure protection. We've already touched on three military-only solutions aligned with the PacketAssure iQ1000: Ultra-Vu, ForceWatch and RockPhone. An additional pair that play an important role in their respective spaces:

- **UltraEagle for RADAR Detection:** UltraEagle uses software defined radio interception to capture, record and analyze enemy radar transmissions. UltraEagle can capture any RADAR technology including those designed to avoid detection such as spread spectrum. Ultra's RADAR detection tool can be augmented by its purpose-built storage and analytics system, TALON-E.
- **Remote Equipment Management System (REMS):** REMS actively monitors and controls unmanned remote sensors or video equipment, providing instant alerts of security breaches or device malfunctions. REMS is widely used for continuous "monitoring of the monitors" in border control and public transportation scenarios, where it helps detect any sensor or camera malfunction. REMS is also a cost-effective alternative to in-person inspections.

Glossary

3GPP – 3rd Generation Partnership Project, a group that unites the standards for wireless and radio communications across other standards-setting bodies including ATIS and ESPI (See ATIS and ESPI below). 3GPP, ATIS and ESPI are all involved in setting standards for electronic surveillance technologies.

ADSL – Asymmetric DSL, the most common form of DSL. Asymmetric because download speed is faster than upload speed.

Active Lawful Intercept – Electronic surveillance technology solutions that reside in a communications service provider's network.

Advanced Persistent Threat – A sophisticated offensive cyberattack that remains unnoticed, even dormant if so designed, for an extensive period of time.

ASIC – Application-specific integrated circuit designed for a specific use.

ATIS – Alliance for Telecommunications Industry Solutions: an organization involved in creating and setting standards for lawful intercept under CALEA and other laws.

BSS/OSS – Automated Business Support Systems (billing) and Operational Support Systems (inventory management systems, order management, provisioning) used by communications service providers (CSPs). BSS produces call detail records and IP data records for billing and is also the basis of metadata collected during intercepts.

BVP – Biometric Voice Print: the means of identifying a suspect by voice regardless of language, device or gender. A BVP is every bit as unique as a fingerprint, and a vital method of identifying and tracking suspects. Biometric technology can isolate a single suspect out of millions of conversations, even if the target switches languages in mid-conversation.

Buffering – Back-up storage of real-time lawful intercept data to prevent data loss in the event of problems or disparities in communications connections.

BYOD – "Bring Your Own Device," the practice of using personal communications and computing devices on the job or for work at home, a practice that if not secured raises the risk of introducing offensive cyber tools into a business enterprise's or government agency's own systems.

CALEA – The Communications Assistance for Law Enforcement Act, enacted by the U.S. in 1994. CALEA outlines the rules of electronic surveillance for criminal cases, the compliance requirements for communications service providers to support lawful intercept, and strict privacy protections for the service provider's customers.

CALEA II – A concept that would extend the domain of CALEA to include all IP, broadband and social media.

CDMA – Code Division Multiple Access, a channel access method used in mobile and other radio services. Improves efficiency of fixed frequency allocation utilization by allowing customers to share a band of frequencies without interference. CDMA was long an access method used in many phone standards; however as of 2010 many service providers such as Verizon have distinued CDMA.

CDR – Call Data Record, data created by a communications service provider of the customer's calls, including originating and terminating numbers, duration of the call and general location by area code.

CIS – Commonwealth of Independent States, a regional organization formed following dissolution of the USSR, primarily comprised of former Soviet Republics.

COMINT – Communications intelligence gathered from people, including voice, text and signaling channel interceptions.

Counter-Terrorism Act of China – China's first comprehensive anti-terrorism law, enacted on December 28, 2015. The law defines terrorism, outlines the responsibilities of communications service providers in assisting law enforcement and government agencies, and the powers of government to use surveillance to detect, monitor, prevent and capture the perpetrators of terrorism. The scope of the Act is broad, extending to all network communications. Following objections by other nations and the technology industry, the Act does not require service providers to maintain all switching and routing equipment in-country. A draft provision requiring "back doors" into communications equipment and devices was dropped after similar objections were made.

CSP – Communications service provider, e.g., a telephone company, Internet provider, cable or satellite dish company.

DCME – Digital Circuit Multiplication Equipment. Traditional voice compression equipment deployed at either end link of a "long distance" oceanic fiber or SATCOM communication.

DHCP – Dynamic Host Configuration Protocol. DHCP is the standard networking protocol used on IP networks for dynamically distributing network configuration parameters such as IP addresses for interfaces and services.

DNS – Domain Name System, for connecting any computing device to the Internet. DNS translates search terms to urls.

DNS Hijacking – A common form of cyberattack. The hacker manipulates and overrides a device's TCP/IP settings to redirect the device to a rogue website, and then hacks the user's device.

DOCSIS – The standard used by cable TV companies to deliver high-speed Internet services over hybrid fiber-coax networks.

DPI – Deep packet inspection, a form of computer network packet interception to examine individual packets as they pass through an inspection point. In lawful intercept, DPI filters packets, identifies those to or from a targeted suspect and creates a mirror image which is then forwarded to law enforcement or intelligence – without the suspect's knowledge or any interruption of the signal.

DRIPA – The United Kingdom's Data Retention and Investigatory Act of 2014. Following BREXIT, the UK passed the Investigatory Powers Act to replace DRIPA, which expired on December 31, 2016. Under DRIPA, communications service providers in Great Britain were required to retain all customers' call and IP metadata records for a period of up to 12 months. In July 2015 the British High Court reviewed and upheld a challenge to DRIPA, wherein opponents charged that portions of the Act violated privacy protections under the European Charter of Fundamental Rights. In October 2015 the UK Homeland Secretary filed an appeal. In December 2015 the Appeal Court referred the case to the Court of Justice of the European Union. DRIPA expired in 2016.

DWDM – Dense Wave Division Multiplexing, a common solution for fiber-based network providers overwhelmed by bandwidth demand, DWDM enables optimization of network loads and speed in the existing fiber optic infrastructure, and is economical versus the cost of deploying more fiber to keep pace with bandwidth expectations. DWDM places data signals from multiple sources onto a single fiber strand in C-Band optical networks, each channel with a separate wavelength transiting fiber infrastructure at multi-GB speeds. A DWDM system can provide up to 96 wavelengths, also known as "lambda" circuits or channels.

E2EE – End-to-end encryption.

ECPA – The Electronic Communications Privacy Act, enacted by the U.S. in 1986. ECPA extended rules on telephone wiretapping to include electronic communications via computer networks, added privacy protections on stored communications through the Stored Communications Act, and added "pen/trap" provisions allowing LEAs to capture the originating and terminating numbers of communications events.

ELINT – Interception of intelligence other than personal communications, such as radio signal analysis and target location with use of electronic sensors.

Ethernet – A computer networking standard originally developed in the 1970s for local area networks (LANs) and now commonly used in metro area networks (MANs) accessing wide area networks (WANs). Metro Ethernet is popular for its low cost versus SONET/SDH MANs.

Ethical Malware – The use of malware, either via a dongle or network connection, to penetrate and take control of a suspect's mobile or other communications device. Also called "offensive cyber."

ETSI – European Telecommunications Standards Institute: creates standards for lawful intercept for members of the European Union. Cooperates with other standards-setting bodies such as ATIS, 3GPP and TIA to ensure international consistency of standards used in lawful intercept.

Femtocell – A small low power-mobile base station typically used by small businesses and in homes, and providing a range of up to 10 meters.

FPGA – Field-programmable gate array: an integrated circuit that can be configured or modified by the customer after manufacture.

FISA – The Foreign Intelligence Surveillance Act (FISA) of 1978. Determines how U.S. agencies may collect intelligence information on foreign nations and agents, including U.S. citizens suspected of espionage.

FISC – Foreign Intelligence Surveillance Court. Oversees requests for warrants to conduct surveillance on suspected foreign agents in the U.S.

FTTX – A generic term for any broadband network using optical fiber to connect to the local loop in the last mile of a telecommunications network.

GCHQ – Government Communications Headquarters, the intelligence and security group in charge of signals intelligence (SIGINT) from telecom and IP networks in the United Kingdom. GCHQ operates listening stations in the UK and abroad, works closely with the NSA and has used NSA-designed programs including PRISM. During World War II, GCHQ's predecessor was responsible for breaking the Enigma code used by Germany. Code-breaking and decryption are still key areas of interest and responsibility at GCHQ.

Geofencing – Use of the satellite-based Global Positioning System (GPS) or radio frequency identification (RFID) to determine the geographic zone of a target's position.

GEOINT – Geospatial Intelligence. Visual identification of natural features and man-made structures on the earth's surface, using SATCOM and/or aerial photography images, infrared and ultraviolet sensors plus analytics to monitor changes in nation state activities, troop movements, the precise location of a target and other intelligence.

GPON – Gigabit passive optical fiber network. Uses gigabit speed point-to-multipoint FTTP (fiber to the premises) economically over a single optical fiber. Considered highly secure, GPON was developed in 2009 to meet the Secret Internet Protocol Router Network (SIPRNet) requirements of the U.S. Air Force, and was adopted by the U.S. Army in 2013.

GPRS – General Packet Radio Service, an IP packet-based method of data communications on 2G and 3G mobile networks.

GSM – A standard developed by ETSI to describe protocols for second generation (2G) digital cellular networks used by mobile phones. It is the de facto global standard for mobile communications available in over 219 countries and territories. Newer mobile standards developed by ETSI include UMTS (Universal Mobile Telecommunications System, for 3G) LTE (Long Term Evolution, for 4G) and emerging standards for 5G.

Gzip – A file format and software application used for file compression and decompression.

IAP – Intercept Access Point. In an active lawful intercept solution, the IAP is an interface built into network hardware. A mediation device programs IAPs across the CSP's network to collect communications data and content specific to a target designated by a lawful intercept court order.

IMEI – International Mobile Station Identity, a unique number assigned to every mobile device on GSM, UMTS, 4G LTE or 5G networks. The IMEI is 15 digits and typically found behind the device's battery. IMEIs are also stored in a mobile operator's Equipment Identity Register (EIR) to validate a device to use the network.

IMSI – International Mobile Subscriber Identity is a second unique code, also 15 digits, that is stored in a 64-bit field on the SIM card in a mobile device and used as the primary identifier of the user, and for validation in home location and visitor location registers kept by the mobile operator. IMSI numbers on 5G smartphones are the first to be protected by encryption.

IMSI Catcher – a surveillance device that works in "active" mode to emulate a legitimate mobile base station, emit a slightly stronger signal than the actual network and capture a device's IMSI and IMEI numbers by making it authenticate with the fake base station. The IMSI catcher then performs a man-in-the-middle attack that intercepts the mobile voice and data communications of targets. The IMSI catcher determines location of the target by triangulating the signal links of his mobile device to local mobile base stations.

Investigatory Powers Act – UK law approved by Parliament on November 29, 2016 requires Internet service providers to retain "Internet Call Records," codify legal right to conduct bulk metadata collection, and establish law enforcement's right to deploy/require "equipment interference" – back doors in network and end user devices that break end-to-end encryption. The law went into effect in 2017, replacing DRIPA, but has faced legal challenges from privacy groups and the EU from the outset.

IPFIX – Internet Protocol Flow Information Export Internet. A standard protocol defining how packet data captured via Flow Monitoring is formatted and sent to a collection device.

IPDR – Internet Protocol Data Record, billing information on the originating and terminating parties of IP communications, including type and duration of the connection.

IP Flow Monitoring – Also known as NetFlow, a system for monitoring and collecting representative packet samples on high speed IP networks. NetFlow was developed by Cisco and is used or mimicked in Flow Monitoring products by Juniper and others.

Iridium – A satellite constellation of low-earth orbital satellites supporting voice and data services from any location on earth.

IoT – Internet of Things.

ISP – Internet service provider.

ISS – Intelligence Systems Support, a term used by the global surveillance industry to define its business and mission.

Lawful Intercept – A term for surveillance of specific individuals by law enforcement agencies by permission of court order. In the U.S., court orders may be obtained only after the requesting LEA has proved probable cause. Previously known as "wiretapping."

LEA – Law enforcement agency.

LI – Lawful intercept, the modern term for "wiretap," but expanded to include all other forms of communications that may be subject to court-ordered surveillance under current laws.

LIDAR – Light Detection and Ranging, a technology that measures distance by illuminating an object with a laser beam. LIDAR is commonly used in facial biometrics, creating a unique map of the human face through thousands of measurements.

LTE – Long Term Evolution. Describes a high-speed mobile network services with peak download rates up to 299.6 MBs and upload rates up to 75.4 MBs.

MAC Address – Media Access Control Address. A unique, typically 15-digit code that serves as the physical address and identifier of a computer or other device allowing transmission of packets from one device to another. A MAC address is typically stored in Read Only Memory (ROM) and is a common target of IMSI catchers for identifying and taking control of targeted mobile devices.

Mediation Device – An appliance that provides centralized management of an active electronic surveillance system that is deployed in a communications service provider's network. The device configures network hardware to intercept targeted suspects' communications, collects, filters and formats the data for buffered transmission to a designated law enforcement agency.

Merchant Silicon – Readily available off-the-shelf chipsets.

Metadata – Record of call data including originating and terminating numbers of a call, time of day and duration of call. Does not include call content.

MLAT – Mutual Legal Assistance Treaties: agreements between nations establishing cooperation between their respective law enforcement agencies on the use of lawful intercept to investigate suspects who operate across borders, or whose communications data is stored outside the boundaries of a nation.

Mobile Location Data – Data gleaned from mobile networks or directly from a mobile device that pinpoints the location of a targeted device. In certain instances, call detail records are used to indicate a suspect's historic location relative to the scene and time of a crime.

Monte Carlo Simulation Engine – Also known as stochastic simulation, the Monte Carlo is a sophisticated computing tool for determining the probability of different future outcomes, taking account of random variables. These tools do not predict the future, but rather, present a panoply of possible futures, which can change randomly with individual probabilities.

NDCAC – The National Domestic Communications Assistance Center, founded in 2012 and located in Fredericksburg, VA. NDCAC is an information and training resource run by the U.S. Federal Bureau of Investigation to provide support and training to state and local LEAs on lawful intercept.

NetFlow – IP Flow Monitoring solution for routers introduced by Cisco in 1996. NetFlow samples packet flows on high-speed networks and singles out anomalies for further investigation. Often powered by field programmable gate arrays (FPGAs) to accelerate performance, IP Flow Monitoring may be used in conjunction with deep packet inspection for full packet examination.

Network Packet Broker – A hardware-based packet monitoring solution that collects, aggregates and copies network traffic at wire speed from switch SPAN ports or network TAPs.

OSI Stack or Model – The Operational Systems Interconnection stack is an abstract model for partitioning a communications network in seven layers, each supporting the layer above it. The seven layers of the OSI stack are: Physical, Data, Network, Transport, Session, Presentation and Application. Packet monitoring systems such as DPI monitor Layers 2 – 7.

OSINT – Open Source Intelligence. OSINT refers to any information that is openly available and in the clear: websites, social media, news sources, blogs, and Deep Web.

Passive Lawful Intercept – A form of lawful intercept that relies on a device called a "probe" that operates independent of a communications service provider network and that "sniffs" designated communications traffic upon activation.

Passive "Off the Air" Monitoring – Direct interception of RF signals from mobile and other radio networks. Passive Off-the-Air RF surveillance intercepts signals from transmitters without interfering with the network.

Pen Register – Lawful intercept of a suspect's call signaling data; does not include content.

PNij – "Plateforme Nationale des Interceptions Judiciaires" is France's integrated platform of domestic surveillance. Conceived in 2010, contracted to Thales, and opened years after the planned launch date, the PNij has been the subject of severe criticism in France.

PRINCE Algorithm Attacks – Probability Infinite Chained Elements cyberattacks using algorithms to efficiently guess and crack passwords.

Probe – A passive surveillance device deployed at the edge of the network and independent of network hardware.

PSTN – Public Switched Telephone Network: the conventional wireline voice network.

RAID - Redundant Array of Independent Disks, a data storage virtualization technology combining multiple physical disk drive components into one logical unit to improve performance, ensure redundancy, or both.

RIPA – The UK's Regulation of Investigative Powers Act 2000, authorizing the technical/legal means for lawful intercept of mass telecom and IP communications in that nation. Subsequently amended in 2003, 2005, 2006 and 2010. A fifth proposed amended version of RIPA was introduced in Parliament in November 2015.

Safe Harbor – Under CALEA, a status of compliance reached by a CSP or its trusted third party when deploying a technology solution that meets the technical standards for lawful intercept.

SATCOM – Satellite communications.

SCA – The Stored Communications Act (Title II of ECPA), a U.S. surveillance law.

SDH/SONET – Standard protocols for transporting multiple bit streams synchronously over fiber optic cable.

SIGINT – Signals Intelligence. SIGINT is the master category defining intelligence gathering of signals whether from human communications (COMINT) or electronic signals not directly used in communications (ELINT). Examples of COMINT include voice, signal and email communications. Examples of ELINT include radar and other signals that indicate types of communications channel and their location.

SOCMINT – Social media intelligence.

SORM – Russian law that sets forth rules of surveillance applicable to of all communications ranging from voice, mobile, texting, Web, TV and streaming video.

Spread Spectrum – A technique using sequential noise-like signal architecture to spread narrowband signals across wideband frequencies. Also known as frequency hopping, the technique is used for a variety of purposes such as securing radio communications, keeping them secret from eavesdroppers, and preventing RF jamming.

SS7 – Signaling System 7, a technology developed in the 1980s to improve the efficiency of networks by creating a data signal pertaining to but separate from each voice or data communications event. More recently, SS7 is used to pinpoint cell towers nearest to a suspect using a mobile device, both for domestic and international surveillance.

Subprobe – In lawful intercept, a subprobe is an intelligent device that is configured to collect targeted communications from a communications service provider network. Subprobes are managed remotely and forward data to a probe or mediation device for aggregation and correct formatting in the protocols specified by a law enforcement agency.

TDMA – Time Division Multiple Access. Provides channel access on shared medium networks, allowing multiple users to share a single frequency channel by dividing the signal into separate time slots.

Thuraya – A satellite communications company based in Dubai, United Arab Emirates. Thuraya provides satellite-based voice and data communications via SATCOM phone in the Middle East, Europe, Central and Northern Africa, Australia and Asia. Thuraya is the SATCOM service most commonly used by Middle Eastern terrorists.

TIA – Telecommunications Industries Association: an advocacy group representing manufacturers of telecom hardware. In the surveillance arena, TIA develops standards pertaining to lawful intercept of traditional telephony (See PSTN).

Title III – That part of The Omnibus Crime Control and Safe Streets Act of 1968 outlining the rules of conventional "wiretapping."

Trusted Third Party – A company that meets the requisite technology standards of the law for providing CALEA solutions on behalf of communications service providers.

USA Freedom Act – Passed by the U.S. Congress and signed into law in June 2015, the USA Freedom Act renewed the Patriot Act but eliminated Section 215 of the Patriot Act, ostensibly banning bulk metadata collection by the National Security Agency. However, the Intelligence Community advocated and Congress subsequently approved a weaker provision – CDR – permitting collection of communications metadata for specified targets,

VPN – Virtual Private Network. Provides the functionality and security of private line service, but over the public network.

WAP Push SL – WAP is the Wireless Application Protocol used for wireless data access on most mobile networks. WAP Push SL is a type of message for sending alerts to mobile devices and giving the option to directly connect to a specific url via the device's Web browser. Government and criminal hackers commonly send targets a bogus WAP Push SL message to guide them to a url that infects and takes control of the target's device.

www.ingramcontent.com/pod-product-compliance
Lightning Source LLC
Chambersburg PA
CBHW071347210526
45465CB00001B/3